Priming the Calculus Pump:

Innovations and Resources

Prepared by the CUPM Subcommittee on

Calculus Reform and the First Two Years

THOMAS W. TUCKER, CHAIR

Mathematical Association of America

MAA Notes Series

The MAA Notes Series, started in 1982, addresses a broad range of topics and themes of interest to all who are involved with undergraduate mathematics. The volumes in this series are readable, informative, and useful, and help the mathematical community keep up with developments of importance to mathematics.

1. Problem Solving in the Mathematics Curriculum,
   Committee on the Undergraduate Teaching of Mathematics, Alan Schoenfeld, Editor.

2. Recommendations on the Mathematical Preparation of Teachers,
   CUPM Panel on Teacher Training.

3. Undergraduate Mathematics Education in the People's Republic of China,
   Lynn A. Steen, Editor.

4. Notes on Primality Testing and Factoring,
   by *Carl Pomerance.*

5. American Perspectives on the Fifth International Congress on Mathematical Education,
   Warren Page, Editor.

6. Toward a Lean and Lively Calculus,
   Ronald Douglas, Editor.

7. Undergraduate Programs in the Mathematical and Computer Sciences: 1985–86,
   D. J. Albers, R. D. Anderson, D. O. Loftsgaarden, Editors.

8. Calculus for a New Century,
   Lynn A. Steen, Editor.

9. Computers and Mathematics: The Use of Computers in Undergraduate Instruction,
   D. A. Smith, G. J. Porter, L. C. Leinbach, and R. H. Wenger, Editors.

10. Guidelines for the Continuing Mathematical Education of Teachers,
    Committee on the Mathematical Education of Teachers.

11. Keys to Improved Instruction by Teaching Assistants and Part-Time Instructors,
    Committee on Teaching Assistants and Part-Time Instructors, Bettye Anne Case, Editor.

12. The Use of Calculators in the Standardized Testing of Mathematics,
    John Kenelly, Editor.

13. Reshaping College Mathematics,
    Lynn A. Steen, Editor.

14. Mathematical Writing,
    by *Donald E. Knuth, Tracy Larrabee, and Paul M. Roberts.*

15. Discrete Mathematics in the First Two Years,
    Anthony Ralston, Editor.

16. Using Writing to Teach Mathematics,
    Andrew Sterrett, Editor.

17. Priming the Calculus Pump: Innovations and Resources,
    Thomas W. Tucker, Editor.

First Printing
© by the Mathematical Association of America
Library of Congress number: 90-070794
ISBN 0-88385-067-2
Printed in the United States of America

Priming the Calculus Pump:
Innovations and Resources

Prepared by the CUPM Subcommittee on
Calculus Reform and the First Two Years (CRAFTY)
Thomas W. Tucker, Chair

Crafty Membership during the period this book was prepared:

RICHARD D. ANDERSON
Louisiana State University

JERRY L. BONA
Pennsylvania State University

MORTON BROWN
University of Michigan

LARRY CURNUTT
Bellevue Community College

RONALD G. DOUGLAS
SUNY at Stony Brook

JEROME A. GOLDSTEIN
Tulane University

WILLIAM HAWKINS
University of the District of Columbia

JOHN KENELLY
Clemson University

PETER D. LAX
New York University, Courant Institute

WILLIAM F. LUCAS
Claremont Graduate School

BERNARD MADISON
University of Arkansas

WARREN PAGE
New York City Technical College

A. WAYNE ROBERTS
Macalester College

SHARON C. ROSS
DeKalb College

MARTHA J. SIEGEL
Towson State University

THOMAS W. TUCKER
Colgate University

ANN E. WATKINS
Los Angeles Pierce College

Table of Contents

Allegheny College • Babson College • Butler University • Chatham College • City College of CUNY • Colby College • The College of Wooster • Colorado School of Mines • Community College of Philadelphia • Cuyahoga Community College • DeAnza College • Dowling College • Drexel University • Eastern Michigan University • Elmhurst College • Franklin and Marshall College • Frostburg State University • Furman University • Grand Valley State University • Gustavus Adolphus College • Harvard Consortium • Humboldt State University • Iowa State University • Ithaca College • Kean College of New Jersey • Knox College • La Trobe University of Australia • Meridian Community College • Montgomery College • Morehouse College • Nazareth College of Rochester • Northeastern University • Northern Virginia Community College • Oklahoma State University • Oregon State University • Penn State Consortium • Polytechnic School • Queen's University • Rensselaer Polytechnic Institute • Rollins College • Rose-Hulman Institute of Technology • San Jose State University • Seaver College, Pepperdine University • Siena College • Simmons College • Spelman College • Tennessee Technological University • Trinity University • University of Arizona • University of California at San Diego • University of Connecticut • University of Hartford • University of Iowa • University of Rhode Island • University of Southern California • University of South Carolina at Aiken • University of Vermont • University of Wisconsin Center—Waukesha Campus • University of Wisconsin, Platteville • Ursinus College • United States Coast Guard Academy • United States Military Academy • United States Naval Academy • Wellesley College • Westminster College • Whitman College • Wilkes University • Worcester Polytechnic Institute.

Preface

The purpose of this book is to disseminate and promote the efforts of the many individuals and institutions who are trying to change calculus instruction. They have responded to the call for a "lean and lively" calculus with energy and imagination—getting grants, setting up laboratories, developing software, assigning student projects, writing their own textbooks, and rethinking from top to bottom what should go into a calculus course. They are changing not only what they teach, but also, what is far more difficult, how they teach. It is one thing to ask for reform in conferences and articles; it is another thing entirely to do it in the classroom, as these people are doing.

The trouble is that this is hard work, hard work that for the most part goes unrecognized. The National Science Foundation has provided seed money, through the Calculus Program begun in 1987 and indirectly though the Instrumentation and Laboratory Improvement Program, to get the effort started; many of the projects in this book have some kind of NSF support. The ideas in this book, however, are for nought if they are not tried elsewhere. It is now up to the mathematical community to provide the widespread support necessary to implement calculus reform. Lip service to undergraduate education is not enough, although even this is welcome after years of silence. What is needed is time, money, space, equipment, travel, positions, and, most of all, professional recognition. Good teaching, like good research, takes talent, dedication, and enthusiasm. Also like research, it is important, it is not easy, and it will not thrive without support.

If mathematicians really are tired of teaching a calculus of algebraic manipulation and "inert material," if mathematicians really are tired of overstuffed textbooks, if mathematicians really are tired of high failure rates and low retention of what is taught, then it is time to try something new. We owe it to our students. We owe it to ourselves. We owe it to mathematics. And it can be done. As one of the projects in this book observes, "There are no obstacles other than our own indifference and the constant lack of fair reward for good undergraduate education."

Acknowledgements

We first wish to acknowledge the ten featured projects for their work in preparing their reports. It is hard enough to organize and teach an experimental course, but to host a site visit and write a forty-page report on a six-week deadline in the middle of the semester is way beyond the call of duty. Clearly this book would not exist without the featured projects, nor would this book have been produced as rapidly as it has without the remarkable cooperation of the individuals associated with those projects.

We wish also to thank the many other projects who sent us abstracts and the candidates for the featured projects who supplied us with so much material on such short notice a year ago. If the rest of the mathematical community approaches the reform of calculus instruction with the same energy, insight, and commitment as these individuals, then the future of mathematics undergraduate education is bright indeed.

To Lynn Steen and co-workers Mary Kay Peterson and Mark Christianson at St. Olaf, we are especially grateful for typing all the abstracts (from very roughly edited copy) and preparing this book for publication from a heterogeneous pile of disks, computer files, and hard copies.

Finally, we wish to thank the National Science Foundation whose grant to the MAA supported the travel of CRAFTY committee members and some of the production costs of this book. Spud Bradley at NSF has also been very helpful in providing information about NSF- supported projects. Even more, we would like to recognize the crucial role that NSF has played in calculus reform both through the Undergraduate Curriculum Development Program in Mathematics and the Instrumentation and Laboratory Improvement Program.

Thomas W. Tucker, Chair
CUPM Subcommittee on Calculus Reform
and the First Two Years (CRAFTY)
Colgate University
July 1990

Introduction

by Thomas W. Tucker, Chair

CUPM Subcommittee on Calculus Reform and the First Two Years

It is now more than four years since the Tulane conference and the "lean and lively" calculus, three years since the symposium on "Calculus for a New Century," and two years since the first round of National Science Foundation grants for new calculus projects. There have been countless articles, contributed paper sessions, poster sessions, talks, and panels. There are also a lot of people doing things in the classroom, some of it innovative, exciting, and even successful. Unfortunately, this activity—the meat rather than the talk—has not been so widely publicized. That is the purpose of this book, to disseminate rapidly to the mathematical community detailed examples of calculus reform in action. These ten featured projects, together with abstracts of more than sixty other projects and a collection of reference materials and resources, should give individuals and departments a concrete idea of what they themselves can do, how to do it, what to use, and who to contact.

On the other hand, we must immediately make some disclaimers. First, with only one or two years' experience, it is too early to assess the long-term effects of calculus reform on many of the pipeline issues that have driven this reform. Projects can compare drop-out and failure rates. They have compared performance on common final examinations. But will there be more mathematics majors? Are scientists and engineers better prepared? Will more women and minorities stay in the pipeline? Are the new calculus courses pumps rather than filters? It is too early to say.

Second, we cannot claim that what we have assembled, in featured projects or abstracts or reference materials, is the "best." The projects we have included, either in featured reports or abstracts, were obtained from open solicitations, from lists of NSF grantees, from lists of speakers at contributed paper sessions, and by word-of-mouth. We are sure that many have been left out. The ten featured projects were selected in September 1989 on the basis of materials we solicited. Since what works at one institution may not work at another, we wanted a wide variety of institutions, large and small, private and public, research and four-year, and a reasonable geographic distribution. (Even then we were not successful; we had almost no two-year colleges to choose from, for example.) We also wanted projects that were well enough developed to write a full report by the beginning of 1990. The result is that many excellent or promising projects were not featured, either because they duplicated the content of other projects or because they were not far enough along. We have been impressed by what we found at the featured projects, but we were also impressed by what we saw of many other projects.

The Featured Projects

With these disclaimers in mind, let us take a look at the ten featured institutions: Clemson University, Dartmouth College, Duke University, Five Colleges (Amherst, Hampshire, Mount Holyoke, and Smith Colleges, and University of Massachusetts), University of Illinois, Miami University (Ohio), University of Michigan at Dearborn, New Mexico

State University, Purdue University, and St. Olaf College. Each of these projects had a site visit by one or two members of this committee. The comments that follow are based on our site visits.

Clemson

Clemson has undertaken a large-scale introduction of hand-held supercalculators into the calculus and other lower-level mathematics courses. This effort currently centers on the HP-28S graphing, symbol-manipulation calculator and involves two terms of one-variable calculus, a term of multi-variable calculus, and courses in linear algebra, statistics, and ordinary differential equations. The project has been underway since 1987 and will eventually encompass most sections of college-level mathematics courses in the first three years of the curriculum. The courses are all based on standard textbooks and the calculator is used as a tool to foster student interest, involvement, comprehension and retention of subject matter. The faculty has developed a considerable collection of calculator enhancement material and this is now available as supplementary manuals.

From our site visit, we were very impressed with the general enthusiasm of both the students and the faculty. The technology did not appear to be driving the course, but rather was being used as an aid to understand concepts and as a tool to solve problems whose size or complexity excludes solution by hand. Because of the scope of the Clemson activities and the space limitations of this volume, only a very small sample of the developed materials is included in this report. Any department considering the introduction of sophisticated calculators into lower-level courses would do well to tap the expertise and experience of the Clemson group.

Dartmouth

The Dartmouth project is for the most part conservative in content, as a quick glance at the syllabi confirms. Day by day mechanics are also standard in the larger details: no laboratories, no major projects, no planned group-learning, even some lecture classes of 100 or more students. On the other hand, this project is unusual in a number of ways. First, the changes are being made across the menu of calculus courses: first, second, and third semester courses; slow, regular, and honors sections; small class and large lecture format. Second, the technological component of the course, which is considerable, is apparently effortless and natural, because the Dartmouth campus is so technologically

friendly. Third, students are occasionally asked to write their own programs or make modifications in existing programs; software is "visible" rather than canned.

The report represents well what we found on our site visit. Our talks with students indicated that the computer usage was not a burden for most students. As one student in the "slow" calculus (Math 1) said: "I'm computer illiterate and the computers are the easiest part of this course." The lectures and materials we saw had more geometrical and numerical emphasis than usual; for example, the section on exponential functions began with the differential equation $\frac{dy}{dx} = y$ viewed as a vector field with solutions plotted by Euler's method. Whether students are absorbing the geometry seems to be, at least from one or two student interviews, a function of how much calculus the students had in high school. The more they had, the more they retain an algebraic and mechanical view of calculus. The greatest impact of the project may be with the weakest and least prepared students. Portability is an obvious question to ask about the Dartmouth project, considering the caliber of students and extent to which the campus is computerized. Classroom computer set-ups are needed, but these are not unusual. Students do homework on computers in their dorm rooms, but there is no reason assignments couldn't be done in a computer center. In fact, the demands made by this course on students and facilities are not excessive, and there is no reason this project could not be duplicated elsewhere.

Duke

Overall the Duke project demonstrates a major rethinking of both subject material and pedagogical presentation; it is refreshingly free of dogmatism in both areas. "Project CALC: Calculus As a Laboratory Course" uses computer lab assignments to drive the classroom work. These assignments require a major writing effort by the students. Students, usually working in pairs, are able to carry out the investigations and write the reports in the same computing environment. The fortuitous availability of a laser printer in the lab has resulted in the students taking great pride in how the reports look and revising them willingly. The student report included is typical of the work we saw produced. Students accept the fact that a lab course is going to require more time than the usual non-lab course. However, this is a very different style math course for most students. The project directors realize that the first few weeks of the course are critical to the student's perceptions and success with the course. As the report indicates, the issues of what students can learn, how they learn it, and how well the concepts of calculus are retained and used are receiving much attention. A barrier test in the second term checks the student's mastery of basic calculus techniques, and a well- developed long-range assessment will begin this year.

This fall will mark a new phase in the Project CALC as it expands to half the freshman calculus sections and is taught by instructors new to the project, and is taught at other schools. The North Carolina School of Science and Mathematics, a public residential high school, will also begin using adapted Project CALC materials this fall. Several of the NCSSM teachers are sitting in on classes at Duke and helping with the development of materials. Now that the bulk of the text materials for Calculus I, II, and III have been written, more time will be available to develop materials for the training and support of faculty who wish to teach the new courses. With the addition of these supplements, Project CALC will be replicable at many other institutions.

Five Colleges

The Five Colleges project is perhaps the most radical reshaping of the calculus curriculum we have seen. The first semester begins on Day One with a system of three-variable, simultaneous, nonlinear, differential equations for an epidemiological model. The first semester also includes some multivariable topics, such as contour plots. Finally, the algebraic viewpoint is so reduced that students had only just learned the quotient and product rules for differentiation at the time of our late November visit in the 10th week of the course. Note, however, that students had been taught the chain rule early on, as well as the derivatives of all the basic transcendental functions, exponential and trigonometric. The technological component of this project, with full scale laboratories using homemade software, is also significant. Laboratories require writing and group learning, making this project different from a standard course in pedagogy as well as content.

The report in this book, especially the excerpts, accurately convey the spirit of the course (see for example the lovely treatment of Fourier transformations in Excerpt 7). Day-by-day mechanics differed amongst the five institutions. Where the work load for students was heaviest, because of lengthy laboratory reports, we heard the most complaints. Gen-

erally, however, students seemed interested, sometimes excited, and even students who had a lot of calculus in high school were, for the most part, remarkably accepting of a course so different from what they expected. The faculty involved in the course were working hard, because exercise sets and some laboratories were being written on the fly (the text had been written the previous summer). The self-assessment given in the report is honest and thoughtful. Overall, this is a dramatically different calculus course that appears to be going very well and that may be much more portable than one might first think.

University of Illinois

The University of Illinois project is the most advanced technologically of any of our featured projects. The core of the course is a dynamic electronic textbook customized by each student; the environment is Mathematica. Classes are one hour a week or less and everything else is in the laboratory (a dedicated room of forty Macintoshes) on a drop-in basis, 10 a.m. to 10 p.m. weekdays. Now in the third semester of operation, the course being taught during our visit was the second course in calculus. In some ways, this is more an individual than departmental project since the staff for the five sections of 10-25 students consists of the two co-directors and three graduate assistants. From our talks with other faculty members, however, it appeared that the department was supportive, as well as people from engineering. During our visit, an economics professor even came in to investigate offering a similar course in his department.

The courses in the Illinois project are not traditional courses plus Mathematica. The content is very different, no doubt influenced by the power and function of Mathematica. The rationale for the content, nevertheless, is sound and historically based. Students have moved from these courses to traditional ones and vice versa, apparently without serious problems, and the intent is to keep these transfers possible. Every student we talked with liked the course. They all agreed it was much more time-consuming than the traditional courses, but they felt they were understanding more. The freedom to investigate and the customized electronic textbook were attractions. Indeed, one strength of the project itself is the willingness to investigate, rethink, redo, reformulate, adapt, and make changes. We know that student involvement in their own learning is crucial; the Illinois project has created classrooms and laboratories alive with involvement, far beyond what we expected.

Miami University

Miami University differs from the other featured projects in its specific focus on students who have had calculus in high school but who have not earned AP credit. Miami offers an alternate course for these students and it is in this course that most of the experimentation takes place. The word "experimentation" is apt, since the project members are trying a number of different instructional methods: group work both in class and out, physical demonstrations in class, computer projects and labs, essay questions on exams. Although the pedagogy is not routine, the content is standard except in one respect: the entire catalog of transcendental functions is introduced at the beginning of the course in some detail. It is interesting that Miami seems to be successful in rearranging material this way, despite using a standard textbook. The Miami project also feels that such a rearrangement is especially effective with their target audience of students who have had calculus before. Another notable feature of this project is the use of technology. This, together with Dartmouth, was the best equipped campus we visited, particularly for classroom use of computers. Every room used for mathematics instruction has a PC locked in a case mounted to the wall, with two monitors in the front two corners of the rooms, easily read from any chair.

Miami University struck us as the place where there was the most uniform support across the entire mathematics faculty for the project underway. By support, we mean awareness (most places have a sizable contingent who need to be reminded that something different is being tried by someone in their own department), lively interest in seeing the project succeed, and either direct involvement or an expectation of being involved in due time. We found the same impression of support in the central administration. Students also supported the course, except in one way. They firmly believed they were learning more in the enhanced course, but they vehemently insisted that they should be getting extra credit because of the extra time they put in (in fact, they got less credit because the alternate course met only 4 times a week versus 5 times for the standard course). We also found some difference in student support depending on gender; students tended to agree that women were more likely to switch back to the traditional course than men (see also our comments on St. Olaf). Overall, this

is a well-conceived, narrowly defined project that seems to be working as planned, and, because of its particular emphasis on students who have had high school calculus, should be of interest to many other institutions.

University of Michigan—Dearborn

This project is perhaps the most carefully planned of any we have encountered. The goal at Michigan—Dearborn was a Calculus I course suitable for department-wide adoption, working within the standard syllabus, involving students more actively in the learning process, encouraging more interaction among students and between students and faculty, engaging students in larger problems, and providing opportunities for students to write mathematics coherently. The means are simple: an 80-minute, once-a-week computer laboratory. What is perhaps most interesting is that the project leaders secured all the initial resources for the laboratory from outside the department but inside the university, with no external support. This is the perfect object lesson for mathematicians about how to decide what you need and how and where to find it.

Our site visit found a very positive response from the students. The few complaints were about lab partners who didn't pull their weight, but for every student who thought it would be easier to work alone, there were at least as many who said they always want to work in pairs. They all agreed that at a campus with such a large commuting population careful scheduling of labs, including evening sessions, was crucial. The few faculty we spoke to who were not involved in the project were well aware and supportive of the project; they also saw that teaching calculus with laboratories would be coming their way soon, and they would have to be ready. For those who feel they do not have the resources or students of a Dartmouth or St. Olaf or Purdue to undertake a major overhaul of their calculus courses, the University of Michigan—Dearborn project may well be, because of its context, scope, and planning, the most instructive and useful in this book.

New Mexico State

Mathematicians at New Mexico State University talk to each other about undergraduate education. Conversation at the conference room table does not center around how bad today's calculus students are. Instead you're likely to hear, "Let me tell you what one of my calculus students did with the 'space-capsule project!' " or "You know, that half-empty Coke can gives me an idea for another project problem. What if we ...?" You might be surprised by the quality of the work their calculus students do when given challenging problems and enough time to do their best. What should come as no surprise, though, is the amount of dedication and time it takes to effect any kind of meaningful change; they're paying the price but loving it. Faculty who have taught in the "Projects" project during the last three years claim that they can't imagine returning to their old ways.

The New Mexico State team does not think that it takes a big budget or a computer scientist's bent to revitalize the learning and teaching of calculus. They would also argue that you need not have a direct line to industry in order to write relevant problems. The essential ingredients of good project problems are: 1) open-endedness; 2) multistage solutions; 3) rich verbal context (some can even be theoretical!); and 4) clear communication of ideas. You don't have to invest in a comprehensive, elaborate, new calculus program. It can be enough to alter the spirit of what you do in the classroom and what you expect of your students. The content of this calculus course is traditional, and it should be easy to use chunks of the materials or to implement home-grown versions. Teaching with student research projects has changed the definition of effective teaching and learning at New Mexico State, as well as the way they view their students' abilities. Teaching calculus with projects is making a difference.

Purdue

This project is more a reform of pedagogy than it is of content. There are those who would say, however, that no aspect of the teaching of calculus is more in need of reform than the way the typical class is conducted. People of this persuasion will want to applaud this project, and there is much here that merits applause. We saw teachers refusing, for considered reasons, to answer questions when they felt that only a hint or a rephrasing or perhaps just a little encouragement would do. We saw imaginative use of group work, not only in computer laboratories, but in classes in which an occasional hiatus allowed groups to talk over some question that had come up. We saw TA's moving around the class and giving to the instructor instant feedback about how some remark was being misinterpreted in the discussion groups. On the other hand, this delivery system appears extraordinarily expensive

and portability may be a problem. We would be remiss if we did not report that anyone emulating this project has a large amount of work ahead of them.

Should calculus students simultaneously be learning a programming language? "No," shout a sizable majority of our colleagues who feel such a burden must rob time from an already overburdened syllabus. "Absolutely," answer the folks in this project. It is in the writing of a program that students come to understand; one does not learn to cook by ordering from a menu. They believe that ISETL offers then the perfect compromise, requiring just enough programming to guarantee understanding. They have made believers of their students. The ones to whom we spoke were absolutely convinced that they were getting more from their course than were their friends in the standard course, more than they got in their high school course (in the case of those who had calculus in high school), and enough to justify the time they spend on computer- laboratory assignments. One well-informed faculty member from another department, father of two Purdue students, praised what one child was getting from this course and lamented the fact that the other child had been allowed to skip over the calculus sequence on the basis of high school work. His conclusion was our conclusion.

St. Olaf

In light of publicity that has surrounded one of the earliest experiments in the use of symbolic computer systems (in this case SMP), we suspect that we may surprise some readers by reporting that mathematics instruction on the campus is not necessarily computer driven. Assignments in a few designated calculus sections, selected by students informed ahead of time about the experimental nature of the course, make use of the computer, but no course has a scheduled laboratory component. Take-home tests allow the use of computers, but few problems require it and many problems ban SMP or remind students that "SMP says so" is insufficient justification for an answer. As the project freely admits, its pedagogy is not "greatly exotic." Rather, the focus is on content. One look at the first problem on the Math 26B Final Take-home Examination in the fall of 1989 should convince one that something different is being taught in this course, and that computers are an integral part of it.

We found that to an unusual extent, the St. Olaf course has strong departmental backing. There are, to be sure, a few members of the faculty who have not and probably will not teach an SMP section of calculus. They claim, however, that awareness of what is being done in the experimental sections, and especially of the nature of the tests being given, has had a definite and they believe positive effect on the way they teach calculus. We were surprised by one point on which we found unanimity among male and female faculty and students in the experimental program. Females are harder to attract into the experimental courses that make use of the computer, are less likely to stay in the program after the first semester, and commonly cite the required use of the computer as the reason for opting out. Whether or not this is a local phenomena is certainly a question that needs to be explored by anyone intending to introduce the computer into their introductory courses.

Abstracts

The purpose of this section is to introduce to the reader a wide variety of calculus projects with the hope that at least one will fit the reader's needs and be near enough geographically that a visit is possible. We make no claim that our list is complete. All that we have assembled are projects that responded to our various solicitations. We would guess that there are at least twice as many institutions seriously involved in calculus reform that do not appear in the abstracts section.

Nor have we attempted any concerted quality control. The projects range from quite modest one-person efforts with only a few students, no funding, and little department support, to significant efforts involving many faculty members and hundreds of students over a number of semesters, sometimes with considerable outside and institutional support. Some have little more than a few laboratory sheets and class notes; others are writing or have written full textbooks or hundreds of pages of supplements. On the other hand, since much curricular reform begins with an individual and only later spreads to a department, it may be the small one-person efforts that are most interesting and most portable. Indeed, a number of the featured projects with widespread departmental involvement began with only one or two individuals. In any case, the abstract given here for each project, large or small, should provide a clear picture of what the project is up to, what materials it has available, and who to contact.

It should be noted that we gave all the projects a specific format for the abstract and a space limitation. Many exceeded that limitation or did not follow our guidelines and their abstracts were edited accordingly. On the other hand, we did allow a small number of abstracts to run over, either because they actually covered a number of projects within the same institution or a consortium of institutions, or because we thought the project might be of particular interest.

Reference Materials

The reference materials we have assembled are meant as signposts pointing the reader in the right direction; they are not exhaustive. All of this material can be found elsewhere. One can write the National Science Foundation to get lists of grant recipients. Our software database is from the *College Mathematics Journal.* Most mathematics departments have a COMAP catalog lying around, or they should. Our brief bibliography of calculus history comes straight from Fred Rickey's MAA minicourse. Our list of minicourses, workshops, and conferences can be gleaned from a few issues of the *AMS Notices* or *Focus.* One can write the College Board or Educational Testing Service for information about the Advanced Placement Program.

All of this can be done, but it usually isn't. We want to remove obstacles and get people started. We hope the reference materials will act as a "whole earth catalog" to inspire some home-made calculus reform.

We would also like to call attention to one particular resource appearing in the section of reference materials, namely the project of the 26 schools in the Associated Colleges of the Midwest and the Great Lakes Colleges Association. We chose not to feature the project in this volume, even though it was recipient of one of the largest NSF grants and is, in terms of the number of participating faculty members, certainly the largest project we know anything about.

There are several reasons for omitting it. It does not aim to develop a particular model of a course that others might wish to emulate; and there is not yet much to describe beyond the goal of putting into the public domain a wealth of materials to be drawn upon by writers, teachers, and the unusually interested student. This project may, however, create one of the most enduring records of the reform effort, and we expect that many readers will want to respond to the call for contributions to the ACM/GLCA project's "Resource Collections for Calculus."

Issues and Observations

The purpose of this book, as we have already stated, is to provide examples and resources for widespread calculus reform and not to assess the efforts presently underway. Nevertheless, having worked on this project for a year, having sifted through a hundred pounds of solicited material to select the featured projects, having visited those projects and read their reports, having edited more than sixty abstracts, we are certainly in a position to make a few observations about various issues surrounding calculus reform.

One obvious issue is the role of technology. Of the ten featured projects, all but New Mexico State have a serious technology component. Most of the abstracts also involve technology. Clearly, computers and calculators are a means of breaking the strangle hold that algebraic manipulation has on most calculus courses and of allowing a more graphical and numerical approach. Just as clearly, technology provides opportunities for students to experiment, to consider larger and more open-ended questions, to work in groups, to justify reasonableness of answers, to encounter realistic problems, to write.

Is then technology necessary for calculus reform? We feel the answer is no. Computers can be catalysts, but most goals of calculus reform can be achieved by other methods, many purely pedagogical. As New Mexico State has shown, the simple introduction of projects—not even "real world" or computer ones—can transform a course. Indeed, we are somewhat concerned by the attention focussed on technology in calculus. Grants and funding are inevitably drawn to equipment, while much less expensive ideas are lost. It may be that, say, take-home exams are the answer to all our woes, but what are the chances that such a proposal would succeed in competition for support or publicity with a proposal for a $100,000 calculus laboratory filled with state-of-the-art equipment and software?

Indeed, technology is sometimes not an answer at all, much less the only answer. Effective use of computers demands reshaping both content and pedagogy. To still teach, say, first or second order linear differential equations as a search for closed form solutions, only letting MAPLE do the algebraic manipulation for the student, is to force a

machine to do what humans can do. Instead, one should exploit the numerical and graphical power of the computer, a power no human has, to study the qualitative behavior of differential equations: draw vector fields, plot trajectories, vary initial conditions, vary parameters, observe stability and chaos. Encasing technology in a traditional, symbol-pushing calculus course reminds us of mounting horse head figures on the hoods of the first automobiles so that the new machines wouldn't scare the old horses still on the roads.

Nor can computers be restrained by standard pedagogy. As anyone can attest who has tried to lecture in a classroom equipped with a computer for each student, within five minutes no one is paying any attention to the lecturer. The University of Illinois project saw this happening and responded accordingly: they did away with nearly all lectures. The computer is entertaining, illuminating, but unruly, and if we cannot accept it as a member of the family, we might not want it around as a guest.

Just as we believe that technology is neither necessary nor sufficient to revive a calculus course, we also do not see for those using technology that a certain level is required for success. While St. Olaf might use SUN workstations, the "laboratory" at Duke is a small room in the basement with fifteen PC's, and Clemson uses HP-28S calculators which cost less than $170 each. Dartmouth has the most advanced network, the most computerized campus, and yet the Dartmouth course was among the most traditional of the projects we visited. In the end, limited resources need not limit the effects of technology in reforming a calculus course.

Another issue in calculus reform, one closely related to technology is the pipeline problem: getting calculus to pump more students, especially women and minorities, into science and engineering. Will a more conceptional calculus course engage more students than it drives away? What effect does the influx of technology have? For example, some students become math majors because they like pencil and paper mathematics and don't like machines and laboratories. Will we lose these students?

Perhaps more troubling are the gender issues relating technology to the pipeline. At St. Olaf, women students seem to be shying away from the computer-based classes, and our site visit team saw this as significant. On the other hand, there was the Dartmouth woman who "hated computers" but found the computer assignments "the easiest part of the course." At Smith and Mt. Holyoke, which admit only women, we found students both pro and con computers. In general, we did not collect data on course enrollments by gender at our featured projects, but the St. Olaf evidence is particularly disturbing because this is a mature project that students know well and are free to choose to take or not to take. The gender-technology question certainly deserves further study.

As for minorities in the new calculus courses, we have almost no information. The only featured project with a large minority population (33% of those taking calculus) is New Mexico State. None of the projects in our pool of candidates focussed on minority involvement in calculus, other than Spelman College (which was more concerned with survey data than with course content and pedagogy). Nevertheless, we do see some encouraging signs. Many of the projects involve group-learning, and as Uri Treisman's work at Berkeley has shown, group-learning can be remarkably effective in improving minority success in calculus. The de-emphasis of algebraic manipulation, common to most of the projects, may also be a help for students whose high school preparation was less than adequate. Finally, the projects we have seen are designed generally for the full audience of calculus, not just stereotypical white male scientists or engineers. These are not "honors" courses, but rather mainstream courses.

The problem of articulation with secondary school mathematics is another issue. Large numbers of students entering college now have "had" some calculus in high school. The fraction who have varies from institution to institution, but in many cases it is well over half. In 1990, about 80,000 students took the Advanced Placement Calculus examinations offered by the College Board (see the reference materials for more information about the AP program). Since fewer than half the students enrolled in the AP Calculus courses actually take the exam, there are probably about 200,000 students taking AP calculus. There may be just as many in non-AP courses, and a larger number still who encounter a few months of polynomial calculus as part of a precalculus course. The result is that the typical college student is not a blank slate as far as calculus is concerned.

The courses taught in the new calculus projects often do not agree with the preconceived notions many of these students have about calculus. At Dartmouth we were told point-blank by one faculty member that this was their biggest problem, and some of our student interviews there seemed to con-

firm it. Although our featured projects appeared well aware of the articulation problem, other than Miami they did not design their projects specifically with such students in mind.

In some ways, radically different calculus courses have an easier time than standard ones when it comes to audiences who have seen it all before, because quite simply they haven't seen it this way before. As one Mt. Holyoke student observed: "There is no comparison between the course I'm taking now and the one I took in high school." On the other hand, it can be difficult persuading students that it really is calculus ("When are we going to stop doing geometry and graphs and start doing calculus?"). We have no easy solution to the articulation problem, but we must remember that school mathematics may be changing more dramatically than college education. The *Curriculum and Evaluation Standards for School Mathematics* recently published by the National Council of Teachers of Mathematics is in the same spirit as calculus reform. The AP program always attempts to reflect what is happening at the college level. It may well be that the articulation problem we see now may be transitory, and that as college calculus changes so will the preparation of incoming students.

Conclusion

Of course, in the end, the main issue is the immediate success of the new calculus projects. We have already disclaimed any assessment of long-term effects, but even for short-term effects, it is not clear what we should measure, and how we should measure it. As for achievement, in every case we know where students from experimental sections took a common final exam with the rest of the calculus students (e.g. Dartmouth, New Mexico State), the experimental students did just as well or better. As for attitudes, most projects report very good, sometimes enthusiastic evaluations by students. How much of this is a Hawthorne effect? It is hard to tell. Instructors are working hard in these courses; students can see that, appreciate it, and work harder themselves as a consequence. Also there is a self-selection process involved—not necessarily among the students, but among the teachers. Which faculty members are the ones most likely to take on calculus reform? We suspect it is the most energetic, most adventurous, most committed. It shouldn't be surprising to find students responding positively to these teachers.

What happens when calculus reform is turned over, as it inevitably must, to the less energetic, less adventurous, less committed? Which parts of calculus reform are robust enough to succeed, no matter who the teacher is? We have some opinions. One thing that seems to work is student projects. It doesn't have to be that many, even two a semester is enough, but anything which asks students to be engaged with one problem for a couple of weeks is good. The other sciences have laboratories, and outside the sciences there are 15-page papers. Mathematics may be the only discipline that bases its instruction on hundreds of exercises of five minutes or less. It is no wonder that students respond by doing without thinking, flushing everything as soon as the test is over, and never seeing the connections. It is a lousy way to teach, and almost anything that breaks that pattern is a good idea, be it projects or laboratory reports or essay questions on examinations.

Another thing that seems to work is group learning. Doing mathematics always alone, as the majority of our students probably do, is not a good idea. Students should learn from each other, as well as from their instructor. It doesn't take much: two students at one computer terminal or lists of names for study groups will do. Many calculus projects have tried consciously to create opportunities for group learning, while others have had it by serendipity. It is not always smooth, since personalities can clash and work load may not be spread evenly in a team, but it does help.

Technology also seems to work, although not necessarily for the reasons one might expect. It certainly can change the content of a calculus course, but its influence on pedagogy is perhaps more important. Classroom demonstrations are not the point. The real impact of technology is the opportunity it provides for students to explore, to work in groups, to write laboratory reports and projects.

What works in terms of content? Most projects give the graphical and numerical viewpoint more emphasis than the standard calculus course. But in general, the content of the featured projects is not radically different, except for Five Colleges, Illi-nois, and Duke. It is not coincidental that these are also the only featured projects not using an off-the-shelf textbook. Amongst the abstracts, the Community College of Philadelphia project, with its "Lagrangian" approach, represents the greatest departure in viewpoint and content, but again most of the abstracts also appear to be staying fairly close to a standard calculus syllabus.

The differences we do see are earlier introduction of the exponential and logarithmic functions and more emphasis on differential equations, especially from a modeling and qualitative viewpoint. We do not see more theory, but rather better motivation and more appeals to intuition. All of this seems to make sense, especially for the mainstream audience.

Even if we know what works for pedagogy and content, there is one significant obstacle to implementing widespread calculus reform: work load! There is no doubt that project members are putting much more time and effort in their calculus classes than they normally would. Does this ever settle down? A number of projects (e.g. New Mexico State) have indicated that the work load does diminish the second and third time around. It is not clear, however, whether it ever returns to the level of effort involved in teaching a standard calculus class on automatic pilot. Maybe it never should, and that is the point. One thing is certain. Individuals involved in the projects feel universally that the extra work is worth it, that there is a payoff, and that the payoff is significant, both to themselves and to their students.

If teaching calculus is going to be more work, then departments and institutions will have to be prepared to expend greater resources on calculus. If English departments can require their classes to have twenty or fewer students, if science departments can require laboratory space and equipment and personnel, then mathematics departments must do the same. It is time for mathematicians, who are notoriously poor entrepreneurs, to begin operating, to lobby for resources, to ask for what they need. Calculus reform will not happen by itself.

Clemson University

CALCULATOR-BASED CALCULUS

by Donald R. LaTorre, Iris B. Fetta, John W. Kenelly,
James H. Nicholson, T. Gilmer Proctor, James A. Reneke

 **Abstract**

Contact:

DONALD R. LaTORRE, Department of Mathematical Sciences, Clemson University, Clemson, SC 29634-1907. E-MAIL: latorrd@clemson.bitnet.

Institutional Data:

Clemson University is a land-grant, state supported institution in northwestern South Carolina offering 69 undergraduate and 87 graduate degree programs in nine colleges: Agricultural Sciences, Architecture, Commerce and Industry, Education, Engineering, Forest and Recreation Resources, Liberal Arts, Nursing, and Sciences. Of the slightly over 16,000 enrollments, approximately 35% of the 12,563 undergraduates are in the College of Engineering or College of Sciences. Clemson also has a well-established honors program, "Calhoun College" with 646 students, and 29% of the current 2,900 freshmen received advanced placement (AP) credit.

The Department of Mathematical Sciences is a large (56 faculty) multidisciplinary department offering innovative degree programs at the bachelor's, masters, and Ph.D. levels. The department maintains somewhere between 160 and 175 undergraduate majors each year and about 100 M.S. and Ph.D. degree-seeking graduate students. Each year, 35-40 undergraduates are graduated with a degree in mathematical sciences. The department provides a major service role, teaching about 1/8 of all credit hours taught at the university each year.

Project Data:

Funded by a three-year grant from the Fund for the Improvement of Postsecondary Education (FIPSE), the goal of the project is to integrate high-level, programmable graphics calculators (to date: the HP-28S) into the basic mathematics courses for science and engineering undergraduates: single-variable calculus (2 courses), multivariable calculus, differential equations and linear algebra. Seven members of the faculty have taught in the project so far, and each of the five courses has been taught at least 4 semesters. All of the courses require regular and substantial use of the HP-28S calculator, both for classwork and homework. Some work on a Macintosh microcomputer is required in the multivariable calculus (software: Master Grapher) and differential equations (software: Phase Portraits) courses.

Project Description:

The project has implemented 5 calculator-based courses: single-variable calculus (two courses), multivariable calculus, differential equations and linear algebra.

All students in the College of Engineering are required to take the full four-semester calculus and differential equations sequence and almost all of the students in the College of Sciences are required to take at least two of the four semesters, with Chemistry, Physics and Mathematical Sciences majors taking all four. Electrical Engineering, Computer Engineering, Computer Science, Computer Information Systems, and Mathematical Sciences majors are also required to take the linear algebra course, and assorted other majors in Engineering and Science take this course on an elective basis. Each course is of a semester's duration and, with the exception of linear algebra, meets 4 days per week. Linear algebra meets only 3 days each week.

The standard departmental-wide texts are used and we basically follow the standard syllabi. But regular, often daily, use of the HP-28S calculator for both classwork and homework has enabled us to

- concentrate on geometrical and graphical aspects,
- focus on essential core theory and methods,
- encourage students' exploration and experimentation,
- require active, in-class participation (a natural for the HP-28S),
- provide interesting and realistic approaches,
- demonstrate advantages of technology.

However, we are not interested in adding substantial amounts of new material, requiring significant calculator expertise, or using high-level calculator routines which deliver "final answers" while obscuring the underlying mathematical processes. *We are primarily interested in increasing student interest, involvement, comprehension and retention of the subject material.*

We have developed our own materials relative to calculator enhancement, and these are now available in a series of course supplement manuals from Harcourt Brace Jovanovich, Inc.

═ Project Report ═

Getting Started

The project began in the 1987-88 academic year when a half-dozen members of Clemson's faculty, recognizing the potential for the newly-released HP-28S calculator (January 1987) to revitalize instruc-

tion and learning in basic undergraduate mathematics courses, committed themselves to the task. Impressed by the high degree of student enthusiasm in an early pilot course in calculus taught by John Kenelly in 1987-88, the project director applied to the FIPSE program for funding. This bid for external support was successful and during the 1988-89 school year the project designed and class tested "pilot" versions of each of the 5 courses (actually, 6 courses—since we have also developed a calculator-enhanced statistics course). Other than some very tentative material from Hewlett-Packard, nothing was available relative to the pedagogical use of this new calculator; thus the pilot year was genuinely a bootstrapping effort. Part of the summer of 1989 was spent refining the pilot versions, and "prototype" courses were taught during the 1989-90 year.

During the first two years, class sizes were held to 30 and 150 HP-28S calculators were loaned to students each semester for the duration of their courses (under signed Calculator Loan Agreements). One senior graduate student acted as a calculator resource person to support the multivariable calculus and differential equations courses.

Each of the calculator-based courses addresses the following key questions.

- Where is calculator use appropriate—or inappropriate—and why?
- What does calculator use cost in terms of time and distraction from standard material?
- Which topics can be more efficiently studied with calculators?
- What is a proper balance between calculator use and hand performance using traditional methods?
- Can calculators genuinely enhance conceptual understanding?
- Which, possibly new topics may be introduced because of the freedom provided by calculators?

The portability, personalized nature and relatively inexpensive purchase price of high-level calculators make them attractive, particularly since they can now model and graphically represent almost all the quantitative models one would normally expect to use in introductory calculus settings. And our success with them so far shows that students see them as especially applicable to their needs. They work equally well in hallways, in the library, at park benches and lab benches, and are a constant companion in their backpacks. Students are able to do serious exploratory work on their own, witness the dynamic nature of mathematical processes, and engage more realistic applications. Their learning has become more active and hence more effective. They need not confine their explorations to central facilities nor individually spend considerable amounts of money on their own workstations.

Reaction from other faculty in the department has been, not surprisingly, mixed. In addition to the original six who made early commitments, another six have asked to teach the calculator-enhanced courses beginning in Fall 1990, and Clemson will teach 16-18 such courses each semester during the 1990-91 year, with no restrictions on class size (which tends to average 40-45 students) and students providing their own calculators. At this writing (Spring 1990) others on the faculty remain either skeptical or indifferent. But our presentations to faculty in Engineering and Sciences (e.g., Physics) have been well-received, and their faculty are generally quite supportive of our use of calculator technology.

Single-Variable Calculus

The HP-28S calculator is in its sixth semester of use in special sections of the single-variable calculus courses at Clemson. The emphasis is on teaching the basic concepts of calculus and their applications to the analysis of functions, not on technology. The current text is *Calculus and Analytic Geometry*, by Thomas and Finney, 1988, Addison-Wesley. Another text was used the first two times the calculator sections were offered, since the calculator material is not text-dependent.

Handouts describing the use of the 28S in connection with particular topics are distributed throughout the semester. The handouts include exercises and programs, but programming by the students is not required. Our intent is to use the calculator for graphing, tedious calculations and certain routine algebraic manipulations. But we take care to avoid push-button solutions to complicated problems.

The calculator allows us to place increased emphasis on the geometric, graphical aspects of calculus. Generally, beginning students are able to sketch by hand only very simple graphs; but with graphs readily available on the HP-28, almost all differentiation problems can be presented effectively in connection with graphs and tangent lines. This helps students to understand differentiation as more than a collection of algebraic rules. From the beginning, functions and their derivatives are sketched

simultaneously for many exercises, not just a few textbook examples. The graphs give a continual reminder of the geometric relations between a function and its first and second derivatives.

The built-in equation-solver routine eliminates some algebra, greatly expands the set of functions that can be considered in curve-sketching and max-min problems, and allows the students to find x-intercepts for all of the functions considered. The calculating and programming aspects of the 28S are also utilized in a series of simple programs which calculate left-endpoint, right-endpoint and midpoint Riemann sums for a function, with interval and value of n chosen by the students. The ability to calculate these sums for fairly large n helps students to understand the limiting process which defines the definite integral. The built-in program for evaluating definite integrals is very useful for evaluating arclength integrals for common functions whose arclength integrals are beyond evaluation by antiderivatives.

The HP-28S' built-in commands are used to find Taylor polynomials, centered at 0, for functions of the user's choice, although students are required to know the formulas well enough to find simple Taylor polynomials by hand. The calculator merely enables them to find and graph those of larger degree. A program is given to find Taylor polynomials centered about any number x_0.

The material covered in the calculator sections is the same as for the standard sections. The first semester course covers basic differentiation and integration, with applications, for polynomial and polynomial-based functions as well as trigonometric functions. The second semester extends the material to exponential, logarithmic, inverse trigonometric and hyperbolic functions. Taylor polynomials and infinite series are then studied, and finally "techniques of integration." Grades are determined by daily (10-minute) quizzes (10%), five hour tests (60%) and the final exam (30%). No exemptions are granted from the final.

Instruction in the use of the calculator takes only several days of total class time during the semester, and this will decrease with our use of a course supplement manual in the future. Our experience is that the calculator saves time in graphing, curve-sketching problems and the presentation of Riemann sums.

A typical 50-minute class includes a 15-20 minute discussion of homework and then a 30-35 minute presentation of new material. Calculator use is ongoing throughout the period and is unrestricted on tests. Students get help with the 28S in class both from the instructor and from other students. The classes are lively and the students are involved. They report that they make considerable use of the machine in their other science and engineering classes.

Student reaction to the use of the HP-28S has generally been very favorable. Most of the good students are enthusiastic, and most of the average students feel that it genuinely helps them. While some of the poorer students feel the calculator is beneficial, others seem to feel that it is just another obstacle to overcome.

Multivariable Calculus

The objective of multidimensional calculus redesign is to *reward geometric thinking* by
- Introducing each topic with an appeal to the students' geometric intuition,
- Stressing geometric definitions of concepts,
- Motivating with geometric applications, and
- Constructing geometric proofs.

The redesign has proceeded through the production of a series of modules, some requiring several lessons to complete. In the fall of 1988, the modules were used for course enrichment. Since then, they have been used to restructure the presentation of part of the core material of the course. Modules for multidimensional calculus include
1. Curves in polar coordinates,
2. Graphs of conic sections,
3. Integration on the HP-28S,
4. Explicit, implicit and parametric curves and surfaces,
5. Limits,
6. Differentials,
7. Classifying critical points for functions of two variables,
8. Pólya's problems (Lagrange multipliers),
9. Applications of integration,
10. Vector fields,
11. Line integrals, and
12. Green's Theorem.

The modules are self-contained and are designed to supplement or replace a section of the book. (Our current text is *Calculus and Analytic Geometry*, by Thomas and Finney, Addison-Wesley, 1988). Each module is broken up into lecture sized units containing a student handout and a set of calculator exercises. Class use of the calculators is largely confined to use with the handouts.

The calculator is the student's handbook, scratchpad and number cruncher, *i.e.*, the student's resource of unstructured computational power. The calculator exercises are designed to
- Build a proficiency with the HP-28S calculator which can be used in courses beyond the Mathematical Sciences Department,
- Lead the student to explore examples beyond the usual pencil and paper textbook problems, *i.e.*, problems which require for their solution some computational power, and
- Lead to a deeper understanding by making connections with previous work.

Elementary Differential Equations

The HP-28S (or HP-48SX) and microcomputer software are used in several sections of the elementary differential equations course which is largely populated by second year engineering and sciences students. A textbook supplement has been created for these sections which gives exercises and programs for the HP-28S (and some modifications for the HP-48SX). The calculator is used during class, on tests, and on homework; microcomputers are used during laboratory time. The calculator and computer programs do not require significant programming time. *The purpose of such technology is to emphasize the concepts encountered in the differential equations class* and to begin adaption to the "workstation environment" now present in scientific careers. Emphasis is given to the graphs of solutions (or the graphs of families of solutions) by using the calculator.

The textbook for this four semester hour credit course is *Fundamentals of Differential Equations* by R. Kent Nagle and Edward B. Saff (Benjamin Cummings, Second Edition 1989). The topics covered are: introduction (includes directions fields, Euler's and improved Euler methods), first order DEs (separable & linear problems, transformation methods), mathematical models (mixing, population, heating, and falling body problems), second order equations (applications to springs, circuits, some nonlinear problems with phase plane graphs), higher order linear differential equations, Laplace transforms, systems of differential equations with applications, and matrix methods of linear systems.

A test is given at the end of the applications involving first order equations, another after linear second order differential equations, a third after the phase plane analysis, the fourth after systems have been introduced. The final exam is comprehensive but emphasizes the matrix methods material. A typical home work assignment to be graded contains ten problems covering about three sections and is given midway between tests.

Homework problems are suggested for each class and assignments to be graded are given in most weeks. In addition to regular homework, there are two *special projects* featuring computation and graphics. The four tests and final examination usually contain some calculator and/or graphics problems but feature many traditional problems.

One or more interactive class problems are presented daily in which the calculator is used as a tool for student participation. The participation takes the form of appropriate input to a "canned" program and making observations/interpretations of the results. The instructor observes individual results by walking through the class as the work proceeds. The remaining time is for questions and presentation of material.

Only a relatively small number of programs have been used in the course; for example, three methods of solving initial value problems are covered; only two are used in class. The algorithms are chosen for pedagogical reason; sophisticated algorithms are left to subsequent courses. Homework and project work include some problems which are not attempted in regular sections, for example, parameter estimation by a least squares fit to data or solution of two or more nonlinear equations using Newton's method.

Graphs are used to illustrate the properties of particular solutions of initial value problems and to show how changes in a parameter or initial value affects the solution. Differences between local and global behavior are noted. Discussion of the asymptotic behavior of solutions is stimulated by appropriate examples. Solutions of problems with chaotic behavior are graphed and references are given to expository popular journal articles. Phase plane solution graphs are constructed, for example, second order differential equations and, in some cases, compared to time history graphs. Numerical integration is used to calculate the time behavior of nonlinear problems, for example, the time required for a cycle in a pendulum problem.

The class was taught for the fourth time in spring semester 1990. Each class has had from 25 to 30 students. Student evaluations were collected at the end of the course. Based on the student reactions (verbal and written) approximately 80% of the students felt the calculators were valuable in un-

derstanding the material and in working the problems, 10% of the students were unable to recognize benefits from their use: the remaining 10% of the students did not particularly like to use the programmable graphics calculators.

Achievement levels for students with good records remain high. Other students show some improvement. All report an increased awareness of the conclusions and questions which result from the use of graphs as part of the solution process.

The differential equations course offers many opportunities for the use of calculation and graphics tools. The HP-28S (or 48SX) has memory and quickly available numerical and graphics capability which is equivalent to combining several microcomputer software packages. We believe it has several advantages over laboratory microcomputers.

Linear Algebra

This is a second-semester sophomore course targeted at the level represented by such popular texts as those of Anton, Barnett and Zeigler, Kolman, and others. The 1989-90 text was *Matrix Methods and Applications*, by Groetsch and King, Prentice-Hall, 1988. Student clientele includes all undergraduates majoring in electrical engineering, computer science, computer information systems, mathematical sciences, and assorted others from engineering and science.

The goal of the course is to present an informal, tangible account of the basic concepts of elementary linear algebra. This is not a rigorous development; instead, we concentrate on explanations and examples in an effort to increase understanding. The main emphasis is on supplying the mathematical tools necessary to solve problems using matrix techniques and in developing a modest theoretical background in linear algebra needed for more advanced courses in science and engineering.

The HP-28S enhances the course primarily by significantly removing the computational burden usually associated with hand performance of matrix algorithms, thus allowing beginning students to focus more clearly on the underlying concepts and theory. The course centers around the following four major themes:
• Gaussian elimination and LU-factorizations,
• Vector space theory associated with matrices,
• Geometrical notions and orthogonality,
• Eigenvalues and eigenvectors.
HP-28S use is regular (almost daily) but we are careful not to require programming skill. Basic

keystroke commands are used when possible (*e.g.,* matrix addition and multiplication, routine computation of matrix transposes, determinants, inverses and norms). Simple programs are provided by the instructor which significantly assist with the computations in

- A variety of matrix editing procedures,
- Gaussian elimination,
- LU-factorizations,
- Checking vectors in R^n for independence-dependence, bases and dimension,
- Executing Gram-Schmidt orthonormalization and producing QR-factorizations, and
- Performing eigenvalue and eigenvector calculations.

However, we avoid high-level calculator routines which deliver "final answers" while obscuring the underlying mathematical processes.

We have resisted the temptation to restructure the course into one emphasizing numerical linear algebra; yet it is certainly within reach of the calculator's technology to examine iterative techniques for solving linear systems (*e.g.,* Jacobi and Gauss-Seidel methods), and the simple power methods for the calculation of eigenvalues.

A typical class format (50 minutes) includes

- Homework discussion (approximately 15 minutes),
- Explanation of new material (approximately 25-30 minutes), and
- Calculator activity (approximately 5-10 minutes).

All quizzes contain calculator-based problems and students may make free use of their calculators on the final exam.

This course has been taught three semesters. Students' reactions are obtained from their responses to an extensive evaluation form prepared by the project's external evaluator. Here is a summary of several key questions, based on a sample size of 70.

Questions	% who Agree or Strongly Agree
1. The graphics calculator helped me understand the course material.	76%
2. The graphics calculator was useful in solving problems.	97%
3. The graphics calculator allowed me to do more exploration and investigation.	83%
4. The graphics calculator helped me have better intuition about the material.	60%

versus

Question	% who Agree or Strongly Agree
5. I could have learned more if I had *not* used a graphics calculator.	86%

Future Plans

Clemson will significantly expand its calculator-based calculus activities over the next several years, both in terms of the number and variety of courses taught and in terms of the level of sophistication of the calculators used. During the 1990-91 academic year, we will teach at least 32 sections of our calculator-based courses, involving 16 faculty. Students will, of course, provide their own calculators.

Also during the 1990-91 year we will conduct a major experiment aimed at integrating the new HP-48SX calculator alongside the 28S in our courses. Released in March 1990, the 48SX is a dramatically upscaled programmable graphics calculator with truly remarkable features. Designed for the engineer/mathematician/scientist, it incorporates the engineering applications of the HP-41CX, the mathematical, graphical and symbolic algebra features of the HP-28S and, most importantly, has the capability of interfacing directly with microcomputers. In short, this affordable, hand-held device is technologically sophisticated to such a degree that it has the potential for changing our approach to instruction in engineering and science oriented mathematics.

As new textual material appears, we will restructure our courses accordingly. We are convinced that the increased use of state-of-the-art calculators will facilitate a new emphasis on the modelling process wherein students will spend substantially more time than now on problem identification, formulation and consideration of possible frameworks for solution.

Though the aforementioned HP-48SX will effectively bridge the gap between microcomputers and hand-held devices, students will also need the expanded versatility provided by microcomputer technology for the more extensive, longer-term projects—projects which involve more substantial engineering and science connections. Thus, we are currently developing a calculus microcomputer lab for such projects, and will give careful attention to the newly emerging interfaces between calculator and microcomputer technologies.

We plan to adopt a graphing calculator approach to our basic precalculus course in 1991-92 and, depending upon our staffing, would hope to do the same in 1992-93 for our two-semester "business calculus" course designed for majors in the College of Commerce and Industry.

═══ Sample Materials ═══

Single-variable Calculus

Seven handouts on using the HP-28 are distributed to students during first-semester course. The first one, on graphing, is given out at the start of the semester; the others are distributed as the topics come up in class. Each is accompanied by a set of exercises. The topics are listed below.
1. Graphing
2. Evaluating functions
3. Finding derivatives (This handout is not distributed until all the basic theorems on the mechanics of differentiation have been covered).
4. Solving equations
5. Linear approximations
6. Curve sketching
7. Integration

Each time the calculator enhanced second-semester course has been given there have been a number of students in the class who did not have the calculator section for the first-semester course. Thus, handouts are distributed again with new exercises appropriate to the material covered in the course. An eighth handout is also distributed during the second semester:
8. Taylor polynomials and infinite series.

Space restrictions prevent the inclusion of these handouts and typical tests in their entirety, so only selected sections and problems from the tests will be given. An excerpt from the handout on graphing:

The Nature of the Graph on the HP-28

The display screen of the HP-28 is a 137 by 32 rectangular grid of picture element dots called "pixels". When the [DRAW]M command is activated, the entire screen is used to display the graph of the expression stored in the storage area designated as EQ.

On the graphing screen, with the default plotting parameters, the pixels in the leftmost column have x coordinates -6.8 and the pixels in the rightmost column have x coordinates 6.8. The highest row of pixels has y coordinates 1.6 and the lowest row of pixels has y coordinates -1.5. Two adjacent columns of pixels have x coordinates differing by 0.1 unit and two adjacent rows of pixels have y coordinates differing by 0.1 unit.

When you activate the [DRAW]M command for a function f stored in EQ, the HP-28 calculates values of $y = f(x)$ for each of the 137 values of x, 0.1 unit apart, from $x = -6.8$ to $x = 6.8$. If the size of the y coordinate puts the point on the display, that is, if $-1.5 \le y \le 1.6$, the appropriate pixel is energized and appears as a point on the graph. So the "graph" of a function on the HP-28 is actually a set of 137, or less, distinct points. When the HP-28 graphs an equation instead of a function, it graphs each side of the equation as a function, and so the graph of an equation consists of 274, or fewer, distinct points.

A few of the graphing exercises: $\sin(\pi x)$, $\cos(10\pi x)$, $\sin(x^\circ)$, $e^{-x/3}\sin(2x)$, $3^{\cos x}$.

From a 10-minute quiz:

On the axes shown below, sketch the graph of $f(x) = x^2 - 3x + 1$.

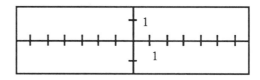

Use the definition $f'(x) = \lim_{\Delta x \to 0} \frac{f(x+\Delta x)-f(x)}{\Delta x}$ to find $f'(x)$. Use this to find $f'(3)$. On your sketch, show the tangent line to the graph of f at the point where $x = 3$.

From another 10-minute quiz:

On the axes shown below sketch the graph of $f(x) = \tan^{-1} x$.

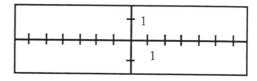

Find $f'(x)$ and $f'(2)$. On your sketch, show the tangent line and the normal line to the graph of f at the point where $x = 2$.

From another 10-minute quiz:

On the axes shown below, sketch the graph of $f(x) = \frac{1}{2}\tan(\frac{1}{2}x)$.

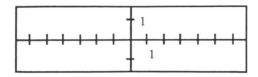

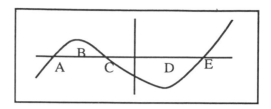

Find $f'(x)$.

Find the linearization $L(x)$ of $f(x)$ at $x = 0$ and show this line on your sketch.

Fill in the table:

x	$f(x)$	$L(x)$
.3		
.1		
.05		

Another hour-test problem:

Sketch the graph of $f(x) = x \ln x$ on the axes given below:

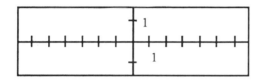

Use l'Hopital's rule to find the limit of $f(x)$ as $x \to 0+$.

Find both coordinates of any extreme points.

Two problems from hour tests which illustrate the geometric emphasis:

1. The graphs of two functions are shown below, one drawn with a heavy line and the other with a light line. One of these functions is the derivative of the other. Which is the derivative? Give reasons for your answer.

2. The curve shown below is the graph of df/dx for a certain function f. The letters represent coordinates on the x axis. For what values of x does f have a local maximum? For what values of x does f have a local minimum? On what closed interval is f increasing? On what closed interval is f decreasing? For what values of x does f have an inflection point?

The following worked-out example from the curve-sketching handout involves several procedures from earlier handouts:

An Example of a Curve-sketching Problem.

Sketch the graph of $G(x) = x^5 + 3x^4 - 2x^3 - 8x^2 + 2x + 6$ showing the coordinates of the important points correct to 3 decimal places.

Enter $G(x)$ into the HP-28 in an evaluation program from which you can both recall $G(x)$ and evaluate it for specific values of x:

$<< \to$ x   'x^5 + 3x^4 - 2x^3 -8x^2 + 2x + 6 ' $>>$

Save this under a name of your choice, for example: 'GEV [STO]

Recall $G(x)$: x [GEV]^M

Graph G: [PLOT] [STEQ]^M ([NEXT] [PPAR]^M [PURGE] [PREV]) [DRAW]^M

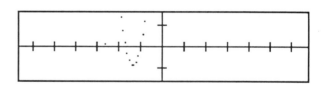

Not much with the default PPAR. Try [NEXT] 5 [*H] [PREV] [DRAW]^M .

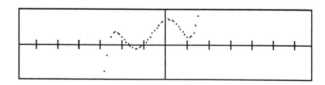

This is better.

To find the coordinates of the extreme points shown, get $G(x)$ on stack level 1; take its derivative: 'x [ENTER], [d/dx] . Store the derivative: 'DG, [STO]. Now recall the derivative and DRAW its graph. Notice that we can see all the x-intercepts.

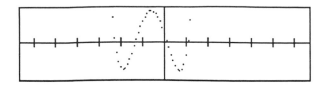

Move the crosshairs along the x axis with the cursor keys to a point close to each root and record the coordinates of each point with INS. Now return to the stack-display screen with ON. For my choice of points, I got:

        4: (1.10000, 0.00000)
        3: (0.10000, 0.00000)
        2: (-1.30000, 0.00000)
        1: (-2.30000, 0.00000)

Solve for the roots of dG/dx: SOLV SOLVR M X M RED X M . This will give you the root corresponding to your point estimate on level 1. Store it: 'DGR1 STO. Repeat this process to get the remaining three roots and store them. You can jot the roots of dG/dx down on a piece of paper if you don't want to store them, However, it is convenient to keep everything in the HP-28. Also, unless you jot down all 12 digits, you lose some accuracy. If you want to keep your USER menu neat, you can store the 4 roots in a LIST. You can now find the y coordinates of the extreme points by recalling each of the roots of dG/dx and finding the value of G there with your evaluation program for G. Store each of these ('GXTR1 STO 'GXTR2 STO etc). Find d^2G/dx^2 and store it: DG, 'x, ENTER d/dx , 'D2G, STO. Now recall D2G and DRAW M its graph:

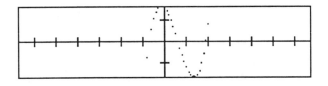

After doing 2*H, I got the graph shown above which is fine since it shows what we want, the roots.

Get initial approximations for the roots by moving the crosshairs close to the x-intercept points and entering these points with INS. Now use SOLVR with these initial approximations to find each of the roots of d^2G/dx^2, storing them as you find them. These are the x-coordinates of the inflection points of G. Now find the values of G at each, using your evaluation program.

Now recall $G(x)$, DRAW its graph, and find its x-intercepts by entering points near them on the stack as initial estimates and then using SOLVR to find the roots of G.

Usually you would now be asked to sketch your own graph of G, with equal unit distances on the two axes if possible, showing the coordinates of important points. This would give a graph of G without the vertical compression usually needed to fit a graph to the HP-28 screen.

Here, however, I will just show the screen graph of G and list the points:

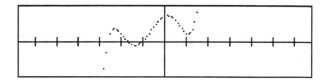

x-intercepts:	-2.607,	-1.668,	-1.000
local maximum pts:	$(-2.275, 3.015)$,		$(0.121, 6.122)$
local minimum pts:	$(-1.335, -0.881)$,		$(1.089, 1.858)$
inflection points:	$(-1.894, 1.333)$,		$(-0.605, 2.627)$,
	$(0.699, 3.692)$		

Curve-sketching problems requiring the use of the calculator are given on tests. The one below is from an hour test:

Sketch the graph of $f(x) = x^3 - 2x^2 - x - 1$, showing the coordinates, correct to 4 decimal places, of any local maximum, local minimum or inflection points .

Curve Sketching

Here is an example of how to use curve-sketching techniques to find highest, lowest, leftmost and rightmost points of a parametric curve

Example. Sketch the parametric curve $x = (1 + \cos t) \cos t$, $y = (1 + \cos t) \sin t$, $0 \le t \le 6.29$, and find both coordinates of the highest, lowest, leftmost and rightmost points of the curve. Since we want to find values of x and y for specific values of t, it is convenient to enter both functions as simple evaluation programs.

        << → T '(1 + COS(T))* COS(T)' >>
        ENTER 'XEVL STO, then
        << → T '(1 + COS(T))*SIN(T)' >>
        ENTER 'YEVL STO.

To plot the parametric curve, return to the default PPAR. Taking $n = 50$ works well here, so enter 50 for n. You can recall x and y to the screen by doing 'T ENTER XEVL M then 'T ENTER YEVL M. Now store these with YSTO M then

$\boxed{\text{XSTO}}$M. Plotting with $\boxed{\text{CLDRA}}$M gives the heart-shaped graph shown below.

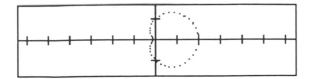

We would now like to find coordinates for the highest and lowest and leftmost and rightmost points on this graph. The high and low points will occur at those values of t for which $y(t) = (1 + \cos t)\sin t$ has a maximum or minimum. The rightmost and leftmost points will occur at those values of t for which $x(t) = (1 + \cos t)\cos t$ has a maximum or minimum. To find the leftmost and rightmost points, you must recall $x(t)$, find its derivative, plot the derivative and find its roots by our previous techniques and then find the x and y coordinates of the points corresponding to these values of t. From the parametric plot above, the rightmost point of the graph is obviously $(0,1)$, which corresponds to both $t = 0$ and $t = 2\pi$ so we can ignore these roots and direct our attention to the other values of t where $x'(t) = 0$. Do T $\boxed{\text{XEVL}}$M 'T $\boxed{\text{ENTER}}$. $\boxed{\text{d/dx}}$, $\boxed{\text{PLOT}}$ $\boxed{\text{DRAW}}$M. This should give you the following graph:

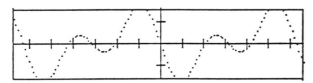

Moving the cursor to each of the three points between 0 and 2π, recording their coordinates on the stack, then using SOLVR with these points as initial guesses gives for the roots the three numbers 2.09440, 3.14159 and 4.18879. The second number is clearly π and the point on the graph corresponding to $t = \pi$ is $(0,0)$ which, from the graph of the parametric curve is a local rightmost point. Evaluating x and y at the first value of t gives $x(2.09440) = -0.2500$ and $y(2.09440) = 0.43301$. Evaluating x and y at the third value gives $x(4.18879) = -.25000$ and $y = -0.43301$. So the two leftmost points of the graph are $(-0.25, 0.43301)$ and $(-0.25, -0.43301)$.

Now you need to get the expression for $y(t)$ on level 1, take its derivative, plot it, get first estimates for the roots from the t intercepts on the graph and use SOLVR to find these roots. Note that the root

at $t = \pi$ gives $y'(\pi) = 0$, but this is neither a max nor a min for y since y' is negative on both sides of it. So we can ignore $t = \pi$. The other two roots of y' are 1.04720 and 5.23599. Finding the points of the parametric graph corresponding to these two values of t gives $(0.75000, 1.29904)$ as the high point and $(0.75000, -1.29904)$ as the low point.

Exercises.

1. DRAW the graph of the given parametric curve: $x = \cos t - \cos(4t)$, $y = \sin t - \sin(4t)$, $0 \le t \le 6.29$

2. DRAW the graph of the parametric curve $x = t^2$, $y = \frac{1}{3}t^3$, $-5 \le t \le 5$, and find the x and y coordinates of the rightmost, leftmost, highest and lowest points of the graph.

Integration and the HP-28

The simplest way to evaluate a definite integral with the 28S is to use the built-in integration routine. But in the student handout on integration, we provide programs which calculate left-endpoint, right endpoint and midpoint Riemann sums. This directory is later expanded to include the trapezoidal rule and Simpson's rule, and helps give insight into the convergence of Riemann sums and to introduce numerical integration. Here are some exercises which exploit these early approximation methods.

Exercises:

1. Approximate $\displaystyle\int_0^3 \frac{1}{1 + x^2}\, dx$

   (a) Using LRECT and RRECT with $n = 50$, 100 and 200. What is the size relation between LRECT, RRECT and the actual value of the integral for a given n? It would probably be helpful to DRAW the graph of the integrand to see the size relations between the Riemann sums.

   (b) Using MID with $n = 50$, 100 and 200

   (c) Using TRAP with $n = 50$, 100 and 200

   (d) Using SIMP with $n = 50$, 100 and 200

   (e) Find an antiderivative for $\dfrac{1}{1 + x^2}$ and evaluate the integral using the fundamental theorem. Compare this answer with those above.

2. Repeat parts (a) through (d) of Exercise 1 for $\displaystyle\int_0^3 \frac{1}{1 + x^3}\, dx$. Can you repeat part (e) for this integrand?

3. Approximate $\displaystyle\int_0^2 \sqrt{1 + \sin^2 x}\, dx$

   (a) Using Simpson's rule with $n = 100$

(b) Using the built-in numerical integration program with error factor .000001.

4. Approximate $\int_1^3 \sqrt{1 + 9x^4}\, dx$ using the built-in program with accuracy factor .0001.

5. Repeat Exercise 4 for $\int_0^{1.5} \sqrt{1 + \sec^4 x}\, dx$.

An hour-test problem which uses these programs:

Sketch the graph of $f(x) = \frac{1}{x}$ on the coordinate axes given below.

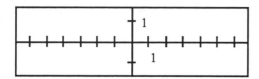

On your graph, show the rectangles of the Riemann sum on the interval [1,4] for $n = 3$ with the function evaluated at the left endpoint of each subinterval.

Evaluate the left-endpoint and right-endpoint Riemann sums on [1,4] for $n = 20$ and $n = 50$.

From an hour test, an arclength problem:

Express the arclength of the curve $y = \sin x$ for $0 \le x \le 1$ as a definite integral and use the integration program on the HP-28 to evaluate it.

Taylor Polynomials and Infinite Series

On the handout concerned with Taylor polynomials and infinite series, we include a program (program INFSM) which will find the sum of a convergent infinite series. The program displays the sum dynamically as a 12-digit number whose last digits change as more partial sums are found. The sum is apparent when the digits stabilize. Here is a problem from an hour test:

Test each series for convergence or divergence, stating the test or theorem you use and showing the details. For those that converge, use the INFSM program to find the sum.

(a) $\sum_{n=1}^{\infty} \dfrac{n}{100n + 1}$

(b) $\sum_{n=1}^{\infty} \dfrac{3^n}{n!}$

From a final exam, a Taylor polynomial problem:

On each set of axes given below, sketch both graphs requested where the Taylor polynomials are for $f(x) = \cos x$, $c = 0$:

(a) sketch $\cos x$ and $P_6(x)$

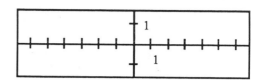

(b) sketch $\cos x$ and $P_{16}(x)$

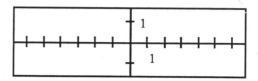

Graphing successively higher degrees of Taylor polynomials for a function is easy to do on the calculator and helps illustrate the transition from the local convergence to the global convergence of the Taylor series.

Multidimensional Calculus

The following sample materials are taken from student handouts and include samples of the text, worked examples and exercises. The materials were selected to illustrate a single sequence of ideas, graphs of conic sections, which is threaded throughout the course. Other sequences of ideas include numerical integration, symbolic manipulation and solutions of equations.

Graphs of Conic Sections

We can create a facility for producing graphs of conics on the HP-28S by entering the following programs.

```
level 1                    level 1
expression ⇒               flag
FNC?:
  <<→ rel                  | Input expression.
    << 'Y' PURGE rel
    'Y' ∂ 'Y' ∂ 0 ==       | Test if function.
    >>  >>
      level 1
      expression ⇒
LDR:
  << 'X' PURGE → rel       | Input expression.
  << rel FNC?              | Test if function.
    << rel 'Y' 1
    TAYLR
    Y' ISOL
    >>                     | Solve for Y.
```

```
<< rel 'Y' QUAD
>> IFTE STEQ      | Store in EQ.
'X' INDEP         | Make X the plotting
                    variable
   >>  >>
```

Apparently DRAW is not well behaved when used in a program; *i.e.*, after an error, something unintended might be left on the stack. We can fix this by using the following program in place of DRAW in a program.

DFix:
```
<<
[1]                | Marks the end of the
                     existing stack
IFERR DRAW         | While executing DRAW,
                     look for an error
THEN DROP ERRM 1   | If there is drop [ 1 ]
  ELSE             | If there is no message
[1]                  return the stack to its
   WHILE SAME
   NOT             | Original state
   REPEAT
[1]
   END             | End WHILE
   END  >>         | End IFERR
```
DRAR:
```
<< −1 1 2 → stp    | Initialize loop
                     parameters.
   << CLLCD
   FOR x x 's1'
   STO DFix stp    | Draw branch of curve.
   STEP
   >> DGTIZ  >>    | Activate cursor.
```

Example. Sketch $4y^2 - 8x - x^2 + 32y + 49 = 0$.
Enter
```
'PPAR' PURGE
'4*Y^2 - 8*X - X^2 + 32*Y + 49' LDR DRAR
```

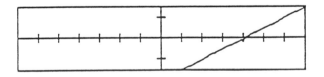

The resulting sketch isn't very staisfying. How can we improve it? Completing squares our equation becomes $(x+4)^2 - 4(y+4)^2 = 1$, so the figure must be a hyperbola with center $(-4, -4)$. Set the center in PPAR as follows:

4 CHS DUP R→ C CENTER

Executing DRAR again produces the following:

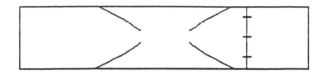

The following material was taken from student handouts. The material was edited to illustrate the use of the symbolic manipulation and graphics capabilities of the HP-28S for exploring a range of problems from multidimensional calculus.

Curves, Surfaces and Functions

In one-dimensional calculus we applied our methods to problems associated with functions: max/min problems for functions, average value of a function, etc. We usually did not distinguish a function and its graph nor make a big deal out of the application of calculus to implicitly defined curves, i.e., implicit differentiation was introduced as a natural extension of differentiation of functions. A consequence of this blurring of distinctions is a heightened emphasis on functions as opposed to geometric curves.

In multidimensional calculus the more fundamental objects of our study are geometric curves and surfaces. Many of the concepts to be introduced are inherently geometric and functions enter as an aid for the study of these concepts. Furthermore, we can usually apply functions to the geometric objects in more than one way, though some particular way will likely be most useful.

Thus the task now is to shift our attention from functions to curves and surfaces, solving some of the same problems introduced in one-dimensional calculus. However, we introduce other problems which have not been studied before. Functions of several variables have more than one kind of derivative, for instance, directional derivatives. Line and surface integrals are not just multidimensional versions of the Riemann integral.

Obviously, in thinking about curves and surfaces we are not going to abandon what we know of differentiating and integrating functions. This leads to two questions. How can we use functions to describe curves and surfaces in higher dimensional spaces? How is calculus applied to those functions to study the curves and surfaces they describe?

A surface **S** is said to be given *explicitly* by a function f, usually from R^n to R, provided **S** is the graph of f. A surface **S** is said to be given *implicitly* by a function f, usually from R^n into R, provided $\mathbf{S}$ is a level set of f, i. e., there is a number k such that $\mathbf{S} = \{\mathbf{x}|f(\mathbf{x}) = k\}$. Note that $\mathbf{S}$ is a subset of the domain of f.

Implicitly defined curves can also be used to study explicitly defined surfaces. Given an explicit surface $z = f(x, y)$ we produce the implicitly defined level curves $k = f(x, y)$. The level curves of a surface, in this case, are a two dimensional representation of a three dimensional surface. Use LDR and DRAR to obtain the level curve representation of the surface $z = 4y^2 - 8x - x^2 + 32y + 49$ given below.

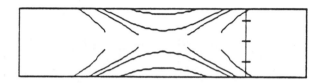

A curve $\mathbf{C}$ (sometimes a surface) is said to be given parametrically by a function $\mathbf{f}$, usually from some subset U of R to R^n, provided $\mathbf{C}$ is the image of U under $\mathbf{f}$, i.e., $\mathbf{C} = \mathbf{f}(U)$. Note that $\mathbf{C}$ is a subset of the range of $\mathbf{f}$.

Line Integrals

Example 4. Compute

$\int \mathbf{F} \cdot \mathbf{dr}$, where $\mathbf{F}(x, y) = xy\mathbf{i} + (x - y)\mathbf{j}$, and C is given in polar coordinates by $r = \cos\theta$. Of course, our first task is to obtain a parametric representation of C in rectangular coordinates. This is easily done since $x(\theta) = r(\theta)\cos\theta = \cos^2(\theta)$ and $y = r(\theta)\sin\theta = \cos\theta\sin\theta$.

```
'X*Y' 'P' STO 'X - Y' 'Q' STO
'SQ(COS(T))' 'X' STO
'COS(T)*SIN(T)' 'Y' STO
'T' PURGE
'X' 'T' d/dx 'XP' STO
'Y' 'T' d/dx 'YP' STO
0 'π' 'P*XP + Q*YP'
'T' IGL
```

Classifying Critical Points

The Taylor series expansion of $f(x)$ about x_0 is given by

$$f(x) = \sum_{p=0}^{\infty} \frac{f^{(p)}(x_0)}{p!}(x - x_0)^p.$$

The first $n + 1$ terms, i.e., for $p = 0, 1, \ldots, n$, is called the nth degree *Taylor polynomial* $P_n(x)$. Of

course, $P_1(x)$ is the line tangent to the graph of $f(x)$ at $(x_0, f(x_0))$. Similarly, each of the polynomials approximates the function. We look for critical points of $y = f(x)$ by finding values for which the graph of $P_1(x)$ is a horizontal line. A critical point x_0 of $y = f(x)$ is classified as a local minimum if the second degree Taylor polynomial $P_2(x)$ for $f(x)$ is a parabola which opens up. Notice that $P_2(x) = f(x_0) + f''(x_0)(x - x_0)^2$, which opens up provided $f''(x_0) > 0$. Similarly, x_0 is classified as a local maximum if $P_2(x)$ opens down. If $f''(x_0) = 0$ then $P_2(x)$ does not give any information for classifying the critical point x_0.

For $z = f(x, y)$ we proceed as follows: let $w(t, x, y) = f(x_0 + t(x - x_0), y_0 + t(y - y_0)$, a section of $z = f(x, y)$ in the direction of (x, y). The second degree Taylor polynomial at 0 for $w(t, x, y)$ (holding (x, y) fixed) is

$$\begin{aligned}P_2(t, x, y) =& f(x_0, y_0) + [f_x(x_0, y_0)(x - x_0) \\ & + f_y(x_0, y_0)(y - y_0)]t \\ & + (1/2)[f_{xx}(x_0, y_0)(x - x_0)^2 \\ & + 2f_{xy}(x_0, y_0)(x - x_0)(y - y_0) \\ & + f_{yy}(x_0, y_0)(y - y_0)^2]t^2\end{aligned}$$

which reduces to

$$\begin{aligned}P_2(t, x, y) =& f(x_0, y_0) \\ & + (1/2)[f_{xx}(x_0, y_0)(x - x_0)^2 \\ & + 2f_{xy}(x_0, y_0)(x - x_0)(y - y_0) \\ & + f_{yy}(x_0, y_0)(y - y_0)^2]t^2\end{aligned}$$

when (x_0, y_0) is a critical point.

Example. Find the second degree Taylor polynomial for $w(t, x, y)$ given $z = (\sin x)(\sin y)$ and $(\pi/2, \pi/2)$ is a critical point . Enter the following:

```
'SIN(U)*SIN(V)' 'Z' STO
'π/2 + T*(X - π/2)' 'U' STO 'π/2 + T*(Y - π/2)' 'V'
STO Z EVAL 'T' 2 TAYLR
```

In order to classify $(\pi/2, \pi/2)$ as either a local maximum, a local minimum or a saddlepoint, we must determine if $P_2(t, x, y)$ is a parabola opening down (or up) for all (x, y) or opening down for some (x, y) and up for others. We can decide by sketching the level curves of $P_2(1, x, y)$. Proceed as follows:

```
1 'T' STO EVAL
'K' PURGE 'K' + LDR
0 'K' STO DRAR DEL ON
1 'K' STO DRAR DEL ON OR
2 'K' STO DRAR DEL ON OR →LCD
```

Clearly, we have produced the level curves of an elliptic paraboloid, i.e., the coefficient of t^2 in $P_2(t)$

must always be either positive or negative. Therefore $(\pi/2, \pi/2)$ must be a local extremum. Furthermore, since the paraboloid looks down $(\pi/2, \pi/2)$ must be a local maximum.

Exercise. Use the method outlined above to classify the critical points $(-1, 11/6)$ and $(1, 1/2)$ of $z = x^3 + y^2 + 2xy - 4x - 3y + 5$ as either local extrema or saddlepoints.

Polya's Problems

If the sum of two numbers is 6, what is the maximum of their product?

Let the numbers be x and y. We are given that $x + y = 6$ and we wish to maximize the function $f(x, y) = xy$ subject to that constraint. From the hand out, we seek a point (x_0, y_0) on $x + y = 6$ where the level curve $f(x, y) = x_0 y_0$ is tangent. We proceed as follows:

Create a program that combines two screen images and leaves the cursor active.

<< OR →LCD DGTIZ >> 'TNGT' STO

We then produce a graph containing both $x + y = 6$ and $xy = 1$, i.e., the constraint and an arbitrary level curve of $f(x, y)$.

'PPAR' PURGE '6 - X' STEQ DRAW DEL ON
ENTER 'K' PURGE 'X*Y - K' LDR 1 'K' STO
DRAR DEL ON TNGT

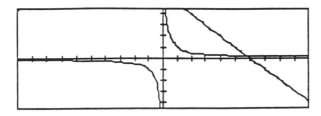

Of course, $xy = 1$ is not tangent to the constraint $x + y = 6$ at any point. We can choose a more appropriate level curve of $f(x, y) = xy$ by moving the cursor to a point on the constraint $x + y = 6$ where we think some level curve is tangent. Capture that point with INS and then continue. (Our guess resulted in $(3.1, 2.9)$.)

ON C→R × 'K' STO

ENTER DRAR DEL ON TNGT
This will produce something like the following:

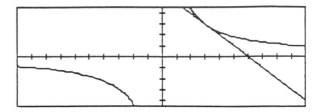

The picture is pretty convincing that either $(3.1, 2.9)$ is a point of tangency or near such a point. To check our answer, we can zoom in by setting the center at $(3.1, 2.9)$ and repeating the process from the top.

Exercise. Find the minimum of $x^2 + y^2$ on the curve $x = y^2 + 1$.

Vector Fields

A vector valued function defined on a subset of R^n, $n > 1$, is called a *vector field*. Similarly, a scalar valued function is called a *scalar field*. We will tend to use a standard notation, for instance, $\mathbf{f}(x, y) = P(x, y)\mathbf{i} + Q(x, y)\mathbf{j}$. Of course, $\mathbf{f}$ is a vector field with component functions P and Q which are scalar fields.

Use the fact that a gradient vector at a point is normal to the level curve through the point to add normalised vectors to the level curves of $z = x^2 - y^2$ given below and recapture the previous picture. Remember which direction is up hill.

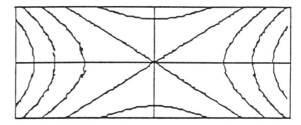

Exercise 4. Produce the gradient fields for $z = y^2$ and $z = x^2 + y^2$.

Green's Theorem

Green's Theorem may be stated as follows: Suppose $\mathbf{F}(x, y)$ is a vector field, i.e., $\mathbf{F}(x, y) = P(x, y)\mathbf{i} + Q(x, y)\mathbf{j}$ where $P(x, y)$ and $Q(x, y)$ are scalar functions (fields). Assume that $P_y(x, y)$ and $Q_x(x, y)$ are continuous in a bounded region R with a piecewise smooth boundary C which is oriented positively. (C is given parametrically by $\mathbf{r}(t) = x(t)\mathbf{i} + y(t)\mathbf{j}$, $a \leq t \leq b$, and $x(t)$ and $y(t)$ are piecewise smooth. Furthermore, as t varies from a to b, $\mathbf{r}(t)$ traces out C keeping R on the left.) Then

$$\int_C \mathbf{F} \cdot \mathbf{dr} = \int_a^b P(x,y)\,dx + Q(x,y)\,dy$$
$$= \int\int_R Q_x(x,y) - P_y(x,y)\,dxdy$$

Exercise. Compute the area of the region R bounded by the ellipse $\dfrac{x^2}{9} + \dfrac{y^2}{4} = 1$. Our standard parameterization of the boundary C of R is given by $\mathbf{r}(t) = (3\cos t)\mathbf{i} + (2\sin t)\mathbf{j}$, $0 \le t \le 2\pi$.

Differential Equations

Programs to calculate and graph solutions of initial value problems using the Euler and improved Euler method for first order, second order and higher order problems are given to the student and used often. A Runge-Kuttta algorithm which adaptively selects step size is presented briefly and used on an earth-moon satellite problem.

The following graphics program (for the HP-28S) named GRAF is used for plotting approximate solutions of $dy/dx = f(x,y)$, $y(x_0) = y_0$:

```
<< CLLCD DRAX 1 N START EULER
DUP2 R→C PIXEL NEXT DGTIZ LCD→ >>.
```

The program first clears the screen, draws the axes and takes the starting values of x and y from the stack, passes them to a loop which takes an input pair to the EULER program for one step using the Euler algorithm, plots the new point and repeats. Finally the cursor is activated and the picture is converted into an object for later display. The EULER program is also short and sweet:

```
<< DUP2 FN H * + SWAP H + SWAP >>.
```

The student must only write a program FN taking the numbers x, y from the stack to produce the value of $f(x,y)$, decide on the step size and the plotting range. The GRAF program can be modified to plot approximate solutions using other solution algorithms or to plot the values of two variables, say $x(t)$, $y(t)$ from a system of differential equations. (A small modification of GRAF is required for the HP-48SX.)

Programs which create the graph of the output of a first and second order linear non-homogeneous system from the variation of constants formula for various forcing functions are given and used with step function, ramp function, and other "natural"

inputs to circuits. The extension to a vector system is outlined.

The asymptotic behavior of solutions of dynamical systems is discussed in several senarios. An example developed in class is shown below. Trajectories are asymptotic to curves of the form $y = n\pi/x$.

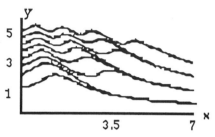

Trajectories for $dy/dx = \sin(xy)$

A second example is to emphasize the steady state solutions to spring/circuit problems with periodic forcing.

A third example is a graphical study of a nonlinear first order discrete system containing a parameter. As the value of the parameter increases the asymptotic behavior changes from a single equilibrium point, to two equilibrium points, which split again and again as the parameter increases to four, eight, sixteen points, etc. A fourth example is a discrete dynamical system in the complex plane; points in the Julia set are graphed. We point out how continuous systems differ from the discrete systems; the importance of equilibrium and periodic solutions in the plane is discussed and an example of chaos in three dimensions is presented.

The calculation of solutions to linear vector systems $dy/dt = Ay + f(t)$ can be accomplished conveniently with the calculator. A program is given which produces the characteristic (eigenvalue) equation when a matrix A is stored. Some solutions may be determined by graphing the values of the polynomial as the value of, say r, is varied. Finally for a characteristic value r, nontrivial vector solutions to the homogeneous equation $(A - rI)v = 0$ may be determined by systematically using a Gauss/Jordan reduction algorithm. In this way, a fundamental set of solutions to $dx/dt = Ax$ is constructed in steps to emphasize the concepts of independance, eigenvector subspaces, etc. (It is feasible to extend the class discussion from vectors of size two in the traditional class to size four or five because of the ease of calculation.) Then a particular solution to the nonhomogeneous differential equation can be determined by using the variation of constants formula or by the determination of coefficient vectors.

The calculator is also used for solving linear and nonlinear equations encountered in the course. Examples include the solution of transcedental equations determining the time when the concentration of a mixture reaches a predetermined value, the time when a parachutists hits "ground," and the solution of some algebraic equations with no rational roots.

Students choose between several suggestions and turn in two projects during the term. A list for the first project contains (1) calculation of the times required by a bead traversing a curve to fall between two points and comparison with the time required when traveling on a cycloid, (2) a study of the trajectories of $dy/dx = \sin(xy)$ (or $\cos(xy)$), (3) a comparison study between the logistic, Gompertz and a third model for constrained population growth. A suggestion list for the second project is (1) a comparison study of trajectories of several spring models, (2) parameter estimation in a population problem using a least squares fit to data and (3) graphs of linear input/output systems.

The following is taken from several such diagrams in a student's project. This particular student was asked to draw a conclusion from his inspection of several periodic input functions: he showed that there would be a periodic output and that other solutions would asymptotically settle toward that solution.

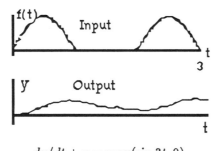

$$dy/dt + y = \max(\sin 3t, 0)$$

The majority of class and homework exercises are taken from the textbook or other available material. Sample questions which reflect calculator usage are provided below.

Problem: Graph $a(x) = -1/(1 + x)^k$ for $k = .95, .5, .01$ and 0 on the interval $0 \le x \le 5$, then graph the solution of $dy/dx = a(x)y, y(0) = 1$ for the same range of x and the same values of the parameter k. What value of $x(k)$ gives $y = .01$?

Problem: A tank initially contains 300 gallons of pure water. Brine containing 1.5 lb/gal of salt enters the tank at 2 gallons/minute and the well mixed solution leaves at 3 gallons per minute. At what times will there be 21 lb. of salt in the tank? This involves finding the solutions of $(1 - t/300)^2 = 1 - 14/(300 - t)$.

Problem: Graph the exact solution of $dy/dx = 1 - x^{1.5}$, $x(0) = .5$ for $0 \le t \le 4$. (This involves finding $x(t)$ from the implicit equation $F(x) = t$ where F is given by the difference of transcedental functions.)

Problem: Suppose a water storage tank has the shape of the lower half of the ellipsoid $(x^2 + y^2)/a^2 + z^2/b^2 = 1$. Water is introduced into the tank at a rate of $f(t) = .25 \sin[\pi(20 - t)/12]$, and evaporation occurs from the exposed surface at a rate $k(t)$ with $k(t) = (10 - \sin \pi(t - 6)/12)/200$. (Here we are assuming the greatest evaporation occurs during the daylight hours.) We want to determine the height $r(t)$ of water in the tank at time t, given $r(0) = 0.5b$. The volume in the tank is given by

$$V = \pi a^2 \int_0^r \frac{b^2 - (b - s)^2}{b^2} \, ds.$$

A balance equation in the scaled variable $u = r/b$ gives the equation

$$b\frac{du}{dt} = \frac{f(t)}{\pi a^2 u(2 - u)} - k(t).$$

Use the following parameters (no physical units intended) and graph the solution for $0 \le t \le 24$ with a step size of 0.08: $a = 2$, $b = 1$ and $r(0) = u(0) = .5$.

Problem: Graph the family of curves $xy = k$ (where k is varied) and members of a second family of curves orthognal to curves $xy = k$.

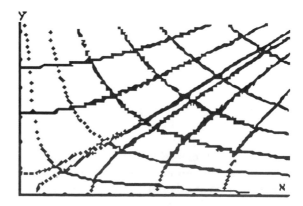

Orthogonal Trajectories

Problem: Use starting values $p = .25$, $q = 2$ to iteratively determine p and q to fit the following (t, y) data:

$$y = 3e^{-pt} - 2e^{-qt}$$

$(0, 1)$, $(.4, 1.89)$, $(.8, 2.01)$, $(1.2, 1.9)$, $(1.6, 1.72)$, $(2, 1.53)$, $(2.4, 1.34)$, $(2.8, 1.18)$, $(3.2, 1.03)$, $(3.6, .903)$, $(4, .79)$, $(5, .57)$. Such problems may occur naturally in this course in trying to determine damping and spring constants in a linear spring motion.

Problem: For what values of s does the problem $dy/dt = ay(t - 1)$ have a solution $y(t) = Ce^{st}$? (Here the student must solve $s = ae^{-s}$ for all real and complex solutions s.) Under what conditions (on a) do all solutions have negative real part? Give examples of functions $\phi(t)$, $-1 \leq t \leq 0$ so that the solution of the differential equation together with the initial condition $y(t) = \phi(t)$, $-1 \leq t \leq 0$ satisfies $y(t) \to 0$ as $t \to \infty$.

Problem: Graph points in the Julia set for the maping $z \to z^2 + (-.123, .745)$ (This graph is called Douady's rabbit: the execise uses complex numbers and is an illustration of the asymptotic set for a dynamical system.)

Problem: Find the period of the solution of $y'' + \sin y = 0$, $y(0) = k$, $y'(0) = 0$ for $k = .5, .75, 1$. (The problem is generalized to $y'' + f(y) = 0$: the student is asked to give examples of functions $f(y)$ so the system has periodic solutions with period which depends on the initial value and which increase (decrease) as as the initial values increase, and to describe other ways in which these systems are different from linear systems.)

Problem: Choose the parameters a, b, c, and d so that the function $C(z) = ae^{-bz} + c + dz$ fits a given data set in a least squares sense. Graph (phase plane and time spacial trajectories) for the system $dz/dx = y/C(z)$, $dy/dx = -C'(z)$, for trajectories satisfying $z(0) = 2000$, $z(xf) = 3000$ where $xf = 24(5280)$.

Problem: A rabbit starts at $(0, 1)$ and runs along $y = 1$ in the positive y direction with speed 1. At the same time a dog starts at $(0, 0)$ and pursues the rabbit with speed 1.3. The dog attempts to point at the rabbit at all times but is constrained by his momentum. That is, for the angle z between the dog's direction and the x axis, dz/dt is restricted. Graph the path of the dog and the rabbit. When does capture take place? The equations of motion are:

$$\frac{dx}{dt} = 1.3 \cos z,$$

$$\frac{dy}{dt} = 1.3 \sin z,$$

$$\frac{dz}{dt} = -(z - \theta(t, x, y))$$

$$x(0) = 0, \quad y(0) = 0, \quad z(0) = 1.8.$$

Here $\theta(t, x, y)$ is the angle between the dog-rabbit vector and the x axis.

Problem: (Lorentz Equations) The problem:

$$\frac{dx}{dt} = \sigma(y - x),$$

$$\frac{dy}{dt} = (r - z)x - y,$$

$$\frac{dz}{dt} = xy - bz.$$

These convective heat transfer model equations were studied by E. Lorentz (1963). An equilibrium or periodic solution to the system represents predictable behavior. The graphs of particular solutions gave surprising results because most trajectories never seem to approach such predictibility. For $r = 1.3$, $\sigma = 10$, and $b = 8/3$ find the three critical points, the linear variational equations at the critical points and the eigenvalues of these systems. Determine stability of the point. Compute and graph a numerical solution for the initial values $x(0) = 13.25$, $y(0) = 19$, $z(0) = 26$ for $\sigma = 10$, $r = 28$, and $b = 8/3$ for $0 \leq t \leq 3$.

Problem: Find the equilibrium solutions of the restricted three body problem modelling the earth-moon-satallite. Determine their stability. (This involves finding the solution of two nonlinear equations in two variables and the eigenvalues of a 4 by 4 matrix.)

Linear Algebra

LU-Factorizations

Example. Attempt an LU-factorization of

$$A = \begin{bmatrix} 2 & 3 & -1 & 2 \\ -4 & -6 & 2 & 1 \\ 2 & 4 & -4 & 1 \\ 4 & 8 & 2 & 7 \end{bmatrix}.$$

Step 1: Enter A onto level 1, press $\boxed{\text{MAK.L}}$M to create an appropriate starting matrix ELL, and press 1, 1 $\boxed{\text{LU}}$M to see

$$\begin{bmatrix} 2 & 3 & -1 & 2 \\ 0 & 0 & 0 & 5 \\ 0 & 1 & -3 & -1 \\ 0 & 2 & 4 & 3 \end{bmatrix}.$$

Step 2: Since the $(2,2)$-entry of this last matrix is 0, we must interchange row 2 with some lower row, say row 3. Thus press 2, 3 $\boxed{\text{RO.KL}}$M to effect the interchange, then bring ELL to level 1 with $\boxed{\text{ELL}}$M, make the same row interchange and store the result in ELL.

Step 3: Now execute 2, 2 $\boxed{\text{LU}}$M to see

$$\begin{bmatrix} 2 & 3 & -1 & 2 \\ 0 & 1 & -3 & -1 \\ 0 & 0 & 0 & 5 \\ 0 & 0 & 10 & 5 \end{bmatrix}.$$

Step 4: Interchange rows 3 and 4 with 3, 4 $\boxed{\text{RO.KL}}$M, bring ELL to level 1 with $\boxed{\text{ELL}}$M and make the same interchange, and store in ELL.

Step 5: See

$$U = \begin{bmatrix} 2 & 3 & -1 & 2 \\ 0 & 1 & -3 & -1 \\ 0 & 0 & 10 & 5 \\ 0 & 0 & 0 & 5 \end{bmatrix}$$

and

$$ELL = \begin{bmatrix} 0 & 0 & 0 & 0 \\ 1 & 0 & 0 & 0 \\ 2 & 2 & 0 & 0 \\ -2 & 0 & 0 & 0 \end{bmatrix}.$$

Get

$$L = \begin{bmatrix} 1 & 0 & 0 & 0 \\ 1 & 1 & 0 & 0 \\ 2 & 2 & 1 & 0 \\ -2 & 0 & 0 & 1 \end{bmatrix}$$

with 4 $\boxed{\text{IDN}}$M $\boxed{+}$, then do $\boxed{\text{SWAP}}$ $\boxed{*}$ to see

$$LU = \begin{bmatrix} 2 & 3 & -1 & 2 \\ 2 & 4 & -4 & 1 \\ 4 & 8 & 2 & 7 \\ -4 & -6 & 2 & 1 \end{bmatrix}.$$

Step 6: (Check) Since $PA = LU$ where P is the product of the elementary permutation matrices

P_{23} and P_{34}, $P = P_{34}P_{23}$, we have $P^{-1}LU = A$ and $P^{-1} = P_{23}^{-1}P_{34}^{-1} = P_{23}P_{34}$. Thus, interchanging rows 3 and 4 of LU, then rows 2 and 3 of the result will show the original matrix A. Try it!

Norms on the 28S

Three matrix and vector norms are provided on the ARRAY menu of the 28S: the Euclidean norm, the ∞-norm and the 1-norm.

- The Euclidean (Frobenius) matrix norm: $\|A\|_F = \boxed{\text{ABS}}$M. Since vectors on the 28S are 1-dimensional arrays sensed as column vectors, $\boxed{\text{ABS}}$M applied to a vector v returns its vector Euclidean norm $\|v\|_2$.

- The row-sum, or ∞-norm: $\|A\|_\infty = \boxed{\text{RNRM}}$M. For a vector v, $\boxed{\text{RNRM}}$M returns its vector max norm $\|v\|_\infty$.

- The column-sum, or 1-norm: $\|A\|_1 = \boxed{\text{CNRM}}$M. For a vector v, $\boxed{\text{CNRM}}$M returns its vector sum norm $\|v\|_1$.

Exercises.

1. Use the matrix A and vector x given below to verify that $\|Ax\| \le \|A\|\,\|x\|$ for the three matrix and vector norms provided by the 28S.

$$A = \begin{bmatrix} -3 & 0 & 0 \\ 2 & 1 & 5 \\ 1 & -4 & 4 \end{bmatrix}, \qquad x = [-1\ 2\ 1]$$

2. Consider the attempt to define a matrix norm by $\|A\| = \max_{i,j} |a_{ij}|$. Experiment with random 3×3 matrices (use 2 FIX mode) to find a pair A, B for which the inequality $\|AB\| \le \|A\|\,\|B\|$ is invalid. (To calculate $\max_{i,j} |a_{ij}|$ for matrix A, separate A into its entries with $\boxed{\text{ARRY}\rightarrow}$M, then reassemble into a vector v with $\boxed{\rightarrow\text{ARRY}}$M. Now apply $\boxed{\text{RNRM}}$M to v.)

Projections and Least Squares

The following program, P.FIT, may be used to create the coefficient matrix A for fitting a polynomial of degree $\le m$ to n data points (x_i, y_i), $i = 1, \ldots, n$. When $m = n - 1$, A will be a Vandermonde matrix.

entries of $A^T A$ may produce very large errors in the solution. Thus, in many real-world problems we seek other ways to solve the normal equations and will return to this point later.

Exercises:

1. (a) With your calculator in 2 FIX mode, generate a random 4×3 matrix whose columns will be called u, v and w. Press REAL NEXT NEXT RND M to convert to 2 digits, then return to STD mode.

   (b) Find $P_v u$ and verify that $u - P_v u$ is orthogonal to v.

   (c) Find $P_w u$ and verify that $u - P_w u$ is orthogonal to w.

2. (a) With your calculator in 2 FIX mode, generate a random 4×3 matrix A and a random vector b in $\Re^4$. Press REAL NEXT NEXT RND M to convert to 2 digits, then return to STD mode.

   (b) Apply Gaussian elimination using G.J.PV M to the normal equations $A^T A x = A^T b$ to obtain a least squares solution to $Ax = b$.

3. (a) Use your calculator to fill-in the following table of values for $y = 2e^{-x}$ (round to 3 decimal places).

x	-1.73	$-.4$	0	$.58$	1.21
$y = 2e^{-x}$					

   (b) Use P.FIT M to get the Vandermonde matrix associated with the interpolating polynomial of degree ≤ 4 for this data.

   (c) Find the interpolating polynomial for this data. Use Gaussian elimination to solve the normal equations.

4. Suppose you wanted to fit a fifth degree polynomial to 20 data points (x_i, y_i), where $x_i = i$. What is the coefficient matrix of the system of normal equations?

Orthonormal bases

Step 1:
$$q_1 = x_1, \; normalized$$
Step 2:
$$q_2 = x_2 - \underbrace{(x_2 \cdot q_1)q_1}_{\text{projection of } x_2 \text{ onto } q_1}, \; normalized$$
Step 3:
$$q_3 = x_3 - \underbrace{(x_3 \cdot q_1)q_1 - (x_3 \cdot q_2)q_2}_{\text{projection of } x_2 \text{ onto } q_1 \text{ and } q_2}, \; normalized$$
$$\vdots$$
etc.

P.FIT (Polynomial Fit Matrix)
Input: level N+2: an integer M
levels 2-(N+1): distinct numbers $x_N, ..., x_1$
level 1: the integer N
Effect: Returns the matrix

$$\begin{bmatrix} 1 & x_1 & x_1^2 & \cdots & x_1^M \\ 1 & x_2 & x_2^2 & \cdots & x_2^M \\ & & \vdots & & \\ 1 & x_N & x_N^2 & \cdots & x_N^M \end{bmatrix}$$

« →LIST DUP SIZE → M L N « 1 N
FOR J L J GET → EX « 1 EX 2 M
FOR I EX I ^ NEXT » NEXT M 1 +
'M1' STO { N M1 } →ARRY 'M1' PURGE
» »

Store variable P.FIT in your ORTH subdirectory, next to PROJ.

Example. Find the interpolating polynomial of degree ≤ 4 for the following 5 data points:

x	1	2	3	4	5
y	.6	1.27	2	2.8	4.1

Key in the number 4, then the x-coordinates of the data, then the number 5 (the number of data points), and press P.FIT M to see

$$A = \begin{bmatrix} 1 & 1 & 1 & 1 & 1 \\ 1 & 2 & 4 & 8 & 16 \\ 1 & 3 & 9 & 27 & 81 \\ 1 & 4 & 16 & 63 & 256 \\ 1 & 5 & 25 & 125 & 625 \end{bmatrix}.$$

Since A is known to be invertible, we may obtain the solution $c = A^{-1}y$ by applying ÷ to A and $y = [.6 \; 1.2 \; 2 \; 2.8 \; 4.1]^T$. Enter y and press SWAP ÷ CLEAN M to show $c = [1.1 \; -1.525 \; 1.321 \; -.325 \; .029]$ to 3 decimal places. Thus the interpolation polynomial is approximately

$$P(x) = 1.1 - 1.525t + 1.321t^2 - .325t^3 + .029t^4$$

Though the above example does not show it, experience has shown that the coefficient matrix $A^T A$ of the normal system, with A given by (*), may be ill-conditioned, in the sense that small errors in the

This is the standard Gram-Schmidt process (there are variations). You should recall that, at each stage, $\mathrm{Span}[x_1,\ldots,x_j] = \mathrm{Span}[q_1,\ldots,q_j]$, so when we're done, $W = \mathrm{Span}[x_1,\ldots,x_k] = \mathrm{Span}[q_1,\ldots,q_k]$ and we have an orthonormal basis for W.

Relative to using the 28S to do the calculations, our position is that for beginning students it is not pedagogically sound to use a program which "does it all" and thus obscures the underlying step-by-step process. We prefer instead to prepare and evaluate a simple program to carry out each step of the construction. Begin with the basis vectors stored as variables X1, X2, ..., XK in USER memory.

Step 1:

<< X1 DUP ABS / ENTER EVAL ' Q1 STO

(Calculates q_1 and stores it as Q1.)

Step 2:

<< X2 X2 Q1 DOT Q1 * - DUP ABS / ENTER EVAL ' Q2 STO

(Calculates q_2 and stores it as Q2.)

Step 3:

<< X3 X3 Q1 DOT Q1 * - X3 Q2 DOT Q2 * - DUP ABS / ENTER EVAL ' Q3 STO

(Calculates q_3 and stores it as Q3.)

$\vdots$

... and so on.

Exercises.

1. (a) With your calculator in a 2 FIX mode, generate a random 4×3 matrix whose columns will be called x_1, x_2 and x_3. Press REAL NEXT NEXT RND^M to convert to 2 digits, then return to STD mode.

   (b) Construct an orthonormal basis $\{q_1, q_2\}$ for $W = \mathrm{Span}[x_1, x_2]$.

   (c) Find the projection vector $P_w x_3$ of x_3 onto W.

   (d) Verify that $x_3 - P_w x_3$ is orthogonal to W by checking that it is orthogonal to both x_1 and x_2.

2. Repeat exercise 1 with a random 5×4 matrix; let $W = \mathrm{Span}[x_1, x_2, x_4]$.

QR-Factorizations

To construct R on the HP-28S

(1) Start with X1, X2, ..., XN and Q1, Q2, ..., QN from Gram-Schmidt in USER memory.

(2) Construct matrix A:

X1 X2 ... XN N ROW→^M TRN^M ' A STO

(3) Construct matrix Q:

Q1 Q2 ... QN N ROW→^M TRN^M ' Q STO

(4) Construct matrix R:

Build R by putting its entries onto the stack in column order, then use →ARRY^M and TRN. Remember that in the upper triangle, $r_{ij} = q_i \cdot x_j$ for $i \le j$, so use QI XJ DOT. Enter 0's for the lower triangle. Store as ' R STO.

(5) Verify $A = QR$:

A Q R * CLEAN^M SAME^M

(SAME is located on the 2nd line of the PROGRAM TEST menu; a 1 indicates $A = QR$, and a 0 indicates $A \ne QR$. If you forget to clean-up QR, you probably won't get $A = QR$.)

Exercises.

1. (a) With your calculator in 2 FIX mode, generate a random 4×3 matrix A and a random vector b in R^4. Press REAL NEXT NEXT RND^M to convert to 2 digits, then return to STD mode.

   (b) Obtain a QR-factorization for A. Check your results.

   (c) Use your QR-factorization to get a least squares solution to the equation $Ax = b$. Use G.J.PV^M when necessary.

2. Repeat Exercise 1 for a random 4×4 matrix.

3. Keep your calculator in 2 FIX mode for this exercise.

   (a) Normalize $v = \begin{bmatrix} 1 & 2 & 3 \end{bmatrix}^T$ to get a unit vector w.

   (b) Use w to build a Householder matrix

$$H = I - 2ww^T.$$

   (c) We have previously noted that H is orthogonal. Look at H. What else is obvious?

   (d) In view of your conclusion in (c), without calculating, what will H^{-1} be?

   (e) Verify your result in (d) with a calculation. Use CLEAN^M at the end.

4. Repeat exercise 3 with a random vector v in R^4.

Eigenvalues and Eigenvectors

Example. Enter two copies of the following matrix

$$A = \begin{bmatrix} 7 & 2 & 4 & 6 \\ -6 & -1 & -4 & -4 \\ 4 & 4 & 5 & -2 \\ -16 & -12 & -14 & -3 \end{bmatrix}.$$

This time [CHAR]M returns

$$\{\,1 \quad -8 \quad 22 \quad -40 \quad 25\,\},$$

so $p(\lambda) = \lambda^4 - 8\lambda^3 + 22\lambda^2 - 40\lambda + 25$. [PROOT]M tells us that the roots are $\lambda = 1, 5$ and $1 \pm 2i$. To find the eigenspace associated with $\lambda = 1 - 2i$ we proceed as before: with $(1, -2)$ on level 1, do [CHS] 4 [IDN]M [*] [+] to see the complex matrix $[A - (1-2i)I]$. Gaussian elimination with [G.J.PV]M and [CLEAN]M shows that $[-1+i \quad 1 \quad -i \quad 1]$ spans the eigenspace.

Example. Make two copies of this 5×5 matrix A:

$$\begin{bmatrix} -.99 & -1.99 & -1.99 & -1.99 & -1 \\ -1.96 & -.96 & -3.96 & -.93 & -2 \\ 3 & 3 & 6 & 3 & 3 \\ -.04 & -.04 & -.04 & -.07 & 0 \\ -2.01 & -1.01 & -3.01 & -1.01 & 2 \end{bmatrix}$$

[CHAR]M returns

$$\{\,1 \quad -5.98 \quad 15.88 \quad -8.678 \quad -.185 \quad .003\,\}.$$

Since this represents a fifth degree polynomial $p(z)$ we cannot apply [PROOT]M directly. First, we locate a real root graphically. With the list of coefficients on level 1, store this list as COEF and then recall it to the stack. Use program PSERS to get an algebraic expression for $p(z)$: Z [PSERS]M. Now graph this expression by using the HP-28S' plotting capabilities (see Chapter 7 of the Owner's Manual if you are rusty about how to do this).

There appears to be a real root near 1, so move the cursor to it and initialize by pressing the white [INS] key. Clear your screen by touching [ON], then hit [SOLV] [SOLVR]M, then [Z]M [RED][Z] (i.e., hit the white key beneath the boxed Z on your screen).

The message appearing above level 1 tells you that $z = .745$ is an approximate root. Touch [ON] to clear the message. You now must divide $p(z)$ by $z - .745$ to get a fourth degree quotient polynomial $q(z)$. Build a list $\{\,1 \quad -.745\,\}$ representing $z - .745$ and call the list of coefficients of $p(z)$ to level 1 by pressing [COEF]M; press [SWAP], then execute the program [PDIV]M. The result shows the divisor on level 1, the remainder on level 2 and the list $\{\,1 \quad -5.235 \quad 11.981 \quad .243 \quad -.004\,\}$ on level 3. This last list represents the quotient polynomial. Drop everything except this list, then use [PROOT]M to show the roots as $.014$, $-.03$, $(2.628, 2.276)$, $(2.628, -2.276)$. You may now tackle the eigenvectors, after storing the 5 eigenvalues as R1, ..., R5.

Program PROOT does not apply to polynomials of degree 5 or higher, but the technique used in the last example may be used on any 5×5 matrix. To graphically locate a real root of the characteristic polynomial with the 28S may require that you adjust the plotting parameters. You may also use this approach on *positive* matrices of higher order, although the work may become a little unwieldly for sizes beyond 6×6. In this connection, we mention the (advanced) theorem due to Perron:

A real, square matrix having only positive entries has a real eigevalue λ, *which is a simple (i.e., not repeated) root of the characteristic polynomial, and all other eigenvalues have absolute value* $< |\lambda|$.

Exercises.

1. Adding a multiple of a row to another row will not change the determinant of a square matrix A. Will this change the eigenvalues? The characteristic polynomial? Use your 28S to investigate these questions for the matrix

$$A = \begin{bmatrix} 6 & 10 & -28 \\ -2 & -3 & 4 \\ 1 & 2 & -7 \end{bmatrix}.$$

2. For each of the matrices A in Examples 4-7:

   (a) Calculate the trace of A ($\text{tr} A$) and $\det A$; record your results.

   (b) Compare $\text{tr} A$ with $\det(\lambda I - A)$; what do you observe? Express your observation as a conjecture.

   (c) Compare $\text{tr} A$ with the sum, $\sum_i \lambda_i$, of the eigenvalues of A; what do you observe? Express your observation as a conjecture.

   (d) Compare $\prod_i \lambda_i$, the product of the eigenvalues of A, with $\det A$; what do you observe? Express your results as a conjecture.

   (e) Ask your instructor about your conjectures.

3. For each of the following matrices, find:

(a) the characteristic polynomial

(b) all eigenvalues

(c) for the eigenvalue λ of maximum absolute value, the associated eigenspace.

$$A = \begin{bmatrix} 4 & -1 & 0 & -1 \\ -1 & 3 & -1 & 0 \\ 0 & -1 & 3 & -1 \\ -1 & 0 & -1 & 2 \end{bmatrix}$$

$$B = \begin{vmatrix} -2 & -8 & -1 & 6 & 5 \\ 1 & 7 & 1 & -2 & -2 \\ -11 & -16 & 0 & 10 & 5 \\ -7 & -8 & -1 & 10 & 4 \\ 7 & 8 & 1 & -6 & 0 \end{vmatrix}$$

$$C = \begin{vmatrix} 8 & 6 & 8 & 5 & -3 \\ -9 & -8 & -10 & -9 & 1 \\ -3 & -2 & -3 & -2 & 1 \\ 11 & 11 & 12 & 12 & 1 \\ -8 & -9 & -8 & -8 & -1 \end{vmatrix}$$

$$D = \begin{bmatrix} -17 & -19 & -12 & -26 & -22 & -26 \\ -5 & 3 & -5 & -5 & -5 & -5 \\ 7 & 13 & 6 & 11 & 9 & 15 \\ -1 & -7 & -1 & 0 & -1 & 7 \\ 3 & 9 & -1 & 7 & 6 & 5 \\ 9 & 3 & 9 & 9 & 9 & 14 \end{bmatrix}$$

Similarity

Example. Consider

$$A = \begin{bmatrix} 0 & 3 & -2 & 0 & -4 \\ -4 & 5 & -4 & 0 & -4 \\ 4 & -3 & 6 & 0 & 4 \\ 4 & -3 & 4 & 3 & 5 \\ -6 & 3 & -6 & 0 & -2 \end{bmatrix}.$$

Program CHAR M returns

$$\{1 \quad -12 \quad 55 \quad -120 \quad 124 \quad -48\}$$

for the list of coefficients of the charactersitic polynomial. Use Z PSERS M to obtain an algebraic expression, then graphically locate an eigenvalue $\lambda = 1$. Use PDIV M to divide the charactersitic polynomial by $z - 1$ to obtain the quotient list $\{1 \quad -11 \quad 44 \quad -76 \quad 48\}$, then PROOT M to show the remaining eigenvalues as $\lambda = 2, 2, 3, 4$.

Since $\lambda = 2$ is the only repeated root, to settle the question as to whether A is diagonalizable or defective we must determine dim$NS[A - 2I]$. Programs G.J.PV M and CLEAN M show two free variables, so dim$NS[A - 2I] = 2$, the multiplicity of 2 as a root of the characteristic polynomial. Thus A is diagonalizable. In fact, a basis for the eigenspace associated with $\lambda = 2$ is $\{[-1 \quad 0 \quad 1 \quad 0 \quad 0] \quad [0 \quad 1.\overline{3} \quad 0 \quad -1 \quad 1]\}$. The eigenspaces associated with $\lambda = 1, 3$ and 4 have $[1 \quad 1 \quad -1 \quad -1 \quad 1], [0 \quad 0 \quad 0 \quad 1 \quad 0]$ and $[-1 \quad 0 \quad 0 \quad 1 \quad 1]$ as bases, respectively. Using these basis vectors as the columns of matrix

$$P = \begin{bmatrix} 1 & 0 & 1 & 0 & -1 \\ 0 & 1.\overline{3} & 1 & 0 & 0 \\ 1 & 0 & -1 & 0 & 0 \\ 0 & -1 & -1 & 1 & 1 \\ 0 & 1 & 1 & 0 & 1 \end{bmatrix}$$

you can verify that

$$P^{-1}AP = D = \begin{bmatrix} 2 & & & & \\ & 2 & & & \\ & & 2 & & \\ & & & 1 & \\ & & & & 3 \\ & & & & & 4 \end{bmatrix}.$$

Exercise.

Determine whether the following matrices A are diagonalizable or defective. For each one that is diagonalizable, find an invertible P and a diagonal D for which $P^{-1}AP = D$.

(a) $\begin{pmatrix} 0 & 1 & -2 & 0 \\ -2 & 3 & -4 & 0 \\ 0 & 0 & 1 & 0 \\ -1 & -1 & 0 & -1 \end{pmatrix}$

(b) $\begin{pmatrix} 10 & -5 & 12 & 0 \\ 9 & -4 & 12 & 0 \\ -5 & 3 & -5 & 0 \\ -7 & 2 & -9 & -2 \end{pmatrix}$

(c) $\begin{pmatrix} 8 & -3 & 8 & 0 & 5 \\ 9 & -4 & 12 & 0 & 9 \\ -5 & 3 & -5 & 0 & -5 \\ -7 & 2 & -9 & 0 & -5 \\ 2 & -2 & 4 & 0 & 5 \end{pmatrix}$

Dartmouth College

TEACHING CALCULUS WITH TRUE BASIC

by James Baumgartner and Thomas Shemanske

≡ Abstract ≡

Contact:

JAMES BAUMGARTNER or THOMAS SHEMAN-
SKE, Department of Mathematics, Dartmouth Col-
lege, Hanover, NH 03755. PHONE: 603-646-2415.
E-MAIL: James.Baumgartner@dartmouth.edu or
Thomas.Shemanske@dartmouth.edu.

Institutional Data:

Dartmouth College is a four-year, private, coeducational institution with an approximate undergraduate population of 4000. Associated with the
College is a medical school, the Thayer School of
Engineering, and the Amos Tuck School of Business
Administration. There are a number of graduate
programs within the Arts and Sciences, but they
are predominantly within the Science division.

The Department of Mathematics and Computer
Science is a joint department supporting separate graduate programs with approximately twenty
graduate students in each. There are 26 full-time
faculty in Mathematics and 13 full-time faculty in
Computer Science. Each year we graduate approximately 25 majors in Mathematics and 35 majors
in Computer Science.

Project Data:

The calculus reform at Dartmouth spans four
courses: two terms of single variable calculus, one
term of multivariable calculus, and a term of differential equations. Virtually all faculty teach some
calculus, and hence virtually all faculty have had
some involvement in the calculus project. It would
be safe to say that more than half of the mathematics faculty are actively contributing new ideas
and software to the project. There is no real de-
marcation for when we began to use computers
in mathematics courses; we have used them since
time-sharing was developed in the 60's, although
the move to Macintosh was done in 1984, and the
Macintosh environment is one that we intend to
use for the foreseeable future. In general, we use
the language True BASIC to generate most of our
software for the classroom, although recently we
have done some experiments with Maple and Mathematica. Three texts are used for the four calculus
courses; their titles are given later in this document.
We have had funding from the PEW foundation,
the NSF, and the Sloan Foundation.

Project Description:

Most of our courses contain fairly traditional material. The focus of all four of the calculus courses
in the calculus project has been on the underlying
geometry of the subject, and to develop the students' ability to interpret the analytic information
geometrically. Typically a class meets for 3 sixty-
five minute periods each week; the format of an
individual class consists of a lecture supplemented

by 5-10 minutes of computer demonstration, if relevant. The computer is routinely used in the classroom and students are expected to use the computer on a regular basis for doing their homework. Since we are not running an experimental section in conjunction with a traditional offering of each course, we find precise assessment of the benefits difficult to obtain. We have modified the examinations to include problems which we believe measure the additional insights that we hope we are giving. What little hard evidence we have suggests that the students who have taken the new calculus course perform as well on a traditional test of skills as those in previous years, and our perception is that the students in previous years would have fared far worse than students in the current courses if examined on more conceptual material requiring a geometric interpretation of analytic data. So, we believe we are succeeding.

Some preliminary materials in the form of handouts or assignments may be available for the cost of reproduction and mailing.

══ Project Report ══

Introduction

Computing in mathematics courses at Dartmouth is pervasive. In the thick of major reform are all the calculus courses (single- and multi-variable as well as differential equations) in addition to the discrete probability course. However, the influence of computing is certainly not restricted to these courses; it reaches broadly across our curriculum touching linear algebra, number theory, logic, and Fourier analysis as well as courses which look at modeling in the social sciences.

The predominant tool is the language True BASIC, supplemented by occasional use of commercially available software. The experience with computing which both students and faculty gain in the calculus sequence serves them well in subsequent courses. After completing the calculus sequence, we can and do assume that students have a reasonable facility with the computer and with True BASIC. As a result, in upper level courses we can introduce students to more intricate or new types of software which will significantly enhance their

understanding of their course subject: simulations in probability, gathering evidence for conjectures in number theory, formal proof-checking in logic.

Day by Day Mechanics

All of our courses meet for the equivalent of three sixty-five minute classes per week for approximately nine or ten weeks. The normal load for students is three courses per term, for a total of nine courses per year, so for purposes of comparison, our courses should be considered equivalent to semester courses at other institutions. We do not conduct any laboratory sessions since the daily lectures and homework typically involve computing. Our classes still cover fairly traditional material in a fairly traditional lecture format supplemented by 5 or 10 minutes of computer demonstration per day—if it is germane to the lecture, or in response to questions about the material from the class.

The format for most of our courses includes daily homework (both written and, if relevant, computer), two hour exams and a final examination, all of which are counted in computing a final grade. Homework usually counts as much as an hour exam and students are encouraged to work together on problem sets. Walk-in tutorial help is available five nights a week, three hours per night.

Getting Started

Dartmouth has a long tradition of computing in mathematics beginning in the early 60's with Kemeny and Kurtz's development of a timesharing system at Dartmouth. Since that time (and until quite recently), we asked students write four or five computer programs as part of the requirements for successful completion of the second calculus course.

In 1984, the College decided to make the Macintosh available to all students and to encourage the use of the computer in all disciplines throughout the College. Dartmouth strongly encouraged incoming freshmen to purchase machines, and at the present time, about 80% of all students own a Mac. To enhance the potential value to students of owning their own computer, the College networked all the dormitories, so that students could use their computers for stand alone applications, to access the mainframe computers for site-licensed software, or to access a public file server which contains course folders for many courses the College offers. Instructors can distribute assignments or handouts as well as send notes to the students in their classes electronically.

Soon after the College made the commitment to use the Macintosh, the Math Department began to reconsider the way in which computing could be used to teach calculus. True BASIC is a very attractive language for its general simplicity, ease of generating excellent graphics, and for its library capabilities. At first, we simply transported programs from the mainframes to the Macintosh, and had students doing the same four or five programs as they did in previous years. However, as we began to realize the potential of the Macintosh, especially in terms of graphics, ideas began to fly.

As in many institutions, these high-level discussions often take place after five in more comfortable surroundings; Dartmouth is no exception. Over drinks at the Faculty club, a half-dozen or so of us began to throw a number of ideas into the hat and tried to think about their utility and feasibility. Having adopted True BASIC whole-heartedly, someone (it seems now inevitably) named the new project "True Calculus", and we were off. After all, what more do you need than a catchy name and lots of enthusiasm. So without a great deal of further thought, we all began, in piecemeal fashion, trying new ideas and telling one another what seemed to work and what didn't. There never was a master plan, nor really is there one now. We certainly had no NSF grant in mind—"Lean and Lively" was still embryonic. Maybe if there had been a grant to apply for, there would have been a master plan, but it has worked out quite well. Our visions now are quite different than they were at the inception of the project, and they continue to evolve. We shall try to pass on our current thoughts as well as the the motivations for our evolution of ideas.

In general, there was reasonable support for "the project" (or the "fourth floor revolutionary group" as we are affectionately called). There were some people with reservations about such an endeavor, a number of people who were indifferent, and some tacitly opposed. The biggest and most pleasant surprise has been the number of converts to the "cause."

One of the obstacles we encountered as we attempted to "reform" calculus was with some of the students. In the early years of our calculus project, we still had the four or five required computer problems, and while the problems were made more relevant to the course material, they were still perceived as an "add-on" feature of the course. This particularly seemed to annoy many students who had studied calculus in high school. Their attitude seemed to be either that they expected a free ride and found that we were now expecting something more from them, that they felt calculus was taking derivatives and finding antiderivatives and again we were trying to tell them that there was something more, or even that computing should not be taught in a mathematics class. Telling them that calculus was a tool with which to study functions, and that the computer could help them understand this tool was not the response they were seeking. We never required much in the way of skills to write computer programs, but we expected them to be able to read a short tutorial handout about True BASIC, work through it, and come out competent to write short programs in True BASIC. This was not a great success. It became clear that we had to spend a few minutes of class time periodically to explain various features of True BASIC, enough to try to teach them in class to write simple programs or to modify more complicated ones that were provided for their use. This approach worked out very well. Students see the instructor write a 10-line program in front of the class, make mistakes typing, hang the system now and then, and take it all in stride. But also, they see how easy it is for someone to gain insight and help answer questions they pose. So we teach them a little syntax, convince them that nobody's program runs the first time, and begin to break down the fear some students have of using the computer.

We have been supported by the Pew foundation, and by the NSF both for a planning grant and for a three-year grant to bring to fruition some of the ideas we have put forth. In a related effort, the Sloan Foundation has funded a workshop to be held at Dartmouth during the summer of 1990 aimed at training faculty to introduce computing into their mathematics courses.

The walk-in tutorials are staffed by graduate students whose job it is to answer questions about the written homework and the computer homework. In general, they have done an excellent job.

Content

Most of our courses contain fairly traditional material. The focus of the calculus courses has been the underlying geometry, and to develop the students' ability to interpret the analytic information geometrically. Ask a random student to draw a picture which corresponds to the sentence $\lim_{x \to 2} f(x) = 3$, and see the number of blank looks you get. On the first day of class in our standard calculus course, we asked students to find $\lim_{x \to 2}[3 + (x - 2)\sin(1/(x - 2))]$. Of course we didn't pose the question in exactly this form. We gave them an intuitive definition of a limit, and drew the graph of the function above. There was no hesitation as to the answer; it was clear. For the rest of the course we struggled to get students to believe that the analytic expression is as clear (in a sense more so) as the geometric one, and to be able to translate the analytic information into geometric terms.

Even in our lowest level of calculus, Math 1 & 2 (which does in two terms what our standard calculus course does in one), significant use is made of the computer. In this course, the students are given each week a program to use or modify in answering homework problems and to gain insight into aspects of calculus such as limits, chain rule, extremum problems, and curve sketching. Math 1 & 2 is designed for students with deficient algebraic skills. The object of the course is to review algebraic techniques as they are required to study the concepts of calculus.

Some of the nicest uses of computing come to the fore in the study of multivariable calculus, simply because a computer can be such a powerful tool for representing mathematical objects geometrically. Curves and surfaces in three-space, level sets, vector fields and the important relations among these objects can all be illustrated both easily and beautifully in a way that gives students confidence that they can do this just as well as their instructors can. Our preference for having students use visible programs here rather than canned ones, is based on our experience that making clear the distinction between, for example, a graph and a parametric image is something that is enhanced by a more "hands-on" introduction. In addition to graphics, the computer makes it possible to take up the gradient method for finding maxima and minima, Newton's method for approximate solutions of systems of nonlinear equations, and numerical evaluation of multiple integrals over fairly complicated regions in two and three dimensions.

The final course in our calculus sequence begins with an overview of ordinary differential equations: first order, second order, and systems thereof, together with appropriate applications. The point here is that in each case numerical methods are introduced so that the class of "appropriate" problems that can be dealt with is expanded way beyond what is possible using pencil and paper methods alone. And of course the graphical representation of a complicated solution then falls well within the grasp of every student, making it possible for us to push harder to get students to be more adept at interpreting the meaning of solutions to problems in applied mathematics. The course ends with two weeks devoted to solution of the simplest second order partial differential equations, using Fourier series. Here again computer graphics are used, this time to display partial sums of Fourier series, and the graphs of solutions to PDEs.

In all of these courses we find ourselves gradually able to ask students to have a better conceptual grasp of the skills that they acquire, although clearly we need more experience along those lines. Our relatively old-fashioned examination format seems so far to be an adequate guard against allowing the hand computation skills to deteriorate.

Pedagogy

Even though the content of the courses we teach is fairly traditional, we do take extreme liberties with the approach to that material. A major problem for us has been the fact that many students come to Dartmouth having seen some calculus in high school, although not enough to gain advanced placement. No matter at what level one begins a course, it is necessary to discuss material which has in some form been seen by a large portion of the students. A common reaction among many fall-term calculus students is that they have seen it all before, and they immediately tune out. By the time one reaches material which is genuinely new (let's ignore the issue that what they thought they saw in high school may not be what we wanted them to have seen), they are in such a stupor they don't recover for a couple of weeks, having lost a great deal of valuable time. Our response is to put them on the edge of their seats on the very first day. On the first day in the first-term calculus course which we offered this past fall, we defined intuitively domain, range, symmetry, asymptotes, the notion of a limit, continuity, differentiability, and extrema, and gave the students the job for homework of applying these terms to several graphs which they could generate on the computer. They all did fine. If calculus was all geometric, they would have passed a final exam on day one. On the second day of the term, we discussed the notion of a function and defined the circular functions. This led naturally to discussing the period and amplitude of sine and cosine, and how these quantities were reflected in the graphs of functions. Then, we took a moment to show them what can happen when certain trigonometric functions are added together: we looked at Fourier series—not seriously, just enough to pique their curiosity and convince them that there just might be a greater purpose for all this machinery.

At this point we had a good set of tools to work with. For a while we settled into a fairly traditional approach to differential calculus, always emphasizing the geometry and how to extract a geometric interpretation from the analytic data. The next major break from tradition was the introduction of integration, which we did by means of solving differential equations geometrically by passing a curve through a field of slopes. We discussed Runge-Kutta approximation (after all, how did the computer draw the pictures?), and writing this geometrically-motivated algorithm down led very naturally to the Fundamental Theorem of Calculus. They thought it was neat! So did we.

In the second semester, the ideas for using the computer for motivation abound, especially when studying infinite series. Prove that the harmonic series diverges, then write and run a 10-line program to print some of the partial sums of this series, and take bets on how large the partial sum will be at the end of class—nobody guesses that it will be less than 14. Rates of convergence and divergence begin to take on meaning. Students are shown that $\sum 1/n^6$ converges. Ask them to give you the smallest number of terms to sum in order to obtain an approximation to the value of the sum which is accurate to 10 places. Then ask them how they can be sure. Also, by graphing successively higher degree Taylor polynomials, students begin to get a feel for what the remainder formula is supposed to tell them.

All of these ideas presented in the beginning calculus course continue to be refined in the subsequent courses. In the multivariable course, one looks at surfaces and vector fields, or extends the techniques for numerical integration to several variables. In differential equations, heavy use is made of the computer to graph numerical solutions of differential equations.

There are two features which are critical to our approach. The first is that the computing which we do is an integral part of the course. We do no demonstration just to do something on the computer. If it's not germane or doesn't enhance the discussion or help to motivate a concept, we don't use the computer. The second feature is the ability to respond to student questions on the fly. The students see that we pick up the computer as they would a calculator, write a quick and dirty program, and get some insight into a problem. Hopefully, they are convinced to do the same.

In the first course, we start with modest demands on the students—usually only the ability to modify in some simple way an existing program. In the second course, we often have them write some simple programs to illustrate various concepts. We always take a few minutes to explain new statements in the language and give the students sample programs which use new statements for them to play with.

Technology

In the classroom, instructors use Macintosh computers modified with video-out plugs. Larger classrooms use projectors, either Sony or Barco CRT projectors, with a Hughes light-valve projector in the largest classroom. Smaller classrooms are equipped with large overhead monitors. All classrooms have network connections. A machine is fully ready for use once it is connected to power, the network, and the video display.

In the mathematics building there are four computers on wheeled carts to service five classrooms. In practice that is usually enough. Classes are frequently taught outside the mathematics building. The College Office of Instructional Services provides machines, carts and portable monitors for those classrooms. The hookup is the same as in the mathematics building, sufficiently simple so that professional mathematicians can usually handle it without difficulty.

Classroom machines are endowed with two megabytes of memory, double the current standard for the Macintosh Plus. System software is installed via a self-starting RAM disk. This removes the need for an external storage device, like a hard disk drive. The RAM disk can also install the True BASIC application as well as communication software for accessing programs, like Maple, on the mainframe computers. Additional software may be supplied by the instructor on a floppy disk.

The biggest obstacle we have encountered has been the projection problem in large classrooms.

At the beginning there were very few projectors capable of handling the Macintosh scan rate. The Hughes projector does a fine job, but it is expensive and in the case of failure, repairs can take months. It also requires as much as an hour of warm-up time. The Sony projector is much less expensive and is easy to use, but it is inadequate in larger classrooms. We hope the new line of Barco projectors will be a better solution. Another nontrivial difficulty has been finding a way to install projection screens so that the blackboard can be used at the same time as the computer. Ideally, the same projection system would be used for computing in a mathematics class and for displaying a videotape in a psychology class. The optimal solution might use more than one screen.

The great majority of students either own their own computer or have ready access to a computer owned by a roommate or friend. In addition there are several public clusters of Macintosh computers, at least one of which is open twenty-four hours a day. All dormitory rooms, as well as all classrooms and faculty offices, are connected to the campus network. Sample programs, homework assignments and important class documents are routinely supplied to the students via the public file server. Communication between the students and faculty is expedited by the campus electronic mail system, the Macintosh software for which is distributed free to all students, whether or not they own computers. Public laser printers are available in the computer center; documents may be directed to these printers over the network. Students who have been asked to write their own programs often find this the simplest way to print out a program and its output for submission in class. Software has been written at Dartmouth that will permit a screen image to be dumped to a laser printer, something often desirable in connection with graphics work in True BASIC. Experiments have been tried in smaller classes with submitting homework via the public server but, except for some computer science courses, this has proved impractical.

Dartmouth makes an effort to supply software free to students whenever possible. True BASIC is site-licensed by the College and is available on the public server, as is all the communications software. Students have free access to several mainframe computers, including a VAX on which Maple can be run. The only computing expense encountered by freshmen in a calculus course is the optional purchase of the True BASIC Reference Manual and

User Guide, for about $25, by those students who have not purchased computers, and the purchase of a few floppy disks.

Perhaps the greatest technological success we can claim is that the system actually works. It may not yet be optimal, but it has been used routinely by ordinary faculty and students. Naturally, there have been occasional failures. A wiring fault can defeat the simplest hookup system. Software sometimes behaves unexpectedly. System updates improve matters but always break something that worked before. But the number of failures has been a good deal lower than anticipated, and the broad experience of faculty and students is that the system is reliable.

Appraisal, Comparisons, Conclusions

Because of the large-scale changes it has undertaken, the Dartmouth project is difficult to assess in standard ways. We have not offered an experimental section alongside several regular ones. Moreover, our regular sections of the first two terms of calculus are quite large (about 100 in each section), making it difficult for the faculty to assess each student's commitment to the new approach. It is traditional for elementary courses to be taught by regular faculty, and the instructors of different sections cooperate in planning the course. Since the majority of the faculty has experience with the use of computing and seems persuaded of its value in mathematics courses, there is nearly always a significant computing component in these courses. Of course, each instructor is responsible for the lectures in his or her own section, and there is no effort to develop computer demonstrations in lockstep. Instructors freely share ideas about computing applications, just as they share ideas about ordinary mathematical pedagogy. It is rare that a traditional approach is used.

In Fall, 1988, the following experiment was conducted. In Mathematics 3, the standard introductory calculus course, exactly the same final exam was given as was given the year before, when much less computing was part of the course. After correcting for a change in the requirements for advanced placement, we discovered that the students' performance in 1988 was indistinguishable from 1987. The exam consisted of traditional material.

So we concluded that we are doing no harm. What we hope, of course, is that our new approach to calculus has succeeded in giving the students insights that they do not get in the traditional mode, but that is exceedingly hard to measure. During the current year, we modified the exams to include problems which we believe measure the additional insights that we hope we are giving. A sample exam is appended. Students found the new questions difficult, but they were able to make progress on many of them. It is our perception that students who took calculus in previous years would not have done nearly as well as the students did this year. But as yet we have no hard data about why certain students found some of the questions difficult. One of the goals of our project is to find better ways to measure the progress that we have made.

We have tried to measure student attitudes, but no firm conclusions can yet be drawn. Anecdotal evidence suggests that some students with high-school experience in calculus feel that the new approach makes the course more work for them. Students without prior experience seem to like the new approach. Possibly this indicates that students with previous exposure to calculus are expecting more of a free ride because of their experience than those without prior experience. Student evaluations of the new material vary. There are those who say they finally understand what their high school instructors were trying to say, and think computing is a great addition to the course. And then there are those who complain bitterly about having to write programs in a mathematics class. What makes the assessment even more difficult is that we as yet have no way of telling which students made any effort to learn a little about True BASIC, and the effort we require is minimal. We give and explain demonstrations in class, and are available to answer questions both in and outside of class. Maybe the students don't attend regularly; maybe they have no interest. It is very difficult and often frustrating to discern from student response the efficacy of our approach. Nonetheless, those students who criticize the computing component don't say that it hinders their learning, and there are many others who say it's the best thing since ice cream. So, undaunted, we continue.

The work load for students is about the same as in previous years. Some students think that the work load has increased, but this is associated to the actual task of learning a little True BASIC or to overcoming their reluctance to learn how to use the computer for more than word processing. Computer homework is not distinguished from any other kind when assignments are made. The total load has not changed much.

The faculty work load has increased, especially in the earlier courses where bigger changes have been made, but the increased load is associated more with curriculum development than with the use of computing in class. The first time a faculty member uses computing in class there is obviously some additional effort. The second time the additional effort is minimal.

One of the continuing challenges that this and other projects have to face is the job of educating newcomers to the educational community. New graduate students and new faculty often arrive with no background in elementary computing. Up to now we have dealt with this on an *ad hoc* basis. We need a more systematic orientation procedure.

What have we learned so far? It is already possible to draw some conclusions. For example, we have learned that if we really want use of the computer to be a routine part of a calculus course then it is unrealistic to demand, as we did for years, that students write four or five substantial programs entirely on their own. With a language like True BASIC the better students can do this easily enough, but the poorer students have some difficulty. This approach also makes the computing seem unduly important in a mathematics class. We offer more support now, in class and in tutorials, and in order to balance the computing homework against the rest, we make more frequent, less substantial assignments. Do we still believe in asking students to write their own programs? The majority of us say yes. Not only does it help shed a new light on the calculus, it confers a kind of independence on the student for later courses that is unobtainable with any closed-ended package.

Publication

Richard Williamson has published a text, *Introduction to Differential Equations* (Prentice-Hall, 1986), that combines the use of numerical methods with the discussion of symbolic solution of each type of differential equation. This text has been used successfully in the fourth calculus course since 1985, at first in preliminary mimeographed form.

Future Plans

Our project already extends to most of the elementary mathematics courses at Dartmouth. The experience students gain in elementary mathematical programming is routinely used in later courses (where we hope the insight they may have picked up will also be of value).

There has been some faculty experimentation with Mathematica. At the present time, running Mathematica requires a Macintosh II with at least four, preferably eight, megabytes of memory. To use it in a class, therefore, would demand a highly restricted laboratory setting, and this is a serious drawback. Over the next year or two advances in hardware, and perhaps in site-licensing agreements, may make it practical for some classroom use.

Reese Prosser is now writing an elementary calculus text. Richard Williamson is revising his differential equations text.

We have had representatives at several workshops and conferences, and we expect to continue to participate in them. The only workshop scheduled for Dartmouth at present is one in Summer, 1990, run by John Kemeny and Thomas Shemanske and sponsored by the Sloan Foundation.

≡ Sample Materials ≡

First Term Calculus

Text (Fall 1989): *Calculus and Analytic Geometry*, Lynn Garner, Dellen Publisher.

Syllabus:

Week 1: Define domain, range, graph, symmetry, asymptote, limit, continuity, differentiability, and extrema all geometrically and give homework to access their comprehension of these geometric concepts. Properties of the real numbers, equation of a line, functions and graphs, definition and graphs of the trigonometric functions.

Week 2: Combinations of functions, limits (one- and two-sided) of algebraic and trigonometric functions.

Week 3: Continuity, tangent lines, rates of change, the derivative.

Week 4: Trigonometric identities, Rules for differentiation including trigonometric functions.

Week 5: Implicit differentiation, related rates, Mean Value Theorem, Newton's Method.

Week 6: Curve Sketching, optimization problems, introduction to integration.

Week 7: Introduction to integration: Differential equations, field of slopes, numerical methods (Euler, Runge-Kutta). Discuss geometrically, develop the numerical methods based on the geometry, write down the sums associated to Euler's method and find yourself staring at the Fundamental Theorem of Calculus. Then formalize and talk about partitions, Riemann sums. Area as a Riemann sum.

Week 8: Properties of definite integrals, Fundamental Theorem, areas by integration.

Week 9: More areas, rectilinear motion, the logarithm and exponential function. Introduce the exponential as a solution to a differential equation, solve it geometrically (field of slopes). Trying to solve the differential equation analytically leads to the logarithm.

Week 10: More on exponentials, logarithms, growth and decay problems.

Second Term Calculus

Texts (Winter 1990): Garner for two-thirds, and Williamson, Trotter *Multivariable Mathematics* (Prentice-Hall) for last third.

Syllabus

Starred topics are additional topics covered in the honors section of the course

Week 1: Improper integrals, completeness of real numbers

Week 2: Integration by parts, the Gamma function*, integrals of trigonometric functions (will use later in course to talk about Fourier series and how to compute their coefficients*) and trig substitution. Begin partial fractions.

Week 3: Partial fractions, Lagrange interpolation*, numerical integration, error estimates for numerical methods, volumes of revolution.

Week 4: More volumes, arclengths, complex numbers, complex exponential function*, finding n^{th} roots of complex numbers*, sequences.

Week 5: Series (motivation—start with power series and evaluate), non-negative series, alternating series.

Week 6: Power series, Taylor polynomials, operations on series.

Week 7: Remainder term of Taylor series, brief look at Fourier series*, vectors in R^n: algebraic and geometric interpretation, parametric equations of lines and planes.

Week 8: Dot product, cross product, other types of equations for planes, systems of linear equations.

Week 9: Matrices, matrix algebra, row reduction, inverses.

Week 10: Linear maps, determinants: algebraic and geometric properties.

First Hour Exam in Math 3

(Approximately 300 students; Median 66%)

1. (Short Answer)
   a. If the graph of $y = f(x)$ intersects the x-axis only at $x = -1, 4$, then the graph of $y = f(x + 1)$ intersects the x-axis only at $x = $ _____.
   b. The functions $\sin(x)$ and $\cos(x)$ have period _____. The function $2\sin(3x)$ has period _____.
   c. If c is an interior point of the domain of a function f and f is continuous at the point $x = c$, then $\lim_{x \to c} f(x) = $ _____.
   d. Suppose that $\lim_{x \to 1} f(x) = M \neq 0$ and $\lim_{x \to 1} g(x) = N \neq 0$, then
   $$\lim_{x \to 1} \frac{f(x) + 2\sin(\pi x/6)}{f(x)g(x)} = \underline{\qquad}.$$
   e. Express the domain of the function $h(w) = \sqrt{w^2 - 1}$ in interval notation.
   f. Show that the slope of the tangent line to the graph of the curve $y = 3x^2 - 2$ at the point $x = 2$ is equal to 12. You must use the definition in terms of limits.

2. Evaluate the following limits (if they exist).
   a. $\lim_{x \to 2} \dfrac{x^2 - 3x + 2}{x - 2}$
   b. $\lim_{x \to 4+} \dfrac{|x - 4|}{x - 4}$
   c. $\lim_{x \to \infty} \dfrac{2x^3 - 5x}{3x^3 - 7}$
   d. $\lim_{x \to \infty} \dfrac{\sin(x)}{x}$

3. Evaluate the following limits (if they exist) and *give reasons for your answers*.
   a. $\lim_{x \to 0} \sin^2(x) \cos(\frac{1}{x})$.
   b. $\lim_{x \to 0} \cos^2(x) \sin(\frac{1}{x})$.
   c. If $-x^2 \leq f(x) \leq x^2$ then $\lim_{x \to 0} f(x) = $ _____ (if it exists).

4. The graphs on the next page (labelled A-E) [*not included here*] have been drawn on the Macintosh using your plotting programs from class. They are (without regard to order) the graphs of the following functions: $\sin(1/x)$, $\sin(x)/x$, $x\sin(x)$, $|\sin(x)|/\sin(x)$, and $1/\sin(x)$. Each function is either continuous at $x = 0$ or has a discontinuity there. The types of discontinuity may include "removable", "jump", "infinite" or "(unclassified) essential" (i.e., none of the previous ones).
   a. Associate one of the functions listed above with each graph.
   b. Determine whether at $x = 0$ the graph is continuous or the type of discontinuity which is exhibited.
   c. Give the reasons for your answers above.

5a. The function $p(t) = \cos(t) - t$ clearly has a root in the interval $[0, \frac{\pi}{2}]$ (see the sketch of the functions $\cos(t)$ and t). Give reasons that allow you to conclude that $p(t)$ has a root in the interval $[0, 1]$.
   b. Suppose that a function f is defined for all real numbers and that $\lim_{x \to \infty} f(x) = 1$ while $\lim_{x \to -\infty} f(x) = -1$. What can you conclude about solutions of the equation $f(x) = 0$? Give a careful explanation of your answer.

6. Consider the functions $g(t) = t$
   and $h(t) = t + \dfrac{1}{t^2}$.
   a. What is the function $h(t) - g(t)$ and what is its range?
   b. Are there any solutions to the equation $h(t) = g(t)$? Why or why not?
   c. Determine:
      i. $\lim_{t \to \infty} (h(t) - g(t))$
      ii. $\lim_{t \to -\infty} (h(t) - g(t))$
      iii. $\lim_{t \to \infty} h(t)$
      iv. $\lim_{t \to -\infty} h(t)$
      v. $\lim_{t \to 0+} h(t)$
      vi. $\lim_{t \to 0-} h(t)$
   d. Notice that it is true that for all values of $t \neq 0$, either $h(t) \geq g(t)$, $h(t) \leq g(t)$ or that neither inequality holds. Which one of these relationships is valid, and which specific piece of information given above justifies your choice?
   e. In the space provided, draw the graph of g. Then, using the information above, sketch the graph of h.

   Hint: Consider the part as $t \to \pm\infty$, then the part as $t \to 0$, and finally complete the graph is a reasonable way. It is not necessary that you plot any points in order to provide an acceptable graph.

Some Homework Problems

1. (Alice's Adventures) An exuberant Math 3 student, flying high over Hanover in a small plane one nice day, decided to take a shortcut back to the dorm. She jumped out of the plane. On the way down she naturally wondered how fast she was going.

"Well," she thought, "since my rate of change of velocity is 32 ft/sec/sec, I need only solve the differential equation $\frac{dv}{dt} = 32$, subject to the initial conditions $v(0) = 0$."

(1) What formula did she write down (she had a notebook and pen in her jump suit) to express her velocity as a function of time?

(2) If the plane was flying at 10,000 feet at the time of her departure from it, how long did it take her to reach her dorm?

(3) How fast was she moving as she passed the window of her first floor dorm room?

Her roommate exclaimed, "My, you're back early. You couldn't possibly have used the right model. Did you take air resistance into account? Air gives a retarding force proportional to the square of your velocity, you know."

"Oh my," said Alice, "I should have solved the equation $\frac{dv}{dt} = 32 - kv^2$, $v(0) = 0$, instead. For that I need my Mac, so it's good that I'm home early."

Alice took the program RungeKuttaPlot and modified it to solve

$$\frac{dy}{dx} = 32 - ky^2, \qquad y(0) = 0$$

on the interval $0 \leq x \leq 50$. She chose the vertical scale to be $0 \leq y \leq 200$. She did not know the value of the proportionality constant k, of course, since she has no security clearance, so she decided to experiment. She did this by putting a statement

```
input prompt ''k = '': k
```

in the program so that she could enter values of k and observe the result.

She tried values $k = 1, 0.1, 0.01, 0.001, 0.0001$. "My goodness," she expostulated, "I didn't know my velocity did that. How clever." She asked her roommate about the weird behavior and he/she

said, "Why, didn't you know, a body reaches a limiting velocity of 185 feet per second in the atmosphere. That's about 125 miles per hour and makes you really think what it must have been like to be in Charleston when Hugo was huffing and puffing at 135 mph."

"Imagine that," said Alice. "I dare say that I can use my model to estimate the value of k. I've wanted to know its value all my life, and now I realize that Math 3 (and my Mac) allow me to discover what it is. I don't even need a security clearance."

She did. She found the value of k to 2 significant digits. Can you?

2. Draw pictures which will convince your instructor that

$$\ln(n + 1) \leq \sum_{k=1}^{n} \frac{1}{k} \leq 1 + \ln(n) \quad \text{for all } n \geq 1.$$

Hint: Consider the upper sum approximating $\int_{1}^{n+1} \frac{1}{x}\, dx$ and the lower sum approximating $\int_{1}^{n} \frac{1}{x}\, dx$ with the partitions of the given intervals consisting of subintervals of unit length. Show that it follows from the above inequality that the *harmonic series* $\sum_{n=1}^{\infty} \frac{1}{n}$ diverges.

Since the harmonic series diverges, it must be the case that the partial sums $\sum_{k=1}^{n} \frac{1}{k}$ get arbitrarily large as $n \to \infty$. Write your guesses to the following questions: If you were to write a computer program to print out the value of the partial sums for the harmonic series and you set this program running (don't do it now—it's tomorrow's homework), how large do you think the partial sums will be after the program has run for an hour, overnight, a day? What do you think the largest number the computer would ever print out for these partial sums if the program were to run indefinitely?

3. Write a computer program which attempts to compute $\sum_{n=1}^{\infty} \frac{1}{n}$. Remember, that this series diverges, so that the partial sums must get arbitrarily large. The idea is to write a program with a do-loop construction. It might be nice to print out the value of n and the partial sum $\sum_{k=1}^{n} \frac{1}{k}$ every ten

thousand terms or so. If you let the program run for an hour, how large is your partial sum? What if you let it run overnight? Does there seem to be an upper bound for the partial sums? Should there be?

4. Now let's figure out how large the number printed in your program will get. Knowing that True BASIC on the Mac can hold 16 significant digits and that the partial sums are certainly greater than ten, there can be at most 14 significant digits to the right of the decimal point. This means that when n satisfies $\frac{1}{n} < 10^{-14}$, the term $\frac{1}{n}$ will no longer be contributing to the partial sum. Approximately how long will this take, and determine how large the partial sum will be when $n = 10^{14}$. *Hint:* Use

$$\ln(n+1) \le \sum_{k=1}^{n} \frac{1}{k} \le 1 + \ln(n) \quad \text{for all } n \ge 1.$$

How does this number compare with your projection from the previous days homework?

5. It is well-known that $\sum_{n=1}^{\infty} \frac{1}{n^2}$ converges, but somewhat less well-known (at least to calculus students) that it converges to $\frac{\pi^2}{6}$. Write a program which determines the number of terms in the partial sums of $\sum_{n=1}^{\infty} \frac{1}{n^2}$ that are necessary to approximate $\frac{\pi^2}{6}$ with an error of less than 10^{-5}. Do the same for $\sum_{n=1}^{\infty} \frac{1}{n^4}$ given that the series converges to $\frac{\pi^4}{90}$.

In each of these series it is easy to determine the number of terms required to obtain a given degree of accuracy, if first we know the answer. On the other hand, if we already know the answer, we don't need the computer. The series $\sum_{n=1}^{\infty} \frac{1}{n^6}$ converges. How many terms do you think might be necessary before the partial sums are within 10^{-8} of the correct value? When relying upon the computer for an answer, after how many terms is it pointless to continue to add terms to the partial sum? Why?

Out of these exercises should come a natural conclusion about what is needed in order to use the computer to effectively compute the value of a convergent infinite series. What do we need?

6. Modify one of your graphing programs (in the PLOTTERS folder) to draw the graph of the function $\sin(x)/x$ using the infinite series

$$\frac{\sin(x)}{x} = \sum_{n=0}^{\infty} \frac{(-1)^n x^{2n}}{(2n+1)!}$$

to define the function $\sin(x)/x$.

This series expression is easily obtained by taking the well-known power series expression for $\sin(x)$ and dividing each term by x. To use the infinite series to draw the graph, you will need to define the function in such a way that given a value of x, you compute enough terms in the partial sum to guarantee that the error between the actual value and the value of the partial sum is less than some reasonable tolerance, say 10^{-6}. That is, you will really only draw an approximation to the actual graph, but one in which each y-value of the approximating curve differs from that of the desired curve by at most 10^{-6}. As the value of x varies, so will the number of terms in the partial sum which are required to achieve the desired degree of accuracy. Note that the series for $\sin(x)/x$ is an alternating series, so the error term is very easy to find. Also, it is not necessary to formulate a factorial function, since a given term in the series is easily computable from the previous one.

7. It often happens that a function that cannot be expressed in terms of "elementary" functions like polynomial, exponential, logarithmic and trigonometric functions, can nevertheless be written as an infinite series. For example, the antiderivative of the function $\sin(x)/x$, more specifically the function

$$\int_0^x \frac{\sin(t)}{t}\, dt,$$

(which Maple calls $Si(x)$) cannot be expressed in terms of elementary functions. It can, however, be written as an infinite series as follows.

To get the series for $\int_0^x \frac{\sin(t)}{t}\, dt$, all we have to do is to integrate the series for $\sin(x)/x$, $\sum_{n=0}^{\infty} \frac{(-1)^n x^{2n}}{(2n+1)!}$, term-by-term from 0 to x.

Modify the program MULTIPLOTTER to compute and graph both $\sin(x)/x$ and its antiderivative, using these series expressions on the interval $[-10, 10]$.

As in the previous part, compute the partial sums of the series to a reasonable tolerance in order to graph the functions. Also compute each term of the series by modifying the previous one, rather than computing them anew each time.

You can check your answer with the results of the program "PlotSeries" in the folder for 2/14.

Calculus with Algebra

First Term (Math 1)

Texts: *Calculus with Analytic Geometry*, by Swokowski

The Math 1-2 sequence is designed to cover in two terms all the topics in our standard first term calculus course, Mathematics 3. The courses are designed for those students whose algebraic skills are not sufficiently well-developed to allow them to keep up with the pace of the standard calculus course. Basic algebraic techniques are reviewed as an integral part of the course as they are required to study the concepts of calculus.

Syllabus

Week 1. 1.1-1.3: Real numbers, Coordinate systems, Lines.
Week 2. 1.4-1.6: Functions, Operations on functions, Trigonometric functions.
Week 3. 2.1, 2.2, 2.4: Introduction to Calculus, Definition of a limit, Techniques for finding limits.
Week 4. 2.4-2.6: Limits of trigonometric functions, Continuous functions.
Week 5. 3.1-3.2: Definition of the derivative, Rules for finding derivatives.
Week 6. 3.2-3.4: More rules for derivatives, Derivative as a rate of change, Derivatives of the trigonometric functions.
Week 7. 3.5-3.7: Increments and differentials, Chain rule, Implicit differentiation.
Week 8. 3.8-3.9: Powers and higher derivatives, Related rates.
Week 9. 4.1-4.3: Local extrema, Rolle's and Mean Value Theorem, First derivative test.
Week 10. 4.4-4.5: Concavity and the second derivative test, Applications of extrema.

Second term (Math 2)

Texts: *Calculus with Analytic Geometry*, by Swokowski

Syllabus:

Week 1. 5.1-5.2: Area, Definition of the definite integral
Week 2. 5.2-5.4: Properties of the definite integral, Fundamental Theorem of Calculus.
Week 3. 5.4-5.5: More on Fundamental Theorem, Indefinite integrals and change of variable.
Week 4. 5.6-6.1: Numerical integration, Computing areas.
Week 5. 6.1, 6.6, 6.9: More on areas, Work, Other applications
Week 6. 6.9, 7.1: More applications, Inverse functions.
Week 7. 7.2-7.3: Natural Logarithm and Exponential functions.
Week 8. 7.3-7.4: Differentiation and Integration of logarithms and exponential functions.
Week 9. 7.6-7.7: Growth and decay problems, Derivatives of inverse functions.
Week 10. 8.1-8.2: Integrals of Trigonometric functions, Inverse trigonometric functions.

Multivariable Calculus

Third Term Calculus (Math 13)

Texts: *Multivariable Mathematics*, by Williamson and Trotter, and *Notes on Multivariable Computing* (Xerox handouts)

The subject of the course is the extension and modification of the ideas and techniques of one-variable calculus to higher dimensions, which is where most of the applications are. Numerical methods will be introduced and applied in areas where traditional calculus techniques are inadequate; in particular, we'll use 3-dimensional perspective computer plotting.

Syllabus:

The course will cover material in Chapters 3, 8, 9, 10, 11 and 12 of *Multivariable Mathematics* and in Xeroxed handouts on numerical methods. Familiarity with simple BASIC programming will be assumed.
Week 1. 3.1-3.2A: Linear functions
Week 2. 8.1-8.3: Functions of one variable, Graphs, Notes on computer plotting, Partial derivatives.
Week 3. 8.4-8.6: Vector partial derivatives, Computer plotting of surfaces, Limits and continuity, Differentiability.

Week 4. 8.7, 9.1, 9.2: Directional derivatives, Gradient vectors, Computer plotting of vector fields, The chain rule.

Week 5. 9.2-9.4: More chain rule, Implicit differentiation, Extreme values.

Week 6. 10.1 & handouts: Gradient method using a computer, Curvilinear coordinates, Iterated integrals.

Week 7. 10.2, 10.4: Multiple integrals, Change of variable.

Week 8. 11.1, 12.1, & handouts: Numerical integration notes: midpoint and Simpson, Line integrals, Green's theorem.

Week 9. 12.2, 12.3, 12.5: Conservative vector fields, Surface integrals, Gauss's theorem.

Week 10. 12.4 & Review: Stokes's theorem.

Sample Supplement

Numerical Integration

A double integral

$$\int_R f(x,y)\,dx\,dy$$

on a rectangle R determined by $a \le x \le b$, $c \le y \le d$ can be written as an iterated integral in either of two orders:

$$\int_a^b \left[\int_c^d f(x,y)\,dx\right]dx \ \text{ or } \ \int_c^d \left[\int_a^b f(x,y)\,dx\right]dy.$$

If evaluating either of these two integrals is possible by finding a succession of two indefinite integrals, then that is probably the best method. (For example, that would give us an answer for lots of values of a, b, c, d.) If we can't find the required indefinite integrals, it may still be possible to evaluate the integral for particular choices of limits, perhaps by some clever change of variable. (See Exercise 4(a).) When all else fails, numerical approximations are available, once again for specific numerical choices of the limits. We'll consider two such approximation methods.

Midpoint Approximations

These are obtained by imposing a grid on R with intersection points at

$$(x_j, y_k) = \big(a + j(b-a)/p,\ c + k(d-c)/q\big);$$

here p and q are the respective numbers of lines in the grid in the x and y directions, while j runs from

0 to p and k runs from 0 to q. Since the dimensions of each rectangle are $(b-a)/p$ by $(d-c)/q$, the midpoint of the grid rectangle with lower left corner (x_j, y_k) is at

$$
\begin{aligned}
(\overline{x}_j, \overline{y}_k) &= \big(x_j + \tfrac{1}{2}(b-a)/p,\ y_k + \tfrac{1}{2}(d-c)/q\big) \\
&= \big(a + (j+\tfrac{1}{2})(b-a)/p,\ c + (k+\tfrac{1}{2})(d-c)/q\big).
\end{aligned}
$$

The **midpoint approximation** to the value of the integral is

$$\int_R f(x,y)\,dx\,dy \approx \frac{(b-a)(d-c)}{pq} \sum_{j=0}^{p-1}\sum_{k=0}^{q-1} f(\overline{x}_j, \overline{y}_k).$$

The main part of a BASIC program to implement the midpoint approximation formula is a double loop of the form

```
LET s = 0
FOR j = 0 TO p-1
FOR k = 0 TO q-1
    LET s = s + f(a+(j+.5)*(b-a)/p,
      c+(k+.5)*(d-c)/q)
NEXT k
NEXT j
PRINT s*(b-a)*(d-c)/(p*q)
```

The rest of the program consists of a definition for f, and an assignment of values to the limits a, b, c, d. To integrate over a region D in $\mathcal{R}^2$ that's more complicated than a rectangle, simply enclose D in a rectangle R. Then define f by its given values for (x,y) in D and define $f(x,y) = 0$ for (x,y) outside D. For example, if D is the quarter disk $x^2 + y^2 \le 1$, $0 \le x$, $0 \le y$, the definition of f might appear in a program format such as

```
DEF f(x,y)
    IF x^2 + y^2 <= 1 AND 0 <= x
      AND 0 <= y THEN
        LET f = cos(x^4 + y^4)
    ELSE
        LET f = 0
    END IF
END DEF
```

Simpson Approximations

If the integrand f in a multiple integral is a fairly smooth function we can take advantage of its smoothness by repeated use of the one-dimensional **Simpson rule** over an *even* number p of intervals:

$$\int_a^b f(x)\,dx \approx \frac{b-a}{3p}\left(f(x_0)+4f(x_1)+2f(x_2)+\cdots\right.$$
$$\left.+4f(x_{p-1})+f(x_p)\right)$$
$$\approx \frac{b-a}{3p}\sum_{j=0}^{p} C_j^{(p)} f(x_j),$$

where $x_j = a + j(b-a)/p$. The pattern for the coefficients $C_j^{(p)}$ is such that the first and last coefficients are $C_0^{(p)} = C_p^{(p)} = 1$, while the intermediate ones are given by the formula $C_j^{(p)} = 3 - (-1)^j$, in other words, alternating 4's and 2's, beginning and ending with 4.

To apply the Simpson formula to a double integral we use a two-stage Simpson approximation to an iterated integral

$$\int_c^d \left[\int_a^b f(x,y)\,dx\right] dy.$$

Thinking of y as held fixed for the moment, start with

$$F(y) = \int_a^b f(x,y)\,dx \approx \frac{b-a}{3p}\sum_{j=0}^{p} C_j^{(p)} f(x_j,\,y),$$

where $x_j = a+j(b-a)/p$. Letting $y_k = c+k(d-c)/q$, with q even, we approximate

$$\int_c^d F(y)\,dy \approx \frac{(d-c)}{3q}\sum_{k=0}^{q} C_k^{(q)} F(y_k).$$

Now replace $F(y_k)$ by the Simpson approximation previously obtained to get

$$\int_c^d \left[\int_a^b f(x,y)\,dx\right] dy$$
$$\approx \frac{(b-a)(d-c)}{9pq}\sum_{k=0}^{q} C_k^{(q)}\left(\sum_{j=0}^{p} C_j^{(p)} f(x_j, y_k)\right)$$
$$\approx \frac{(b-a)(d-c)}{9pq}\sum_{j=0}^{p}\sum_{k=0}^{q} C_j^{(p)} C_k^{(q)} f(x_j, y_k).$$

Note that in this formula we use the values of f at all grid points in the rectangle R, including those on its boundary. The double loop in the implementing program now looks like

```
LET s = 0
FOR j = 0 TO p !  Odd number of values,
   with p even.
      FOR k = 0 TO q !  Odd number of
      values, with q even.
         LET s = s + Coef(j,p)*Coef(k,q)*
            f(a+j*(b-a)/p, c+k*(d-c)/q)
      NEXT k
   NEXT j
PRINT s*(b-a)*(d-c)/(9*p*q)
```

Before the loop we need to include the definition of Coef(j,p):

```
DEF Coef(j,p)
   IF j = 0 OR j = p THEN LET Coef = 1
   IF 0 < j AND j < p THEN LET Coef =
      (3-(-1)^j)
END DEF
```

Sample Exercises

(Complete programs are in the public file server: MIDPT2 and SIMP2)

1. Write a program to implement the midpoint rule for an $f(x,y)$ definable in BASIC and test it on the example $f(x,y) = x^2 + y^4$ over the rectangle $R:\ 0 \le x \le 1, 0 \le y \le 1$. Having computed the correct value by hand, you can find our how small p and q can be while still producing 4-place accuracy.

2. Write a program to implement Simpson's rule for $f(x,y)$ and test it on the same example as in the previous problem to find minimal values for p and q for 4-place accuracy.

3. Find an approximate value for

$$\int_R e^{\sin \pi(x+y)}\,dx\,dy$$

where R is
(a) The rectangle $0 \le x \le 2,\ 0 \le y \le 3$.
(b) The rectangle $0 \le x,\ 0 \le y,\ 3x + 2y \le 6$.
(c) The circular disk $x^2 + y^2 \le 1$.
(d) The quarter disk $0 \le x,\ 0 \le y,\ x^2 + y^2 \le 1$.
(e) The parabolic region $x^2 \le y \le 2$.

4. The double integral

$$G(a,b,c,d) = \int_a^b \int_c^d e^{-x^2-y^2}\,dx\,dy$$

can't be evaluated in terms of elementary functions of a, b, c, d. Nevertheless, we can still find such a value when the rectangular region is replaced by a circular region of radius a centered at the origin. The trick is first to make the change of variable

$x = r\cos\theta$, $y = r\sin\theta$, $dxdy = rdrd\theta$, then to compute the resulting double integral by iterated integration over the region $0 \le r \le a$, $0 \le \theta \le 2\pi$.

(a) Compute the value of the integral as suggested above, and find the limit as a tends to ∞. [Ans. $\pi(1 - \exp(-a^2)); \pi$.]

(b) Compute approximations to π by finding Simpson approximations to $4G(0, a, 0, a)$ for suitable values of a.

(c) Estimate how small you can make the positive number a in part (b) while still maintaining four-place accuracy.

5. The Simpson formula for a triple integral over a three-dimensional rectangle R is

$$\iiint_R g(x, y, z)\, dxdydz$$

$$= \int_e^f \left[\int_c^d \left[\int_a^b g(x, y, z)\, dx \right] dy \right] dz$$

$$\approx \frac{(b - a)(d - c)(f - e)}{27pqr}$$

$$\cdot \sum_{j=0}^p \sum_{k=0}^q \sum_{l=0}^r C_j^{(p)} C_k^{(q)} C_l^{(r)} g(x_j, y_k, z_l).$$

Extend your program for Simpson approximations from 2 to 3 dimensions, and test it by integrating $g(x, y, z) = x + y + z$ over the rectangle $0 \le x \le 1$, $0 \le y \le 1$, $0 \le z \le 1$.

6. Find approximate values for the following integrals.

(a) $\int_R \ln(xyz)\, dxdydz$, $R : 1 \le x \le 2$,

$1 \le y \le 3$, $2 \le z \le 3$.

(b) $\int_R \sqrt{x^2 + 2y^2 + z^2}\, dxdydz$

$R : x^2 + y^2 + z^2 \le 1$, $0 \le z$.

(c) $\int_R \sqrt{x + y} \sin(x + z)\, dxdydz$,

$R : 0 \le x \le 1$, $0 \le y \le 2$, $0 \le z \le 3$.

7. (a) Sketch the region in R^2 bounded by the four lines $x + y = 1$, $x + 2y = 4$, $x - 2y = -1$ and $x - 3y = 1$.

(b) Find an approximate value for the area of the region in part (a). Can you find the exact value, 187/120, by elementary geometry?

(c) Find an approximate value for the integral of $f(x, y, z) = x^3 + y^3$ over the region described in part (a).

8. (a) Sketch the region in the positive octant of R^3 bounded by the planes $x + y + z = 3$, $x + y + 2z = 6$, $z = 1$ and $z = 2$.

(b) Find an approximate value for the volume of the region in part (a). Can you find the exact value by elementary geometry?

(c) Find an approximate value for the integral of $f(x, y, z) = x^4 + y^4 + z^4$ over the region described in part (a).

9. A cube of side-length 1 is made of material of density at a given point equal to the distance of that point from one corner of the cube. Estimate the total mass of the cube.

Differential Equations

Fourth Term Calculus (Math 23)

Text: *Introduction to Differential Equations* by Richard E. Williamson, published by Prentice-Hall.

If you've passed Math 13, Multivariable Calculus, your preparation for the course should be adequate. More particularly, you should be familiar with the most elementary aspects of the following: infinite series, calculus of one or more variables, vector geometry, matrix operations and BASIC programming.

Syllabus

The course will cover the material in Chapters 1, 2, 4 and 8 of the text, together with Ch. 5, Sec. 2 and Ch. 9, Secs. 6 & 7. There will also be a handout for a special computer project. The essential background material on numerical solution of differential equations is included in the text .

Week 1. 1.1AB, 1.2ABC, 1.3AB: First order differential equations, Direction fields

Week 2. 1.5ABCD, 1.6C, 1.7: Linear equations, Application to rocket flight, Numerical methods.

Week 3. 2.1AB, 2.2ABC, 2.3AB: Higher-oder equations, exponential solutions, Complex solutions, Nonhomogeneous equations.

Week 4. 2.3C, 2.4ABC, 2.5: Dynamical applications, Pendulum, oscillations, Numerical methods.

Week 5. 4.1, 4.2, 4.3AB: Vector equations, Linear systems, Mixing, oscillations.

Week 6. 4.3CD, 4.4, 5.2AB: Inverse-square law, phase space, Numerical methods, supplement on springs, Eigenvalues and eigenvectors.

Week 7. 5.2B, 4.5AB, 9.5AB: Stability, linear systems, nonlinear systems, Review of power series.

Week 8. 9.6, 9.7, 8.1AB, 8.2AB: Differential equations, series solutions, Linear partial differential equations, Fourier series.
Week 9. 8.3AB, 8.4AB: Adapted Fourier expansions, Heat equation, Wave equation.
Week 10. 8.4C, 8.5: Laplace equation

Sample Supplement

Nonlinear Spring Analysis

Analysis of linear motion of physical bodies linked by springs satisfying Hooke's law leads naturally to a system of second order linear differential equations with dimension equal to the number n of bodies under consideration. If the motion is allowed to take place in a plane, or else in 3-dimensional space, it takes two, or else three, coordinates to specify the position of each body. Hence the relevant system of differential equations will then involve $2n$, or else $3n$, coordinates if we're keeping track of n bodies. At first sight it might seem that just setting up the system of differential equations is a formidable problem. There is a simple way to proceed, however, using vector ideas and notation. The next figure shows two objects with masses m_1, m_2 linked by three springs to each other, and to two points $\mathbf{a}$ and $\mathbf{b}$. Depending on how the motion proceeds, it might in fact be confined to the line joining $\mathbf{a}$ and $\mathbf{b}$, or to some plane containing those two points, or in the most likely circumstances the motion of each body might be three-dimensional in an essential way. Our derivation of the differential equations will include all these cases at once.

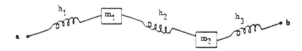

Along with the stiffness constants h_1, h_2, h_3 we need to take into account the relaxed (i.e. unstressed) lengths of the springs, which we'll denote by l_1, l_2, l_3 respectively. Denoting the position at time t of the body of mass m_1 by $\mathbf{x} = \mathbf{x}(t)$, the total force acting on the body can be expressed as $m_1\ddot{\mathbf{x}}$, that is, mass times the acceleration vector. On the other hand, this same total force can be expressed as the sum of the two forces acting on the body *via* the two springs attached to it. The first spring will have extended, or maybe compressed, length $|\mathbf{a} - \mathbf{x}|$. The difference $|\mathbf{a} - \mathbf{x}| - l_1$, if it's positive is just the extension of the spring beyond

its relaxed length, and, if it's negative it measures the amount of compression of the spring below its relaxed length. The numerical value of the force due to the first spring is then $h_1(|\mathbf{a} - \mathbf{x}| - l_1)$. Since an extended spring pulls the body toward $\mathbf{a}$, we can express the force vector by its numerical value times the vector of length one pointing from the body toward the point $\mathbf{a}$. This unit vector can be written as the vector $\mathbf{a} - \mathbf{x}$ divided by its length: $(\mathbf{a} - \mathbf{x})/|\mathbf{a} - \mathbf{x}|$. The resulting force vector is then the first of the two vectors

$$h_1(|\mathbf{a} - \mathbf{x}| - l_1)\frac{(\mathbf{a} - \mathbf{x})}{|\mathbf{a} - \mathbf{x}|},$$

$$h_2(|\mathbf{y} - \mathbf{x}| - l_1)\frac{(\mathbf{y} - \mathbf{x})}{|\mathbf{y} - \mathbf{x}|}.$$

Denoting the position of the second body by $\mathbf{y} = \mathbf{y}(t)$, we similarly get the second of these two vectors for the other spring force acting on the first body. The differential equation expressing $m_1\ddot{\mathbf{x}}$ as the sum of these forces is then

$$m_1\ddot{\mathbf{x}} = h_1(|\mathbf{a}-\mathbf{x}|-l_1)\frac{(\mathbf{a} - \mathbf{x})}{|\mathbf{a} - \mathbf{x}|}+h_2(|\mathbf{y}-\mathbf{x}|-l_2)\frac{(\mathbf{y} - \mathbf{x})}{|\mathbf{y} - \mathbf{x}|}.$$

A parallel derivation yields the equation for the force on the second body:

$$m_2\ddot{\mathbf{y}} = h_2(|\mathbf{x}-\mathbf{y}|-l_2)\frac{(\mathbf{x} - \mathbf{y})}{|\mathbf{x} - \mathbf{y}|}+h_3(|\mathbf{b}-\mathbf{y}|-l_3)\frac{(\mathbf{b} - \mathbf{y})}{|\mathbf{b} - \mathbf{y}|}.$$

For motion in three dimensions, each of these two vector equations is equivalent to three real coordinate equations, making up a system of six equations governing the motion of the entire mass-spring system. One striking feature of this system is that it is in general nonlinear. The nonlinearity is due to the presence of the expressions $|\mathbf{b} - \mathbf{y}|, |\mathbf{y} - \mathbf{x}|, |\mathbf{a} - \mathbf{x}|$. For example, letting $\mathbf{y} = (y_1, y_2, y_3)$ and $\mathbf{b} = (b_1, b_2, b_3)$, we note that

$$|\mathbf{b} - \mathbf{y}| = \sqrt{(b_1 - y_1)^2 + (b_2 - y_2)^2 + (b_3 - y_3)^2},$$

an obvious nonlinearity in general.

If it seems strange that the two-body mass-spring system is in general nonlinear when we've already seen that for motion of two bodies on a line the system is linear, consider the next example, which explains this apparent anomaly.

Example 1

For two bodies moving on a line we may as well take the line of motion to be a coordinate axis. We place the fixed points $\mathbf{a} = a$ and $\mathbf{b} = b$ and the the moving points $\mathbf{x} = x$ and $\mathbf{y} = y$ in order on the line: $a < x < y < b$. Then the vector lengths can all be interpreted as absolute values, and we can use the order relations among the points to compute their values. For example $|a - x| = x - a$ since $a < x$. In particular, the unit vectors that we use to determine direction of force all turn out to be either 1 or -1. For example, $(a - x)/|a - x| = -1$, since $a < x$ implies $a - x < 0$. Using the relevant relations of this sort, the general two-body vector equations reduce to

$$m_1 \ddot{x} = -h_1(x - a - l_1) + h_2(y - x - l_2)$$
$$m_2 \ddot{y} = -h_2(y - x - l_2) + h_3(b - y - l_3) \, .$$

The system is certainly linear. The only reason these equations don't look exactly like the ones we derived earlier for this special case is that then we used two different locations on the line for the zero-values for x and y and we assumed there was no tension on the springs when they were in equilibrium. Thus the new equations are really more general than what we had earlier. (You can find the equilibrium values for x and y from the new equations by setting $\ddot{x} = \ddot{y} = 0$ and solving for x and y.)

Example 2

The linear case treated in the previous example is the only one that has solutions that can be expressed in terms of elementary functions. If we want to resort to numerical approximations to the true solutions, it's sometimes desirable to display the differential equations in coordinate form rather than vector form. Written out, the six real coordinate equations for a two-body mass-spring system might use the following coordinates: $\mathbf{x} = (x_1, x_2, x_3), \mathbf{y} = (y_1, y_2, y_3), \mathbf{a} = (a_1, a_2, a_3), \mathbf{b} = (b_1, b_2, b_3)$. To avoid repeating the square-root expression for vector length more than necessary, we rearrange the equations slightly, as follows:

$$m_1 \ddot{x}_j = h_1 \left(1 - \frac{l_1}{|\mathbf{a} - \mathbf{x}|}\right)(a_j - x_j)$$
$$+ h_2 \left(1 - \frac{l_2}{|\mathbf{y} - \mathbf{x}|}\right)(y_j - x_j), \ j = 1, 2, 3.$$

$$m_2 \ddot{y}_j = h_2 \left(1 - \frac{l_2}{|\mathbf{x} - \mathbf{y}|}\right)(x_j - y_j)$$
$$+ h_3 \left(1 - \frac{l_3}{|\mathbf{b} - \mathbf{y}|}\right)(b_j - y_j), \ j = 1, 2, 3.$$

Linear damping would simply add terms of the form $-k_1 \dot{x}_j$ and $-k_1 \dot{y}_j$ respectively to the two right sides of our equations for the undamped motion, where k_1, k_2 are positive.

Example 3

There is nothing in the derivation of the vector equations of motion that requires $\mathbf{a}$ and $\mathbf{b}$ to be fixed in space, as we tacitly regarded them at the time. For example, we could hold one of them fixed at $\mathbf{a}(t) = (0, 0, 0)$ and specify a periodic motion for the other one by $\mathbf{b}(t) = (2 \sin t, \cos t, \sin 2t)$.

Sample Exercises

1. Suppose two bodies moving on a line are each joined to the other and to a fixed point by springs of equal stiffness, so $h_1 = h_2 = h_3$. Find the equilibrium points for the system in terms of the unstressed lengths l_1, l_2, l_3 of the springs and the distance l between the fixed points. Assume no forces are acting except that of the springs. (The answer is intuitively obvious in case $l_1 = l_2 = l_3$, and your general answer should be consistent with this special case.) [Ans. $x_e = (l + 2l_1 - l_2 - l_3)/3, y_e = (2l + l_1 + l_2 - 2l_3)/3.$]

2. Derive vector equations of motion for two bodies of mass m_1, m_2, respectively, if they are joined sequentially by springs to two fixed points and to each other, and assuming a constant gravitational acceleration of magnitude g in the direction of a fixed unit vector $\mathbf{u}$.

3. Find a first order system in standard form that is equivalent to the second order system for two bodies and three springs.

4. Derive the equations of motion for a single body of mass m attached by springs to two fixed points.

5. Adapt your numerical system solver so that you can plot a graphical solution of the equations of motion for each of the following. Then run the program with graphic output with some specific parameter and initial values.

(a) Mass m joined to two fixed points $\mathbf{a}$, $\mathbf{b}$ by springs with unstressed lengths l_1, l_2, and Hooke constants h_1, h_2.

(b) Masses m_1, m_2 joined to each other and to two fixed points **a**, **b** by springs with unstressed lengths l_1, l_2, l_3, and Hooke constants h_1, h_2, h_3, as in Example 2.

(c) What is the effect of making $h_3 = 0$ in part (b)? Run your system solver with that assumption included.

(d) Assume initial conditions that will restrict the motion in part (b) to a line and run your system solver under that assumption.

Duke University

PROJECT CALC

by David A. Smith and Lawrence C. Moore

Abstract

Contact:

DAVID A. SMITH and LAWRENCE C. MOORE, Project CALC, Department of Mathematics, Duke University, Durham, NC 27706. PHONE: 919-684-8124. E-MAIL: das@math.duke.edu.

Institutional Data:

Duke University is a private university with many programs of high national and international stature. It includes Trinity College of Arts and Sciences, the Graduate School, the Fuqua School of Business, the Divinity School, and Schools of Engineering, Medicine, Law, and Forestry and Environmental Studies. It has some 9,000 students enrolled in degree programs, just under two-thirds of whom are undergraduates. Approximately 47% of the freshman are female and 11% are (non-Oriental) minorities; these percentages are also approximately the same for enrollments in freshman calculus.

The Duke Department of Mathematics numbers 25 in tenured and tenure track positions and 10 in visiting, adjunct, part-time, and term positions. It has 41 graduate students in residence, almost all of whom are pursuing the Ph.D. degree. Each undergraduate class at Duke has about 30 first majors and another 30 second majors in mathematics.

Project CALC is a joint effort with the North Carolina School of Science and Mathematics (NC-SSM), a state-supported residential school for high school juniors and seniors located about two miles from Duke. It enrolls approximately 475 students from diverse geographic, social, economic, and educational backgrounds. The student-faculty ratio of the School is about 11 to 1; nine of the full-time faculty are in the Department of Mathematics and Computer Science.

Project Data:

The first two semesters of the Project CALC course were offered (for the first time) to two sections at Duke in the 1989-90 academic year; Co-Directors Smith and Moore were the instructors. In the Fall of 1990, seven sections of Calculus I and one of Calculus III will be offered at Duke; these will be taught by six faculty members and two graduate teaching assistants. In addition, two sections of an adapted Project CALC course will be offered at NCSSM.

The two 1989-90 sections at Duke enrolled 42 students in the Fall term, 33 in the Spring. The corresponding conventional sections had 456 students in the Fall and 325 in the Spring. (Many entering freshmen at Duke place into accelerated, advanced, or honors sections of calculus that are not counted in these figures.)

In addition to the usual three hours in the classroom, Project CALC students are scheduled for a two-hour laboratory each week. The laboratory accommodates up to 16 students working at eight IBM PS/2 Model 30-286 computers. The student stations are part of a Novell network that also includes faculty, secretarial, and classroom computers; the file server is a Zenith 386 Model 25 with a 150-megabyte hard disk. The principal software packages used in the labs are EXP (for technical word processing), *MathCAD*—Student Edition (for numerical and graphical computation and for discovery experiments), and *Derive* (for symbolic and graphical computation).

All text materials for the course are being prepared by the participants in Project CALC. These include a "lean and lively" textbook, a set of laboratory projects, a set of classroom projects, and a small reference volume. The Project will also develop materials for training and supporting faculty and graduate students who will eventually teach the course.

Major funding for Project CALC has come from a National Science Foundation curriculum development grant (1989-93) and an NSF-ILI grant (1989-92) to equip three-fourths of the labs and classrooms we will need as the program expands to include all calculus taught at Duke. NSF also provided a planning grant for 1988-89. In addition, the Project has received support from a major grant to Duke from the Howard Hughes Medical Institute, the purpose of which is to improve access for women and minorities to careers in science and medicine. During our preliminary phase (1988-89), we were assisted by the publishers of our selected software packages, Wadsworth & Brooks/Cole, Addison-Wesley, and the Soft Warehouse. Our network software and communications cards were provided at no cost through a grant to Duke from the Novell Corporation.

Project Description:

Project CALC is developing a three-semester calculus program based on a laboratory science model. Its key features are real-world problems, hands-on activities, discovery learning, writing and revision of writing, teamwork, and intelligent use of available tools. Our intended audience includes all students of calculus.

Our course meets for three 50-minute periods in a classroom equipped with one computer for instructor demonstrations. Each section (maximum of 32 students) splits into two lab groups; each group has a scheduled two-hour lab each week. Each lab team (two students) submits a written report almost every week; after receiving comments from the instructor, the team revises and resubmits the report for a grade.

Lecturing is limited to brief introductions to new topics and responses to demands for more information. Teams of four work in the classroom on substantial problems that lead to written reports approximately every other week. In weeks with no report, each student has a week-long assignment of computations that will be used in the next team project. As with the lab reports, each project report is revised and resubmitted. One class period each week is "group office hours"; the instructor responds to student problems but does not initiate new material. Every student must own and be able to use a scientific calculator; no particular brand or model is required.

The Project has a University Advisory Panel that includes one representative from each department or program that requires calculus; panel members have all been very supportive. Support within the Department is strong and visible, as is support from the University administration. The decision to expand the program to all calculus sections has not been made yet; it will depend on demonstrated success of the program.

Instruments have been developed and tested for evaluating students in both the conventional and new courses in the areas of basic skills, concepts, non-routine problem solving, attitudes, and writing. During 1991-92, when we have students one year and two years beyond each course, those four groups will be evaluated to determine long-range effects.

Preliminary versions of the textbook, lab projects, and classroom projects are available at no cost. A small number of selected test sites will use these materials in 1990-91. By the following year, a stable set of materials will be available for any school wanting to be a test site. The Project publishes a newsletter about three times a year; anyone may request to be on the mailing list at no charge.

Project Report

How It All Began

In one sense, this project has been going on for 15 years. For at least that long, Duke has offered a calculus sequence with an attached "laboratory." This course was taken by students willing to meet an extra hour a week to discuss computer-based problems and to do extra work on their own, using the public computer facilities on campus. However, this course always followed the standard syllabus, the teachers did the standard things in three class periods per week, and the students took the standard examinations. For a number of reasons that are not important here, the computer-supplemented standard calculus course has now died for lack of interest.

Many members of our department have long been aware of the importance of having students write, but this activity was largely restricted to junior and senior courses, populated mostly by majors. An important precursor of Project CALC was David Smith's work during 1984-86 at Benedict College during a leave of absence that was supported by the United Negro College Fund and the NIH Minority Access to Research Careers program. Among other things, Smith experimented with having math students at all levels write papers on a word processor instead of taking the usual tests. He also required students to work in teams, and he set up and ran a microcomputer lab in which they could do open-ended experiments and learn by discovery.

While these devices were seen as ways to cope with the very weak preparatory skills of students in an open admissions college, he realized that what he was learning about learning would be applicable throughout the educational spectrum. Upon return to Duke, he started experimenting with the use of writing assignments as a required part of freshman and sophomore courses: the computer-supplemented calculus course and the honors-level third semester calculus course.

At about the same time, Duke's present University Writing Program was brought into being by Professor George Gopen, one of the inventors of Reader Expectation Theory. Every undergraduate at Duke is required to take the University Writing Course in his or her first semester, so all undergraduates can be expected to have a common knowledge of Reader Expectation Theory. Furthermore, the Program offers workshops for faculty and graduate students to learn the essentials of the theory, the language in which students talk about language, and the techniques for responding effectively to stimulate better writing and for grading writing in a reasonable time frame. This support from the Writing Program is now an integral part of Project CALC.

We (Moore and Smith) became aware of the emerging calculus reform movement with the publication of MAA Notes No. 6, the "Lean and Lively" conference proceedings, in 1986. We read it carefully and wrote a review of it for *The College Mathematics Journal.* Our Provost, Phillip Griffiths, as soon as he learned of the impending NSF Special Program in Calculus, proposed to our (then-) chair, Michael Reed, that someone at Duke should get in on that program. Mike said he knew just the people and promptly approached us with the suggestion that we think hard about what should be done with calculus and write a proposal to NSF. The rest is, as they say, history.

Well, not quite. We started work on the proposal during the summer of 1987 and submitted it in early 1988. In the meantime, we had established contact with our colleagues at the North Carolina School of Science and Mathematics and found them very receptive to a joint project. They had already written and were using a precalculus course that was consistent in spirit with the "Lean and Lively" recommendations (and with the subsequent NCTM Standards). Their own students, taking this new course in their junior year and calculus in their senior year, were now complaining about how *dull* calculus was, after they had had such an interesting year in precalculus. Thus, the NCSSM faculty were eager to start on a new calculus course that would fit with their highly successful precalculus offering—if they could find a way to support the effort. Both their motivation and their experience were clearly valuable to us.

And we were valuable to them as well: As a secondary school, they were not eligible to apply for the new NSF calculus program, but they could be partners in our effort. Even though secondary schools were not eligible to apply, NSF (through the then-director of the program, Louise Raphael) strongly encouraged cooperative efforts between colleges and schools. Official policy also supported regional and statewide consortia: Louise initiated contact between us and every college in

North or South Carolina that inquired about a calculus grant. In the end, no one we talked to except NCSSM was interested in as radical a reform as we wanted to pursue.

The 1988 proposal, which requested major funding for a five-year project, resulted in a small planning grant for 1988-89, to be shared equally with NCSSM. The money was probably well spent by NSF, because we worked very hard that year refining our ideas and the proposal itself, in preparation for the 1989 round of proposals. We also developed roughly one-fourth of the labs we would need for the first-year course, and we got a small number of volunteer students to work through the labs in return for a credit that Duke calls "small-group learning experience." Addison-Wesley gave us copies of the student edition of *MathCAD*, so we were able to give a copy to each of our volunteers. We were also given very favorable arrangements by Wadsworth & Brooks/Cole Advanced Books and Software and by The Soft Warehouse that, respectively, enabled us to use *EXP* and *Derive* for our experimental labs.

We made enough progress in 1988-89 that the revised proposal asked for only four years support to complete the project. NSF gave us almost everything we requested for three years (1989-92) and modest additional support for 1992-93. What we didn't get for the fourth (or is it fifth?) year we expect will be supported by the eventual publisher of our materials.

At about the same time as the major curriculum development grant, we received an Instrumentation and Laboratory Improvement grant from NSF to equip our labs and classrooms. We had requested enough money (including the required matching money from Duke) to equip four labs and three classrooms (in addition to two already in place), a number that would be sufficient (with careful scheduling) to extend our program to all calculus instruction at Duke. NSF cut the budget by the equivalent of one lab and one classroom; we hope to make up the difference by donations from one of the computer manufacturers.

We have also received support from two grants to Duke that were not initiated by us. One is a major grant from the Howard Hughes Medical Institute, the purpose of which is to improve access for women and minorities to careers in science and medicine. Our share of the grant is supporting a graduate student who is assisting us this year in preparation for teaching one of our sections in 1990-91. The other

grant is from the Novell Corporation, and it provides network software and communications cards for any network on campus that is used primarily for educational purposes.

The most serious impediment to getting our laboratory course started was the series of problems we encountered in getting the first lab set up. The ILI grant was not funded until October, 1989, and our course had to start in August. We asked the Dean of Arts and Sciences to advance two-thirds of the eventual matching money *in advance* of the ILI award. We knew by that time that the calculus grant would be funded and that the ILI grant was "likely," but it wasn't exactly a bird in the hand. Because it took some time for the matching money to be made available, we were late ordering equipment, so we didn't actually have PS/2's when the term began.

We borrowed eight AT&T 6300's, owned by Academic Computing, from a public lab in the Chemistry Building that was undergoing renovations. They were frustratingly slow for our first *MathCAD* experiments, but we got by for two weeks until the IBM's arrived. Then it took additional time to get the network installed and operating, so we continued to operate with floppy disks. We have been lucky not to have any major disasters with equipment, but we would recommend planning to have all hardware and software in place at least a month before starting a lab course and to use that month for a thorough shakedown, testing, and training period. Four to six weeks into a course is no time to be learning to run a network for the first time.

The most important item in the budget for our curriculum grant is our half-time secretary. The Department had one-quarter time available from another secretarial position, which enabled us to offer a more attractive position: 30 hours per week qualifies for University benefits. However, it took us longer than we anticipated to find a suitable person, so we had to start the course without this assistance. In the meantime, the Department committed extra secretarial assistance to the Project. When we finally found the person we wanted, she was overqualified for the salary and position we could offer, but she was attracted to the part-time nature of the job. We got additional support from the administration to upgrade the position slightly, and our candidate accepted.

We have already mentioned some of the ways in which the University and the Department have supported Project CALC. The most important com-

mitment we have received is *space*. Running calculus as a laboratory course requires new space for
labs, and space is probably the scarcest commod-
ity in any college or university. We got the space
we needed for the first lab as soon as the grant
was announced, and space for the second lab has
been committed in an adjacent room. We don't
know yet where the third and fourth labs will be or
which other classrooms will be equipped with computers, but the Dean has assured us we will have
the space when we need it. Furthermore, on his
own initiative, the Dean saw to it that our first lab
was renovated beyond our expectations.

The eventual expansion of Project CALC to all
calculus sections at Duke is a *plan*, not an accomplished fact. In particular, we still need to convince
our colleagues that this is the way to go. Nevertheless, almost everyone, in our department, in
other affected departments, and in the administration, has been very supportive. When our present
chair, Bill Pardon, requested additional funds to
staff Project CALC sections for 1990-91, the President of the University made the money available as
an increment to the Arts and Sciences budget. This
will enable us to staff most of the Project CALC
sections with faculty rather than TA's and to shift
the TA's to lab supervision duties that did not previously exist. Furthermore, we had more faculty
volunteers for the 1990-91 sections than we could
accommodate.

Nuts and Bolts

Our course meets for three 50-minute periods in
a classroom equipped with one computer for instructor demonstrations. Each section (maximum
of 32 students) splits into two lab groups; each
group has a scheduled two-hour lab each week. In
addition, the lab is open and staffed by undergraduates 20 hours a week (evenings) and by instructors and graduate assistants 12 hours a week (weekdays). Each lab team (two students) submits a
written report almost every week; after receiving
comments from the instructor, the team revises and
resubmits the report for a grade.

Student Preconceptions

Our students come to us with some very firm
beliefs about mathematics. (Your students come to
you the same way.) These beliefs have little to do
with anything communicated directly by teachers

or textbooks, but they have a lot to do with the
indirect messages that have been hammered home
in 12 years of school mathematics. Teachers and
textbooks can alter student *behavior* on a temporary basis (to get past the next test), but they have
little impact on student *beliefs*. Here are some of
them, expressed in words the students would not
use themselves:

• Only "special" people can really understand
  mathematics; attempts by "ordinary" people are
  counterproductive.

• The purpose of academic mathematics (i.e., of
  making students take math courses) is to provide
  a series of hurdles (some low, some high) that one
  has to get over in order to be admitted to study
  in a variety of other disciplines.

• Mathematics has *no* connection with anything
  else, real or academic. This is especially true
  of the so-called "word problems," which are an
  especially difficult topic designed to stymie "ordinary" students like themselves. (Possible exception: Some of mathematics looks a little like
  what they see in a physics class, but they *know*
  that has nothing to do with the way things really
  work.)

• The fundamental object of mathematics is *formula*; synonyms include "function" and "equation." If the word "expression" were in their
  vocabulary, it too would be a synonym—but it
  isn't.

• Successful performance in mathematics requires
  setting aside any attempt to *think* and replacing
  that activity with mechanical adherence to *rules*.

• The object of a problem is *the answer*. If you
  get the answer, it makes no difference how you
  get it. If you don't get it, the secondary object
  is to write down enough symbols to get partial
  credit. The only ways to *know* if an answer is
  right are (a) it's in the back of the book or (b)
  teacher says so. (This is actually a logical de-
  duction from the observation that the rules are
  completely arbitrary and can be changed at any
  time by the teacher or the textbook author.)

• Reading and writing have nothing to do with
  mathematics. Assignments can be completed
  and tests passed without actually reading the
  book, and it would be unfair for the teacher to
  ask that anything be communicated in complete
  sentences.

This list could go on, but the point is this: Unless we recognize that this is where our students
are, we have no hope of inducing them to learn, to

understand, as we ourselves understand mathematics. We can change their behavior, but only they can change their beliefs.

What our students have actually learned from school mathematics—more precisely, what they have invented for themselves—is a set of "coping skills" for getting past the next assignment, the next quiz, the next exam. When their coping skills fail them, they invent new ones. The new ones don't have to be consistent with the old ones; the challenge is to guess right among the available options and not to get faked out by the teacher's tricky questions. At Duke, we see some of the "best" students in the country; what makes them "best" is that their coping skills have worked better than most for getting them past the various testing barriers by which we sort students. We can assure you that that does not necessarily mean our students have any real advantage in terms of understanding mathematics.

The bottom line is that we have to induce students to invent real mathematics by making sure that that's the only kind that will work. "Work" in this context means "lead to good grades." Our students will go to great lengths to get good grades; we have to use that to our (and, ultimately, their) advantage.

What's in the Course?

We are often asked, "What topics do you leave out of the syllabus?" The only sensible answer we have found is "All of them." The SYLLABUS is part of our problem with calculus; therefore, it is not likely to be part of the solution. We have a syllabus—a copy is provided with our sample materials—but you may have trouble finding in it any of your favorite "topics." The course is driven by "prototype problems" from other disciplines—physics, biology, chemistry, economics—that students can recognize as important, at least for someone, if not personally for them. These problems are selected to lead into the need to develop most of the usual "topics," but as necessary tools for solving the problems, not as ends in themselves.

In our view, the *raison d'etre* for the study of calculus is differential equations. Thus, that's where we start. The need to describe information about rates of change leads to the definition of the derivative. But some rates of change are inherently discrete, and we move back and forth frequently between discrete and continuous models.

The course has a single axiom: Every initial value problem has a unique solution. This statement is intuitively obvious (after the students have seen enough direction fields, both discrete and continuous), and it's also false. We tell them that it's false and that we will take up the conditions that make it true when that becomes important. It hasn't yet. The only real problem we have with our single axiom is that it contains five words most of the students do not know how to use correctly. Much to our students' surprise and dismay, mathematics and language are intimately linked.

A number of threads, both mathematical and non-mathematical, run through our course and give it structure. For example, the entire first semester (and part of the second) centers around "slope equals rise over run"—in many different guises. In addition to the obvious ones, this leads to early introduction of Euler's method, which reinforces approximation of derivatives by difference quotients and simultaneously provides a tool for generating solutions of initial value problems. Thus, students are enabled to see their unique solutions, graphically and numerically, long before they know formulas for any of them. Formulas are also convenient tools; the quest for them then makes sense.

A second (closely related) thread is *scaling*, both of independent and dependent variables, both linear and nonlinear scaling. This leads us, for example, to log-log and semilog graphing and the detection of power and exponential functions by looking for straightness. The first introduction of the chain rule arises from proportional scaling of the independent variable; the general case arises by considering "instantaneous scaling."

A third thread is the role of inverse problems. Problems come in pairs; one of the pair is easier, and the other is interesting. We learn something about the interesting one by working on the easier one first, then turning our results inside out. Studying derivatives to get at antiderivatives is an obvious example; here's a less obvious one. "Partial fractions" is not a topic in our course, but we need to solve the logistic differential equation. Our approach is to suggest that a fraction with two linear factors in the denominator is the "answer" to a problem they have already solved—what problem? There are two possibilities, one of which (the more obvious one) leads nowhere; the other leads to a solution by "guessing" the form of the "problem."

Our non-mathematical threads are the prototype problems. We start with population dynamics, which leads naturally to exponential and logarithmic functions as our first objects of study. We return to this topic several times, introducing immigration, logistic constraints, and superexponential growth. All of these concepts are readily introduced and supported by real data, so students can see at once that we are dealing with important problems of the real world. When they discover that the growth of world population is superexponential, with an asymptote before 2030 AD, they get a very vivid picture of the population problem. Eventually, the discrete logistic model (equivalently, an Euler solution of the continuous logistic model) leads us into the modern study of chaos. Thus, our students find that some of the content of the course has been discovered in their own lifetimes.

(An aside: Most of our students have had a calculus course in high school, although we are concentrating for now on the ones who did not place out of anything. They quickly see that this course is not a repeat of what they studied in high school.)

One of our physical threads, the one represented by most of the sample materials that follow, is that of motion in a gravitational field. The falling body problem (without air resistance) is our introduction to polynomials. We later add air resistance (which ties into another mathematical thread, exponential decay), then expand to projectile motion (with and without resistance) as our introduction of parametric representations, and eventually add wind (see Lab 13, Fall Term).

Teaching and Learning

Teaching is irrelevant. What's important to us is *learning*. We are learning every day how and why students do and don't learn. In particular, it makes little difference what the "teacher" *says* or how well he or she says it. What matters is what the students *do*. Thus, we try to keep all the students actively engaged whenever they are in the classroom.

We keep lecturing to a bare minimum: brief introductions to new topics and responses to student demands for more information. Students work in teams of three or four in the classroom on substantial problems that lead to written reports approximately every other week. (In our first semester, we tried to make this every week, but the workload was too heavy for the students.) In weeks with no report, each student has a week-long assignment of computations that will be used in the next team

project. As with the lab reports, each project report is revised and resubmitted. Only then do we give it a grade, and all members of the team get the same grade.

The acronym in our name stands for Calculus As a Laboratory Course. Our computer lab is central to the learning experience, not a peripheral "add-on." We provide software "tools" and somewhat structured environments in which student teams carry out open-ended experiments that lead to discovery of fundamental principles of mathematics and its representations of the real world. As much as we can, we use these experiences to drive the classroom and out-of-class activities.

The classroom often functions as a laboratory also. With a computer in the classroom, we can carry out "group" experiments, whether planned in advance or in response to student questions. We have the capability on the classroom computer (although we seldom use it) to pull up student files from the labs. And we often prepare demonstration files to support new topics. We try to avoid "show and tell" demonstrations, which may be entertaining, but usually have little lasting impact. Rather, our demonstrations require active involvement of the students, for example, in selecting parameters or examples and in guiding the course of the exploration.

We also have the students do non-computer experiments in the classroom. They measure the period of a pendulum (a doorknob on a string), the height of a bouncing ball and the time until it stops bouncing (to illustrate geometric series), the lengths of their arms and of the blackboards (for studies of the normal distribution), and the balance points of plane figures. They take great interest in these activities, and their theoretical calculations become more meaningful when they can compare them with data they *know* are real.

We have written elsewhere at some length about the importance of *writing* as a learning tool in mathematics: G. D. Gopen and D. A. Smith, "What's an Assignment Like You Doing in a Course Like This?: Writing to Learn Mathematics," *The College Mathematics Journal 21* (1990), 2-19. Rather than repeat ourselves (at the expense of more trees), we refer the interested reader to that article.

Technology

Our laboratory (the first of an eventual four) accommodates up to 16 students working at eight

IBM PS/2 Model 30-286 computers. A Novell network links the student stations with faculty, secretarial, and classroom computers; the file server is a Zenith 386 Model 25 with a 150-megabyte hard disk. The lab contains one Hewlett Packard Laser Jet Series II and one IBM Proprinter.

The principal software packages used in the labs are *EXP* (for technical word processing), *MathCAD Student Edition* (for numerical and graphical computation and for discovery experiments), and *Derive* (for symbolic and graphical computation). We have also used *Feedback*, a public domain program for exploring the dynamics of discrete dynamical systems.

Our classes meet in one of two computer-equipped classrooms installed in 1986. The computer is an IBM XT, and students view its CGA output on one of two 19 inch Electrohome RGB monitors mounted from the ceiling. The details of this installation were reported in D. P. Kraines and D. A. Smith, "A Computer in the Classroom: The Time is Right," *The College Mathematics Journal* *19* (1988), 261-267.

Every mathematics student at Duke is expected to own and know how to use a scientific calculator; we do not specify a particular brand or model, nor do we specify capabilities beyond having the standard elementary functions. We call on students frequently to calculate things in class, and some of our projects specifically require the use of a calculator. On occasion, we see students in the computer lab take out a calculator to find a particular number, perhaps because it would be more hassle, say, to exit *EXP* and bring up *MathCAD*. We often ask them, in response to their first submissions of project reports, to focus on their selection of tools for solving a given problem—for example, to explain why they used a computer instead of a calculator for a given task.

Appraisal of First Semester

We taught two sections of the new course in 1989-90 to a total of 42 students in the Fall, 33 of whom continued in the Spring. Our students were initially selected at random from among those signed up for the standard course. During the first few weeks, students had the opportunity to switch sections, and approximately one-third of our original students did; they were replaced by an approximately equal number of students transferring in from regular sections. Most of these students had taken a calculus course in high school but had not done well enough to place beyond the first semester course.

The students found the first few weeks of the course stressful. Predictably, they were nervous about working in teams and about writing reports. However, we failed to anticipate that they would have such a difficult time with the concept of a *function*. These were, after all, very good students from very good secondary schools, and they should have been ready for a course in calculus. Their notion of function was synonymous with that of *formula*; they found it extremely difficult to think of a function in terms of its graph or in terms of data. Because of this early difficulty, we plan to revise our introductory materials to give students more experience with thinking about functions from a variety of points of view.

On the other hand, the students took to the laboratory easily. They accepted *MathCAD* and *EXP* readily and were soon using both programs for projects in other courses. (*Derive* was introduced in the second semester and was learned just as easily.)

In addition to the clear pedagogical advantage of working in teams of two, such teams significantly reduce the level of frustration with software that "won't do" what the student wants it to do. It always helps to have someone else to discuss problems with. For the same reason, we found it important to have an instructor readily available for help with software or clarification of what was expected of the student. We remain convinced that eight workstations is close to the optimal number for one lab instructor to keep track of.

We discovered a number of "things that work" by serendipity rather than be conscious design. For example, since our students worked in teams both in the lab and in the classroom, some of the first friends they made at the university were fellow calculus students—and their first joint activity was discussing calculus.

Another example: The laser printer in the lab was a greater asset than we had anticipated. Not only is it quiet and reliable, but it also makes the work the students submit look important. This, in turn, encourages them to spend more time on the write-ups. Some of this time went into cosmetics, but a considerable portion went into deciding what they wanted to say about what they had done. We had originally planned to allow use of the laser printer only for final drafts, but now we have them use it for everything unless we lack an appropriate driver for the software being used.

Student Evaluation

By the time students evaluated the course at the end of the semester, the consensus was that they had worked hard and that it was a good course. The following are typical quotes from their evaluations:

- The projects were a very positive approach to calculus and taught people to really work together and exchange ideas. The lab pulled things together.
- I liked the idea of the teacher going over our reports, giving suggestions for improvement, and resubmitting them. This way I was able to learn from and use the teacher's comments rather than not learning until after the fact—and after I received a grade.
- ... wonderful job helping me understand what the equations, symbols, constants, etc., mean and how they relate ... for the first time I feel as if I have actually learned something.

Numerical ratings on student evaluation forms are notoriously unreliable, but we will report them anyway. In the following table, we abbreviate the questions asked and report the average ratings for both sections combined. After the first two questions, responses were on a five point scale, with "1" meaning "poor" or "not at all," and "5" meaning "excellent" or "extremely." It is not unusual for students at Duke to rate most of the faculty as "above average" (the Lake Wobegon effect), but it is unusual for an entire class to rate a calculus course as any better than "fair" (2.0).

Average number of hours per week outside class and lab	7.5
How demanding, intellectually?	(Very) Difficult
How well presented?	3.3
Instructor's style, enthusiasm	4.0
Instructor's approachability	4.4
Grading fairness	4.2
Clarity of expectations	4.0
Overall rating of instructor	4.2
Text materials	3.1
Class discussions and projects	4.1
Lab projects	4.2
Increased ability to discuss and apply concepts	3.8
Increased knowledge	3.6
Increased interest	3.3
Overall rating of course	3.8

As the final draft of this document is being prepared, we have just read the student evaluation forms for the second semester; some of the numbers are slightly higher, some slightly lower, but, in general, students rated the second semester very much like the first.

Second Semester Barrier Tests

In the second semester we require each student to pass tests on basic differentiation and integration techniques; "passing" means getting essentially everything right. A student must keep taking versions of each test until he or she passes it at the required level. The tests are open book and open notes, with as much time as is required to work eight problems. The tests do not affect course grades, except that each student must pass them to complete the course.

Around a third of the students passed the differentiation test on the first try, and the rest averaged three tries. As we anticipated, most of the problems in differentiation centered on the proper use of the chain rule, particularly its repeated use. When a student took several versions of the test, we could see the results focus his or her attention on a shrinking range of difficulties. Under this attention, the problems were understood and eliminated. In the end, the student had a feeling that he or she really did know how to do these calculations.

There was little correlation between the curiosity and creativity of a student on the one hand and his or her barrier test performance on the other. Several of the most creative students were too impatient to get all the problems right on the first or second try. Many of them had survived all their mathematical lives on partial credit on such tests; a requirement that all problems be done correctly was a shock. Those who took a large number of tries to pass the test were a little embarrassed, but, in general, they did not see this as reflecting on their overall performance in the course.

Future Plans

In the Fall of 1990, Project CALC will offer seven sections of Calculus I and one of Calculus III at Duke; these will be taught by six faculty members and two graduate teaching assistants. In addition, two sections of an adapted Project CALC course will be offered at NCSSM. A small number of selected test sites will be using our materials during 1990-91. By the following year, a stable set of materials will be available for any school wanting to be a test site. If everything proceeds on schedule, and if the Department agrees, 1991-92 will be the year that the new course becomes the only calculus course at Duke; NCSSM plans to make the same transition.

Our project evaluator, Jack Bookman, has developed and tested instruments for evaluating students in both the conventional course and the Project CALC course in the areas of basic skills, concepts, non-routine problem solving, attitudes, and writing. During 1990-91, when there are roughly the same number of students in each course, we will evaluate their relative progress and capabilities. A year later, when we have students one year and two years beyond each type of course, those groups will be evaluated to determine the long range effects of each course.

≡ Sample Materials ≡

On the following pages we offer samples of the materials we have prepared for our students. As these materials are from the first offering of the course, we expect to change most of them, in small ways or large, before the Fall Term of 1990. Our samples include:

- A general information handout for the first class
- The syllabus for two semesters
- The Introduction and part of Chapter 5 from our textbook, *The Calculus Reader*
- Instructions and *MathCAD* worksheets for two lab assignments
- Instructions for an in-class/take-home project
- A student report on that project
- The first semester final examination
- The second semester midterm examination
- Barrier tests on differentiation and integration.

The samples of text material, project, and lab assignments have been selected to trace parts of the "physics" thread running through the course. We are preparing a report for the forthcoming MAA Notes volume, *Calculus as a Laboratory Course* (L. Carl Leinbach, et al., eds.), that will include sample materials tracing the "biology" thread.

Spacing and formats of some materials have been changed to conserve space in this document; otherwise, what you see here is what our students saw. *Our inserted comments appear in italics.*

General Information and Procedures

Project CALC at Duke University is part of a nationwide effort to revitalize the teaching of calculus. Supported by grants from the National Science Foundation and the Howard Hughes Medical Institute, we are developing a new approach to the course. Our approach is based on an interactive computer lab and emphasizes writing and student cooperation as integral parts of class activity.

Each of you is scheduled for a weekly laboratory session in our new computer lab. Here you will work in teams of two, exploring "real world" problems involving population growth, the spread of epidemics, the fall of objects through the atmosphere, the oscillation of a pendulum, electrical circuits, price dynamics in economics, and more. In addition to the laboratory explorations, there will be group activities introduced and explored during class. Both in the laboratory and in the classroom, each group will be expected to write a report of their investigations.

As is clear from the remarks above, writing is an important part of this course. The standard for your written work will be the same as that expected of you in your University Writing Course. (You will be introduced to a friendly technical word processor, *EXP*, in the second laboratory session.) Also, it should be clear that cooperation with your fellow students is an important part of the course. To make this easier, we will provide you with a list of phone numbers and dorm room numbers of all students in the Project.

This course will be entirely separate from the other sections of Mathematics 31; we will use different text materials, take different exams, and cover ideas in a different order and from a different point of view. For those students planning on taking both Math 31 and Math 32, we strongly suggest that you plan on taking a Project CALC Math 32 section in the Spring. There will be a Project CALC Math 103 course offered in Fall 90.

Materials Required for the Course

1. You do not need to buy a textbook; we will supply you with text materials as the course progresses. You will need to buy a large three ring binder to keep the materials in. In fact, later in the semester, you may need to buy more binders to keep track of laboratory handouts, class handouts, the class journal (see below), etc.

2. We are asking you to buy a copy of the *Student Edition of MathCAD*. This is available at the

bookstore for about \$40. It is the major piece of software used in the course and will be useful in future mathematics, science, and engineering courses. Having your own copy of *MathCAD* will give you a personal copy of the manual and enable you to work on projects at various MS-DOS machines around campus.

Tests and Grades

There will be an open book, essay-type midterm exam in class on Friday, September 29. The final exam will be of the same type; it will be scheduled at the same time as the other calculus exams. (A university final exam schedule will appear later in the semester.)

Letter grades will be assigned to each laboratory report and to in-class projects. The course grade will be determined by the grades above using the following weights:

laboratory reports	35%
class projects	35%
mid-term	10%
final exam	20%

General Procedures

Lab reports for labs on Wednesday or Thursday are due the following Monday. Lab reports for labs on Tuesday are due the following Friday. After an initial reading by us, the reports will be returned to you for revision. These must be resubmitted within a week. Only the second submissions will receive a letter grade. Class project reports will be handled in a similar manner. The submission and revision dates will be announced later.

In each class, one of each lab report and one of each report on a class project will be distributed to the class, for a class journal recording what we have done. The standard for receiving an A on a report is that it be good enough for "publication" in the class journal. (Of course, several reports may be good enough, even though only one will be published.)

Lab and project teams are encouraged to consult with others (in or out of the course), but the report must be written by the team members, and each member must sign it.

Nature of the Text Materials

You must read the assigned text materials and attempt the activities requested there before coming to class. We will spend little time in class going over what you have read; class time will be used for questions, discussions, and group activities based on the reading. We stress that this must be active reading. Work the exercises; write comments in the margin. What you learn depends on what you do, not on what you passively read in your room or drowsily hear in class. **This is an active course; most of the time you will be participating in the solution of a problem or conducting an experiment, not sitting back watching others.**

1989-1990 Syllabus

The syllabus that appears on the next two pages was revised between the first and second semesters. Thus, it reflects what we actually did in the first semester and what we planned for the second.

The Calculus Reader

We reproduce here parts of the Introduction and Chapter 5 of our textbook; the latter is the transition from the end of the first semester to the start of the second. The section presented here supported roughly one week's work at the end of the first semester.

Introduction

WHAT'S A STUDENT LIKE YOU DOING
IN A COURSE LIKE THIS?

What Is This Book About?

Most math textbooks are about "answers" (and how to get them). This book is about *questions*. Just like the world around us, this book has far more questions than answers. Indeed, our questions *are* the questions of the world around us. Almost everything of importance in our world is moving or changing, and *calculus* is the mathematical language of motion and change.

To give you some idea of the importance of our subject, we will pose some sample questions that calculus might help us answer. You won't find the answers in the back of the book; indeed, answers that fit neatly in books are seldom real solutions to real problems. Here we go.

Are we in the midst of a global population explosion? If so, what resources will we exhaust first: food, fuel, or terrestrial space? As a response to such a crisis, should we colonize outer space? If so, what does it take to do that, and how do we go about it? Can people survive in large numbers on the moon or Mars? Can we move enough of them

Project CALC Syllabus, 1989–90, First Semester

Textbook	Classroom Take-home	Laboratory
0. Functions of time • Tables and graphs • Linear functions • Discrete vs. continuous	1a. Rates of change with linear and quadratic data • Speedometer/odometer	
1. Population models • Discrete Malthusian model • Rates of change, slopes • Math notation vs. words • Instantaneous rates • Delta process, derivative • Discrete population growth • Solution of discrete IVP • Continuous IVP, dir. field • Limit notation • Limit experiments • Natural base • Change of base, time scale • Change of population scale • Solution of Malthus model • Semilog and log-log graphs • Summary	1b. Exponential growth • Fruit fly data 2a. Derivatives of various functions (delta process) • Discussion on writing (in place of student-driven class) 2b. Derivation of e • Properties of exponential function 3a. Resolve fruitfly model • Radioactive decay 3b. Puppy growth • Planetary years 4. Review 5a. Symmetric differences 5b. Midterm exam	1. MathCAD tutorial • Falling body data (no resistance) 2. *EXP* tutorial • Rate of change of common log 3. Logistic model of fruitflies with crowding • Euler solution 4. Log graphs in *MathCAD* • College cost data 5. Falling body with resistance • Find proportionality constant • Symmetric difference quotient
2. Initial Value Problems • Effects of scaling • The homicide detective • Immigration model • Problem-solving strategies • Bisection method • Immigration (reprise) • Falling bodies • Polynomials, power rule • Second derivatives, graphs • Newton's method • The product rule • Solution of air resistance model	6a. Polynomials, graphs, derivatives 6b. Find zeros by bisection 7a. RL circuts 7b. Falling body with resistance 8. Newton's method	6. Derivatives of polynomials • Zeros by zooming 7. Direction fields • Uniqueness of solutions of IVP's 8. *(No lab; Fall Break)*
3. Balloons, epidemics, and prices • Weather balloons • Chain rule • Differential • Epidemic models • Vector functions • Supply and demand models • Continuous response to imbalance • Functions of several variables • Introduction to partial derivatives • Geometric sums • Instability	9a. Related rate problem (air traffic control) 9b. Minimum distance 10a. Supply and demand • Dependence of stable price on parameters 10b. Functions of two variables	9. Spread of epidemics • Vector Euler method 10. Discrete price stability model • Instability, exact solution • Comparison with continuous model
4. Oscillations • RL circuit with AC generator • Derivatives of sine and cosine • The equation $y'' + y = 0$ • External Drive force • Phase shift • Pendulum motion • Linear approximation to DE • Period vs. Length	11a. Sine and cosine demonstration • Limit of $(\sin t)/t$ 11b. Undetermined coefficients 12a. Forcing function • Superposition 12b. Pendulum timing experiment • Solution of linearized equation	11. Parametric curves with sine and cosine • Conics • Lissajous figure 12. Pendulum motion • Euler solution, period vs. length
5. Projectile motion • Winning the peace • Initial angle: how many solutions? • Escape velocity • Rocket propulsion • One stage or two?	13a. Parametric curves • Velocity and acceleration 13b. Projectile with no air resistance • How to hit a given target 14. Honesty day: Where we've been and where we're going	13. Projectile with air resistance • Firing into the wind *(End of First Semester)*

Project CALC Syllabus, 1989–90, Second Semester

Textbook	Classroom Take-home	Laboratory
	0. (partial week) Honesty day	
	1a. Indefinite Integral	1. Rocket propulsion
	1b. Review rules for differentiation and antidifferentiation	
	1c. Barrier test on differentiation	
6. Population growth revisited • Logistic model • Another inverse problem: If the answer is $1/x(a-x)$, what's the question? • Coalition model • Combining models • Dicrete logistic model • Graphical analysis and chaos • Feigenbaum's constant	2a. U.S. population, 1840–1940 2b. Solution of the logistic model *Note:* Classes and lab on this material postponed to Week 8 before Spring Break.	2. World population • Superexponential growth
7. Average function values • Area and Riemann sums • Area as solution of an IVP • Definite integral • Speedometer and odometer (again) • Fundamental Theorem (discrete) • Fundamental Theorum (continuous) • Moments and centers of mass • Trapezoidal and midpoint rules • Simpson's rule • Substitution techniques • Use of integral tables • Undoing the product rule • First order linear DE's • RL circuits (reprise)	3a. Graphical estimates of error in left and right sums 3b. Integral of function defined by data 4a. Experiments with balance beam, plywood shapes 4b. Centers of mass 5a. Graphical comparison of TR, MR 5b. SR as weighted average 6a. Symbolic computation 6b. Midterm exam 7a. Substitution techniques 7b. Use of tables	3. Trapezoidal and midpoint rules 4. Derivative of integral and vice versa 5. Simpson's rule 6. Tutorial on *Derive* 7. Experiments with *Derive*
	8a. Demonstration of bifurcation, cycles, onset of chaos 8b. Calculate Feigenbaum's constant	8. Chaotic behavior of logistic equation • Discover Feigenbaum's constant
(Spring Break)	9a. Integration by parts 9b. First order linear IVP's	9. Comparison of Euler's method with exact solution via *Derive*
8. Approximation and error • Normally distributed data • The error function • Polynomial approximations • Taylor's Theorem • Power series • Interval of convergence • What happens at the end points? • Improper integrals • The error function (reprise) • New methods from old: weighted averages • Acceleration of convergence	10a. Measure heights of class • Are they normally distributed? 10b. Review integration techniques 10c. Barrier test of integration 11a. Polynomial approximations 11b. Taylor polynomials 12a. Approximations to $1/(1-x)$, $\ln(1-x)$ 12b. End point behavior 13a. Improper integrals	10. Normally distributed data • Distribution function 11. Taylor polynomials and convergence 12. Summing alternating series • Corrected partial sums 13. Numerical methods for improper integrals
	13b. More improper integrals 14. (partial week) Honesty day: Where we've been and what's next	

there to make any difference? If so, what are the scientific, engineering, economic, political, sociological, theological, and biomedical problems we have to solve? How do we solve them?

Suppose we find there *is* a population crisis, but there is *no* viable solution to the problem of space colonization. What problems do we have to solve to continue our existence in relative peace on Earth? Population control? Waste management? Pollution control? Technological advances in computers, consumer goods, weapons, communications? Arms control or reduction? Management of international relations? Peace through strength or strength through peace? Economic growth or economic stability?

Suppose there is no impending population explosion—population may be self-limiting. What then? Will we see world population "level off" at some stable number? If so, how big can we expect that number to be? Would its sheer size lead us to grapple with a host of other problems, such as extreme scarcity of resources and drastically lowered standard of living?

If there is no leveling-off point, will there be oscillations in the population level? If so, will these be wild swings between very high and very low levels, or will they be modest variations at manageable levels? If the latter, which would suggest that population problems need not be high on our priority scale, what are the *important* problems of a society and a world that appear to be changing ever more rapidly?

These are challenging questions about *change*, more precisely about the *rates* at which dynamic quantities change and about the consequences we can determine from those rates. Calculus provides us the conceptual framework and many of the computational tools for the quantitative and qualitative study of rates of change, and *that's* what this book is about.

Why Study Calculus?

Each semester we ask our beginning calculus students why they are taking the course. Here are a few of the most common answers to this question:

1. It's required for my major.
2. I have always had a mathematics course, and this was the next one in line.
3. My parents said I had to take it.
4. I like mathematics.
5. Everyone says mathematics is important; I just felt that I ought to do it.

6. Calculus is central to understanding the development of philosophy and science in the last three centuries; without a thorough grasp of this fundamental branch of mathematics, one cannot be considered an educated person.

Well, honesty compels us to admit that no one has actually given response number 6, but our hope springs eternal.

Why do you, your major department, and/or your parents think that mathematics (calculus, in particular) is important? In large part, this rests on the belief that much of what we experience in the world around us can be *understood*—i.e., that our experience can predict what will happen in the future or explain what happened in the past. In other words, mathematics helps us *solve problems*.

Calculus is particularly important for solving problems involving change, problems involving limiting behavior, problems involving stasis in the midst of change. Calculus is essential for understanding the motion of atomic particles, automobiles, satellites, galaxies. Calculus is basic to the study of flow in rivers, currents in the oceans, and air over airplane wings.

What Is This Course About?

Often a problem comes to us in the form of data: An object falls through the air; we have data consisting of distances fallen at ten different times. How far had it fallen at some time not measured? Can we predict how a similar object will fall in the future? In this case, the theory comes to us from physics; the language of the theory is a mixture of English and calculus, and the calculations necessary to answer the questions require the same mix.

This course concentrates on the use of calculus to solve problems. It might seem that any calculus course ought to do that, but not all do. Traditional courses emphasize pencil-and-paper calculations, often without revealing how these calculations are used to solve real problems. In many cases, the calculations take on a life of their own and are performed mindlessly, without any purpose or application in mind.

Calculators and computers can now do much of the numeric, algebraic, and graphic manipulations that are the focus of a traditional course. These tools empower us to solve problems involving "messy" data and large numbers of computations, exactly the kind of problems we find in the real world. We can also use these tools to experiment with different ways of attacking problems and thereby obtain the intuition that comes from experience. For these reasons, the heart of this course is the *laboratory experience.*

Three paragraphs back we said, "the language of the theory is a mixture of English and calculus, and the calculations necessary to answer the questions require the same mix." English—reading and writing—what has that to do with mathematics? We want to solve problems; solving problems requires deciding what should be done, executing the calculations, and interpreting the results. The environment for this intellectual activity is language, English in our case. Until you can describe what you have done, why you did it, and what it means, you have not solved the problem. For this reason, our course is a *writing course*; we expect you to write up the laboratory experiments and the in-class and take-home projects on which you will work.

Chapter 5. Projectile Motion

Winning the Peace: A Parable For Our Time

Relations between the two nations sharing the Island of Paradise have become increasing belligerent over the last few months. Schwartz, the crazed dictator of South Paradise, is determined to inflame the patriotic passions of his people and lead them in an attack on North Paradise in order to gain control of the entire island.

Schwartz's daughter, Maria, is a playwright who is much beloved by her fellow South Paradisians. Her new play, *Chaos in Paradise*, opened two weeks ago in Belmo, the capital of North Paradise. The next morning the headlines of the papers in South Paradise read

> ## 'CHAOS' PANNED IN BELMO!!
>
> "UTTER MADNESS ... CRAZINESS" REVIEWER STATES
>
> SCHWARTZ DECLARES STATE OF EMERGENCY, CLOSES BORDER WITH NORTH PARADISE, URGES CITIZENS TO AVENGE INSULT TO MARIA

As Schwartz whipped up the war fury, all communication with the North was broken off.

However, the review in the Belmo paper had actually been quite favorable: "This is a delightful farce with beautifully balanced examples of utter madness. The author has the touch of divine craziness that enables us to both laugh at our failings and view others in a new light. Don't miss it!"

Kept in ignorance of the favorable reception of Maria's play, the Southerners prepared for war. The Northerners indignantly mobilized to meet the attack. Fighting seemed inevitable.

A small group of Northerners decided on a daring plan to try to avoid the bloodshed. They had one old mortar shell and a mortar launcher. Under cover of darkness, they planned to paddle an inflatable boat around the border and carry the shell and launcher on a seldom-used trail to the top of a hill overlooking the Southern capital of Ergo. They removed most of the explosive from the shell and replaced it with copies of the Belmo review. At dawn the shell would be launched; the small explosive charge would burst the shell at a height of 100 meters and allow the copies of the review to flutter down and enlighten the war-mad Ergons.

The hill is 400 meters above the plain on which Ergo is situated and 600 meters (horizontal distance) away from the center of Ergo. If the muzzle velocity of the shell is 100 meters per second, at what angle should the brave peacemakers aim the launcher to have the shell travel to a point 100 meters above the capital?

The small explosive device is activated by a timed fuse that is started when the shell is launched. For what length of time should the fuse be set so that the shell will burst when at the desired point, 100 meters over Ergo?

We now assume the role of members of the peacemaking band and attempt to solve this two-part problem. We start by attempting to describe the position of the shell as a function of time since launch. In other words, we want a parametric description, with time t as the parameter, of the curve traced out by the shell. Recall that we have seen such parametric descriptions before in connection with curves described by the sine and cosine functions.

We choose a coordinate system (see Figure 1) with origin at the launch site, the x-axis horizontal and pointing in the direction of Ergo, and the y-axis vertical with the positive direction upward. As usual in problems of motion, we obtain the description of the motion from Newton's Second Law. We call the mass of the shell m, the horizontal and vertical velocities v_x and v_y, respectively, and the corresponding accelerations a_x and a_x. If we assume that the only force acting on the projectile in flight is gravity, then we have

(1) $m \cdot a_x = 0$ and $m \cdot a_y = -m \cdot g$.

(Why?) Since acceleration is the derivative of velocity, we may divide by m and write

(2) $\dfrac{dv_x}{dt} = 0$ and $\dfrac{dv_y}{dt} = -g$.

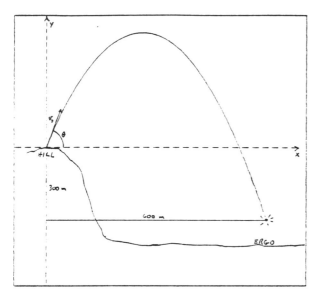

Figure 1.

To determine v_x and v_y, we need the initial conditions for these quantities. If the shell is launched at a speed of 100 meters per second and at an angle of θ with the horizontal, we have $v_x(0) = 100 \cdot \cos\theta$ and $v_y(0) = 100 \cdot \sin\theta$. (Again see Figure 1.) Thus, we have two initial value problems:

(3) $\dfrac{dv_x}{dt} = 0$, with $v_x(0) = 100 \cdot \cos\theta$;

(4) $\dfrac{dv_y}{dt} = -g$, with $v_y(0) = 100 \cdot \sin\theta$.

Exercise 1. Show that the initial value problems (3) and (4) have the solutions

(5) $v_x(t) = 100 \cdot \cos\theta$

and

(6) $v_y(t) = 100 \cdot \sin\theta - g \cdot t$,

respectively.

Exercise 2. Show that the following formulas for $x(t)$ and $y(t)$ follow from Exercise 1:

(7) $x(t) = 100(\cos\theta)t$;

(8) $y(t) = 100(\sin\theta)t - \dfrac{g \cdot t^2}{2}$.

Now that we have explicit formulas for x and y as functions of t, our problem reduces to finding an angle θ and a time t such that $x(t) = 600$ and $y(t) = -300$ (why?). If we substitute those values into (7) and (8), we get

(9) $600 = 100(\cos\theta)t$;

(10) $-300 = 100(\sin\theta)t - \dfrac{g \cdot t^2}{2}$.

Equations (9) and (10) are a pair of equations in the unknowns t and θ, just the quantities we want to determine for placing the mortar and setting the fuse. However, they are not linear equations, so it may not be obvious how to proceed on the algebraic problem of finding their solutions.

[The symbol θ is a constant throughout this calculation; the mortar is going to be placed and fired only once. However, it is still an unknown, because we haven't determined yet what value or values will meet our requirements for passing through a particular target point. The symbol g is also a constant, but its value is known: 9.807 m/sec^2 (but varying slightly with location on the surface of the Earth). We have left g in symbolic form up to this point because it is easier to write "g" than to write "9.807."]

In general, the method for algebraic solution (as opposed to *numerical* solution—say, by Newton's Method or something like it) of a system of nonlinear equations is "whatever will work." Sometimes, what works is to solve one equation for one of its variables, then substitute the result into all the other equations, thereby "eliminating" that variable from the problem. We can easily do that here, because t appears linearly in equation (9). When we solve for t, we get

(11) $t = \dfrac{6}{\cos\theta}$.

Now we substitute for t in equation (10), and we get

(12) $-300 = \dfrac{600(\sin\theta)}{\cos\theta} - \dfrac{36g}{2\cos^2\theta}$.

Equation (12) is now a single equation in the single unknown θ, but solving for θ doesn't look very promising. However, we can simplify the equation by observing that (a) $\dfrac{\sin\theta}{\cos\theta} = \tan\theta$, (b) $\dfrac{1}{\cos^2\theta} = \sec^2\theta$, and (c) $\tan\theta$ and $\sec\theta$ are related by the formula $\sec^2\theta = 1 + \tan^2\theta$.

Exercise 3. Use the hints just given to rewrite equation (12) in terms of $\tan\theta$ only. Manipulate the terms and cancel a common factor to show that (12) can be rewritten as

$$(13) \qquad 3g\tan^2\theta - 100\tan\theta + 3g - 50 = 0.$$

Now we're getting somewhere. If we substitute another variable, say, u, for $\tan\theta$, then equation (13) will be a quadratic equation in u. This is a good point at which to also give g a numeric value. When we make those two changes, (13) becomes

$$(14) \qquad 29.42u^2 - 100u - 20.58 = 0.$$

Somebody told you (and somebody else presumably told our Paradisian peacemakers) in high school that you would need the quadratic formula someday. The truth is, you don't need it often, but when you do, it will probably be for an equation like (14), not one with small integer coefficients, as is customary in high school textbooks.

Exercise 4. Solve equation (14) for u. Recall that u is the tangent of the angle θ we are looking for. Use the "inverse tangent" button on your calculator to find one or more values of θ that satisfy equation (13). Be careful about the distinction between radian and degree measure; find θ both ways. Substitute your value or values for θ into equation (11) to find the corresponding value or values of t. We get only one try with the mortar, and it absolutely has to be right; a war hangs in the balance. Check your work by substituting your (θ, t) pair or pairs into equations (7) and (8) to see if placing the mortar at the angle θ and setting the timer for t seconds will send the shell to the right spot.

Exercise 5. It is not unusual for nonlinear equations (singly or in a system) to have multiple solutions. In particular, a quadratic equation such as (14) often has two solutions, if it has any at all. How many solutions did you find? Are they all really solutions of the peacemakers' problem? Explain carefully your reasons for accepting or rejecting each of the solutions.

Your solution to Exercise 4 may or may not be adequate to prevent a war. This solution started with the assumption that air resistance was not a factor, and, indeed, that *nothing* affected the path of the projectile other than gravity. In a laboratory accompanying this Chapter, you will have an opportunity to explore what happens when air resistance is considered and then to see the effect of firing into a headwind.

Instructions for Lab 5

1. In Lab 1 you investigated the motion of an object falling under gravity with negligible air resistance. We return to this problem to consider the effect of air resistance. When you load Lab 5 and calculate the worksheet (press <F9>), *Math-CAD* will read in data on the distance an object has fallen as a function of time. This data reflects the force of gravity, the retarding force and a small random measurement error. We reproduce half of the data here.

time (in seconds)	distance (in meters)
0	0
1	4.8
2	18.5
3	40
4	69
5	104
6	146
7	193
8	244
9	300
10	361

We want to find a function $s(t)$ that approximates this data. In the case without air resistance, we found that the rate of change of the velocity, the acceleration, was essentially constant. If we assume that dv/dt is constant we are led to an approximating function that is a quadratic polynomial. In the present case we'll begin the same way, by trying to approximate the velocity, $v = ds/dt$. Since we have data at a number of discrete points, it is reasonable to approximate the derivative by difference quotients.

Suppose we label the times at which we have the data as $t_0, t_1, \ldots, t_{20}$ and try to estimate the derivative at t_1. Our first idea is to use the *forward difference quotient* $\dfrac{s(t_2) - s(t_1)}{t_2 - t_1}$, since it is the limit of difference quotients of this type (as t_2 approaches t_1) that defines the derivative. We argue that a better estimate to the derivative is obtained by using the *symmetric difference quotient* $\dfrac{s(t_2) - s(t_0)}{t_2 - t_0}$. If we think of the difference quotient as representing the slope of the line segment between the two points on the graph of s, the following illustration shows why this is the case.

[A computer-drawn figure is inserted at this point. It shows a segment of an exponential curve, a tangent line at the center of an interval, and the chords whose slopes are the symmetric difference quotient and the forward difference quotient. The slope of the longer chord is visibly different from, but very close to, the slope of the tangent line; the slope of the shorter chord is not even close.]

In general, we estimate the derivative of s at t_k by the symmetric difference quotient

$$\frac{s(t_{k+1}) - s(t_{k-1})}{t_{k+1} - t_{k-1}}$$

for $k = 1, 2, \ldots, n-1$. Here n is the index of the last data point (10 in the table above, 20 on the *MathCAD* worksheet). We do not estimate the derivative at the end points, t_0 and t_n. Why can we not use symmetric difference quotients for these points?

2. If you have not done so, load Lab 5 into *MathCAD*. Press <F9> to read in the data and plot the graph. Move down in the worksheet; calculate the symmetric differences and plot them.

It is often assumed that the retarding force is proportional to the velocity. Let's see if we can decide whether this is a reasonable assumption for the data presented. The physical law involved is Newton's Second Law of Motion which says that the force on the object is the product of the mass and the acceleration. In symbols, this is

$$F = ma$$
$$= m\frac{dv}{dt}.$$

We are considering two forces acting on the object; the force of gravity F_g and the retarding force F_r. Here $F_g = m \cdot g$, where g is the acceleration due to gravity, approximately $9.8 \, \text{m/sec}^2$. Our assumption is that the retarding force has the form $F_r = -k \cdot v$. (Where did the minus sign come from?) Thus under our assumption, the total force F on the object has

the form $F = m \cdot g - k \cdot v$. Putting this back into Newton's Law, we have

$$mg - kv = m\frac{dv}{dt}.$$

Dividing by the mass m, our assumption leads to the following expression for $\dfrac{dv}{dt}$:

$$\frac{dv}{dt} = g - \frac{k}{m}v.$$

Our question may now be rephrased: Is it reasonable to assume that the quantity

$$c = \frac{g - \frac{dv}{dt}}{v}$$

is constant? Well, we have an estimate for v at $t_1, t_2, \ldots, t_{19}$. Use *MathCAD* to calculate symmetric differences of the approximations to v to obtain approximations to $\frac{dv}{dt}$. (For what times t_k will the approximations be defined?) Use these numbers to decide whether it is reasonable to assume c is constant. If it is, give an approximate value for c. (You may find that you are running out of space; the Student Edition of *MathCAD* has a two-page limitation. When this happens, remove something, e.g., the graph of the approximation to v, to obtain more space.)

3. Remember that our goal is to find a function $s(t)$ that approximates the distance s the object has fallen as a function of t. Use your results in Part 2 to write down an initial value problem which should be satisfied by v, i.e., write down the differential equation for v and the initial condition.

A Preview of Future Work

At a later time we will finish this investigation by carrying out the remaining two steps:
  (a) We will solve the initial value problem for $v(t)$.
  (b) Using our expression for $v(t)$ from (a), we will find an expression for $s(t)$.

The MathCAD *worksheet for Lab 5 is shown in Figure 2, exactly as the students first see it. We have filled in those items we don't want them to spend time on at this point, and we have left to them the items they need to think about.*

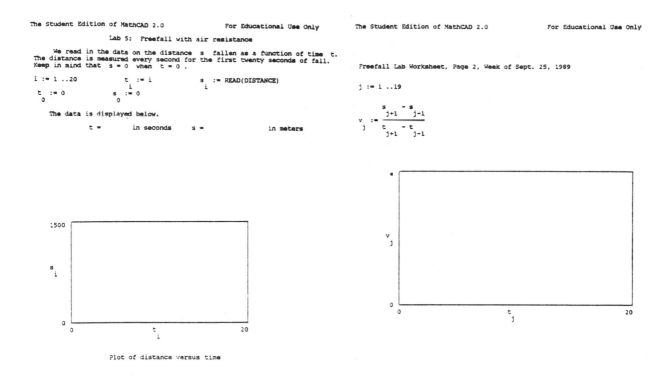

The Student Edition of MathCAD 2.0   For Educational Use Only   The Student Edition of MathCAD 2.0   For Educational Use Only

Lab 5: Freefall with air resistance

We read in the data on the distance s fallen as a function of time t. The distance is measured every second for the first twenty seconds of fall. Keep in mind that s = 0 when t = 0 .

Freefall Lab Worksheet, Page 2, Week of Sept. 25, 1989

$i := 1 ..20$ $t_i := i$ $s_i := READ(DISTANCE)$

$t_0 := 0$ $s_0 := 0$

$j := 1 ..19$

The data is displayed below.

$t =$ in seconds $s =$ in meters

$$v_j := \frac{s_{j+1} - s_{j-1}}{t_{j+1} - t_{j-1}}$$

Plot of distance versus time

Figure 2

Instructions for Project 9

It is a calm autumn day in southeast Iowa at the Ottumwa air traffic control radar installation. Only two aircraft are in the vicinity of the station, American Flight 1003 from Minneapolis to New Orleans is approaching from the north northwest and United Flight 366 from Los Angeles to New York is approaching from the west southwest. Both are on paths which will take them directly over the tower. There is plenty of time for the controllers to adjust the flight paths to insure a safe separation of the aircraft.

Suddenly the tower is rocked by an earthquake. The conventional power is lost and the auxiliary generator fails to start. In desperation a mechanic rushes outside and kicks the generator; it sputters to life. As the radar screen flickers on, the controllers find that both flights are at 33,000 feet. The American flight is 36 nautical miles from the tower and approaching it on a heading of 171° at a rate of 410 knots. The United flight is 41 nautical miles from the tower approaching it on a heading of 81° at a rate of 455 knots.

- How fast is the distance between the planes decreasing?

- How close will the planes come to each other?
- Will they violate the FAA's minimum separation requirement of 5 nautical miles?
- How many minutes do the controllers have before the time of closest approach?
- Should the controllers run away from the tower as fast as possible?

This project was the first introduction in class to the Chain Rule. It was also the only "related rate" problem we considered, and it happens to also be an optimization problem. We are deliberately trying to remove "problem type" clues that allow students to revert to their "plug into a formula" mode of problem solving. They already knew the connection between derivatives and max/min, a very easy connection to assimilate through graphical experiments in the lab. We had not said anything about the Chain Rule or about related rates when they started on this project.

Teams of three or four worked on the project in class, trading ideas, calculating, struggling to relate the problem to something familiar. What they didn't accomplish in class became a take-home project leading to a paper that would explain their solution. Some teams got discouraged with their symbolic computations and went to the lab to solve the problem numerically with MathCAD. *If that was the only solution submitted, we suggested they go back to their hand and calculator work, with gentle pushes in the right direction, to see that they could solve the problem more quickly and more accurately without the computer.*

A common stumbling block in this problem was an intuitive certainty that the minimum distance must be achieved when one of the planes is over the tower.

The following document is the second submission from one team of three, warts and all. The authors can't make sense of the phrase "Pythagorean Theorem," but they know what it is and how to use it. At the time of this writing, they still couldn't embed their calculations in sentences, but most of our students can do that now. These authors are still erring on the side of overkill in their explanations of easy details; judgement comes later. The paper also contains one flat-out error, which fortunately does not affect the result; we leave its detection as an exercise for the reader.

On the positive side, note the <u>appearance</u> of this work. Our students take pride in making their work look good, and EXP *and the laser printer make that relatively easy. Notice the evident <u>involvement</u> of the students in their work. And finally, observe that these students are <u>having fun</u>.*

Student Report on Project 9

In Project 9, we were presented with an interesting series of problems caused by a couple of airplanes and an earthquake. Our situation involved two planes, each approaching the infamous Ottumwa air traffic control tower. The tower was located in Southwest Iowa, an area known to be extremely dangerous due to frequent earthquake activity, and so the controllers were not too surprised when a quake rocked the building, causing it to lose all power. The controllers, however, were used to this sort of situation, so they immediately sent Fred the mechanic out to the rescue. Fred gave the generator a powerful swift kick and restored life to the tower. The controllers quickly checked their radars after the electricity had returned and discovered, to their dismay, that the two planes - an American and a United flight - were headed towards the tower on paths that could mean very close flying in the friendly skies. Immediately, panic engulfed the entire control tower.

Our job here at Project CALC headquarters was to take over for the controllers of the tower since they were too disturbed to perform the calculations which could determine how much danger they were in. Our first task was to find the speed at which the distance between the planes was decreasing. From the radar we found that the American flight was approaching the tower from 36 nautical miles away at a heading of 171° and a speed of 410 knots. The United flight was approaching the tower from 41 nautical miles away at a heading of 81° and a speed of 455 knots. From this information we quickly sketched this picture of the situation:

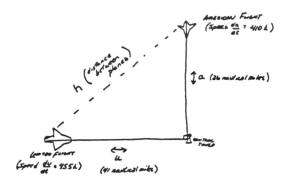

From the degree measures and our sketch, we saw that the planes were approaching each other along the legs of a right triangle. Thus, the distance between them was equal to the hypoteneuse of our right triangle. What luck! It was as if this problem had been found in a calculus book and was being acted out in mid-air! Anyway, after a short celebration, we quickly got to work setting up an equation for dh/dt, the rate at which the two planes were approaching each other. We recalled Pythagorous' theory and applied it to our right triangle to find that h^2 (the hypoteneuse squared) was equal to a^2 (the American flight's distance from the tower squared) plus u^2 (the United flight's distance from the tower squared). We used this information to calculate dh/dt:

$$h^2 = a^2 + u^2$$

(Pythagorous' theory)

$$2h\frac{dh}{dt} = 2a\frac{da}{dt} + 2u\frac{du}{dt}$$

(We took the derivative $\frac{d}{dt}$ of both sides of the equation.)

$$h\frac{dh}{dt} = a\frac{da}{dt} + u\frac{du}{dt}$$

(We divided both sides
of the equation by 2.)

$$\frac{dh}{dt} = \frac{a\frac{da}{dt} + u\frac{du}{dt}}{h}$$

(We divided both sides
of the equation by h.)

Now that we had dh/dt isolated, we needed to replace the known symbols with their equivalent numerical values. The symbols that we could replace, from information given above, were:

$$a = 36 - 410t \qquad u = 41 - 455t$$

$$\frac{da}{dt} = -410 \qquad \frac{du}{dt} = -455$$

Our equation, with the known values, was

$$\frac{dh}{dt} = \frac{(41 - 455t)(-455) + (36 - 410t)(-410)}{h}.$$

By expanding, the equation became

$$\frac{dh}{dt} = \frac{-18655 + 207025t - 14760 + 168100t}{h}.$$

And finally, when we combined terms, the equation became

$$\frac{dh}{dt} = \frac{375125t - 33415}{h}.$$

Thus, we knew that the rate at which the distance between the two planes was decreasing was

$$\frac{375125t - 33415}{h}.$$

We wanted to know how fast the planes were approaching each other at the initial time $t = 0$. To solve this problem we needed a value for h at the initial time. Again, we recalled Pythagorous' theory and used it to find h.

$$h^2 = a^2 + u^2$$
$$h^2 = (36)^2 + (41)^2$$
$$h^2 = 2977$$
$$h = 54.56$$

Then we used this value to solve for dh/dt at $t = 0$:

$$\frac{dh}{dt} = \frac{0 - 33415}{54.56}$$
$$= -612.445 \text{ knots}$$

So, at the initial time $t = 0$, the planes were approaching each other at 612.445 knots.

We assumed that the planes would have a point of closest approach, at which the $h(t)$ function would have a definite minimum. From our past experience with second-order polynomials, we knew that this minimum would correspond with a zero in the derivative. Therefore, to find this zero, we set dh/dt equal to zero:

$$\frac{dh}{dt} = \frac{375125t - 33415}{h} = 0$$

Since h is a constant, multiplying both sides by h is algebraically legal. Thus:

$$375125t - 33415 = 0$$

We added 33415 to both sides and then divided the equation by 375125 to obtain the value of t when dh/dt equals zero:

$$t = \frac{33415}{375125} = .0890769$$

So, at the point t equals .0890769 hours, the two planes come closest together. To find this distance, we need only to evaluate h at .0890769. We recalled that the value of h is equal the square root of $a^2 + u^2$. Since we knew the values of a and u as functions of t, we substituted these values into Pythagorean's Theorem to find the following:

$$h(t) = \sqrt{(36 - 410t)^2 + (41 - 455t)^2}$$

We expanded and combined the terms to obtain $h(t)$ equal to $\sqrt{375125t^2 - 66830t + 2977}$. When we evaluated .0890769 for t, we obtained $h(.0890769)$ to be equal to .70207 nautical miles. This is the point of closest approach of the two planes.

This distance violates the FAA's minimum separation requirement of 5 nautical miles. To find how long it would take before the planes were at their point of closest approach, we converted .0890769 hours to minutes and obtained a time of 5.34 minutes. Since the planes weren't going to crash, the controllers didn't have to clear the tower. However, we here at Project CALC headquarters think that it might behoove the controllers to evacuate the tower since it was just hit by a very brutal, punishing, and tyrannic earthquake. This is Project CALC headquarters signing off. See you Thursday, same time, same place.

Instructions for Lab 12

Pendulum Motion

For the background to this lab, see the handout entitled "Text Material for Pendulum Motion, Week of Nov. 13." You should have read this material and worked on the exercises at the end of the handout before coming to the lab!

Bring up *MathCAD*, load LAB12, and scroll through the worksheet. Notice that we have written the iteration conditions for both θ and v in a single vector equation:

$$\begin{bmatrix} \theta_{k+1} \\ v_{k+1} \end{bmatrix} = \begin{bmatrix} \theta_k + v_k \cdot \delta t \\ v_k - \frac{g}{L} \cdot \sin(\theta_k) \cdot \delta t \end{bmatrix}.$$

Note also that the constants L, N, and δt are given as global variables under the first graph plot.

We want to estimate the period of the pendulum for various values of length L. An easy way to do

this is to estimate the length of a half-oscillation by finding the value of t_k, where k is the first value of the index such that v_k is positive. (Why is this reasonable?) Enter your estimate for the period in the table started for you. Repeat this for a variety of L's between 0 and 100. Try to estimate periods for at least 10 values of L, and include a couple for L's of 10 or under. Change the limit on the index j from 20 to the number of periods you have estimated, and plot the periods versus the lengths. What sort of function describes the relationship between the period and L?

Compare your numerical observations with the experimental observations made in class and with your estimate in Exercise 3 of the handout on pendulum motion.

The MathCAD worksheet for this lab is displayed in Figure 3.

The Student Edition of MathCAD 2.0   For Educational Use Only

Lab 12
Pendulum Oscillation

NOTES: 1. Distances are measured in centimeters and time in seconds.
        2. Acceleration due to gravity is   g := 980.7

The following constants below will be changed frequently during the lab.
They are given as global variables under the first graph plot.

The length L of the pendulum
The number N of time steps
The length δt of each time step

The variables are:

θ: the angular displacement from vertical in radians
v: the angular velocity in radians/second
t: the time in seconds

The initial conditions are:

θ := π/6   The initial displacement is π/6, positive to the left.
 0

v := 0   The initial angular velocity is 0.
 0

k := 0 ..N - 1   The iteration step is given below.

$$\begin{bmatrix} \theta_{k+1} \\ v_{k+1} \end{bmatrix} := \begin{bmatrix} \theta_k + v_k \cdot \delta t \\ v_k - \frac{g}{L} \cdot \sin(\theta_k) \cdot \delta t \end{bmatrix}$$

The time at step k is
t := k·δt
 k

θ ,0
 k

0   t
                     k

L = 100   N = 600   δt = .005

The Student Edition of MathCAD 2.0   For Educational Use Only

Estimate the period by taking as an estimate for the half period
the value of t where v changes from negative to positive.

m := 204

v   2·t
 m   m

j := 1 ..20

L :=   Period :=
┌─────┐   ┌─────┐
│ 100 │   │ 2.04 │
└─────┘   └─────┘

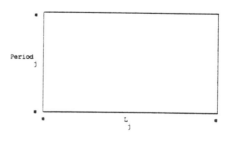

Period
 j

               L
               j

Figure 3

Final Examination

Fall Semester, 1989

For this exam, write your responses in complete sentences and paragraphs. We will grade primarily on the quality of your explanations. You may use your text, project, and lab materials and your notes.

Of the four problems on the exam, you should select two to be graded. We suggest that, at the beginning of the exam period, you select three of the problems and work on them in the space provided. Then select two to write up using your own paper. If you are satisfied with the write-ups on the two problems and there is still time left in the examination period, you may work on one of the two remaining problems for extra credit.

1. Consider the doorknob pendulum that was used for our experiment in class. Suppose we attach new string to it so that it now has a length of two meters. If we pull it to one side to an angle of $\frac{\pi}{7}$ with the vertical and give it an initial upward velocity of $d\theta/dt = 1.2$ radians per second, find the approximate location of the doorknob after 2 seconds.

2. Suppose the rubber ball that we experimented with on the last day of class is dropped onto the classroom floor from a height of two meters.

a. Estimate how far the ball travels up and down before it comes to rest.

b. Estimate how long it takes for the ball to come to rest.

3. Dr. Moore likes to run in the evening when he gets home after work. On this day he had planned an eight mile run at a pace of eight minutes per mile. When he walked in the door, he smelled a turkey cooking; in the kitchen he found a note from his wife: "I have to go to Raleigh. Take the turkey out when it is done."

The turkey is done when the meat thermometer reads 185°F. At 6:05 Dr. Moore pulled the turkey out of the 350°F oven; the thermometer registered 125°F. He put the turkey back in and went upstairs to change clothes. Then he did stretching exercises, glanced at the mail, and went outside to see why the dog was barking. At 6:23 he was ready to start his run. Another glance at the turkey showed that the thermometer was reading 145°F. Does Dr. Moore

have enough time to complete his run before he has to be back to take the turkey out?

4. Suppose we assume that an economy is modeled by the continuous price evolution model given in the "Handout for Economics Unit" and that we have determined that the demand and supply functions are:

$$D(p) = 400 - 10 \cdot p \text{ and } Q(p) = 20 + 9 \cdot p.$$

Time is measured in weeks and prices are recorded at the beginning of the year and at the end of the week for the first four weeks of the year (each Friday at 5:00):

time	price
0 weeks	$37.0
1	32.0
2	28.5
3	26.0
4	24.3

• What is the equilibrium price?

• Write down an initial value problem to model the price as a function of time.

• Estimate the price after 8 weeks; after 11 weeks.

Midterm Examination

Spring Semester, 1990

*Select **one** of the following two problems, and write your solution and explanations on your own paper. When you have solved the problem, write a complete explanation in essay form; you will be graded primarily on the quality of your explanation. You may use text, project, and lab materials, notes, calculator, and anything else you have with you.*

1. You are making a mobile, and one of the suspended parts is to have the shape of the figure below. You are going to cut the figure from a sheet of a thin material. At what point will you attach the thread so that the figure will hang in a horizontal plane?

If the origin of the coordinate system is placed at the "nose" of the figure, the upper half of the figure is bounded by the graphs of $y = \sqrt{x}$, $y = x - 2$, and $x = 7$.

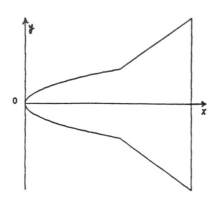

2. If the population P of the world is measured in billions and time t is measured in years, the historical data suggests that the growth rate of P can be modeled by $dP/dt = 0.005P^2$. Given the fact that the population reached 5 billion about the start of 1987, this growth rate predicts that P becomes infinite (as we saw in Lab 2) about the start of 2027.

We consider now a fanciful scenario in which the world population problem is complicated by a small but steady immigration from outer space: Alien beings that are identical to humans in every way start arriving in 1987 at the rate of 5 million (0.005 billion) per year. Since the immigrants are travelling from a distant solar system (in a state of suspended animation, so they arrive with normal ages in Earth years), there is no way to turn off this steady flow—the travellers for the next several hundred years are already en route.

Thus, from 1987 on, the growth rate for world population has an added component to account for immigration. That should mean that P is predicted to become infinite even earlier. How much earlier?

Differentiation Barrier Test

January 19, 1990

Find the following:

$$\frac{d}{dt}(10t^6 - 4t^3 + \pi) \qquad \frac{d}{dx}(cos x)^3$$

$$\frac{d}{dt}\ln(\sqrt{3t - 9}) \qquad \frac{d}{d\theta}\frac{\sin(\theta)}{\cos(2\theta)}$$

$$\frac{d}{dx}\frac{1}{x^3} \qquad \frac{d}{dt}t^{\frac{7}{5}}$$

$$\frac{d}{dx}xe^{-x} \qquad \frac{d^2}{dx^2}xe^{-x}$$

Integration Barrier Test

March 23, 1990

Evaluate the following integrals. Your work should show how you would work the problems without a using a table of integrals. You may use the table for checking.

$$\int (t^2 + \sin 3t)\, dt \qquad \int \frac{7}{\sqrt{s + 1}}\, ds$$

$$\int ze^{7z}\, dz \qquad \int \frac{x + 1}{x^2 + 4}\, dx$$

$$\int_3^7 \frac{4}{2y + 1}\, dy \qquad \int_{-\frac{\pi}{4}}^{\frac{\pi}{4}} \cos 2x\, dx$$

$$\int_0^2 e^{-0.3t}\, dt \qquad \int_0^{\frac{\pi}{4}} \arcsin t\, dt$$

The Five Colleges

CALCULUS IN CONTEXT

by James Callahan, David Cox, Kenneth Hoffman
Donal O'Shea, Lester Senechal, Frank Wattenberg

Abstract

Contact:

JAMES CALLAHAN, Mathematics Department, Smith College, Northampton, MA 01063. PHONE: 413-585-3863. E-MAIL: jcallahan@smith or jcallahan%smith.bitnet@cunyvm.cuny.edu.

Institutional Data:

The Five Colleges are Amherst, Hampshire, Mount Holyoke and Smith Colleges and the University of Massachusetts at Amherst. The colleges are private, four-year, liberal-arts institutions enrolling between 1000 and 2600 undergraduates each. Mount Holyoke and Smith are women's colleges. Hampshire was created by the other four institutions about twenty years ago to be a site for educational innovation. The University is a public institution with 18,000 undergraduates and strong graduate and professional programs. This project is sponsored by Five Colleges, Inc., a small and effective administrative body set up to promote collaboration between the institutions.

There are 32 mathematics faculty at the four colleges, ranging from 2 at Hampshire to 12 at Smith. Each year the colleges graduate about 75 majors all together. The University mathematics department has 60 faculty and graduates about 40 majors each year.

Project Data:

Twenty-three faculty members from the five institutions have taught versions of Calculus in Context—from one at Amherst College to nine each at Mount Holyoke and Smith. At Hampshire, Mount Holyoke, and Smith, *all* calculus students take the new curriculum; this amounts to 370 enrollments in Calculus I during 1989-90. At Amherst, 25 students take a year-long course that combines pre-calculus material with the new calculus. At the University, three of the twenty-five sections of the calculus course for mathematics and physical science students (75 from a total of about 650) used a new curriculum in the fall of 1989.

A precursor of the new course was taught at Hampshire College for ten years before the Five College project was created. Versions of the new course have been taught there, and at Mount Holyoke and Smith Colleges, since the fall of 1988: Calculus I for four semesters, Calculus II for three, and Calculus III for two. In the fall of 1989 some of the Calculus I material was introduced into the Amherst course that includes a review of pre-calculus. A different curriculum was introduced at the University at the same time: Calculus I and Calculus II have each been offered there for one semester.

Computing is an essential part of Calculus in Context. Computers are used on a daily basis in the classroom, as tools for instruction and exploration. Students also work on assignments in computer laboratories and workshops. All the software used is homemade, designed to fit the needs of the new curriculum. There are programs to plot the graphs, contour lines, and Fourier transforms of functions; others plot solutions of systems of differential equations—both as graphs and as trajectories in phase space. The computers are IBM PS/2 Model 30 286s with math co-processors and VGA color monitors.

Members of the project are currently writing texts, exercises, and laboratory assignments. A 360-page first draft of a text for Calculus I, produced in LaTeX, is being used in 1989-90 at the colleges. The project will revise this text and produce drafts for Calculus II and III for the fall of 1990. A different text was written at the University.

Principal funding for Calculus in Context is a five-year National Science Foundation grant awarded in August 1988. The principal investigators are Jim Callahan (Smith), David Cox (Amherst), Ken Hoffman (Hampshire), Don O'Shea and Lester Senechal (Mount Holyoke), and Frank Wattenberg (University of Massachusetts). We are aided by an advisory board consisting of Solomon Garfunkel (COMAP), Peter Lax (Courant), John Neuberger (University of North Texas), Barry Simon (Caltech), Gilbert Strang (MIT) and John Truxal (SUNY Stony Brook). The evaluation of our project is directed by Hariharan Swaminathan of the University of Massachusetts.

In addition to our NSF grant, various project activities have been funded by the Pew Charitable Trusts, supporting the New England Consortium for Undergraduate Science Education, and by the Sloan Foundation. Computers have been provided by IBM and by the Instrumentation and Laboratory Improvement program of the National Science Foundation.

Project Description:

Calculus in Context is a fundamental restructuring of the three-semester calculus sequence, in which the major mathematical concepts grow out of substantial problems from the sciences. Computers are used to explore ideas and to answer questions that would otherwise be inaccessible. For example, systems of differential equations are introduced early because they are common in scientific work, and a beginning student can understand how to solve them numerically. Furthermore, since numerical solutions are only approximate, and errors are reduced by successive approximations, students see limit processes introduced in a compelling way. Our ultimate goal is to expose the underlying power, beauty, and integrity of calculus as a coherent intellectual achievement.

The new curriculum is intended for all students. Each course is taught by regular faculty in a 13-week semester with three or four class meetings per week. Computers are used in the classroom as the need arises, and laboratories are offered in addition. There are daily homework assignments, lab reports, occasional quizzes, and two or three exams. Some questions on assignments are complex, subtle, and ill-posed; often students can resolve them adequately only by using a computer and by working in groups.

Since the new courses and the existing ones have different goals, it is not possible to compare directly the performance of students in the two. One of the purposes of the project is to enhance the ability of students to use calculus in their subsequent work in science and mathematics. To assess this, the progress of students will be followed over several years, and their instructors will be interviewed.

Beginning in the summer of 1991, the project will offer workshops to demonstrate course materials and software and to discuss strategies for transporting and implementing the new curriculum. Texts and software are currently available at cost.

═══ Project Report ═══

Getting Started

The Five Colleges consist of Amherst, Hampshire, Mount Holyoke, and Smith Colleges and the University of Massachusetts at Amherst. There are substantial differences between the institutions. Amherst, Mount Holyoke and Smith are well-established, traditional liberal arts colleges; Mount Holyoke and Smith are women's colleges. The University is public institution with strong graduate and professional programs as well as a large undergraduate program. Hampshire was created in 1970 by the other four colleges to serve as a site for innovation and experimentation. The five schools lie within twelve miles of one another and have always enjoyed close relationships. An umbrella organization, Five Colleges Inc., was set up in 1965 to coordinate joint ventures by the colleges, and our calculus project is administered by this organization.

The history of our project goes back to the late 1970's and early 1980's, when Hampshire's Kenneth Hoffman and Michael Sutherland developed an unusual calculus course for students in the life and social sciences. The course introduced differential equations at an early stage and made heavy use of the computer. Other experiments with the calculus curriculum were tried from time to time, but matters didn't come to a head until 1987-88. During that year, the University developed a calculus course with a computer lab, and funding from the Sloan Foundation's New Liberal Arts program made it possible to teach a section of the Hampshire course at Mount Holyoke. This brought various ideas of calculus reform to a wider audience, and people who for years had been concerned about calculus now began collectively to think about alternatives. At just this time, the NSF calculus program was announced, and we decided to make a formal proposal. Our grant application was heavily influenced by the Hampshire course, and our basic theme was that its ideas could be expanded to create a course that served the needs of all calculus students.

After receiving the NSF grant in the summer of 1988, the process of developing course material began immediately. That fall, Mount Holyoke and Hampshire taught sections based on the Hampshire course, while Smith embarked on a different approach that emphasized functions of several variables. During the summer of 1989 these approaches were synthesized into a common text covering the first semester (what we call "Calculus I") that is now used at Hampshire, Mount Holyoke and Smith. Development of material for the second semester (Calculus II) began at Smith in the spring of 1989, and a complete version will be written during the summer of 1990. Work on the third semester (Calculus III) started in the fall of 1989 and is still underway. The rapid pace of writing forced us to seek additional funding. The budget of our NSF grant didn't fund Calculus II until 1989-90, and we were fortunate to get support from NECUSE (New England Consortium for Undergraduate Science Education, funded by the Pew Charitable Trusts) so the writing of Calculus II could begin a semester early.

The transition from old to new calculus was handled differently at different schools. Hampshire had the easiest time since students had been using versions of the new calculus for a decade. Mount Holyoke began the transition in the fall of 1988 by teaching a version of the Hampshire course to students in the life and social sciences, and a short "bridge course" was set up in the January term for those rejoining the usual calculus sequence in the spring. The next year, all students took the new Calculus I. At Smith, the process was more abrupt: starting with the fall of 1988, all beginning calculus students encountered a new calculus curriculum. The standard calculus sequence no longer exists at any of these three schools.

At Amherst College and the University of Massachusetts, the situation is different. Amherst is keeping the standard sequence for most students. However, there is a year-long course that combines a semester of precalculus with a semester of calculus, and some of the Calculus in Context material is used in this course. At the University, a different version of Calculus I and II was taught during 1989-90 in three sections of the calculus sequence for scientists and engineers. The University course was structured so that students could switch to the standard sequence after either semester. Six sections will be offered in 1990-91.

At the colleges, the small size (of both the departments and the number of students involved) made it easier to commit to changing the calculus curriculum, although different schools adopted different strategies for doing so. At the University, a more gradual course of action was needed.

The Philosophical Setting

We believe that all calculus students should have a solid understanding of the interplay between the tools, language, and techniques of calculus and some of the real-world contexts from which they arise. This is different from teaching either "calculus with applications", where one typically develops the mathematics autonomously and then casts about for a couple of applications, or the "applied calculus" courses where the mathematics is viewed principally as a tool. An essential component of our approach is the fact that scientific problems are sometimes poorly posed, noisy, described by incomplete or ambiguous data, and too rich to be solved by a single technique. The role of mathematics is often more that of an experimental tool for probing the nature of the problem, rather than a method for solving it in the sense that students are used to seeing problems solved. There may be a continuum of good answers rather than a "right" answer.

Getting our students to think about problems in this way calls for a major shift—they need to go beyond the technique-and-formula conception of mathematics which has characterized most of their mathematical training so far. An ability to play with mathematical ideas, to try out a variety of approaches to see what works rather than looking in a book for the "right" technique, is a basic skill for all scientists and mathematicians.

Another basic feature of our course involves the role of computers, which have fundamentally changed the way we handle information and do science and mathematics. The computer is a wonderful teaching aid which allows us to better demonstrate traditional concepts like limits and Taylor series. But it is much more than this: the existence of computers substantially affects the choice and presentation of mathematical concepts as well. For instance, it is now possible to treat manifestly more realistic, and hence more interesting, problems in the classroom. There are a number of computationally intensive techniques which are nevertheless conceptually simple and can now be introduced early in the curriculum. Computers have created a paradigm shift (in the sense of Kuhn) in how mathematics is used and, in some cases, in the way it is conceived. We want to reflect the impact of this revolution in what we teach and how we teach it.

There are a number of more specific principles which derive from the broad general ideas laid out above:

- Differential equations should be treated as fundamental objects. Dynamical systems are an integral feature of many branches of science and mathematics, so an ability to handle systems of simultaneous (often non-linear) differential equations should be developed early. This is possible because the concept of derivative—as rate of change—can be approached initially as an intuitively clear concept independent of the process of differentiation.

- Modelling natural systems lies at the heart of many parts of calculus. Students therefore need to be adept at constructing and analyzing such models.

- The process of successive approximation is fundamental as both a theoretical concept and an analytical tool. Many ideas should be developed by numerical methods before closed-form solutions for well-behaved cases are presented.

- Many natural processes involve more than two interacting variables. A familiarity with functions of several variables should be developed early.

- Geometric visualization greatly enhances our power to develop a sophisticated understanding of concepts, techniques, and problems. The geometric viewpoint should be an integral part of the way students are taught to think about calculus.

Content

In designing the new calculus curriculum, we started with the question "What do we want to put in?" rather than "What can we afford to take out?". The philosophy described above guided us in deciding what topics to include and how to treat them. We also wanted to take into account the fact that some students—often social scientists and biologists—take only one semester of calculus, while others take two, and mathematicians and physicists take all three semesters. We therefore designed the courses so that Calculus I contains the topics likely to be of greatest interest to those students for whom this would be their terminal calculus course, leaving for Calculus III topics primarily of interest to physical scientists.

In contrast, the Calculus I and II taught at the University are intended to be compatible with the standard courses. Thus, our discussion of content has two parts: first, we will describe what was in the courses at Hampshire, Mount Holyoke and Smith, and second, we will discuss what was taught at the University. In both cases, it should be kept in mind that there will undoubtedly be substantial changes in the next several years as we revise the courses in light of classroom experience.

Hampshire, Mount Holyoke, and Smith

In 1989-90, the chapters of Calculus I as taught at the three colleges were:

    I. The Contexts of Calculus
   II. Functions
  III. Rates of Change
  IV. Differential Equations
   V. Integration
  VI. Differentiation
 VII. The Fundamental Theorem of Calculus

The chapters of Calculus II were:
VIII. Periodicity and the Fourier Transform

IX. Dynamical Systems
X. Series and Approximations

The chapters of Calculus III were:
XI. Newton's Gravitation Model
XII. State Spaces and Functions of Several Variables
XIII. Integration over Curves and Surfaces

Since Calculus I is the most fully developed of the three courses, we will direct most of our subsequent discussion towards it. There are a number of unusual features of the course which are apparent just in the way the topics are arranged. Most striking, perhaps, is the separation between the concept of derivative and the process of differentiation. This reflects the fact that real-world problems often come to us in the form "Can you find a function whose rate of change behaves in the following way?" rather than "Here's a function—what's its derivative?" The concept of rate of change is taken initially as being intuitively clear. In fact, the first two class meetings are spent developing a simple epidemiological model involving a population that consists of three groups: the Susceptibles, the Infecteds, and the Recovereds. A discussion of the way individuals move from one group to another leads to a model where the rates of change are described by the following equations:

$$S' = -aSI$$
$$I' = aSI - bI$$
$$R' = bI$$

Here a and b are parameters determined by the specific disease being modelled. We then look at some of the elementary implications of this model (e.g., What determines whether or not an epidemic takes hold? When does the infection hit its peak?) and have the students work out simple variations on the model. (Excerpts from this portion of our textbook are included in the sample materials at the end of this report.)

We return to this S-I-R model and its variations as the prototypical dynamical system at several points during the first semester. Dynamical systems are introduced in more detail in Chapter IV, and developed at length in Chapter IX in the second semester.

A second point is that we leap right into the subject in the first week. In addition to the S-I-R model mentioned above, Chapter I considers (and solves, using computers) the problem of finding the arc length of an ellipse, and introduces functions of several variables using topographic maps as the motivating example. There is no pre-calculus review as such. Review material, such as the equation of a line, the distance formula, or the quadratic formula, is covered as needed during the consideration of the problems.

A key theme of Calculus I is local linearity. Through the use of computer software with a "zoom" feature, students are led to the conjecture that all the standard functions "look like" straight lines when examined close up at almost all points. The derivative of the function at the point is then defined to be the slope of this local linear approximation. This leads naturally to the approach taken in Chapter IV, where systems of differential equations are solved numerically using piecewise linear approximations.

Throughout the course basic concepts are developed through numerical approximations before closed-form solutions are derived. This is because many important problems lack closed-form solutions. Nevertheless, students can attack such problems and gain considerable insight using numerical methods. For example, Chapter IV uses Euler's method to solve systems of differential equations before any analytic solutions are given; Chapter V uses numerical methods to compute integrals, with the Fundamental Theorem deferred until Chapter VII; and rates of change are computed numerically long before differentiation rules are developed in Chapter VI.

It needs to be emphasized, however, that this is not a numerical methods course in that conceptual simplicity is given precedence over power and efficiency. Thus, while we eventually discuss the virtue of using the midpoint rather than either endpoint in calculating Riemann sums, this is not dwelt upon at length. Simpson's rule is mentioned only briefly. Similarly, while some of the software written for exploring dynamical systems uses Runge-Kutta methods, the students are only expected to understand and use the more straightforward Euler's method for approximating solutions to differential equations.

A related feature of our course is that students tend to develop a strong feeling for the limit concept. There are several reasons for this. First, in many of the settings in which limits arise the value of the limit itself is never known. For instance, in the arc length problem, the students can say that the length of one quarter of the ellipse $x^2 + 4y^2 = 4$

is 2.42211..., and they are convinced that they could add more digits if they had to, but they never have "the answer". Second, students see the process of limit in many different settings—straight lines as the local limits of curves, curves as limits of broken line segments, solutions of differential equations as limits of approximations from Euler's method, areas as limits of sums of rectangles, numbers as limits of sums of other numbers.

The University

In 1989-90, the chapters of Calculus I as taught at the University were:

    I.   Discrete Dynamical Systems, Approximation, and Limits
    II. Differentiation
    III. Differential Equations and Transcendental Functions
    IV. Riemann Sums and the Integral

The chapters of Calculus II were:

    V.   Two-Dimensional Dynamical Systems
    VI. Techniques and Applications of Integration
    VII. Complex Numbers and Infinite Series

This course is compatible with standard calculus, so that students were able to transfer from the new course to the standard one at the end of either semester. We will discuss only Calculus I, and our comments will be most detailed for those portions of the course that differ most from standard calculus. Nonetheless, even when the course covers material straight out of a standard course in the same order as a standard course, it frequently does so with a different perspective and emphasis.

The first two chapters of the course cover basic topics like continuity and differentiation, but from a rather different point of view. Chapter I begins with discrete models of population change governed by an equation of the form $p_{n+1} = f(p_n)$. Students see the exponential and logistic models, and the homework problems include simulations done on calculators. Using the experimental results of this assignment, the ideas of limit and equilibrium point are introduced, first as descriptive concepts, then as careful definitions. An example involving water quality (When would a polluted lake be safe for swimming? When would its water be safe for drinking?) is used to introduce the epsilon definition of limit. The bisection method is used as a lead-in to the concept of continuity.

Attracting and repelling equilibrium points are studied for logistic models and exponential models with immigration. Graphical techniques give students a strong feeling for when an equilibrium point is attracting or repelling. In the linear case $p_{n+1} = mp_n + b$, the unique equilibrium point is shown to be attracting if $|m| < 1$ and repelling if $|m| > 1$. This result is applied to examine the stability of an economic model of price change with linear supply and demand functions. The homework asks practical questions like "If the manufacturing cost for a particular product rises, how much of the rise will be passed on to consumers?" or "How long will it take the market to adjust?". These questions recur later in the course when we discuss models in which manufacturers set prices to maximize profit.

The development of the derivative is motivated by a desire to extend the theory of attracting and repelling equilibrium points to the non-linear case. The derivative is discussed as both the tangent to a curve and as the rate of change. Examples include motion, waterflow, sensitivity of demand to prices, etc. This is followed by sections that cover the usual differentiation formulas, curve sketching, optimization and Newton's method. Transcendental functions are covered in a later chapter.

We discuss geometric ray-tracing using Fermat's principle at several points during the course. The homework includes problems asking students to find the apparent image of an object reflected in a mirror. These problems illustrate one of the underlying ideas of the course: students do longer problems in which the most recently-learned techniques are only part of the solution.

In the last section of Chapter II we answer the question that motivated our development of the derivative. We show that a nonlinear dynamical system $p_{n+1} = f(p_n)$ may have several equilibrium points and that an equilibrium point p_* is *locally* attracting if $|f'(p_*)| < 1$. Students have already seen examples (e.g., population models for a species that hunts in packs) that have local but not global attracting equilibria. This theorem illustrates the way that the derivative gives the same kind of *local* information that the slope does. Together with Newton's method we now have two nice examples illustrating the derivative as local linearization.

Finally, Chapters III and IV follow the material in a standard calculus class on integration and the transcendental functions quite closely. The main difference is the early introduction of differential equations. Another difference is that the exponential function is developed as the solution of the initial value problem $y' = y$, $y(0) = 1$.

Pedagogy

The pedagogy and the content of Calculus in Context are bound together by our general philosophy. Each course topic is both a specific set of useful definitions, techniques, and insights, and an example of how to think and work mathematically. Our curriculum is constructed as a coherent whole; it does not readily translate into modules that can be selectively introduced into the traditional course.

We try to get our students to view the activity of mathematics as a way of thinking about problems, of exploring possibilities, and not just as a collection of rules and techniques to be memorized. This is different from the way mathematics has often been presented to them in the past, and we feel it is important to address the difference right from the beginning. We want to enable our students to think constructively about real-world problems which are typically messy, unsolvable, and for which there is often no obvious technique at hand.

Problems are a much more important part of the student's work than is customary, and much class time is given to discussing them. We have tried to design classes and assignments so that general principles emerge from exploring problems. This reverses the traditional order in which the general principles are first laid out in class lectures, some examples are worked, and the students are then given some homework problems applying these principles. There is less time than usual spent in formal lectures.

Computers are used extensively as a basic tool for experimentation, for probing questions, and for arriving at general principles inductively. The course integrates computers into daily classroom activities as the need arises: to sketch a graph, to plot the contours of a function of several variables, to calculate an integral, or to plot the solutions of a system of differential equations. The students' work with computers is an essential component to the development of the concepts discussed in class.

Students naturally work in twos and threes at a computer; this is efficient and productive, as well as accommodating the need to share scarce resources. We encourage students to work in groups outside the class as well. This helps them to work harder and more effectively, and it lets us assign more complex problems than would be reasonable if they were working alone.

Fundamental concepts like rate of change and limit are used freely for quite a while without precise definitions. We believe that students must acquire a strong intuitive feel for such ideas before formal definitions will have much meaning. As concepts evolve over time, students become more sophisticated in using them. This enables us to introduce precision and rigor as tools for better understanding what the concepts mean.

Mathematics is presented as a subject grounded in current topics, not just as a collection of immutable laws handed down over the centuries. By the second semester students are given articles from journals like *Science* or *Nature* as a part of the development of topics like dynamical systems and the Fourier transform. This is done to make them aware of the extent to which mathematical ideas grow out of problems in other disciplines.

Technology

Calculus in Context uses, in an essential way, the computational and visual power of the current generation of microcomputers. The computers are IBM PS/2 Model 30 286 machines with math coprocessors and VGA color monitors. A typical computer classroom is equipped with a dozen machines that are arranged on tables around three sides of the room, all facing the center. At the University all calculus students use HP-32S programmable calculators, and next year they will use IBM PCs with math coprocessors.

We use computers in several distinct ways: to help students visualize concepts, to make accessible many important but difficult problems, and to promote the students' understanding of algorithms. Computational efficiency is *not* the goal; a method that is transparent and intuitive is always preferred. We want students to see they can tackle complex and difficult problems with tools they can understand.

Two sorts of computer programs are used in the course. First, we have a variety of graphics programs that graph functions of one variable, draw contours for functions of two variables, and plot solutions of differential equations, either as graphs over time or as trajectories in phase space. Several of the programs have a "zoom" feature that allows students to magnify a portion of the picture on the screen. These programs were written locally for the Calculus in Context project.

The second set of programs used consists of "template" programs written in Pascal or a version of Basic. These programs duplicate the computations students have already done with paper and pencil (such as calculating Riemann sums or stepping through Euler's method). While we do not teach programming, students learn to read code well enough to alter the programs to get more accurate answers and to solve other problems.

We plan to have machine-independent software that runs our programs; it remains to be decided whether we will try to adapt existing commercial packages to our needs as well.

Day by Day Mechanics

As already explained, the Calculus in Context material is the only calculus taught at Hampshire, Mount Holyoke, and Smith Colleges. In addition, parts of Calculus I are used in a year-long course at Amherst College that also covers pre-calculus material. A separate curriculum has been developed at the University of Massachusetts; this year it is offered in three sections of the regular calculus course for students in science and engineering. The day-to-day details of how things were done varied from one institution to the next.

Each course is offered for a 13-week semester. At Hampshire, there are three 80-minute meetings per week. At Mount Holyoke, there are four 50-minute meetings: two large lectures and two computer laboratories. At Smith, there are also four 50-minute meetings: three regular classes and one laboratory. By the fall of 1990, all classes at the three colleges will meet in computer classrooms. At the University and Amherst, classes meet for four hours per week.

Assignments and tests vary, but typically there are daily homework assignments, weekly lab reports, occasional quizzes, and two or three exams, plus a final. Exams may have both an in-class, closed-book component, and a take-home, open-book component. Students may use computers for homeworks and exams. Laboratory reports may be either simple worksheets or more demanding two- or three-page documents written by the student, and graded for clarity of presentation as well as mathematical accuracy. Overall, the work load for students is greater than in the traditional course.

Several methods are used to give students support in learning the new curriculum. Group work is encouraged, and a group may submit a single homework assignment (thereby reducing the grader's

work load). Several hour-long optional workshops are offered each week, conducted by students familiar with the new curriculum. They are held in the computer rooms. These rooms are also kept open during most of the day and staffed during the afternoons and evenings. This is essential, because much of the computing in our course requires high-resolution color graphics, not likely to be available on whatever computers individual students may have in their rooms.

Each section counts as one course in a faculty member's total annual load of four or five courses. During the development period, the actual time spent by faculty teaching the new courses has been two to four times the normal amount. The additional effort that has gone in to composing exercises and laboratories has been compounded by the need to create material that can be used by other instructors in other sections and at other institutions. Instructors at a particular institution have met weekly to plan and analyze their work; instructors from the different institutions have met a few times a semester to exchange ideas.

Appraisal, Comparisons, Conclusions

At the end of the fall semester of 1989, students in the ten sections of Calculus I at Hampshire, Mount Holyoke, and Smith were surveyed by questionnaires, and personal interviews were conducted by evaluators not involved in teaching the course. Students in the related courses at Amherst College and the University of Massachusetts also filled out the questionnaire. In addition, the evaluators interviewed some instructors, not all of whom had been involved in developing the materials. The general sense was that the course had accomplished much of what we hoped for, but, not too surprisingly, there are a number of areas we need to think about. Here are some of the key issues that came up.

The students appreciated the effort to develop the mathematical concepts within the context of specific problems. They felt that this gave them a sense of the connection of mathematics to broader kinds of inquiry which was missing in much of their previous mathematics work.

The material on differential equations seems to be working well, by and large. The students come away with a good feel for the nature of dynamical systems and the equations which describe them.

Overall, the way computers are used in the course seems to be effective in developing the kinds of insights we want our students to have. While a few students wondered whether this was a math course or a computer science course, most seemed to feel that the computers were integral to the mathematics. We clearly need to spend more time on writing documentation for the software and on writing handouts to lead the students more smoothly into elementary programming. We plan to continue the current balance between writing our own software and having students modify simple programs in Pascal or a version of Basic. Some commercially available software, such as Maple or Mathematica, may allow a similar balance. This is an area we will explore as time goes on.

Students found working in groups to be effective and generally quite satisfying. It is our perception, borne out by the responses of the students, that group work often helped them through difficult topics, and it certainly contributed to a general sense of collective engagement with the material.

Despite spending a lot of energy and time on functions of several variables, we have still not integrated the topic into the first-year curriculum. There is still too much of a sense that they are an optional add-on. At the elementary level, the role of functions of several variables seems to be quite different from functions of one variable: one can argue that, initially, the language of functions of several variables is more important than the analytical tools we apply to them.

One frequent criticism from both students and other mathematicians is "Where's the math?" The relative weight given to general principles, modelling and numerical methods at the expense of formulas and rules goes against the sense of many people's perception of the fundamental nature of mathematics. We need to be clearer in explaining to our students why we approach things the way we do. It may well be, too, that a slightly greater emphasis on formulas would be appropriate at several points.

A related concern is that our focus on the cultivation of general skills and attitudes rather than on specific (and named) techniques leaves students unsure about what they know at the end of the course. It is also more difficult for us to assess the success of this program, since in many ways the real tests will only come over the next two or three years as our graduates encounter problems needing the kinds of general tools we have tried to develop.

Many students were uneasy with the open-ended nature of the problems they were assigned. They were accustomed to homework that consisted of minor variations of examples solved in class or in the textbook. Since there is relatively little of this kind of work in our course, we must help students reappraise the way they learn mathematics. Needless to say, many of them were frustrated, including some who were particularly proficient at the previous system. On the other hand, most of the students seemed to adapt to this mode by the end of the first semester, and a number came to find it exhilarating.

Many of our students arrive having already had a year of calculus in high school. These students find our calculus sequence different enough from what they have seen that they can profitably start with Calculus I. While some students can move directly into Calculus II or Calculus III, there will be a number of things they don't know, and some fast catching up will be needed at the beginning. We have begun to consider a concentrated course for these students which covers the material of Calculus I and Calculus II in a single semester.

There are clearly major questions of portability of this curriculum which we are only starting to address. Over the next year we hope to be talking with mathematicians at a wide variety of institutions to get a clearer sense of the effect of constraints like large class size, courses taught primarily by graduate students, limited computer facilities, a large engineering clientele, and the need of students to be able to transfer to and from standard calculus sequences.

Future Plans

In the next two years, we will revise Calculus I and complete first drafts of Calculus II and III. Next spring we will start developing an accelerated version of Calculus I and II aimed at students who have had a year of calculus in high school. We will also refine the software and start working on versions that run on different computers. The evaluation process will continue, and we will keep track of what Calculus in Context students do in subsequent mathematics and science courses.

In the summers of 1991, 1992 and 1993, we plan to host a number of one-week workshops for interested faculty members outside the Five Colleges. We expect that these workshops will result in additional feedback and will provide a forum for discussing problems that may arise in adapting our materials for use at different institutions.

≡ Sample Materials ≡

In what follows, Calculus in Context material will be in normal type, with our comments in italics. We have included seven excerpts from the version used at Hampshire, Mount Holyoke and Smith, and two excerpts from the version used at the University of Massachusetts.

Hampshire, Mount Holyoke, Smith

1. Table of Contents

We begin our selection of course material with the table of contents of Calculus I. This will give a better idea of what we cover in the first semester and will help the reader to see how subsequent selections fit into the course.

2. The Spread of Disease

This selection from §2 of Chapter I shows what is covered during the first two class meetings.

Many human diseases are contagious: you "catch" them from someone who is already infected. But contagious diseases are of many kinds. Smallpox, polio, and plague are severe and even fatal, while the common cold and the childhood illnesses of measles, mumps, and rubella are usually relatively mild. Moreover, you can catch a cold over and over again, but you can have measles only once.

. . .

At first glance, it seems hopeless to try to understand something so complex as the behavior of a disease as it spreads through a population. ... But the dangers posed by contagion—and especially by the appearance of new and uncontrollable diseases—compel us to learn as much as we can about the nature of epidemics.

. . .

Let's suppose the disease we want to model is like measles; that is

• it is mild, so anyone who falls ill eventually recovers;
• it confers permanent immunity on every recovered victim.

In addition, we will assume that the affected population is large but fixed in size and confined to a geographically well-defined region. ... At any time, that population can be divided into three distinct classes:

Susceptible: those who have never had the illness (but can catch it);

Infected: those who currently have the illness and are contagious;

Recovered: those who have already had the illness and are now immune.

Let S, I, and R denote the number of individuals in each of these three classes, respectively. ... If the disease is active and spreading, then the quantities S, I, and R are not static; they change over time. So if we are going to study an active disease, we need a language to talk about change. We'll use S', I', and R' to denote the rates at which S, I, and R are changing, measured in units of persons per day. For example, if the school health service reports that 20 children have recovered on a given day, we would write $R' = 20$ persons per day.

Using this intuitive notion of rate of change, we derive the equations $S' = -aSI$ and $R' = bI$, where a and b are parameters determined by the disease and population being modelled. To help the students visualize the situation, we use a compartment diagram as part of our discussion:

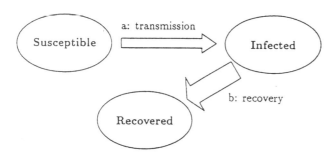

Our compartment diagram is now complete; there is an arrow from S to I of size a, and another arrow from I to R of size b. The final rate equation we need—the one for I'—just reflects what is already clear from the diagram: every loss in I is due to a gain in R while every gain in I is due to a loss in S.

$$S' = -aSI$$
$$I' = aSI - bI$$
$$R' = bI$$

The sum of these three quantities is the overall rate of change of the whole population. It is zero. Do you see why?

The three rate equations we have derived (along with the accompanying flow diagram) make up the basic S-I-R model of epidemiology. The fundamental question we ask of the model is this: if the transmission and recovery coefficients are known, can S, I and R be determined over time? That is, can we actually use the model to follow the course of the epidemic?

Let us look at an example. Suppose that

$$a = .00001 \text{ (person-days)}^{-1}$$
$$b = .08 \text{ day}^{-1}$$

and, furthermore, that the current sizes of the three populations are

$$S = 35,400 \text{ persons}$$
$$I = 13,500 \text{ persons}$$
$$R = 22,100 \text{ persons.}$$

Then the three rates have current values

$$S' = -.00001 \times 35400 \times 13500$$
$$= -4779 \text{ persons/day}$$
$$R' = .08 \times 13500$$
$$= 1080 \text{ persons/day}$$
$$I' = -S' - R'$$
$$= 3699 \text{ persons/day}$$

What will happen between tomorrow and the next day? In particular, will there be an identical increase in the value of S? If there were, then it would be a simple matter to follow the values of S, I and R into the future. Alas, matters are not simple. The reason is that tomorrow's values of S' and I' have also been altered, because they depend on tomorrow's values of S and I. In fact, in the exercises you will compute that the new values of S' and I' are -5266.5 and $+1375.9$ persons per day, respectively. (The appearance of "fractional" people here reminds us that we are dealing with a mathematical model! To be useful, the model's numbers need only *approximate* the true numbers, so we can tolerate them being fractions.) The model tells us to expect about 5266 people to fall ill between tomorrow and the next day, but only 4779 between today and tomorrow.

$\cdots$

Here are some specific questions we can ask about our model. Notice that some of the questions are rather broad and imprecise. This is often the case when we first looks at a problem; one of the purposes of a model is to help us sharpen questions and make them more precise.

1. Is S eventually 0; that is, does everybody get sick?

2. If not, how many people avoid the illness? Does that number depend on the transmission coefficient a, and, if so, how?

3. Sometimes an *endemic* disease (i.e., one always present but at low levels) becomes an *epidemic*

(i.e., affecting a significant fraction of the population). Can the model show how or why such a change might happen?

4. It seems reasonable to say the disease is spreading so long as the number of infected is increasing. When does it stop spreading and begin to decline? Does the infection peak early, when most people are still in the susceptible category, or later?

5. Can the model be modified or expanded to describe a disease that does not confer lifelong immunity? How might the modifications be made?

The exercises for this section range in difficulty from straightforward (Exercise 1) to challenging (Exercise 13). The first 8 exercises use the S-I-R model

$$S' = -aSI$$
$$I' = aSI - bI$$
$$R' = bI$$

with transmission coefficient $a = .00001$ (person-days)$^{-1}$ and recovery coefficient $b = .08$ day^{-1}. As in the text, we assume that

$$S = 35,400 \text{ persons}$$
$$I = 13,500 \text{ persons}$$
$$R = 22,100 \text{ persons.}$$

on "day 0". Exercises 9 through 13 then ask the students to change the model in various ways.

1. Compute the values of S, I, and R on day 1, 2, and 3, by recalculating the rates S', I', and R' daily. (The text has already done part of this.)

2. Recompute the values of S, I, and R on day 1 by recalculating the rates S', I', and R' every 8 hours. Compare the day 1 values of S, I, and R calculated in questions 1 and 2.

3. We say that a disease is spreading as long as the number of infected increases. Explain why there are 8000 susceptibles left when the the disease stops spreading and begins to decline.

The answer to question 3 depended on the fact that the transmission and recovery coefficients had definite values, namely $a = .00001$ (person-days)$^{-1}$ and $b = .08$ day^{-1}. For the remaining questions suppose that a and b have arbitrary values.

4. Determine how many susceptibles are left (in terms of a and b) when the disease stops spreading and begins to decline. (Does your answer have the right units?)

5. Under what conditions is the infected population increasing? Decreasing? What about the susceptible and recovered populations? (To see what sort of conditions are involved, you might first answer these questions when $a = .00001$ (person-days)$^{-1}$ and $b = .08$ day^{-1}?)

6. Under what conditions is the infected population increasing at exactly the same rate as the recovered population?

7. Under what conditions is the infected population increasing more rapidly than the recovered population? Less rapidly?

8. Suppose that, in the middle of an epidemic, the sizes of the three populations on a certain day were

$$S = 475,000 \text{ persons}$$
$$I = 170,500 \text{ persons}$$
$$R = 210,000 \text{ persons.}$$

Estimate the sizes of these populations two days earlier. Explain how you made your estimate. What could you do to make a more accurate estimate? (Don't do it; just say what you would do.)

9. Create an *S-I-R* model to describe the course of a disease which lasts an average of 5 days. (Assume the transmission coefficient a is still .00001.) Draw a compartment diagram, with suitable "compartments" and "arrows," and then write rate equations that describe your diagram.

10. Modify the *S-I-R* model to describe the course of a disease (like the common cold) where immunity is not permanent. Specifically, assume a recovered person loses immunity after N days. How does this change affect the compartment diagram on page 7 in the text? Draw a new compartment diagram, and then write rate equations that describe your diagram.

11. In question 10 take $N = 20$ days, and assume the transmission and recovery coefficients have the definite values $a = .00001$ (person-days)$^{-1}$ and $b = .08$ day^{-1} we have used earlier. Can you determine how often each person might expect to catch a cold? Does your answer depend on the size of the the whole community or the individual populations?

12. Modify the *S-I-R* model a different way, to describe the course of a *non*-infectious disease like glaucoma. Here the rate at which people catch the disease depends not on the number of contacts between infected and susceptibles, but only

on the number of susceptibles. Draw the compartment diagram and find the rate equations.

13. Modify the *S-I-R* model to describe a disease like the one in the text, but which kills half the people who get it (i.e., a disease closer to smallpox than to measles). For simplicity, assume that the infected people who die expire on the last day of the disease. Include both a diagram and the corresponding rate equations.

3. Constructing Linear Models

An important topic of Chapter II is the way functions are used to construct models. Instead of the standard review of straight lines, we emphasize linear functions and linear models. The selection below is taken from the exercises to §5 of Chapter II.

The following excerpt is from Mark Twain's *Life on the Mississippi* (1884). It comes from a chapter called "Cut-offs and Stephen." The Lower Mississippi River meanders over its flat valley, forming broad loops called ox-bows. In a flood, the river can jump its banks and cut off one of these loops, getting shorter in the process. The Connecticut River just south of Northampton has an ox-bow that has been cut off in exactly this way. Here is a bit of what Twain has to say about ox-bows and cut-offs:

> In the space of one hundred and seventy six years the Lower Mississippi has shortened itself two hundred and forty-two miles. That is an average of a trifle over a mile and a third per year. Therefore, any calm person, who is not blind or idiotic, can see that in the Old Oölitic Silurian Period, just a million years ago next November, the Lower Mississippi was upwards of one million three hundred thousand miles long, and stuck out over the Gulf of Mexico like a fishing-pole. And by the same token any person can see that seven hundred and forty-two years from now the Lower Mississippi will be only a mile and three-quarters long, and Cairo [Illinois] and New Orleans will have joined their streets together and be plodding comfortably along under a single mayor and a mutual board of aldermen. There is something fascinating about science. One gets such wholesome returns of conjecture out of such a trifling investment of fact.

Suppose we let t be the time, in years, from when Twain wrote the book, and $L(t)$ the length of the river at time t.

1. What is the rate at which L is changing, in miles per year? In what form does Twain give us this information? Is the rate of change constant?

2. According to Twain, the river will be only $1\frac{3}{4}$ miles long in 742 years from when the book was

written. How long must the river have been when he wrote the book?

3. Can you give a formula for $L(t)$? What kind of function is $L(t)$?

4. According to your formula for L, was the river "upwards of 1,300,000 miles long" a million years ago? In other words, does your formula confirm Twain's assertion?

5. Was the river ever 1,300,000 miles long; will it ever be $1\frac{3}{4}$ miles long? What, if anything, is wrong with the "trifling investment of fact" which led to such "wholesome returns of conjecture" that Twain has given us?

4. Local Linearity

Although the concept of rate of change is used freely in Chapters I and II, only in Chapter III does it take center stage. The derivative is treated numerically and geometrically, but instead of using the tangent line, we emphasize the notion of local linearity. The following selection is taken from §3 of Chapter III.

In the previous section, we found the slope of a curved graph by zooming in on the graph. This worked because zooming gave us a virtually straight line—whose slope could be found by standard methods. Zooming produced a picture of the immediate neighborhood of a point on the graph; *in each neighborhood the graph appeared linear.* This property of the function is called **local linearity**.

How common is this property: are there other locally linear functions? Fortunately, there are many; even very "wiggly" graphs can look straight if we zoom in on them enough.

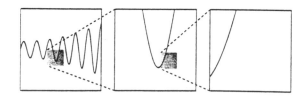

We hasten to add that not all functions are locally linear at all points. Look what happens when we zoom in on the graph of $y = x^{2/3}$ at the origin.

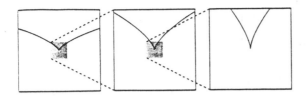

However, even this function turns out to be locally linear at all other points. You can see on the top of the next figure what happens if we zoom at $x = 1$, for instance.

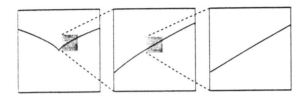

In spite of all these examples, it is important to realize that local linearity is a very special property. ... **Calculus is the study of functions that are locally linear almost everywhere; these are the functions that have derivatives.**

...

So let us return to locally linear functions. Recall from Chapter II §4 that a *linear* function defines a particularly simple relation between changes in input and changes in output:

$$\text{if} \quad y = mx + b \quad \text{then} \quad \Delta y = m \cdot \Delta x.$$

Thus, a change in output is just a multiple of the change in input that produced it, and the multiplier m is the slope of the graph of the linear function. As the following graph shows, local linearity insures that a version of this result carries over to a *non-linear* function $y = f(x)$. The non-linear version says

$$\text{near} \; x = a \quad \Delta y \approx f'(a) \cdot \Delta x.$$

It differs from the linear version in being approximate rather than exact, and in being local rather than global. That is, the relation holds approximately inside a small window, the degree of accuracy improving as the size of the window decreases. The multiplier is the value of the derivative $f'(a)$ at the center of the window; if the window moves, the multiplier changes. In particular, the effect a small increase in x has on y depends on the sign and magnitude of $f'(a)$:

- $f'(a)$ is large and positive $\Rightarrow$ large increase in y
- $f'(a)$ is small and positive $\Rightarrow$ very small increase in y
- $f'(a)$ is large and negative $\Rightarrow$ large decrease in y
- $f'(a)$ is small and negative $\Rightarrow$ very small decrease in y

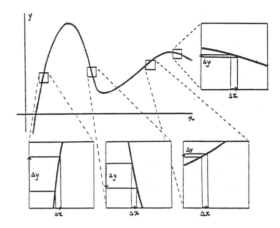

5. Euler's Method

Numerical solutions of differential equations may sound like a sophisticated topic for students in their first semester, but many of the more common methods use little more than the basic idea of derivative. Thus Chapter III gives our students exactly what's needed in order to start solving differential equations. The following excerpt is from §3 of Chapter IV.

In Chapter III we saw that graphs of functions are locally linear. This means that in a small neighborhood of any point on the graph the graph looks very much like a straight line. Moreover, the slope of this line is simply the derivative of the function at that point. But if we know a point on a line and the slope of a line, we can find the coordinates of other points on the line. These simple observations were developed by the Swiss mathematician Leonhard Euler (1707-1783) into a powerful technique for approximating solutions to differential equations. While in actual practice various refinements are added to make the technique more efficient, the underlying concept remains the same. Let's see how it works.

The text then discusses a cooling cup of coffee. The motivating question is "How long will we have to wait until the temperature of the coffee is 30° C?". We use Newton's law of cooling to set up an initial value problem, which is solved using Euler's method with step-sizes of 1 minute and .1 minute. The computations are presented in both tables and graphs, so that students can see how better and better approximations to the solution are being generated. This example concludes as follows:

Let's summarize the essence of Euler's method. Suppose we want to evaluate a function f. Suppose further that we somehow know f' everywhere (typically f' will be given in terms of t and/or the values of $f(t)$), and that we know $f(t_0)$ for some point t_0. We proceed as follows:

1. If Δt is a "small" change in t, since we know $f'(t_0)$ we can approximate the change this would produce in the function by $\Delta f = f'(t_0)\Delta t$. Thus the point $(t_0 + \Delta t, f(t_0) + \Delta f)$ will will lie near the graph of f.
2. Using this approximate new point, calculate f' there. Use these values to estimate the value of of f another "short" distance further along.

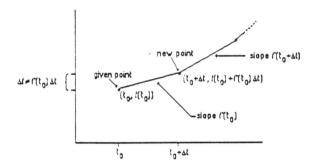

3. Repeat step 2 until you get to where you want to go.
4. By redefining the meaning of "short" to be smaller and smaller, the approximations for the function values obtained via the above method can be made arbitrarily close to the true values of the function.

Here are three of the exercises that accompany the section on Euler's method.

2. Find approximate solutions $y = f(x)$ to the differential equation

$$\frac{dy}{dx} = y - x$$

on the interval $[0,1]$ using Euler's method. Consider two different boundary conditions

(a) $f(0) = 1$

(b) $f(0) = 0$

In particular, find the value $f(1)$ in each case. Since we want to have some idea of how accurate the solutions are, use two different schemes for chopping up the interval $[0,1]$ in each of (a) and (b), and comment on how well the different schemes agree with each other in giving $f(1)$. Present your solutions in the form of tables with headings x, y, dy/dx, and compare your results with the text description of this equation starting on page 111.

3. For each of the following differential equations

i) $\dfrac{dy}{dx} = y^2 - x$;

ii) $\dfrac{dy}{dx} = \sin y$;

a) Use a computer program to help you sketch typical solutions. (Some suggested starting points are $(0, 0)$, $(0, 0.5)$, $(0, -0.5)$, $(0, 1)$. From each starting point be sure to "step left" as well as "step right"—that is, find y for values of x *less than* the starting value of x.)

b) Point out where your solutions have maxima or minima. Compare your answers here with what you found in exercise 4 of §1.

4. With a suitable computer program, use Euler's method to estimate what will happen to a population R of 2000 rabbits in a meadow after 6 months, when the growth rate is given by the differential equation

$$R' = 0.1\,R \quad \text{rabbits per month.}$$

The computation is made at intervals of 1 month. Modify the program so that the computations are made at intervals of 3 days (that is, 1/10-th of a month) instead of 1 month. Does this change the calculated value? If so, how? Recompute the population several more times, using step sizes of 1/100-th, 1/1000-th, and 1/10000-th of a month. (Each time assume the initial population is $R = 2000$.) What is the point of using smaller step sizes? What do you see happening in this sequence of estimates? What would you expect the calculated value to be, in whole numbers, if you used a step size of 1/100000-th of a month? Were you right? Finally, can you use the model $R' = 0.1\,R$ to predict the population after 6 months, *independently of the step size used in Euler's method?*

6. Riemann Sums and the Integral

The chapter on integration begins with the problems of estimating energy consumption and distance travelled, and Riemann sums are introduced as a language to unify the estimates obtained. The definition of Riemann sum is given in full generality. After some examples, the text of §2 of Chapter V continues as follows:

In practice, Riemann sums are rarely this arbitrary. For example, when dealing with functions given by formulas, such as the velocity function $v(t) = 60t^{1.1}(2-t)^{1.1}$, it pays to be systematic: one usually works with subintervals of equal size, and picks the "same" point from each subinterval (i.e., always pick the midpoint or always pick the left endpoint). This is especially useful when using a computer to evaluate Riemann sums. As an example, let's apply this strategy to the above function $\sqrt{1+x^3}$ on $[1,3]$. We'll use subintervals of equal length $\Delta x = 2/n$, and x_i will be the midpoint of the ith subinterval. Here is the picture of the n subintervals:

Note that the midpoints are Δx apart. Then the following algorithm computes the Riemann sum for n subintervals, evaluating the function at the midpoint of each subinterval:

```
input:  n {the number of subintervals}
output: S {the Riemann sum}

delta_x := 2/n
x := 1 + delta_x/2
S := 0
For i := 1 to n do
   S := S + sqrt(1+x^3)*delta_x
   x :  = x + delta_x
Print S
```

If this algorithm were programmed to run on a computer, the input $n = 100$ would give the output $S = 6.2299345$, and other Riemann sums could be obtained just as easily. Thus 6.2299345 is a Riemann sum for $\sqrt{1+x^3}$ on the interval $[1,3]$.

This algorithm also makes it easy to change the point we use in each subinterval. For example, if we always use the left endpoint, then we get the picture:

The left endpoints are Δx apart and the first one is $x_1 = 1$. Thus, if we replace the above line '$x = 1 + \Delta x/2$' by '$x = 1$', then we get an algorithm that computes Riemann sums using the left endpoint of each subinterval. It is similarly easy to make a version of the algorithm that uses right endpoints.

The above algorithms can used for any function given by a formula. However, there are other ways to describe functions. For example, suppose that $f(x)$ is given as the solution of a differential equation with certain initial conditions. Then the methods for approximating $f(x)$ given in Chapter IV can be easily adapted so that one simultaneously generates the function and the Riemann sum (the stepsize used in Euler's method is used as the width Δx in the Riemann sum).

When a function is given by data, the situation is a bit different. For example, the power graph in §1 represents data collected over a 24 hour period. Unlike the velocity case, we can't pick t_i and get $p_i = p(t_i)$ by "plugging in." Rather, our Riemann sum was computed using values $p_1, \ldots, p_n$ taken from the power data. More precisely, the data was broken up into subsets corresponding to different time intervals, and then representative values were chosen from each subset of the data. This is related to the idea of *sampling* a data set.

In the case of the power graph, we were working with a potentially infinite amount of data (one can imagine a plotter that continuously records the amount of power used). Another situation that often arises is when there is only a finite amount of data to work with. In this case it makes sense to use *all* of the data to make a Riemann sum. Here, the Δx_i's would be the distances between successive data points. Furthermore, when the data is available in a computer file, it is not difficult to write a computer program to calculate the Riemann sum.

This section defines both Riemann sums and the integral, and there are a lot of problems that the students can do. Here are 6 exercises, which range from the computational (Exercise 16) to the conceptual (Exercises 17 and 21).

16. Consider the integral $\int_2^4 1/u\,du$.

a) Make a table of the approximations of this interval, and for each approximation, indicate the number of subintervals and whether you used right or left endpoints or midpoints.

b) Which of the approximations from part a) do you think is the best. How accurate is it (i.e., to how many decimal places is it accurate)?

17. In this problem, we will study waste production in a yeast colony. Let $P(t)$ be the mass, in grams, of living yeast in a vat of fermenting grape juice. We will measure time t in hours. We will assume that the waste (alcohol and carbon dioxide) is proportional to both the yeast population $P(t)$ and the amount of time Δt that has passed. As we learned at the very end of Chapter II (see pages 56-58), this means that if W is the amount of waste, then

$$W = kP(t)\Delta t .$$

However, this formula only holds if the population $P(t)$ is constant over the time interval Δt. When the population is not constant, use the ideas of this section to express the total waste produced for the time interval $t_0 \le t \le t_1$ as an integral. Be sure to fully explain your reasoning.

18. Assume that the population of the yeast colony is growing exponentially according to the formula

$$P(t) = 300e^{.2t} .$$

We will also assume that the constant k in problem 8 is .1 gram waste per hour per gram yeast (do you see why these are the correct units for k?).

a) Use the integral formula from the previous problem to get an integral that expresses the waste generated in the first four hours (which means $0 \le t \le 4$).

b) Approximate the integral of part a).

19. Here are some more problems concerning waste production.

a) If we double the yeast population in problem 8, what effect does this have on the waste produced for $0 \le t \le 4$? Of the properties of the integral mentioned on pages 158-160, which one is being used here?

b) If we instead double the growth rate (which means $P(t) = 300e^{.4t}$ rather than $300e^{.2t}$), then compute the waste produced for $0 \le t \le 4$.

c) Which has a greater effect on the waste production: doubling the population or doubling the growth rate?

20. In this problem, we will again study waste production, but this time, suppose that the population $P(t)$ satisfies the differential equation

$$P' = .2P(1 - .005P) .$$

As before, assume that the initial population is $P(0) = 300$. As in the previous problems, we want to compute the waste production for the four hour period $0 \le t < 4$. If the constant of proportionality from problem 8 is still $k = .1$, then as usual the waste production is given by the integral $\int_0^4 .1P(t)\,dt$. The problem is that we don't have a formula to use when computing Riemann sums— here, we are dealing with a function defined by a differential equation.

The key observation is that if we compute $P(t)$ numerically using Euler's method, we can easily compute a Riemann sum for $\int_0^4 .1P(t)\,dt$. For concreteness, suppose we the differential equation using Euler's method for $0 \le t \le 4$ with step size $\Delta t = .1$. This gives us approximations to $P(0)$, $P(.1)$, $P(.2)$, ..., $P(3.9)$, $P(4)$, and then we can compute a Riemann sum with 40 equal subintervals of length $\Delta t = .1$ using either left or right endpoints.

a) Write a computer program that simultaneously does Euler's method and computes the Riemann sum using right endpoints. Here are some hints. Begin with a program that implements Euler's method, and suppose that P is the variable that stands for the value of the function (thus P has 300 as its initial value). Now introduce an new variable RS (for Riemann sum), which has initial value 0. Then the program should have a line something like

```
P = P + PPRIME*DELTA
```

that computes the next value of P from the previous one. This line should be immediately followed by the line

```
RS = RS + .1*P*DELTA .
```

Be sure you understand why this generates a Riemann sum using right endpoints.

b) Modify your program so that it generates a Riemann sum that uses left endpoints. Hint: Interchange the above two line in the program. Do you see why this has the desired effect?

21. Suppose that the amount of food required by a certain population is proportional to both the population and the length of time interval considered. Let the population at time t, measured in years, be $P(t)$, and assume that one person requires 3 lbs of food per day. Express the pounds of food required over a one year period as an integral. Explain your reasoning. Hint: This problem is similar to problem 17.

7. Periodicity and the Fourier Transform

Our final selection comes from Chapter VIII, which is part of Calculus II. This chapter is entitled "Periodicity and the Fourier Transform", and its purpose is to give a significant application of integration that highlights the scientific context. Here is the introductory section of the chapter:

One of the tantalizing problems in many branches of science is determining whether or not a certain phenomenon is *periodic*—that is, whether it repeats itself over some fixed period of time. Here are some situations where the question of periodicity arises naturally:

Example 1. In Chapter IV we studied the Lotka-Volterra model for the interaction between predator and prey species. One of the predictions of this model is that the numbers of predator and prey species should be periodic. How can we tell if this is happening? Ecologists have examined data for a number of species. Some of the best documented cases come from the records of Hudson's Bay Company, which trapped fur bearing animals in Canada for almost 200 years. The graph below gives the data for the numbers of lynx pelts harvested in the Mackenzie River region of Canada during the years 1821 to 1934. (This data is from "The ten-year cycle of numbers of lynx in Canada" by C. Elton and M. Nicholson, *Journal of Animal Ecology 11* (1942), 215-229. Used by permission of the publisher, Blackwell Scientific Publications, Ltd.) It is obvious that the values go up and down, but are they periodic? And if they are periodic, how do we determine the exact period?

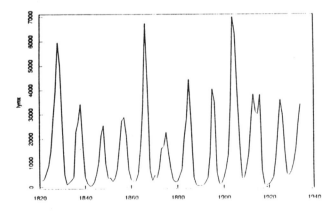

Example 2. Different features of the earth's orbit about the sun exhibit periodicity. For instance, the ellipse traced out by the earth's orbit rotates slowly

in space (this movement is called **precession**) with a period of roughly 23,000 years. Other changes in the orbit have periods of 41,000 years (obliquity cycle) and 95,000, 123,000, and 413,000 years (eccentricity cycles). In 1941 the Serbian geophysicist Milutin Milankovitch hypothesized that these periodicities in the earth's orbit affected global weather patterns and should therefore cause periodicities in a number of quantities measureable in the geological record. Scientists have looked at a number of phenomena—thickness of the annual layers in shale beds, the thickness of annual ice layers in the Antarctic ice cap, the fluctuations of CO_2 concentrations in the ice caps, changes in the O^{18}/O^{16} ratio in deep-sea sediments and ice caps, etc.—to check this hypothesis. At the end of this chapter we will look at the results of one such study.

Example 3. There have been several theories about behavior of the stock market which claim that it is cyclic, both over short periods and over periods many years long. How can such claims be tested?

Example 4. Over the years a number of claims have been made that sun spots affect a variety of phenomena on earth, particularly weather patterns like rainfall and temperature, and CO_2 concentrations in the atmosphere. The basis for such claims often lies in noticing that the phenomena in question have an 11 year or a 22 year cycle, which correlates with the length of the sun spot cycle.

What makes many of these problems difficult is that there are typically many other forces at work as well, so that the underlying periodic behavior may be so masked by the effects of the other forces that it isn't immediately obvious. This leads to the related problem of distinguishing periodic behavior from random fluctuations.

The above introduction is followed by a short section that reviews the basic concepts of period and frequency. Students also learn how to create trigonometric functions with preassigned frequencies. Then we discuss the Fourier transform:

To introduce the basic idea of the Fourier transform, we will study an example of a signal embedded in some noise. ... Here are the graphs of the three functions $\sin(5x)$, $g(x)$ and $h(x) = \sin(5x) + g(x)$. Notice that we have made the noise $g(x)$ three times as loud as the signal $\sin(5x)$:

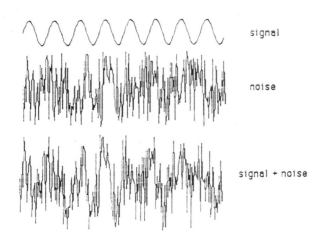

At first glance, the last two pictures look pretty much the same; there is nothing about the third picture which leaps to the eye to suggest that it contains a strong periodic signal.

Key insight: The function $\sin(5x)$ is fluctuating up and down between positive and negative values. If we multiply $\sin(5x)$ by an unrelated function, we are just as likely to get positive as negative values, so if we integrate this product the positive and negative contributions will roughly cancel each other, and the total integral will be small. The noise function $g(x)$ is an example of such an unrelated function, and we have by numerical integration that

$$\int_0^{10} g(x)\sin(5x)\,dx = .77 \,.$$

On the other hand, if we multiply by a function which is strongly correlated with $\sin(5x)$, such as $h(x)$ or $\sin(5x)$ itself, and integrate, then we should get a much larger integral. For instance

$$\int_0^{10} \sin(5x)\sin(5x)\,dx = 5.03$$

and

$$\int_0^{10} h(x)\sin(5x)\,dx = 5.80 \,.$$

(Note that the last integral is simply the sum of the first two, as we would expect since $h(x) = \sin(5x) + g(x)$.)

If we had multiplied $h(x)$ by other sine or cosine functions, we would not have gotten such a large value. For instance, one can compute that

$$\int_0^{10} h(x)\sin(x)\,dx = .744$$

$$\int_0^{10} h(x)\sin(9x)\,dx = -.391$$

$$\int_0^{10} h(x)\sin(13x)\,dx = -.105$$

$$\int_0^{10} h(x)\sin(17x)\,dx = -.858$$

$$\int_0^{10} h(x)\cos(2x)\,dx = -.624$$

$$\int_0^{10} h(x)\cos(5x)\,dx = -.274$$

$$\int_0^{10} h(x)\cos(8x)\,dx = -1.144$$

$$\int_0^{10} h(x)\cos(11x)\,dx = -1.119$$

In other words, it appears that the $\sin(5x)$ signal hidden in the function $h(x) = \sin(5x) + g(x)$ can be detected by the fact that the integral of its product with $h(x)$ is relatively large.

Based on this discussion, the text defines the sine, cosine and power transforms to be

$$F_s(\omega) = \int_a^b f(x)\sin(2\pi\omega x)\,dx$$

$$F_c(\omega) = \int_a^b f(x)\cos(2\pi\omega x)\,dx$$

$$P(\omega) = F_s(\omega)^2 + F_c(\omega)^2$$

Here $f(x)$ is a function defined on $[a,b]$. To illustrate these definitions, the text returns to the above example.

For example, the function $h(x)$ we have been considering has the following power spectrum:

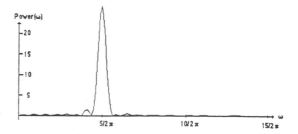

Note the very conspicuous spike in the power spectrum centered at the frequency $\omega = 5/2\pi$ corresponding to the frequency of the signal $\sin(5x)$ embedded in the white noise.

The text then considers some of the classic integrals (like $\int \sin(\alpha x)\sin(\beta x)\,dx = (\alpha\cos(\alpha x) \sin(\beta x) - \beta\sin(\alpha x)\cos(\beta x))/(\beta^2 - \alpha^2)$ when $\beta \neq \alpha$) and explains their relevance to the Fourier transform. The section concludes by applying this theory to some real scientific data.

Let's apply the insights we've gained here to examine some of the examples that led off this chapter. For instance, if we take the Mackenzie River Lynx data, the corresponding power spectrum looks like

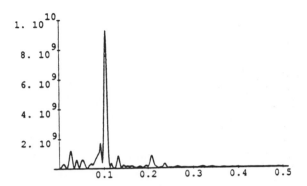

Note the very strong spike centered just beyond a frequency of .1. This corresponds to a period of slightly less than $1/.1 \approx 10$ years, suggesting that the lynx population is strongly periodic, with a period of 9.6 years.

The text also considers the power spectra of other species, and then discusses data on the thickness of sediment layers. In this example, the spikes in the power spectrum correspond to the eccentricities of the earth's orbit.

Computing the Fourier transform of a trigonometric function or a linear function leads to nice exercises involving integration by parts. Other exercises come from questions concerning the way phase shift affects the Fourier transform. In doing the exercises, students used locally written programs that graphically represent the spectrum of a data function or a function given by a formula.

When using a computer to find the spectrum of a function given by a formula, one must be careful, because numerical integration can lead to incorrect conclusions about the spectrum. Here are some exercises we wrote to deal with this issue:

The purpose of the Fourier transform of a function is to make visible the periodic patterns contained with a given function. However, the *method of computing* the transform used by the program

Spectrum can introduce spurious information, too. It can tell us there are periods that are not really present in the function. So we must take the Fourier transform with a grain of salt. The purpose of these exercises is to point out the spurious information, show why it arises, and how we can get rid of it.

1. Using the program Spectrum, analyze the function $\sin 2\pi x$ on the interval $0 \leq x \leq 3$, using an ω range of $0 \leq \omega \leq 3$. (Note: the program recognizes the typed expression pi as the value of π, so you can enter the function as sin(2*pi*x). Describe the graph of the power spectrum and the Fourier sine and cosine transforms. Make a sketch of the power spectrum.

[*Answer*: The power spectrum has a single peak at $\omega \approx 1$; at that point the sine transform is large and positive, while the cosine transform is zero.]

2. Now change the x-interval to $0 \leq x \leq 30$, but leave the ω-interval as it was in question 1. How does the power spectrum change? Sketch the new power spectrum on the same coordinate plane you used in question 1.

[*Answer*: The peak at $\omega \approx 1$ becomes much "sharper", and other peaks appear at $\omega \approx 2/3$, $7/3$, and $8/3$. The one at $\omega \approx 7/3$ is almost as tall as the one at $\omega \approx 1$.]

These new peaks convey spurious information: the function $\sin 2\pi x$ has no components whose frequencies are $2/3$, $7/3$, or $8/3$. We now consider why this should happen.

3. To get a close look at the transforms near $\omega \approx 7/3$, change the ω-interval to $2.25 \leq \omega \leq 2.4$. What is the value of the power spectrum at the point $\omega \approx 7/3$ where it has its peak? Describe the sine and cosine transforms there.

[*Answer*: The cosine transform is zero, and the sine transform has a large negative value. The value of the power spectrum is ≈ 225, so the value of the sine transform is ≈ -15.]

4. According to exercise 3, the peak in the power spectrum at $\omega \approx 7/3$ is due to the integral

$$\int_0^{30} \sin(2\pi x)\sin\left(2\pi\frac{7}{3}x\right)\,dx.$$

Using one of the sine and cosine integrals on page 14 of Chapter VIII (see also page 4 of the exercises), determine the *exact* value of this integral.

The program Spectrum calculates transforms numerically, using 100 steps. For the integral in exercise 4, the step size was therefore $\Delta x = .3$. The following exercises mimic the numerical work that Spectrum carries out on this integral.

5. Make a sketch of the graph of the function

$$h(x) = \sin(2\pi x)\sin\left(2\pi\frac{7}{3}x\right)$$

on an appropriate interval. What is the period of this function?

6. Determine the value of $h(x)$ at $x = 0$, .3, .6, .9, 1.2, and 1.5, and use these values to construct a Riemann sum for the integral

$$\int_0^{1.5} h(x)\,dx$$

using left endpoints and a step size of $\Delta x = .3$ Mark these values of h on the sketch you made in exercise 5.

[*Answer:* The Riemann sum is $-.3(2\sin^2(2\pi/5) + 2\sin^2(\pi/5)) = -.75$.]

7. Construct a Riemann sum for the integral

$$\int_0^{30} h(x)\,dx$$

using left endpoints and a step size of $\Delta x = .3$ How can exercise 6 be used to answer this question?

[*Answer:* -15. Since $h(x)$ is periodic with period $x = 1.5$, the interval $[0, 30]$ contains 20 periods of h. The integral of h over $[0, 30]$ is therefore 20 times its integral over $[0, 1.5]$.]

8. Compare the Riemann sum from exercise 7 with the exact value of the integral from exercise 4 and with the value given by Spectrum in exercise 3.

As exercises 3-8 demonstrate, the exact and computed values of the Fourier transform can be quite different, essentially because the steps in a Riemann sum can pick out very special values of the integrand. A possible solution is to change the step size. Spectrum will do this if you type I (that is, type i while you hold down the *Shift* key).

9. Refer back to exercise 2 and the graph of the power spectrum of $\sin 2\pi x$ you sketched there. When you have the graph in Spectrum, press I and note how the power spectrum changes. Do any peaks diminish or disappear? Which ones? Press I again; is there any further change? What is it? Press I a third time? What has happened to the spurious frequencies?

The University of Massachusetts

While the topics covered in the course taught at the University were chosen primarily on mathematical grounds, their organization and presentation are heavily influenced by the Calculus in Context philosophy. The following excerpts all deal with one context, population models, to illustrate the way the context is treated in depth and recurs throughout the text. Population models form one of several major contexts treated in the course.

1. Some Simple Models

The first day we introduce students to the concepts of mathematical modeling and, in particular, to discrete time models of population growth.

We share the earth's finite resources with bumblebees and elephants, with turnips and our fellow human beings. From day to day and year to year, our numbers and the numbers of our earthly neighbors vary. This section is about mathematical models of population change. We want to study the way in which populations vary. This is a very complex field of study. There are many, many different species interacting in very complex ways. Weather and a myriad of other factors can cause population changes. We will be able to investigate only some very simple models. But even these simple models will be able to give us some insight into population changes at the same time that they illustrate the ways in which calculus can help us understand our world.

We will begin by studying single-species models, models with only one species in an enclosed habitat. Even these simple models have some very surprising complexity. For example, we will see that it is possible for the population of a single species to vary as shown in Figure 1. You might guess that the kind of cyclical variation shown in this graph would necessarily be caused by some external factor, perhaps sunspots or harvesting by farmers, but we shall see that this kind of cyclical variation can actually occur without any outside interference.

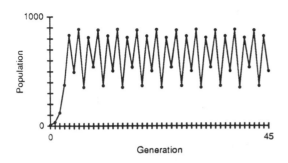

Example of Single Species Population Growth

Figure 1

One of the most important themes of this course is the use of mathematics to solve real problems. The emphasis here is on the word *real*. Real problems are often extremely complex. They cannot be solved completely in a single day. Many of them will not even be solved completely in our lifetime. One must frequently start with a very simple approximation to the real problem, an approximation which includes just a few of the most important factors involved in the real problem. Part of the art of using mathematics is identifying the most important factors involved in a given problem so that one can build a model that is simple enough to be tractable and yet close enough to reality to give at least some insight into the original problem. For example, suppose one wanted to study what happens when one drops an object from a tall building. If the object were a heavy brick, the most important factor would be the force of gravity and a model including just the force of gravity would be a reasonable model to use. However, if the object were a feather, then air resistance would be extremely important. A model that neglected air resistance and wind would be completely useless.

In practice, one often builds a series of models, beginning with a simple model including only a few of the most important factors and then adding more factors to get more complicated and better models. The process of building such a series of models can be extremely helpful. In some ways building mathematical models is very much like other everyday model building activities—building verbal models (describing things in words) or building visual models (painting pictures). The very act of describing something verbally or painting a picture forces one to think carefully about reality and to isolate the most important features.

The purpose of this section is twofold. First, we begin immediately using mathematics for modeling. Second, we want our students to gain some experience. The most important part of this section is the exercises. They ask students to simulate a variety of models. This past year they did these simulations using programmable calculators, and next year they will use computers. The models simulated include exponential models

$$p_{n+1} = Rp_n,$$

exponential models with immigration

$$p_{n+1} = Rp_n + m,$$

logistic models

$$p_{n+1} = ap_n(1 - bp_n)$$

and a model for species that hunt in packs. In the course of running these simulations the students will encounter sequences that go to infinity, sequences that converge to a point, sequences that are bounded but do not converge to a point, sequences that converge to cycles of various lengths, chaotic sequences and sequences whose longterm behavior depends on the initial value.

The students come to the second class with a wealth of experience. In the second meeting we introduce the notions of equilibrium point, boundedness and limits as descriptive concepts, useful for describing their experience. We also introduce one of the most important mathematical topics of the course, the classification of equilibrium points. We return to this topic later in the semester when we classify the equilibrium points of linear dynamical systems. Then the derivative is developed in order to generalize the results for linear systems to the nonlinear case.

2. Interacting Species

In the second semester we look at two-species (Lotka-Volterra) population models and use these to motivate our study of the character of equilibrium points for two-dimensional dynamical systems. By looking at different interspecies interactions (predator-prey, competitive, aggressive, weak-strong), the students see different kinds of equilibrium points. The mathematical behavior is strongly related to the underlying biology, which makes the mathematics more comprehensible and more meaningful.

The following section begins this study qualitatively by having students construct rough direction fields.

Our analysis so far has allowed us to identify four equilibrium points for the dynamical system

$$p' = p(1 - .005p - .001q)$$
$$q' = q(1 - .001p - .005q)$$

Based on our simulations we can see that one of them, $(166.67, 166.67)$, appears to be attracting. This fits very nicely with the biology. In this situation it appears as if the two species coexist very nicely. This is a reasonable result since the two species compete for some things but not for everything.

One of our main themes in this chapter will be analyzing the long term behavior of dynamical systems like this one. In particular, we would like to be able to determine when an equilibrium point is attracting. The computer-based graphical methods we have discussed so far are extremely powerful tools for investigating this question. In fact, we can also find out a great deal about the equilibrium points using graphical methods by hand without the computer. We proceed as follows.

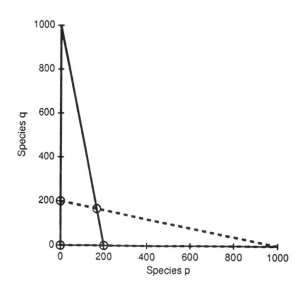

$$p' = 0 \text{ and } q' = 0$$

Figure 2

First, look at Figure 2. The equation

$$p' = p(1 - .005p - .001q) = 0$$

is satisfied at all the points indicated by the heavy solid lines. These lines divide the region into two pieces—the triangular region to the left of the

slanted heavy line and the other region to the right of the slanted heavy line. Using the Intermediate Value Theorem one can show that in each of these two regions

$$p' = p(1 - .005p - .001q)$$

is either always positive or always negative. We can determine which of these two alternatives holds in each region by evaluating

$$p' = p(1 - .005p - .001q)$$

at any one point in each region. For example, in the triangular region to the left of the slanted heavy line we can pick the point $(100, 100)$. At this point we see that:

$$\begin{aligned} p' &= p(1 - .005p - .001q) \\ &= (100)(1 - .005(100) - .001(100)) \\ &= 40 \end{aligned}$$

is positive. Hence, p' is positive in this whole triangular region. In the other region we can pick, for example, the point $(1000, 1000)$. At this point we see that:

$$\begin{aligned} p' &= p(1 - .005p - .001q) \\ &= (1000)(1 - .005(1000) - .001(1000)) \\ &= -5000 \end{aligned}$$

is negative. Hence, p' is negative in this entire region.

This means that the currents are flowing to the right (because p' is positive) in the triangular region and are flowing to the left (because p' is negative) in the other region. We can represent this information by arrows pointing to the left and right in the appropriate regions as shown in Figure 3.

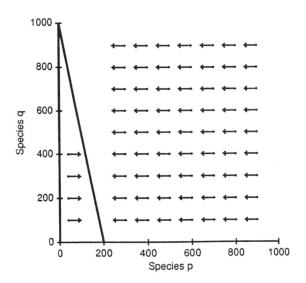

Left and Right Arrows According to p'

Figure 3

A similar analysis of the equation for q':

$$q' = q(1 - .005q - .001p)$$

leads to Figure 4. The arrows point up and down because q is graphed on the vertical (y-) axis.

Putting the information from Figures 3 and 4 together we see that the two equations:

$$p' = p(1 - .005p - .001q) = 0$$

and

$$q' = q(1 - .001p - .005q) = 0$$

divide the region into four pieces and we can see *generally* which way the direction arrows point in each of the four regions. Figure 5 summarizes all this information in one graph. This figure is not as precise as the computer-generated graphs we saw earlier but it does contain the same general information. In particular, one can see that the currents will carry corks towards the equilibrium point at (166.67, 166.67).

We arrived at Figure 5 by putting together the information from Figures 3 and 4 as follows. First, consider the small region in the lower left. This region is the intersection of the triangular regions from Figures 3 and 4. In Figure 3 we see that the arrows in the triangular region go to the right. In Figure 4 we see that the arrows in the triangular region go upwards. Thus, in the intersection of these two regions the arrows go upward and to the right as shown in Figure 5. The general directions of the arrows in the other three regions are determined in the same way.

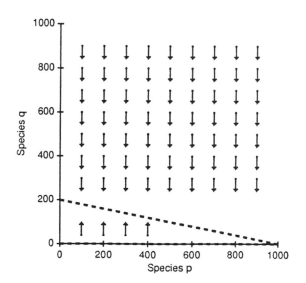

Up and Down Arrows According to q'

Figure 4

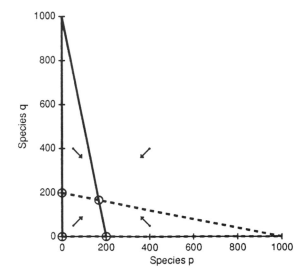

General Direction Field

Figure 5

University of Illinois

CALCULUS AND MATHEMATICA

by Don Brown, Horacio A. Porta, and J. Jerry Uhl

≡ Abstract ≡

Contact:

DON BROWN, H. PORTA, AND J. J. UHL, University of Illinois, 1409 West Green Street, Urbana, IL 61801. PHONE: 217-244-8312. E-MAIL: horacio@math.uiuc.edu.

Institutional Data:

The University of Illinois is a large land grant institution with a long tradition of excellence in science and mathematics. There are about 25,000 undergraduate students at Illinois, a large percentage of which are in the sciences, engineering, and commerce. Illinois is a residential university with steadily increasing minority enrollment. Illinois has a well-known program for physically handicapped students.

The Mathematics Department has about 100 full-time faculty. About 100 students receive undergraduate degrees in mathematics each year, and another twenty receive undergraduate degrees in mathematics education.

Project Data:

There have been two faculty members, five teaching assistants, and one academic professional involved in teaching at Illinois. This will be expanded next fall. In addition, several faculty have helped with advice and encouragement. Each semester there have been 50-80 students completing the new courses. This represents slightly less than 1/10 of the total enrollment in corresponding classes at Illinois. It should be pointed out that the course is also running at Ohio State, Vanderbilt, Knox College, and Westmont College. Reports from these test sites have been invaluable. Additional schools that have signed on to be test sites are Randolph-Macon, Fresno State, Stevens Tech, St. Bonaventure, University High. Actual instruction at Illinois began in January 1989.

The computer lab is the center of the course. No lectures are given, but the students meet with their instructor once a week to discuss matters of calculus literacy.

The only software used is Mathematica running on a selection of Apple Macintoshes (IIcx's, II's, SE/30's, and SE's). There is no text as such; all text materials reside in the computers with students printing what they choose to take home.

Project Description:

Calculus&*Mathematica* is an electronic calculus course based on the principle that calculation and plotting set up theory which in turn sets up more calculation and more plotting which in turn sets up more theory which .... In this way, the students learn instead of being taught.

To be issued in preliminary version by Addison-Wesley later this year, Calculus&*Mathematica* is a sequence of electronic Mathematica Notebooks comprising an electronic text with live calculations and graphics that bring students through calculus of one variable (a multivariable sequel lies in the not-too-distant future). The format is that of problems and solutions. Each notebook opens with Basic problems which introduce many of the new ideas in the material under study. Second comes the Tutorial problems which introduce techniques and applications. Full electronically active solutions are provided to each Basic and Tutorial problem. Closing each notebook is a section of "Give-it-a-try" problems. Here no solutions are given, but the student, through word processing and calculation, adds his or her solution directly to the notebook and electronically turns in the completed notebook for comments and grading.

As taught at Illinois, discussion sessions meet once a week with students spending 6-10 hours a week at the machines in a rather loosely organized lab. There is an effort to provide quality lab assistance at all times, and there is one teaching assistant whose sole job is to keep a lot of pep in the lab. Students turn in assignments electronically about once a week. These assignments count for 1/3 of the grade. There are also three calculus literacy tests for 1/3 of the grade, and a final lab project for the other 1/3.

We have been quite pleased with the backing given to us by our department chair and the university administration. Also we are pleased that by next fall we expect to be field testing at 15-20 other schools.

══ Project Report ══

Day by Day Mechanics

The course is taught as a laboratory-recitation course. Lectures are kept to a minimum. Experimentation led us to reduce the lecture period to half an hour a week, with another half an hour for organized discussion, quizzes, etc. The rest of the instruction is provided at the lab by faculty, TA's and lab assistants (recruited among the students who took the course previously). This requires more contact hours than a regular course and a careful monitoring of supporting staff (we still feel like beginners in the area of staff supervision, as the traditional mathematics course is usually a solo

performance). On the other hand, the schedule is less rigid than a traditional course and preparation time is minute.

As the center of activities, the lab is the classroom. Since we allow only calculus students in our lab, considerable mutual support and camaraderie develops among the students. Soon we found that work teams appeared with personalities determining the individual roles. We feel that cooperation at this level is a very positive characteristic of our course since it encourages these future professionals to work in teams, the normal way of organizing any enterprise.

The decision to reduce the length and prominence of the lectures led us to consider also the role of the instructor in the lab, and we adopted the "Illinois" model suggested by George K. Francis: The computers are arranged in a large "U" formation with the screens facing towards the interior of the "U." The instructor's place is at the top of the "U" looking over the students' shoulders.

From this vantage, the instructor can monitor the work being done on each station and quickly spring to the aid of a student in trouble. This allows the instructor to make direct contact with the student at the "optimal teachable moment". Thus the professors and assistants are placed in a strong supporting role behind the students, pushing them instead of pulling them. We believe that in this way learning becomes the primary activity, while teaching is relegated to a secondary position and supplied only in the necessary dosage. Winston Churchill once said that he was always ready to learn but seldom ready to be taught; we like to think Churchill would have been comfortable in our lab.

Each section consists of 10-25 students who meet in a standard classroom (without machines) for discussion and literacy quizzes. Aside from this formal meeting, there is no other set time that the students must appear in class or the lab. The rest of the time is spent on the lab with the students going through the electronic text and working their assignments. These assignments include calculations, plotting and fully word-processed explanations. The assignments are turned in electronically slightly more often than once a week.

At the beginning, faculty and teaching assistants devote more time to this course than they would to the traditional course, but with experience the time decreases. Teaching assistants are used interchangeably with faculty, but some of our best teaching is done in the lab by undergraduate lab assistants who have already done well in a previous section of the course. Currently we are experimenting with a position called the "lab pep person" whose duties are confined to roving about the lab and promoting a lively mathematics atmosphere.

Getting Started

The original idea for the course was Steven Wolfram's. He described the possibilities for an electronic text in the form of *Mathematica* Notebooks.

As two colleagues who always liked to talk about calculus, we (H. P. and J. J. U.) were intrigued with the idea and met with Wolfram, Dan Grayson and Theo Gray a few times and then we decided to give it a try in partnership with one of our students (D. B.).

Following Wolfram's suggestion we contacted Allan Wylde, of Addison Wesley's Advanced Books Program, who suggested that we send a prospectus to the College Division of Addison-Wesley. The College Division quickly rejected the idea and Wylde said that his Advanced Books Program would back the project as a low-overhead experiment in electronic publishing. Wylde presented our case to Apple Computer and soon we had three Macintoshes and a printer and the promise of modest funding from Wylde's division.

We are pleased to deal with Wylde because his division of Addison-Wesley does not seem to have the marketing pressures that in part have led to the current situation in the traditional calculus texts.

We worked very hard for six months, writing lessons, making proposals and trying to arrange for some more computers so that we could begin instruction. The result was the first set of Calculus&*Mathematica* Notebooks. At this point, Ward Henson, chair of the Illinois mathematics department, gave us decisive support and soon the university administration and Apple Computer decided to buy twenty machines in time to start instruction in January, 1989. In addition, Wolfram Research contributed the necessary copies of *Mathematica*. We were on our way.

But not everything went so smoothly. Our NSF ILI proposal was returned with exceedingly negative reviews. Only then did we understand that many mathematics teachers were firmly committed to other software and did not share our vision of how *Mathematica* Notebooks can be used. Almost the same thing happened to our NSF proposal for the Calculus Initiative, but luckily one or two of the referees wanted to give us a chance; so we received a small one year award.

Getting other professors in the department interested has been a challenge. We have considerable expression of support from some colleagues, but many keep their distance. A recent presentation to the Illinois mathematics faculty and graduate students attracted a large enthusiastic audience; but, before it was over, one of the star members of the faculty walked out in a dispute about whether computer produced plots are ever accurate. Many research mathematicians do not want their images soiled by expressing any interest in calculus and choose to put themselves above the current calculus crisis. We believe that the future health of mathematics departments depends on turning this attitude around. We are trying and maybe we can succeed.

On the other hand, we were able to attract interest outside our university. Ohio State, Vanderbilt, Westmont and Knox agreed to test the course with their students starting in fall, 1989. What we learned from these test sites proved invaluable. Additional schools that have signed on to be test sites are Randolph-Macon, Fresno State, Stevens Tech, St. Bonaventure, University High School (Urbana), Niles West High School and Kennesaw State.

And the future looks good: last summer, the university administration announced that they would set up a forty station lab devoted to our project and couple it to another forty station open lab over the next two years. Our experience is that support from the administration is typical. It may be hard to explain but university administrations are often more interested in technology and calculus revision than are mathematics departments.

By fall, 1990, we expect to be in the classroom, in a limited experimental way, at fifteen to twenty universities, colleges and high schools. At the same time, we hope that a preliminary version of the course will be offered by Addison-Wesley.

Content

Let's begin with what's not in the course. Peter Lax once said:

> Calculus as currently taught is, alas, full of inert material which will remain there as long as the teaching of calculus is controlled by ... the group presently entrusted with teaching it.

We take Lax's words as part of our credo. Missing are topics like limit at a point, relative max-min,

Rolle's theorem, second derivative test, Riemann sum definition of the integral, definition of logarithm via integration, shell method vs. disk method, many of the mindless applications of the integral like surface area of solids of revolution, Simpson's rule, serious treatment of L'Hopitals rule, formula for the error term in Taylor Series. These are all filler topics in one-variable calculus. They make for stock test questions that allow teachers to fool their students into the belief that they have learned something.

Henri Lebesgue warned us off topics like these when he said:

> Unfortunately, competitive examinations often encourage one to commit this little bit of deception. The teachers must train their students to answer little fragmentary questions quite well, and they give them model answers that are often veritable masterpieces and that leave no room for criticism. To achieve this, the teachers isolate each question from the whole of mathematics and create for this question alone a perfect language without bothering about its relationships to other questions. Mathematics is no longer a monument but a heap.

Our view is that calculus is nothing more or less than a course on how to use the tools of differentiation, integration and approximations to make precise measurements. The venerable Granville course had moderate success in this endeavor. Then came the new math and Johnson and Kiokemeister. In the sixties, it became the fashion to study the tools of calculus without actually using them. Instead of using a hammer, it became fashionable to put it on the table and to study it from all angles. By the eighties, the reaction was to pick up the hammer and use it as a rote procedure by driving nail after nail into the same piece of lumber. The tool was used, but nothing was built. In the Calculus&*Mathematica* course we attempt to pick up the hammer, get its feel, drive a few nails and then begin to build (tables, ladders, houses ...)

Calculus is related closely to applied mathematics because applied mathematics is a great consumer of exact and approximate calculation and measurements, but it is a mistake to say that calculus is an introduction to applied mathematics. Calculus is related to real analysis because calculus uses many of the tools of real analysis. But calculus, as a course in exact and approximate calculation and measurement, is a positive endeavor; real analysis with its emphasis on counterexamples is in many ways a negative endeavor. So calculus is not an introduction to real analysis either. If anything, calculus is an introduction to complex analysis, at

least in the sense that they share a spirit of positive thinking. But when the dust settles, calculus should be viewed as the first course in a newly emerging discipline called "scientific computation" by James G. Glimm, chair of the SIAM Committee on Science Policy. With the definition of calculus nailed down, we can move on to a discussion of what *is* in the course.

If a topic contributes to measurement, then the topic is not tossed out; if a topic is purely mathematical and may serve to whet the students' interest in mathematics, then it is not tossed out. Otherwise the topic is probably a goner. All calculus cottage industries like L'Hopital's rule, convergence at endpoints, etc. are de-emphasized or missing.

Our goal is to elicit the proper mathematical response to a given situation. Traditional courses with their emphasis on counterexamples and limitations do just the opposite. They feign an axiomatic course, but the students never learn what the axioms are. Counterexamples and limitations are the stuff of real variable theory, but in calculus we should "accentuate the positive."

Our course takes the view that the quantities under measurement like instantaneous growth rate, tangent line, area, volume, arc length, e^x, etc. do not need to be defined because they are already there. For instance, we *find* the natural base for logarithms instead of using the integral to *define* natural logarithms. The tools of calculus are important insofar as they can make measurements—either mathematical or scientific.

Here is a sampler of innovations and revisions:

- Feeling for limits and convergence is set up through plots showing the curves $(f[x + h] - f[x])/h$ crawl onto $f'[x]$ as $|h|$ gets small.

- The chain rule forms the keystone of our treatment of differentiation. We get at the logarithms by using the chain rule to differentiate the functional equation of the logarithm. The power rule and product rules are obtained by logarithmic differentiation.

- Continuity and limits do not sit at the front of the course but emerge in a natural way throughout the course. Students work on continuity by being able to report how many accurate decimals of x are needed to calculate, say, eight accurate decimals of $f[x]$.

- The mean value theorem is studied as a consequence of something we call the Race Track Principle. This principle says that if $f(a) = g(a)$ and $f'(x) \geq g'(x)$ for $x \geq a$, then $f(x) \geq g(x)$ for

$x \geq a$. Another version of the principle says that if $f(a) = g(a)$ and $f'(x)$ is close to $g'(x)$ for x near a, then $f(x)$ is close $g(x)$ for x near a. The Mean Value Theorem is a corollary of the Race Track Principle.

- Early in the course we confront the problem of plotting $f[x]$ given only $f'[x]$ and one value of $f[x]$. Differential equations appear liberally in this course, but maybe not so much as in some other calculus revision projects.

- Lots of emphasis on curve matching and approximation.

- Our course is not in the business of defining slopes, areas, arc lengths and volumes. We are in the business of measuring these quantities. Newton never heard of a Riemann sum, but Newton did teach us via the fundamental theorem $\int_a^b f'[x]dx = f(b) - f(a)$ that we can measure any quantity once we calculate its derivative. Finding the derivative of area, volume, arc length and the like is good geometric mathematics. Traditional courses take the mathematics out and bring the dogma in when they insist on using Riemann sums to define these concepts. Courant introduced the integral in terms of area. Emil Artin, in his Princeton courses, defined integral in terms of area. We follow their precedent.

- Most of the usual techniques of integration are present. Substitution is important and integration by parts is emphasized. Partial fractions is in the course, but it is buried in a section on integration by the method of undetermined coefficients.

- Conventional wisdom has it that pointwise convergence is easier than uniform convergence. Yet what our students observe via plots—and what they call "cohabiting functions"—is uniform convergence.

- We build feeling for series via plotting and calculations in the context of power series and other approximations. Series of numbers come up at the end as a short digression and are looked at from the standpoint of series of functions evaluated at a point. Most of the usual "convergence tests" have been scrapped.

- Expansions in powers of x are obtained algebraically by the method of undetermined coefficients and change of variable. Students are quite experienced with expansions before they study Taylor's formula for the coefficients.

- Complex numbers and the complex exponential are introduced and play a big role in the study of Taylor expansions. The main criterion for convergence of Taylor expansions is by complex singularities.

- Empirical plots of power series and power series solutions to differential equations are studied. The question is: over what interval should we be able to trust the plot of a truncated expansion to mimic the plot of the true solution?

- Non-differentiable functions and divergent series make only rare appearances in the course and the existence of a non-integrable bounded function is never mentioned.

- Throughout the course, we ask for student opinions, explanations and conjectures. We are not so much interested in whether a student can give an air-tight proof, but whether the student has the right mathematical reaction to a given situation and can discuss the reasons for his or her reaction.

- Accompanying each lesson is a "Literacy sheet" which is a series of questions meant to be answered orally.

- The course operates at two levels. Informal, but correct, arguments flourish in the computer lessons. For precision and rigor, we are working on an accompanying "theory book" in the style of Artin.

Pedagogy

The course is presented through the medium of *Mathematica* Notebooks. Very simply, *Mathematica* Notebooks allow fully word-processed text to be inserted in the middle of active *Mathematica* code. It was this medium that convinced us to undertake the project and to develop the course around Calculus&*Mathematica* Notebooks (abbreviated C&M Notebooks).

Imagine a mathematics text in which each example is infinitely many examples because each example can be redone immediately by the student with new numbers and functions. Imagine a symbolic or numerical computer routine into which fully word-processed descriptions can be inserted at will between lines of active code. Imagine a text whose paragraphs can be modified and added to as the teacher sees fit. Imagine a text that has better graphics and plots than any available in any standard mathematics book and imagine that the amount of graphics is limited only by computer memory instead of the cost and weight of printed pages. Imagine that all the graphics can be in color and that the three-dimensional graphics are all prefectly shaded and can easily be viewed from any desired viewpoint. Imagine a text in which a student can launch his or her own graphic and calculational explorations with graphics and calculations appearing as the student desires. Imagine a text in which the student can find as much space as he or she needs to solve the assigned exercises. Imagine that a student, in a matter of seconds, can copy his or her own homework and turn it in while retaining the original.

If you can imagine all of this, then you have imagined an electronic text in the form of *Mathematica* Notebooks. We are convinced that the medium of *Mathematica* Notebooks will change forever the way undergraduate mathematics is taught. About this point, Thomas Morley, of Georgia Tech, agrees saying that this will be the case even for those teachers who choose not to use this resource, as the style adopted by the whole mathematical community is very likely to evolve in the direction of *Mathematica* Notebooks. As retroactive support of this rather bold prediction, we mention the striking similarity between some pages of Euler's *Analysis of the Infinite* and a bare bones *Mathematica* Notebook on expansions of rational functions in powers of the variable.

Traditional courses proceed by releasing measured amounts of theory. Then the theory is illustrated by the exercises. Courses based on *Mathematica* Notebooks do not have to be this way. In our project we have followed Euler's lead (*loc. cit.*) by extensively using calculations and plotting as a dynamic device to explore and discover underlying concepts. The students seem quite excited to get answers almost instantaneously and to explain them in their own terms. With a little guidance (found in the text cells of the notebook) they soon enough announce correct results based on their experience. Granted, they cannot always give mathematical proofs of these facts, but they are quite convinced, being the originators of the statements and fully committed to their validity.

Main Pedagogical Principle

Early success in this regard and an important discussion with Francis Sullivan of the National Institute for Science and Technology, helped us to articulate our main pedagogical principle:

> Calculation and plotting set up theory which in turn sets up more calculation and more plotting which in turn . . .

This is the way research mathematicians work and it works equally well with students.

A bonus of this approach is that because of *Mathematica*'s graphic and calculational power, students make computations that lead to a mathematical experience far richer than students taught the traditional way, who are limited by rigged examples where all coefficients are small integers, all quadratic forms have integer roots, all maxima are attained at values with simple descriptions, etc. An analysis of the Predator-Prey model, for instance, is out of the question in most traditional courses; it is routine and almost painless in Calculus&*Mathematica*.

Presentation of Notebooks

Deciding on how to present the Notebooks to the students is no small problem.

The typical traditional calculus class opens with a lecture on the theoretical points of the day's material. The students politely sit with blank expressions on their faces allowing the professor to indulge himself with his lecture until the point at which the lecture moves into the problem solving phase. At this point the classroom becomes alive with note taking and questions while the students program themselves with the templates for the problems they will solve for the assignment. The assignments will be corrected so the student can find the bugs in the way he or she programmed his mind to do the problems.

Often the student has no idea whether one of his or her individual solutions is right or wrong. Neither does the student usually have an idea of the purpose of a problem. The main goal is to get it the way the teacher wanted it. Traditional calculus instruction resembles traditional instruction in religion with the teacher serving as the curator of the dogma, the students reciting the litany of rote procedures and the grader being the arbiter of truth.

At the beginning, we held to the traditional lecture system. We used *Mathematica* to prepare beautiful classroom demonstrations and animations. But the students quickly made us change our approach. Their method was simple: they quit coming to the lectures. When we asked why, they said that everything in the lecture was already available in the *C&M* Notebooks that we had installed in the computers; so there was no reason for them to attend our lectures. They were right. The result of this experience was the laboratory-recitation approach that was discussed above.

Classroom

Our experience sets up a friendly warning to those who plan to append *Mathematica* onto a course taught with traditional text materials. Although this sounds good in principle, it may be a step backwards due to the disconcerting imbalance between lecture and laboratory that will result. In the classroom, *Mathematica* is not a quiet guest. It can significantly alter a standard calculus course because most of the questions usually asked in a test or assignment can be answered by simply evaluating *Mathematica* instructions.

This arrangement may result in bored students with confused attitudes about what calculus is: on the one hand, the textbook material with lots of caveats and terminal timidity regarding calculations or on the other the formidable exploits of *Mathematica* crunching calculations and graphs that often explain by themselves more than the textbook. Our belief is that *Mathematica* should be brought into the classroom successfully only at the same pace as new courses are designed.

Our students turn in their work in the form of *Mathematica* Notebooks. We ask the students for the same care in writing calculus as they apply to their rhetoric or social science courses. Originally we thought that getting good writing would be difficult, but this has not been much of a problem.

Our students generally work one student per machine mainly because their work is their own. But we encourage and get lots of huddles in the lab. Often we see three neighboring students grouped around one machine discussing the points of a solution. Then the huddle breaks up and the students return to their own machines to go to work. This is group learning at its best.

The result of our design is a different outlook for the calculus course. One of our students remarked that this was the only class in which he did his homework. When asked why, he replied: "Because homework is the class."

Technology

This project probably uses technology to a greater extent than most other calculus revision projects; yet the role of technology in it is so natural that it does not stand out in the day-to-day operation.

In order to prepare students for challenging careers, we try to train them in the use of legitimate tools. Custom "user-friendly" classroom interfaces are not steps forward if the result is that the only place that the students can undertake serious calculations is the calculus lab. We want our students to be able to use *Mathematica* wherever they find it. Therefore we do not customize *Mathematica* in any way; instead we use *Mathematica* directly as it comes off Wolfram's shelf. Even all the grisly graphics instructions are in the open for our students to use and modify. This decision (hard as it was at the moment), more than any other, resulted in our students becoming competent calculators very quickly.

Our current lab consists of forty Macintoshes (IIcx's, II's, SE/30's and SE's) and one NeXT networked together with two SE's acting as file servers and for submission of homework. Each machine has its own copy of *Mathematica* hidden from greedy fingers. The computers sit in a large room near the mathematics library. The lab is open 9:00 a.m to 10:00 p.m. Monday through Thursday, 9:00-6:00 on Fridays, 12:00-5:00 Saturdays and 12:00-10:00 Sundays.

As indicated above, the laboratory is the class, the homework and the exercises. The only software in the lab is *Mathematica*. No non-mathematical activities are permitted in the lab.

We estimate that a ratio of one computer for four or five students is adequate. One of our computers will serve many more students when *Mathematica* becomes available at locations outside our lab.

The lab is staffed with undergraduate laboratory assistants who are graduates of one of our courses. They have full responsibility for security, opening and closing. They are extremely valuable to the operation and they are cheap.

Students do a lot of writing and calculating in the lab and the question comes up about whether the students write their own programs and who writes home-made programs. The truth is that there is very little traditional programming in the course in the sense of programming in Fortran or C. There are no homemade programs. These is a lot of interactive writing with *Mathematica* as the students do their homework in *Mathematica* Notebooks which they prepare.

Appraisal and Conclusions

The first question usually asked of us is whether the medium interferes with the message. The answer is no. The medium is *Mathematica*, the message is mathematics.

The second, related, question is whether *Mathematica* would be hard for the students to learn. Originally we thought *Mathematica* syntax would be hard for the students to learn. We were wrong. In the lessons the students saw *Mathematica* commands and *Mathematica* routines in context and picked them up very quickly. After the second week, *Mathematica* syntax was not much of a problem at the level needed for successful performance. Professors with background in Basic seem to have more trouble adjusting to *Mathematica* than the students. The old-timers tend to emphasize *Mathematica* as a programming language, while the kids emphasize it as a calculation and communication medium.

This is the time to experiment with *Mathematica* courseware. The students are ready. The universities and high schools are hoping to turn around the current math (and science) teaching malaise and will support serious proposals. The cost of hardware is not that overwhelming to put reasonably sized labs out of reach. There are no obstacles other than our own indifference or exhaustion and the constant lack of fair reward for good undergraduate instruction. Those difficulties will have to be fought on an individual basis.

But already our students share the joy of working out problems; they experiment and discover—thereby feeling that their own exploits are not too different from those of research scientists—and are beginning to develop the correct reflexes to attack technology problems with power and elegance. Also, they make fewer mistakes (and fewer desperate attempts at getting half-credit through the old technique of blurring part of a formula). Very possibly their verbal skills are improving to more acceptable levels. They seldom submit incorrect answers because (as we have seen in the lab) they detect the errors by themselves and seek advice on time, and they cooperate with one another. This last point has several interesting implications pertaining to special education, continuing education, teacher training, and similar out of the ordinary programs.

Our downside is our initial drop rate. Our course requires more commitment and time than does the standard course. Some students bail out saying that the standard course is easier and requires much less time. Others note that we are not doing the same thing as they see in the standard course and they believe that the standard course must be doing things right. All this means that at Illinois, our transfer-to-the-standard course rate is about thirty percent. We think this is too high, but are somewhat relieved by the fact that our test sites at other universities to have a lower drop rate.

Quite often we are asked about comparing our students with students from the standard course on common exams. We have not yet done this because such tests will necessarily be on the standard calculus cottage industries that we are trying to replace. In fact, we go so far as to say that a genuinely new calculus course should not try to evaluate itself this way.

At last count there were about one million students taking one form or another of calculus instruction. Perhaps we have no right to extrapolate from the sixty or so we had each of the last three semesters that a great number will benefit from our course. But we are confident that, for a sizable number of students, our course is a much more rewarding, attractive and challenging undertaking than the ordinary course, and that many students are willing to spend the extra time and effort to participate in the first calculus adventure in 300 years.

References

Emil Artin, A freshman honors course in Calculus and Analytic Geometry taught at Princeton University, notes by G.B. Seligman, CUPM publication, Mathematical Association of America, 1957.

Richard Courant, Differential and Integral Calculus, (Translation by E.J. McShane), Interscience, New York, 1937.

Leonard Euler, Introduction to Analysis of the Infinite I, (English translation by John Blanton), Springer-Verlag, 1988.

William Anthony Granville, Percey F. Smith and William Raymond Longley, Elements of the Differential and Integral Calculus, 1929, Gin and Co.

Richard E. Johnson and Fred L. Kiokemeister, Calculus with Analytic Geometry, Second edition, 1960, Allyn and Bacon, Boston.

Peter Lax, On the teaching of calculus in "Toward a Lean and Lively Calculus", 61-69, Ronald G.

Douglas, Editor, Mathematical Assn. of America, Washington, 1986.

Henri Lebesgue, Measure and Integral (Translation of "La Mesure des Grandeurs", edited by Kenneth O. May), Holden-Day, San Francisco, 1966.

Thomas D. Morley, The Georgia Tech Calculus Project, Presentation at the Second Conference on Technology in Collegiate Mathematics, Columbus, Ohio, October 1989.

Jacob T. Schwartz, Computed-Aided Instruction in "Discrete Thoughts: Essays in Mathematics"; by Mark Kac, Gian-Carlo Rota and Jacob T. Schwartz, Harry Newman, Editor, Birkhauser, Boston 1986.

═══ Sample Materials ═══

Powers and Products

Here is how a Calculus&Mathematica *Notebook opened to the Basic Problem section looks on the screen; one solution is displayed:*

▦ **Basics**

☐ **B. 1) Constants and sums.**

☐ **B. 2) Using the logarithm to calculational advantage: Powers and products.**

• **B.2.a.i)** It may not be clear how to calculate the derivative of

$$f[x] = x^4 \sin[x]^6.$$

But it is very clear how to differentiate

$$\log[f[x]] = \log[x^4 \sin[x]^6]$$
$$= 4\log[x] + 6\log[\sin[x]].$$

In fact the chain rule tells us that

$$D[\log[f[x], x] = 4/x + 6(1/\sin[x])\cos[x]$$
$$= 4/x + 6(\cos[x]/\sin[x]).$$

How does this tell us how to calculate $f'[x]$?

• **B.2.a.ii)** Why is $f[x]D[\log[f[x]], x] = f'[x]$ and what calculation advantage does this give us?

▦ **Answer:** The chain rule tells us that

$$D[\log[f[x]], x] = f'[x]/f[x].$$

Multiply both sides by $f[x]$ to see that

$$f[x]D[\log[f[x]], x] = f[x]f'[x]/f[x] = f'[x].$$

Try it:

```
>> Clear [f,x]
```
This is a live executable instruction which is executed by the student:

```
>> Expand [f[x] D[Log[f[x]],x]]

f'[x]
```
Right on!

This is a welcome calculational advantage in the cases, as in part i), in which that $f[x]$ is hard to differentiate but $\log[x]$ is easy to differentiate. Some old timers call this technique logarithmic differentiation. Let's try this out. If

$$f[x] = x^{-5}e^x,$$

then $f[x]$ is hard to differentiate, but

$$\log[f[x]] = -5\log[x] + \log[e^x] = -5\log[x] + x$$

is easy to differentiate. The formula

$$f[x]D[\log[f[x]], x] = f'[x]$$

tell us that $f'[x]$ is given by:
This is another live executable instruction which is executed by the student:

```
>> Expand [x^(-5) E^x ((-5/x) + 1)]

        x      x
  -5 E     E
  -----  + --
     6      5
    x      x
```
Check:
This is another live executable instruction which is executed by the student:

```
>> Expand [D[x^(-5) E^x,x]]

        x      x
  -5 E     E
  -----  + --
     6      5
    x      x
```
Right on the money.

• **B.2.a.iii)** By this time most of us believe that if t is any non-zero exponent, then the derivative of

$$f[x] = x^t$$

is

$$f'[x] = tx^{t-1}.$$

Use the identity $f[x]D[\log[f[x]], x] = f'[x]$ to explain this basic formula which some folks call the **power rule**.

- **B.2.b)** Give the derivatives with respect to x of the following functions. Check with *Mathematica*.
- **B.2.b.i)** $x^{2/3}$.
- **B.2.b.ii)** $(e^{-7x} - 3e^{2x})^{1/5}$.
- **B.2.c.i)** Use the identity

$$f[x]D[\log[f[x]], x] = f'[x]$$

to calculate the derivative of $f[x] = x^5 \log[x]$.

- **B.2.c.ii)** Use the identity

$$h[x]D[\log[h[x]], x] = h'[x]$$

to explain why the derivative of the product of two factors is the sum of two terms each of which consists of the derivative of one factor times the other factor.

- **B.2.c.iii)** Calculate the derivative of $x^4 e^{3x}$. Check with *Mathematica*.
- **B.2.d)** We can also handle quotients. Calculate the derivative of

$$\sin[x]/(3 + x^4).$$

Check with *Mathematica*.

Tutorials

Here is how the same Calculus&Mathematica Notebook opened to the Tutorial Problem section looks on the screen; one solution is displayed:

▨ **Tutorial**

☐ **T.1) Derivatives.**

☐ **T.2) Using the derivative to help to get a good representative plot.**

☐ **T.3) Describing the growth of the function by looking at the derivative.** In each case, take the derivative of the given function and use the information contained in the derivative to describe (in words) the behavior of the function. Then confirm your description with a representative plot of the function.

- **T.3.a)** xe^x.

▨ **Answer:** Factor the derivative:
This is a live executable instruction which is executed by the student:

```
>> Factor [D[x E^x,x]]
```

```
 x
E  (1 + x)
```

The instantaneous growth rate ($=$ the derivative) is positive for $x > -1$ and the derivative is negative for $x < -1$. (Remember e^x is never negative or 0.) Therefore the function is going up for $x > 1$ and going down on for $x < -1$. Confirm with a representative plot:
This is another live executable instruction which is executed by the student:

```
>> Plot [x E^x,x,-4,1,PlotRange->All]
```

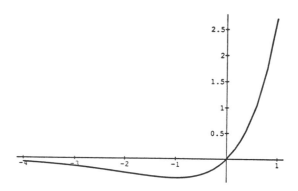

```
--Graphics--
```

The description is right on the money.

- **T.3.b)** $\log[x]/x$.

Here is how the same Calculus&Mathematica Notebook opened to the "Give it a try" Problem section looks on the screen:

▨ **Give it a try**

☐ **G.1) Taking derivatives.**

☐ **G.2) Functions and their derivatives.**

☐ **G.3) Going up or down.**

☐ **G.4) Up then down.**

- **G.4.a.i)** Plot x^3/e^x for $0 \leq x \leq a$ where a is chosen to be large enough to see the curve go up and then down on $[0, a]$. Factor the derivative to find the exact x at which the curve changes direction.

The student enters the Notebook here and solves the problem mixing word processing and calculations.

- **G.4.a.ii)** Plot x^6/e^x for $0 \leq x \leq a$ where a is chosen to be large enough to see the curve go up and then down on $[0, a]$. Factor the derivative to find the exact x at which the curve changes direction.

The student enters the Notebook here and solves the problem mixing word processing and calculations.

- **G.4.a.iii)** Plot x^{12}/e^x for $0 \le x \le a$ where a is chosen to be large enough to see the curve go up and then down on $[0, a]$. Factor the derivative to find the exact x at which the curve changes direction.

The student enters the Notebook here and solves the problem mixing word processing and calculations.

- **G.4.a.iv)** Fix a positive number t. Factor the derivative to explain why the curve $y = x^t/e^x$ first goes up as x advances from 0 and grows until x reaches a point $x[t]$ after which the curve goes down. Find the exact value of $x[t]$ as a function of t.

The student enters the Notebook here and solves the problem mixing word processing and calculations.

- **G.4.b.i)** Plot $\log[x]/x$ for $0 \le x \le a$ where a is chosen to be large enough to see the curve go up and then down on $[0, a]$. Factor the derivative to find the exact x at which the curve changes direction. Don't forget that $\log[x] = a$ means $x = e^a$.

The student enters the Notebook here and solves the problem mixing word processing and calculations.

- **G.4.b.ii)** Plot $\log[x]/(x^{1/8})$ for $0 \le x \le a$ where a is chosen to be large enough to see the curve go up and then down on $[0, a]$. Factor the derivative to find the exact x at which the curve changes direction.

The student enters the Notebook here and solves the problem mixing word processing and calculations.

- **G.4.b.iv)** Fix a positive number t. Factor the derivative to explain why the curve $y = \log[x]/x^t$ first goes up as x advances from 0 and grows until x reaches a point $x[t]$ after which the curve goes down. Find the exact value of $x[t]$ as a function of t.

The student enters the Notebook here and solves the problem mixing word processing and calculations.

- ☐ **G.5) Good representative plots.**

- ☐ **G.6) Oil slicks.**

- ☐ **G.7) Efficient design.**

- ☐ **G.8) Finding the derivative of x^x.**

- ☐ **G.9) Higher derivatives.**

- ☐ **G.10) Products.**

Literacy Sheet

Here is a sample of part of a Calculus&Mathematica *Literacy Sheet:*

What you should know away from the computer:

- **1.** Differentiate the following functions with respect to x by hand and simplify:
- **1.a)** $\cos[x]$
- **1.b)** $\arctan[x]$
- **1.c)** $-\log[\cos[x]]$
- **1.d)** $2x^{5/2} - 5x^{2/5}$
- **1.e)** $\sec[x]^2 = 1/\cos[x]^2$
- **1.f)** x^e
- **1.g)** e^x
- **1.h)** $x^2 \log[x]$
- **1.i)** $x^3 \sin[x]^2$
- **1.j)** $e^{-2x} \cos[7x]$
- **1.k)** xe^{-3x}
- **1.l)** $(1 + 8x - x^3)/(1 + x)^2$
- **1.m)** $1/\sqrt{[x]}$
- **1.n)** $1/(1 - x)$
- **1.o)** $1/(2(1 - x)^2)$
- **1.p)** $e^x/(1 + e^x)^2$
- **1.q)** x^x
- **1.r)** $-(2 + 2x + x^2)e^{-x}$
- **1.s)** $-\arctan[1/x]$
- **1.t)** $\arctan[\sqrt{[x]}]$
- **1.u)** $x \log[x] - x$
- **1.v)** $(e^x \cos[x] + x^4)^3$

- **2.** Examine the derivative of $f[x] = xe^{-x^2}$ to give a reasonably good hand sketch of the curve $y = f[x]$. What is the maximum value $f[x]$ can have?

- **3.** Examine the derivative of $f[x] = x \log[x]$ to give a reasonably good hand sketch of the curve $y = f[x]$. What is the minimum value $f[x]$ can have?

- **4.** Examine the derivative of $f[x] = x - \cos[x]$ to see whether the curve $y = f[x]$ ever goes down.

- **5.** What familiar function is approximately equal to $(\arctan[x+.000001]-\arctan[x]/.000001$?

- **6.** Suppose $f[x]$ and $g[x]$ are two functions with $f'[x] > 0$ and $g'[x] > 0$ for all x's. This means both $f[x]$ and $g[x]$ go up as x advances from left to right. Does it also mean that the product $f[x]g[x]$ also goes up as x advances from left to right? To help you form your opinion, try $f[x] = x$ and $g[x] = x$. What happens if in addition $f[x] > 0$ and $g[x] > 0$ for all x's?

- **7.** If x is measured in radians, then the derivative of $\sin[x]$ with respect to x is $\cos[x]$. Use the formula

$$\sin[x \text{ degrees}] = \sin[(2\pi/360)x \text{ radians}]$$

to calculate the derivative of $\sin[x \text{ degrees}]$ with respect to x. Why does the resulting formula make calculus difficult if we insist on working with degrees instead of radians?

- **8.** To say that $y[t]$ is proportional to $f[t]$ means that

$$y[t] = K f[t]$$

for some constant K. To say that $y[t]$ is proportional to both $f[t]$ and $g[t]$ means that

$$y[t] = K f[t]g[t]$$

for some constant K. To say that $y[t]$ is inversely proportional to $f[t]$ means that

$$y[t] = K/f[t]$$

for some constant K.

   a. If $y[t]$ is proportional to t^2 and $y[1] = 7$, then give a formula for $y[t]$.

   b. If $y[t]$ is proportional to e^{4t} and $y[0] = 8.1$, then give a formula for $y[t]$.

   c. If $y[t]$ is proportional to t^3, then what function is $y'[t]$ proportional to? What function is $y'[t]$ inversely proportional to?

   d. If $y[t]$ is proportional to t^3 and $y'[1] = 7$, then give a formula for $y[t]$.

   e. If $y[t]$ is proportional to e^{3t}, then why is $y'[t]$ proportional to $y[t]$?

   f. If money is invested at interest $100\ r\%$ compounded continuously and left untouched, then why is the instantaneous growth rate of the account proportional to the amount in the account? Do large accounts grow faster than small accounts?

   g. If $h[t]$ is proportional to $f[t]$, then is $h'[t]$ proportional to $f'[t]$?

   h. If $h[t]$ is proportional to both $f[t]$ and $g[t]$, then is $h'[t]$ proportional to both $f'[t]$ and $g'[t]$?

- **9.** Write in mathematical symbols:
   a. In a certain controlled expansion of a gas, the pressure is inversely proportional to the square of the volume.
   b. The number of fruit flies is proportional to the food supply but is inversely proportional to the temperature.
   c. The strength of a beam is proportional to both its width and the square root of its depth.

- **10.** Explain why:
   a. The area of a equilateral triangle is proportional to the square of one of its sides.
   b. The area of a circle is proportional to the square of its diameter.
   c. If rectangular box with length x inches, width y inches, and height z inches is to contain a certain fixed volume V cubic inches, then area of the top is inversely proportional to z.

Student Writing

Here is a selection of lightly edited samples of student writing based on problems from the laboratory final project in first-semester calculus turned in by Calculus&Mathematica students at Illinois in December 1989. All students were freshmen in various majors including engineering, biology, general liberal arts and sciences, etc. Not everything these students said is correct or mathematically precise, but they did arrive at correct answers. Our goal is for the students to be able to solve the problem correctly and explain themselves to the best of their ability. Note that student writing is a strong component of Calculus&Mathematica, and that a clear organization of the solutions is strongly emphasized. To achieve these goals the authors of these solutions spent considerable amounts of time.

We have added some comments: they are contained in closed cells marked by "COMMENT" and can be displayed by opening them and in some cases, executing them.

□ *A standard optimization problem. The answer is by Ginny Mark.*

- **Problem:** Find the <u>highest and lowest</u> points on the graph of $f[x] = x^9 \exp[-x^2]$. (Recall $\exp[x]$ is another way of writing e^x.)

□ **Student's answer:**

```
>> Simplify [Exp[x] - E^x]
```

0

Oh, okay, thanks.

COMMENT: *Notice the "dialogue!" This is not unusual in this and other students who felt very much a part of the project and the real owners of the lab.*

```
>> f[x_]=x^9 Exp[-x^2]
```

$$\frac{x^9}{E^{x^2}}$$

```
>> Plot [f[x],{x,-5,5}, PlotRange->All,
   AspectRatio->Automatic]
```

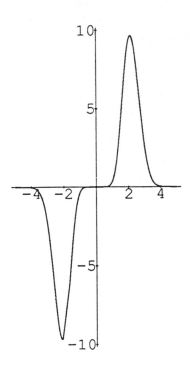

--Graphics--

The highest point appears to be around $(2.5, 9.5)$ and the lowest point $(-2.5, -9.5)$.

```
>> Plot [f'[x],{x,-5,5},PlotRange->All,
   AspectRatio->Automatic]
```

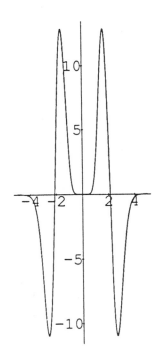

--Graphics--

```
>> FindRoot [f'[x],{x,-2}]
```

{x -> -2.12132}

```
>> FindRoot [f'[x],{x,2}]
```

{x -> 2.12132}

```
>> f[x]/.x->2.12132
```

9.66343

```
>> f[x]/.x->-2.12132
```

-9.66343

COMMENT: *Notice how the student gains confidence and goes from "appears" to "is." Also the whole thing is done without much apparent suffering, which is quite an accomplishment in itself.*

Our high point is $(2.12132, 9.66343)$ and our low point is $(-2.12132, -9.66343)$.

□ *Here is a first-class calculus problem suggested to us by Rod Smart of the University of Wisconsin. Answers to all parts by Cody Buchman.*

- **Problem:** You own the Calculus&*Mathematica* Steel Plate Company. In comes an order for 750 square steel plates each measuring 12 feet wide and 12 feet long.

```
>> border = Graphics[Line[{{-6,0},{-6,12},
   {6,12}, {6,0},{-6,0}}]]

>> Show [border, Axes->{0,0}, AspectRatio->
   Automatic, AxesLabel->{''x,'' ''y''}]
```

A drill (or router) is to be used to drill out a parabolic arch bounded by the parabola

$$y = -4(x - 3/2)(x + 3/2).$$

```
>> arch = Plot[{-4 (x-3/2) (x+3/2)},
   {x,-3/2,3/2}, DisplayFunction->Identity]

>> Show[border,arch,
   PlotRange->{{-6,6}, {0 ,12}},
   Axes->{0,0},Aspect-Ratio->Automatic,
   AxesLabel->{''x,'' ''y''},
   DisplayFunction->$DisplayFunction]
```

- **Part a:** You have a selection of router bits to do the job. The bits available to you come in the following cutting diameters: 2 inch, 3 inch, 4 inch, 5 inch, 6 inch. Which of these bits can be used to cut out the arch? Discuss how you arrived at your answer.

▢ **Student's answer:** First: to cut out such an arch, the bit must be no wider or narrower than the diameter of the smallest osculating circle. Why? We saw already that the smallest osculating circle of a parabola is always found tangent to the vertex of said parabola, and always fills the bottom of the parabola. If a bit larger than the diameter of this circle was used, it would cut a whole different, wider parabola, with possible ugly results (depending what the plates were being used for). Using a bit of smaller diameter would produce similar results. This will be illustrated later. For now, let's find the radius by setting the $f[x]$ equal to $-4(x - 3/2)(x + 3/2)$, and running it through the mill:

COMMENT: *The student is using here a Tutorial problem that describes how to find the osculating circle at any point.*

```
>> Clear [x,y,h,k,r,f,x0]

>> f[x_] := -4 (x-3/2)(x+3/2)

>> circleeqn = ((x-h)^2+(y[x]-k)^2 == r^2);

>> eqn1 = circleeqn/.{x->x0,y[x]->f[x0]};
```

```
>> firstderiveqn = D[(x-h)^2 + (y[x]-k)^2
   == r^2,x];

>> eqn2 = firstderiveqn/.{x->x0,y[x]->f[x0],
   y'[x]->f'[x0]};

>> secondderiveqn = D[(x-h)^2 + (y[x]-k)^2
   == r^2, {x,2}];

>> eqn3 = secondderiveqn/.{x->x0,y[x]->f[x0],
   y'[x]->f'[x0],y''[x]->f''[x0]};

>> hkrsolved = Solve[{eqn1,eqn2,eqn3},{h,k,r}]
```

$$\{\{r \to \frac{\text{Sqrt}[1+192\ x0^2 +12288\ x0^4 +262144\ x0^6]}{8},$$

$$h \to -64\ x0^3, \quad k \to \frac{71 - 96\ x0^2}{8}\},$$

$$\{r \to \frac{-\text{Sqrt}[1+192\ x0^2 +12288\ x0^4 +262144\ x0^6]}{8},$$

$$h \to -64\ x0^3, \quad k \to \frac{71 - 96\ x0^2}{8}\}\}$$

The first "r" is the correct one. Now let's solve for "r".

COMMENT: *What the student means to do here is to find the smallest radius to guarantee that the bit will fit everywhere.*

```
>> radius = hkrsolved[[1,1,2]]
```

$$\frac{\text{Sqrt}[1+192\ x0^2 +12288\ x0^4 +262144\ x0^6]}{8},$$

```
>> Solve [D[radius,x0] == 0,x0]
```

$$\{\{x0 \to 0\}, \{x0 \to \frac{I}{8}\}, \{x0 \to \frac{I}{8}\},$$

$$\{x0 \to \frac{-I}{8}\}, \{x0 \to \frac{-I}{8}\}\}$$

This could also be seen from the illustration, but the numerical proof is here. Now let's find the diameter by plugging back into r:

```
>> diameter =2 (N[radius/.x0->0])
```

0.25

Hmm. The bit that I need must be 0.25 feet—3 inches—in diameter.

- **Part b:** Choose a bit that will work and give a plot of the path of the center of the router.

▢ **Student's answer:** Let's start by solving for its slope:

```
>> y = -4 (x-3/2) (x+3/2)
```

$$4 \ (-(\tfrac{3}{2}) + x) \ (\tfrac{3}{2} + x)$$

```
>> yprime = D[y,x]
```

$$4 \ (-(\tfrac{3}{2}) + x) - 4 \ (\tfrac{3}{2} + x)$$

This is the slope for the tangent line at any x along the parabola. Now, let's say a single point on the bitpath is (s, t). This point is on a line whose equation is given above, and which lies a distance a from a point on the parabola, $\{x, y[x]\}$. This gives us the equation

$$\{x, y[x]\} + b\{-1, 1/y'[x]\} = (s, t),$$

where b is some multiple that increases as distance a increases. Now, we know that distance $a = 0.125$ (the radius of the osculating circle). Also, we can simplify the right term of the equation to $\{-b, b/y'[x]\}$, which represents the distance we must travel from the point $\{x, y[x]\}$, along the perpendicular line, to reach the point (s, t). Since we know this $= 0.125$, we get the equation $\sqrt{[b^2 + b^2/y'[x]^2]} = .125$, and can solve for b:

```
>> Clear [b]
```

Solving for b in the above equation gives:

$$b^2 = .125^2/(1 + 1/y'[x]^2),$$

or

```
>> b = .125 /Sqrt[1+1/yprime^2]
```

$$\frac{0.125}{\text{Sqrt}[1 + (-4 \ (-(\tfrac{3}{2}) + x) - 4 \ (\tfrac{3}{2} + x))^{-2}]}$$

Now, let's plug in "b" and plot:

```
>> pointt = {x,y}+ Sign[x] b {-1,1/yprime}
```

$$\{x - \frac{0.125 \ \text{Sign}[x]}{\text{Sqrt}[1 + (-4 \ (\tfrac{3}{2}(-)+x) - 4(\tfrac{3}{2} + x))^{-2}]},$$

$$-4 \ (-(\tfrac{3}{2}) + x) \ (\tfrac{3}{2} + x) + (0.125 \ \text{Sign}[x]) \ /$$

$$((-4 \ (-(\tfrac{3}{2}) + x) \ (\tfrac{3}{2} + x))$$

$$\text{Sqrt}[1 + (-4 \ (-(\tfrac{3}{2}) + x) - 4 \ (\tfrac{3}{2} + x))^{-2}])\}$$

```
>> par = ParametricPlot[{x,y},{x,-3/2,3/2},
   PlotPoints->75,DisplayFunction->Identity]
```

```
--Graphics--
```

```
>> router = ParametricPlot[pointt,
   {x,-3/2,3/2}, PlotPoints->75,
   DisplayFunction->Identity]
```

```
--Graphics--
```

```
>> Show[par,router,DisplayFunction->
   $DisplayFunction]
```

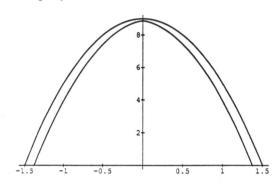

```
--Graphics--
```

Note that glitch at the top. This is caused by a round-off error in the plotting mechanism. To solve this, let's go in for a close-up of the top:

```
>> Show[par,router,PlotRange->{{-.2,.2},
   {8.8,9}}, DisplayFunction->
   $DisplayFunction]
```

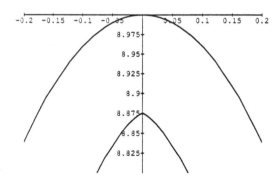

```
--Graphics--
```

Note the peak at the top of the drill path. This is logical, for when the drill reaches the point of maximum curvature of the parabola, it must abruptly shift directions to maintain the cut's shape.

□ *A problem dealing with accuracy of computation. Answer by Anne Bierzychudek.*

- **Problem:** If x is in $[-12, 12]$, then how many accurate decimals of x are needed to guarantee k accurate decimals of $f[x] = x^4 - 3x^3 + 5x^2 - 2x + 8$?

□ **Student's answer:**

```
>> Clear[f,x,y,z]
>> f[x_] = x^4 - 3x^3 + 5x^2 - 2x + 8
              2     3    4
8 - 2 x + 5 x - 3 x + x
>> Plot[Abs[f'[z]],z,-12,12,PlotRange->All]
```

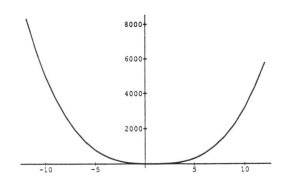

```
--Graphics--
```

The plot makes it clear that for z in $[-12, 12]$,

$$|f'(z)| < 10,000 = 10^4.$$

To get k accurate decimals of $f[x]$ for x in $[-12, 12]$, we need at least $k + 4$ accurate decimals of x.

COMMENT: *Notice that the student trusts the graph and what she sees. This allows her to obtain the correct solution without delay.*

□ *A standard problem on tangent lines. The answer is by Mabel Chiu.*

● **Problem:**

Describe all constants a, b and c such that the parabola

$$y = ax^2 + bx + c$$

passes through $\{0, 1\}$ and is tangent to the line $y = x$. Plot several of these parabolas and the line on the same axes.

□ **Student's answer:**

```
>> Clear[c,a,x,b,y]
```
Let's set the equation first:

```
>> y=a x^2 + b x + c
              2
c + b x + a x
```
Since the equation y and $y = x$ are tangent to each other they should share the same slope and touch at a certain point. So, the two equations are set equal to one another.

```
>> eqn1 = a x^2 + b x + c == x
```

```
        2
c + b x + a x == x
```
Since the two equations are tangent to each other their derivatives should be equal to one another. So the derivative of the two equations are set equal to each other.

```
>> eqn2 = D[(c + b x + a x^2),x]==D[x,x]

b + 2 a x == 1
```

There are three unknowns that must be found a, b, c. So, one more equation must be generated in order to get what we want. We have three unknowns so we need three equations in order to solve for a, b, c.

COMMENT: *Notice the comfort in deciding the correct number of needed equations. The student simply adds an equation assigning to the unknown variable c the only value it can take (because the parabola passes through $x = 0, y = 1$), and proceeds with the solution. This is usually lacking in standard calculus courses.*

```
>> eqn3 = c==1

c == 1
```
Now let's solve those three equations.

```
>> solved1=Solve[eqn1,eqn2,eqn3]

         -Sqrt[4 - 4 (1 - 4 a)]
{{x -> ----------------------,
                 4 a
       2 + Sqrt[4 - 4 (1 - 4 a)]
 b -> -------------------------, c -> 1},
                 a
        Sqrt[4 - 4 (1 - 4 a)]
 {x -> --------------------,
                4 a
       2 - Sqrt[4 - 4 (1 - 4 a)]
 b -> -------------------------, c->1}}
                 a
```
Since the variable "a" is first in the alphabet, the computer chose a as the variable that everything else should be in terms of since it wasn't specified. So, x, b, c is solved in terms of "a."

```
>> y

          2
c + b x + a x
```
Now, the solved values of x, b, c are plugged back into the equation "y" which is above. However, the plot will need the variable x to remain in the equation in order for it to be plotted since $f(x)$ is dependent on the variable x. So, x remains to be x. Do not solve for x in terms of a. It is not necessary. Look down. Only b and c is plugged into the equation.

COMMENT: *Here the student discards the possibility of using a as a parameter, or any fancy arrangement, and simply sticks with x as a variable. The successful answer justifies the choice.*

```
>> alt1 = y/.{b -> (2 + (4 - 4*(1 - 4*a))^
   (1/2))/2,c -> 1}
```

$$1 + \frac{(2 + \text{Sqrt}[4 - 4\ (1 - 4\ a)])\ x}{2} + a\ x^2$$

```
>> alt2 = y/.{b -> (2 - (4 - 4*(1 - 4*a))^
   (1/2))/2,c -> 1}
```

$$1 + \frac{(2 - \text{Sqrt}[4 - 4\ (1 - 4\ a)])\ x}{2} + a\ x^2$$

There are solutions for this equation. The values b and c are part of one solution. They go together. Let's try to plot both the solutions.

```
>> Plot[{alt1/.a->3,alt2/.a->3,alt1/.a->1,
   alt2/.a->1,alt1/.a->.4,alt2/.a->.4,x},
   {x,-2,2}]
```

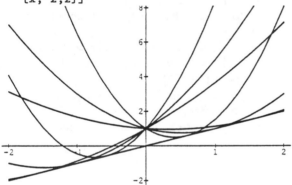

```
--Graphics--
```

Answer: Here is the plot of the two solutions along with the tangent line $y = x$ and random values for the variable a. All of the curves for equation "alt1" are on the left side of the graph. All of the curves for equation "alt2" are on the right side of the graph.

COMMENT: *The student talks about "two solutions" which is not really correct, but the construction is quite satisfactory and the two families of parabolas are neatly distinguishable.*

□ *A problem on Fourier approximation. The answer is by Rod Johnson, a student in the first semester calculus who had never heard of trigonometric polynomials.*

• Problem:
Find the constants a, b and c that make

$$\int_0^\pi (x(\pi - x) - (a\sin[x] + b\sin[2x] + c\sin[3x]))^2\,dx$$

as small as possible. After you have found a, b, and c, plot

$$x(\pi - x) \text{ and } (a\sin[x] + b\sin[2x] + c\sin[3x])$$

on the same axes for $0 \le x \le \pi$ and assess the quality of the fit.

□ Student's answer:
We will integrate this equation first.

```
>> Clear[a,b,c,x]

>> int = Integrate [(x(Pi - x) - (a Sin[x]
   + b Sin[2x] + c Sin[3 x]))^2,{x,0,Pi}]
```

$$\frac{\text{Pi}^2}{30} - 8\ a + \frac{\text{Pi}\ a^2}{2} + \frac{\text{Pi}\ b^2}{2} - \frac{8\ c}{27} + \frac{\text{Pi}\ c^2}{2}$$

All we have to do is take the derivative of the integral with respect to each constant, set it equal to 0, and solve for that constant.

COMMENT: *First semester calculus students do not normally consider maxima and minima of functions of several independent variables. We found, however, that this does not seem to be a conceptual difficulty, and here is an example.*

```
>> derva = D[int,a]

-8 + Pi a

>> Solve[{derva == 0},a]
```

$$\{\{a \to \frac{8}{\text{Pi}}\}\}$$

Now let's find b.

```
>> dervb = D[int,b]

Pi b

>> Solve [{dervb == 0},b]

b -> 0
```

Finally we can get c.

```
>> dervc = D[int,c]
```

$$-(\frac{8}{27}) + \text{Pi c}$$

```
>> Solve [dervc == 0,c]
```

$$\{\{c \to \frac{8}{27\ \text{Pi}}\}\}$$

To get

$$\int_0^\pi (x(\pi - x) - (a\sin[x] + b\sin[2x] + c\sin[3x]))^2 dx$$

as small as possible we will need to use $a = 8/\pi$, $b = 0$, and $c = 8/(27\pi)$.

Now let *Mathematica* know the values for a, b and c:

```
>> a = 8/Pi;

>> b = 0;

>> c = 8/(27 Pi).
```

We are going to plot the equation

$$(a\sin[x] + b\sin[2x] + c\sin[3x])$$

with these values for a, b and c so let's find out what that will be.

```
>> eqn2 = (a Sin[x] + b Sin[2x] + c Sin[3x])
```

$$\frac{8\,\text{Sin}[x]}{\text{Pi}} + \frac{8\,\text{Sin}[3\,x]}{27\,\text{Pi}}$$

Here is the plot of

$$x(\pi - x) \text{ and} (8\sin[x])/\pi + (8\sin[3x])/(27\pi)$$

from 0 to π.

```
>> Plot [eqn2,x(Pi - x),x,0,Pi]
```

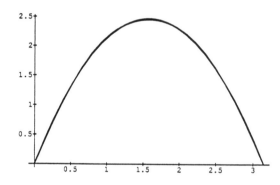

--Graphics--

Now that is a hell of a fit. That is looking real sweet the whole way from 0 to π.

□ *A problem from The Waterloo Maple Calculus Workbook. The answer is by Ann Brzozkiewicz.*
Problem: The Waterloo Tile Company is designing 1 foot by 1 foot ceramic tiles. On each tile, two fourth degree polynomial curves are running from the lower left hand corner to the upper right hand corner. They are positioned so that they trisect the square's right angles at the lower left hand and the upper right hand corners. A shade of red paint is to be applied above the top curve, white paint is to be applied between the curves, and a shade of blue is to be applied below the bottom curve. It is required that the red area = white area = blue area

= 1/3 square feet. Determine the equations of the polynomial curves and plot.

□ **Student's answer:**

• There are two curves and therefore two equations. One I will call $f[x]$ and the other will be $g[x]$.

```
>> Clear [a,b,c,d,f,g,k,l,m,n]

>> f[x_] := a x^4 + b x^3 + c x^2 + d x + e

>> g[x_] := k x^4 + l x^3 + m x^2 + n x + o
```

COMMENT: *In the next paragraph the student (engineering) makes a little inventory of the data and conditions of the problem. The handling of "number of equations = number of unknowns" is quite adequate and the explanations are remarkably clear.*

There are five variables so I will need five equations to solve each one. The things that I know are:

a. The equations move from $x = 0$ to $x = 1$, so for x equal to zero the function equals zero, and for x equal to one, the function equals one.

b. The slope of the tangent lines to the curves at 0 and 1 are given (because they trisect the corners of the tile).

c. The area between the curves must be equal to 1/3 square feet.

The following equations are therefore true.

$$f[0] = g[0] = 0$$

$$f[1] = g[1] = 1$$

$$f'[0] = g'[1] = \tan[\pi/3]$$

$$f'[1] = g'[0] = \tan[\pi/6]$$

$$\int_0^1 f[x] - g[x]\,dx = 1/3$$

This last equation and the information given also indicate that the area under the lower curve will be equal to 1/3, and the total area under the higher curve will be 2/3. Keeping this in mind I start out to make $f[x]$ the lower curve.

```
>> f1 = f[0] == 0

e == 0

>> f2 = f[1] == 1

a + b + c + d + e == 1

>> f3 = f'[0] == Tan[Pi/6]
```

$$d == \text{Tan}[\frac{\text{Pi}}{6}]$$

>> f4 = f'[1] == Tan[Pi/3]

$4 a + 3 b + 2 c + d == Tan[\frac{Pi}{6}]$

>> f5 = Solve[Integrate[f[x],x,0,1] == 1/3]

$\{\{a \rightarrow \frac{5}{3} - \frac{5 b}{4} - \frac{5 c}{3} - \frac{5 d}{2} - 5 e\}\}$

>> newf5 = a == $\frac{5}{3} - \frac{5 b}{4} - \frac{5 c}{3} - \frac{5 d}{2} - 5 e$

$a == \frac{5}{3} - \frac{5 b}{4} - \frac{5 c}{3} - \frac{5 d}{2} - 5 e$

>> N[Solve[{f1,f2,f3,f4,newf5},{a,b,c,d,e}]]

{{a -> -2.11325, b -> 4.5359, c -> -2.,
d -> 0.557735, e -> 0.}}

COMMENT: *Again, the student knows where she is
going and the plan is followed with a lot of elegance
and confidence.*

These are all my variables, I just need to substitute
them into my fourth power polynomial.

>> goodf = f[x] /.{a -> -2.11325, b -> 4.5359,
c -> -2, d -> 0.57735, e -> 0.}

$0. + 0.57735 x - 2 x^2 + 4.5359 x^3 - 2.11325 x^4$

>> Plot[goodf,x,0,1,PlotRange->0,1]

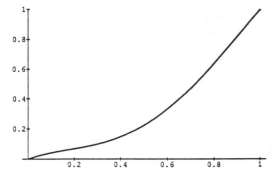

--Graphics--

It looks like it might be ok, let's proceed to g[x].

COMMENT: *A partial assessment, quite in tune
with the experimental character of the course.*

>> g1 = g[0] == 0

o == 0

>> g2 = g[1] == 1

k + 1 + m + n + o == 1

>> g3 = g'[0] == Tan[Pi/3]

$n == Tan[\frac{Pi}{3}]$

>> g4 = g'[1] == Tan[Pi/6]

$4 k + 3 l + 2 m + n == Tan[\frac{Pi}{3}]$

>> f5 = Solve[Integrate[g[x],x,0,1] == 2/3]

$\{\{k \rightarrow \frac{10}{3} - \frac{5 l}{4} - \frac{5 m}{3} - \frac{5 n}{2} - 5 o\}\}$

>> newg5 = k == $\frac{10}{3} - \frac{5 l}{4} - \frac{5 m}{3} - \frac{5 n}{2} - 5 o$

$k == \frac{10}{3} - \frac{5 l}{4} - \frac{5 m}{3} - \frac{5 n}{2} - 5 o$

>> N[Solve{g1,g2,g3,g4,newg5},{k,l,m,n,o}]]

{{k -> 2.11325, l -> -3.9171, m -> 1.0718, n
-> 1.73205, o -> 0.}}

>> goodg = g[x]/.{k -> 2.11325, l -> -3.9171,
m -> 1.0718, n -> 1.73205, o -> 0.}

$0. + 1.73205 x + 1.0718 x^2 - 3.9171 x^3 + 2.11325 x^4$

>> Plot[{goodf,goodg},{x,0,1},
PlotRange->{0,1},AspectRatio->Automatic]

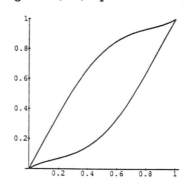

--Graphics--

COMMENT: *The next comment sums it up: the
student found an answer but she wants to verify
that it does what she wants to do. This is a very
healthy reflex that seldom appears in the standard
courses where the objectives are usually the repeti-
tion of techniques with little regard for the value of
the answers obtained.*

That's rather a funny looking tile, so to check my
answers I will integrate the different areas to see
if they meet the specifications. First, the area between the curves should be equal to 1/3.

>> Integrate[goodg - goodf,x,0,1]

0.333333

Got it. Another thing to check is that the area from
the top of the tile to the top function g[x] is also
equal to 1/3.

```
>> Integrate [1 - goodg,x,0,1]
```

0.333333

Got it again! Lastly, the area under the bottom curve should be 1/3.

```
>> Integrate [goodf,x,0,1]
```

0.333333

COMMENT: *A great final sentence:*

Right on the money! Now the ratio of the three areas red to white to blue is 1:1:1 and the tangent lines to the curves trisect the right angles.

□ *A problem on chemical reactions. All answers are by Maria Gonzales. Notice how she uses the word processor to do interactive algebra with Mathematica.*

• **Part i)** Two chemicals A and B react with each other and in the process one molecule from A bonds with one molecule from B to form a new compound. Let a be the number of molecules of A present at the start, and let b be the number of molecules of B present at the start. Let $y[t]$ be the number of molecules of the new compound present t seconds after the reaction starts.

At any time t, there are $a - y[t]$) molecules of A available for reaction, and there are $(b - y[t])$ molecules of B available for reaction. Since reaction requires collision of an $A-$ molecule and a $B-$ molecule, it makes sense to assume that $y'[t]$ is proportional to both

$$(a - y[t]) \text{ and to } (b - y[t]).$$

What differential equation does this lead to?

□ **Student's answer:**

COMMENT: *Here the answer is given with no justification. This is part of the course: the students should be able to communicate facts from their previous work, very much as in tests. This component of the course requires a little discipline because it is rather tempting to start each problem from scratch. The students got used to this type of question, but at the beginning they sometimes felt that they had not "answered enough." Several times we heard the question: Is that all we have to say?*

$$y'[t] = K(a - y[t])(b - y[t]) \text{ with } y[0] = 0.$$

• **Part ii)** Find a formula for $y[t]$ in terms of a, b and K. What is the limiting behavior of $y[t]$ if $a > b$?

□ **Student's answer:**
Rewrite the equation to get:

$$y'[t]/((A - y[t])(B - y[t])) = K.$$

```
>> Clear[y,t,x,K,a,b]
>> left = Integrate[y'[t]/((a-y[t])(b-y[t])),
   t,0,t]/.y[0]->0
```

$$-\left(\frac{\text{Log}[a]}{a-b}\right) + \frac{\text{Log}[b]}{a-b} + \frac{\text{Log}[a-y[t]]}{a-b} - \frac{\text{Log}[b-y[t]]}{a-b}$$

```
>> right = Integrate[K,t,0,t]
```

K t

Simplify the left side of the equation to get:

$$(1/(a - b))(\log[b(a - y[t])/(a(b - y[t])])$$

Now solve for $y[t]$:

$$(1/(a - b))(\log[b(a - y[t])/(a(b - y[t])]]) = Kt$$

$$\log[b(a - y[t])/(a(b - y[t]))] = (a - b)Kt$$

$$b(a - y[t])/(a(b - y[t])) = e^{((a-b)Kt)}$$

```
>> Solve [(b (a - y[t]))/(a(b - y[t]))
   == E^((a - b) K t),y[t]]
```

$$\{\{y[t] \to -\left(\frac{-(E^{K\,a\,t}\,a\,b) + E^{K\,b\,t}\,a\,b}{E^{K\,a\,t}\,a - E^{K\,b\,t}\,b}\right)\}\}$$

To find the limiting behavior of $y[t]$ if $a > b$, then simplify $y[t]$ by multiplying the top and bottom by $e^{(-Kat)}$:

```
>> y[t] = ((a b)(1-(E^(K b t))/(E^(K a t))))/
   (a - b(E^(K b t))/(E^(K a t)))
```

$$\frac{(1 - E^{-(K\,a\,t) + K\,b\,t})\,a\,b}{a - E^{-(K\,a\,t) + K\,b\,t}\,b}$$

Now if $a > b$ then the term $e^{(Kbt)}/e^{(Kat)}$ will be 0 leaving:

$$y[t] = ab(1 - 0)/a$$
$$y[t] = b$$

(since a will cancel). Therefore, the limiting behavior of $y[t]$ if $a > b$ will be approaching b.

Miami University

AN ALTERNATIVE CALCULUS

by *Tom Farmer and Fred Gass*

Abstract

Contact:

TOM FARMER or FRED GASS, Department of Mathematics and Statistics, Miami University, Oxford, OH 45056. PHONE: (513)-529-5818. E-MAIL: tfarmer@miavx1.bitnet or fsgass@miamiu.bitnet.

Institutional Data:

Miami University is a selective, residential, public institution in southwestern Ohio with about 16,000 students on the main campus at Oxford, Ohio. Undergraduate students choose to major in Arts and Science, Business and other programs in roughly 40:35:25 percentages; there is no school of engineering.

The Department of Mathematics and Statistics has flourishing programs at the undergraduate and graduate levels with about 40 majors per year graduating with Baccalaureate degrees and about 10 with Master's degrees. In addition, approximately 25 students per year earn the Bachelor of Science degree in Mathematics Education.

Calculus at Miami is taught primarily by regular PhD faculty (numbering about 40 on the Oxford campus) and by several part time and full time instructors who hold Master's degrees in Mathematics. Only a few sections of Calculus I per year are taught by graduate students.

Project Data:

In the second year of experimentation, during Fall term 1989, the alternative Calculus I course had 260 students in seven sections taught by seven different PhD faculty. The regular Calculus I course that term was taken by 1240 students in 28 sections. Computers tend to be used more in the experimental course than in the regular course, although computers are not the main focus of this project. Because of a state grant in 1986, each calculus classroom is equipped with an IBM PC and monitors and there is a PC lab for mathematics students. Graphical and numerical calculus software has been locally written and is available on all the computers. DERIVE is increasingly available but is not yet playing a significant role in calculus experimentation. The textbook currently used in all the calculus courses is Purcell/Varberg (5th edition). Planning and implementation of the alternative calculus course was facilitated by an internal grant during the summer of 1988 and by an NSF planning grant in 1988-89. The project currently has no external funding.

Project Description:

A calculus course which is conceptual, makes effective use of technology, and gives students a range of problem solving experiences is apt to be a difficult course for poorly prepared students. Therefore, this project targets an intended audience which is relatively well-prepared, consisting of first year college students who have had calculus in high school but who did not earn advanced placement or college credit. Because of the continued growth of high school calculus enrollments, this is a large and important segment of the college calculus population. Our approach to calculus reform is to experiment with an alternative course for this special audience with the expectation that successful experience with this course will lead to improvements in the other calculus courses.

The alternative Calculus I is a 4 credit course and has four 50-minute class periods per week as compared with the regular course which is 5 credits and meets 5 times per week. The rationale for the difference is that the students in the alternative course do not need as much coaching on the techniques of differentiation and integration so not as much class time is needed. The major syllabus change is to make early use of transcendental functions. As was suggested in "Toward a Lean and Lively Calculus", the students can be presented with a catalog of all the functions in the course right at the start. With some review of the precalculus properties of the trigonometric, inverse trigonometric, logarithmic and exponential functions in the first few days of the course, students are ready to focus on calculus concepts for the rest of the term.

One of the goals of the alternative course is to provide a more conceptual treatment of the subject. Since the students have typically practiced the techniques of differentiation and integration they seem to be willing to focus more on understanding the concepts and applying them to non-routine problems. Of course, each instructor has a unique view of what constitutes a conceptual treatment of calculus.

A significant part of the project concerns the uses of computers in teaching and learning calculus. A number of computer projects have been written and used by many of the instructors in both the alternative course and the regular sections of Calculus I and II. In general, each of these computer projects consists of a one or two page handout which includes some exposition together with several problems that can be solved with the help of any calculus software package. The usual practice is to have a computer demonstration in class to prepare the students for each assignment. In addition to the computer work needed for solving the problems, the assignments also require some understanding of calculus concepts, interpretation of the computer

output and some written explanation.

Materials available upon request include supplementary exercises on the calculus of transcendental functions, sample computer assignments, and calculus software for IBM PCs.

═══ Project Report ═══

Calculus at Miami University

The first semester of calculus at Miami University covers limits, derivatives, applications of derivatives, the definite integral with application to area, and the calculus of transcendental functions. The decision to cover transcendental functions instead of applications of integration in the first course is based on the fact that the Business students are required to take one semester of mainstream calculus and it is important for them to do logarithms and exponential functions. The standard course, numbered MTH 151, is a 5 credit course meeting 50 minutes each day. The rest of the standard calculus sequence consists of MTH 251 and MTH 252, both of which are 4 credit courses. The first covers applications and techniques of integration, improper integrals, sequences and series, conics, and two dimensional vectors, and the second is multidimensional calculus.

The alternative Calculus I course, MTH 153, is a 4 credit course meeting for four 50-minute periods per week. The students coming out of this course join the mainstream students in MTH 251. While the operation of the class may not be extraordinary (it varies with the instructor just as in all courses), what is unusual is that the prerequisite for this course is high school calculus with no AP credit. We intend to take advantage of what the students bring to the course and we want to give them a challenging, interesting course. It is the case that MTH 153 instructors tend to expect more of the students in this course than in the standard course: routine techniques should be learned with little time spent in class, students should be able to explain their solutions on exams, and computer assignments during the semester require some written analysis.

In addition to the alternative Calculus I course for students with high school calculus backgrounds, there is another option available in the Fall term for first year students. This is an alternative Calculus II course for incoming students who have AP credit for Calculus I. It is a 5 credit course (as compared with the usual 4) which, roughly speaking, spends one credit hour reviewing key ideas from Calculus I before covering the Calculus II material. In this way, we can make sure that these students have no serious gaps in their preparation for the Calculus II material and we can address their special needs as first term college mathematics students. To summarize, incoming students elect all or part of one of the sequences

    (a) 151-251-252
    (b) 153-251-252
or (c) 249-252

depending on whether they have (a) little or no calculus background, (b) high school calculus but no AP credit, or (c) AP credit for Calculus I.

Getting Started

This project began in 1986 when the Department of Mathematics and Statistics won a grant of $125,000 from the Ohio Board of Regents for the purchase of computers and monitors to equip 11 classrooms for demonstrations and to set up a student laboratory. Together with locally written calculus software, this equipment gave our calculus program computer capabilities on a large scale almost unequaled at other universities at that time. The challenge was to find out how to make effective use of the equipment in teaching and learning calculus.

The "Calculus for a New Century" meeting in October, 1987, was a catalyst which brought the authors together and convinced us to commit ourselves to calculus reform. Miami was in a unique position to be one of the nation's leaders in the effort and not only because of computer resources. The calculus program at Miami is staffed mostly by PhD faculty and by experienced instructors all of whom are uncommonly interested in the success of the courses (TAs taught 4 out of 35 sections of Calculus I in the fall term of 1989). And the calculus reform movement coming to campus was particularly timely since there has been a massive effort going on to redesign and improve the liberal education requirements of Miami University. Thus, there was the promise of widespread support in the department and throughout the campus for efforts to improve the teaching of calculus. Institutional support was evidenced by internal grant awards for the Summer of 1988 for the authors to write computer projects and other materials to be used in an experimental course in the fall term. The project was also given a boost and increased credibility on campus by receiving an NSF planning grant for 1988/89.

An important aspect of the Miami project is the emphasis on the high school calculus backgrounds of college students. The number of students taking calculus in high school has been growing rapidly in recent years and the percentage of students in beginning college calculus courses who studied calculus in high school but did not earn AP credit is significant and expanding. The usual reaction of college calculus teachers has been to wish that students would spend their high school years mastering precalculus instead of rushing to get to calculus. But rather than wishing for something that is not likely to happen, it is time to find ways to take advantage of the calculus background of these students—to make the best of the current trends. Our idea was that this group of students would be a perfect match for a conceptual, lean calculus course.

One of the goals of the NSF planning grant was to get input from faculty in other departments regarding calculus content. Calculus user departments were surveyed to determine the following.

1. How is calculus used in other undergraduate courses?
2. What applications in other disciplines could be presented in the calculus program?
3. Are there some calculus topics currently taught that are no longer used by anyone on campus?

On the basis of responses from several departments which indicated strong interest, follow-up interviews were arranged so that the authors could learn in detail about the ideas and concerns of individuals in Physics, Systems Analysis, Economics and Decision Sciences. In each case, the interview discussion covered calculator and computer usage, teaching methods, calculus content and calculus applications. Some conclusions derived from the survey and interviews include the following points.

- Suggested applications were in the areas of optimization, continuous probability, marginal concepts in economics and the modeling of physical properties. Few ideas were suggested which would be new to the calculus program and would be feasible. Colleagues in other departments expressed less interest than expected in having applications from their disciplines included in the calculus courses.
- There was concern about the abilities of students coming out of the calculus courses to analyze problems and represent them mathematically.
- There was considerable agreement that it would

be a good idea to reduce the number of topics covered in order to spend more time on understanding the underlying concepts.

• No user departments cared much about inverse trig functions; no other topics were particularly identified for exclusion from the courses.

Several valuable by-products of the process of surveying and interviewing faculty members from other departments can be identified. First of all, the process served to publicize the existence of a reform effort to people who are potentially interested in becoming involved in some way. In particular, these faculty members from other departments play an important role in advising their incoming students about the choice of calculus courses. Secondly, a number of survey responses indicated gratitude (and surprise) that they had been consulted; so a significant amount of good-will and support may have been generated. And finally, a few individuals were identified who showed the enthusiasm and knowledge necessary to be valuable consultants as the project continues.

Content

It was decided at the beginning of the project to operate under a constraint preventing the new Calculus I course from being radically different from the standard course. The constraint was that students coming out of the alternative course would be channeled into the mainstream Calculus II course. This meant, of course, that the syllabus for MTH 153 had to include that of MTH 151. Any topics deleted must be of minor importance as the calculus sequence continues. Also, there is not a free choice of text; it is assumed that all calculus courses will use the same text and a strong case would have to be made to do otherwise. Despite these disadvantages, the constraint was adopted because it was felt that conservative changes would be widely accepted by the department and would lead to evolutionary improvements in the whole calculus program.

There is just one significant difference in the syllabus of MTH 153 as compared with MTH 151 and it is that transcendental functions are introduced early and are used throughout the course. By contrast, in the standard course, these functions appear only in the last two weeks of the semester during which time the students face a large number of new facts including properties and differentiation formulas for exponential, logarithmic and

inverse trig functions. The results are often not very satisfactory. In particular, some students come away from the standard course with confused notions about what are the concepts of calculus and what are the properties of peculiar functions. If we do nothing else, we should leave students with a clear picture of the central concepts of calculus. In the alternative course, we begin the semester with a brief, graphical, precalculus review of all the algebraic and transcendental functions that will appear in the course. To facilitate the review, we hand out and briefly discuss a document which is a short catalog of functions. It lists samples of all the algebraic and transcendental function types, gives a computer-generated graph of each one, and lists one or two important observations about the graph. From that point on, with the introduction of each calculus concept, we choose examples and exercises involving transcendental as well as other functions.

The seven faculty members who were scheduled to teach sections of MTH 153 in Fall term, 1989, met several times during the preceding Spring to make plans and determine what supplementary materials should be produced. One of the goals of the meetings was to decide upon what was referred to as a minimal syllabus. The idea was to find a minimal path through Calculus I; that is, for each topic in the course, the question was asked: must students necessarily be exposed to this topic in order to understand the concepts of calculus and to succeed in Calculus II and beyond? It was hoped that there would be agreement that certain topics were not central to the course and could be treated as optional, thereby allowing for more class time to be spent dwelling on the remaining central topics. However, only these few optional topics were identified: arccos and arcsec functions, functions of the form $f(x)^{g(x)}$, logarithmic differentiation, and derivation of functional properties of logs and exponentials (the properties themselves are part of the prerequisites of the course).

Pedagogy

In order to achieve a Calculus I course which is more conceptual and captivating for students who have taken a calculus course in high school, pedagogy may be more important than syllabus content, use of technology, or any other factor. In all good mathematics courses from about fourth grade on, we must lead students to be more thoughtful about concepts, more adaptable to non-routine questions, and more tenacious in analyzing and solving complex problems. To accomplish this change in attitudes, instructors will surely have to change their habits too and, perhaps, spend less time lecturing and commit more time to planning appropriate classroom activities, writing and grading conceptual exams, and grading homework.

What are we trying to accomplish by our choices of teaching and grading methods in calculus? We want students to have a clear understanding of the main ideas of calculus. We want them to improve their problem solving abilities and to build confidence in problem solving. We hope that our students will progress in the area of critical and logical thinking. And it would be very valuable for them to practice working together and communicating their mathematical thoughts.

MTH 153 has offered an opportunity for faculty members to try out instructional methods different from the ones which have become the routine. Experimentation of this sort is particularly valuable because successful methods will often be transportable to other courses. Some pedagogical areas of interest to one or more of the instructors of the course are listed below.

- Collect homework. We all know this is valuable; the only question is how to find the time to correct it. We are trying all kinds of schemes including spot grading, randomized grading, and the use of student workers.

- Encourage working together on homework. The potential benefits for the students who combine calculus with socialization and who practice the language of calculus with each other completely outweigh the possible unfairness in homework grades.

- Do group-work in class. Well-planned group activities can be very satisfying and, if nothing else, a welcome break from the usual lecture. We have seen students eagerly stay overtime to complete group assignments, and not for the sake of grades. Students get to know each other, they talk about problem solving strategies, and the instructor gets a good chance to interact with the students. We have found that group-work is more feasible with stronger students and with class sizes of 30 or less.

- Use essay-style questions on exams and increase writing expectations. When the students write sentences they reveal their understanding and misunderstanding of concepts more clearly than they do in solving more traditional calculus prob-

lems.

- Do physical demonstrations in class. This is another way to break away from the lecture routine and give the students a memorable class period. As one example, we have demonstrated Torricelli's Law regarding the rate at which water drains from a hole in a tank. The students learned a little about modeling in the context of solving separable differential equations and it was one of the highlights of the course.

- Have students solve problems at the blackboard.

- Use conceptual questions on exams.

- Encourage learning from the text.

- Make computer related homework assignments and require written reports.

Technology

Our intention is to make good use of computers and calculators as tools which can aid in understanding the graph of an equation and can quickly carry out numerical computations (we are not yet using symbolic manipulation software in any significant way). We have written a dozen or more computer assignments for use in Calculus I and II. We hand out these one- or two-page assignments to the students and then give a short demonstration on the computer in class. The hand-out material contains notes regarding the classroom demonstration and several related exercises for the students to work out on a computer outside of class. The exercises often require the students to ponder and react to the computer results. What is to be turned in is invariably some hardcopy together with written commentary. For the most part, the computer assignments are independent of the calculus software. This is reasonable and desirable partly because the goal is to learn calculus concepts rather than to become adept in the use of particular software.

Since 1986, all of our calculus classrooms have been equipped with a built-in IBM PC with color graphics, a hard disk, and color monitors and we have very good, locally written, calculus software which is menu-driven and does most of the tasks that one would expect (but not symbolic manipulation). Having the equipment readily available for use in the classroom is important but it still takes careful planning to use the computer effectively. For one thing, it is very difficult for students to comprehend and to take notes on what happens so quickly in a computer demonstration. That is why we have chosen to prepare written materials to accompany the demonstrations corresponding to each computer assignment. The extent of use of the computer varies a great deal from one calculus instructor to another; some have the machine on every day and use it at least briefly to graph a function while others use it only for two or three topics in the course. The MTH 153 instructors tend to use the computer somewhat more than is done in the standard course; in part this is due to our agreement that the significant use of the computer would be one way that our students who had calculus in high school would get a different experience in college.

Despite our careful preparation and presentation of computer assignments, many students have found these assignments to be troublesome. The trouble is usually not in the use of the computer but is more likely to be in understanding the calculus or in finding the way through a multi-step problem or in writing up a suitable report. Mathematics students are unaccustomed to having problem assignments outside the usual setting within a section of the text, and they are certainly unaccustomed to assignments which require them to interpret and explain information obtained from a computer. The fact that students have trouble with these kinds of assignments is probably a good reason to do them.

Appraisal and Conclusions

Seven sections of the new MTH 153 course were offered during Fall term 1989. There were seven different PhD faculty instructors and about 35 to 40 students per section. The students selected this course on the basis of the catalog description and an advisory placement exam. The catalog description indicated simply that this is a Calculus I course which is intended for students who studied calculus in high school but do not have AP credit. There was no indication that this was in any way an experimental course. The placement exam has been in place for many years as a part of our summer orientation program for incoming students. There is an algebra-trig part and a calculus part and students were expected to have a certain range of scores on both parts in order to choose MTH 153.

It is difficult to list successes and failures in this project because there was so much variation across sections. However, it seems clear that the concept of offering a special course for this band of students (those with high school calculus but no AP credit) has been worthwhile and will be continued. Also,

the early introduction of transcendental functions, which was used by all the instructors, was clearly successful. That is, all the instructors found this to be an improvement over what happens in the standard course. We have no data to support a claim that the students benefit from this approach, but we do believe that to be the case. It is interesting and not surprising to find that many of the students who had high school calculus are quite weak in their understanding of transcendental functions; the review at the beginning of the course must be done carefully, and these functions have to be used frequently both in classroom examples and in homework.

In order to determine whether changes might need to be made before offering MTH 153 again next year, several sections of the 1989 course were surveyed. We asked students the following questions.

Please comment on the following aspects of the course. Do you find them very helpful? ... somewhat helpful? ... a waste of time? What could be done to improve them?
- Class lectures and discussion
- Homework assignments
- Computer assignments
- Working in small groups
- Quizzes and exams

To what extent has the calculus that you studied in high school helped you in MTH 153? How would you characterize the difference between MTH 153 and MTH 151? If you had it to do over, in what mathematics course would you enroll for the first semester? If you were the professor, what would you do to improve MTH 153?

Survey responses from one section of the course may not be representative of the other sections because of differences due to the instructor. Even so, some of the expressed opinions seem to be valid for the MTH 153 course as a whole. Almost all the students in one section indicated that class lectures, homework assignments, quizzes and exams were helpful or somewhat helpful. But nearly a third of the class responded that the computer assignments were a waste of time. Comments to the effect that the problems were confusing, irrelevant or impossible were typical. This class had 5 computer assignments during the term each of which should require no more than 2 or 3 hours to complete. In view of the student remarks, it appears that it would be an improvement to spend some class time in the computer lab or have an assistant in the computer lab to work with the students helping them see the relevance of the assignments and removing some of their frustration in understanding the computer output.

The students were practically unanimous in pointing out that the calculus they learned in high school was vital in preparing them for this course. Students tended to have the opinion that MTH 153 was harder and more theoretical than MTH 151 and required more effort. Nearly half the students in this section claimed that if they had it to do over again they would take MTH 151. The common reasons were that they could then get a higher grade or that they could get one more hour of credit with the same grade. On the other hand, (as the optimist would say) more than half the students said they would again choose MTH 153.

Future Plans

The near future of the calculus program at Miami University is now in the hands of a calculus committee that was formed in response to the Miami liberal education revision as well as to the calculus reform movement. To help bring the teaching of calculus into the mainstream of these currents, the committee has begun to prepare special materials for all instructors and students. Among the things to be included are a guide to writing mathematics and many recommended examples of computer assignments, group work, conceptual problems and logic exercises. We expect some of the suggested ideas to become a standard part of the calculus syllabus in all the calculus courses. Also, the early use of transcendental functions is well accepted in MTH 153 and is supported by supplementary material developed this past year. We anticipate that faculty will continue to experiment with teaching methods in MTH 153 and that the successful results will gradually be adapted to the other courses in the program.

═══ Sample Materials ═══

MTH 153—Brief Syllabus

Classes

1-4 Structure of the real number system; catalogue of functions; review symmetry, translations and other graph properties; pretest on functions and derivatives.

5-9 Limits; use of graphs, calculators and informal reasoning; ϵ/δ definitions; one-sided, two-sided limits; limits involving ∞.

10-24 Derivatives; rate-of-change and slope interpretations emphasized; notation, rules, higher-order derivatives, differentials, implicit differentiation, related rates, Newton's method.

25-36 The shape of a graph; max-min; graph-sketching; Mean Value Theorem.

37-52 Antiderivative; introduction to differential equations; area problems to motivate the definite integral; Fundamental Theorem of Calculus (both versions); numerical integration; area of a plane region.

53-56 Focus on transcendental functions while reviewing course topics.

Notes

1. Transcendental functions will be discussed throughout the course, and several computer assignments are made.

2. The classes numbered above include hours given over to exams (usually 3 or 4), as well as an occasional day free for an enrichment topic.

3. A full semester includes 60 meetings of MTH 153 (in practice, 59). The final exam is scheduled separately for a 2-hour period during final exam week.

Graphing: Adjusting the Window.

Objective: To learn how to explore a graph by adjusting the window that the computer uses to display a portion of the graph.

Prerequisites: None

Example: When you graph a function on a sheet of paper or on a computer screen, you necessarily restrict the coordinates to certain intervals $x \in [a, b]$ and $y \in [c, d]$. In computerese, this rectangle of values is sometimes called the window. What you draw or see on the screen is the portion of the graph that happens to lie in this window. The problem is to adjust the window to gain a good view of the interesting features of the graph. Figure 1 shows four different windows on the graph of $y = \sin(x)$; clearly some views are more informative than others depending on the type of information that is needed.

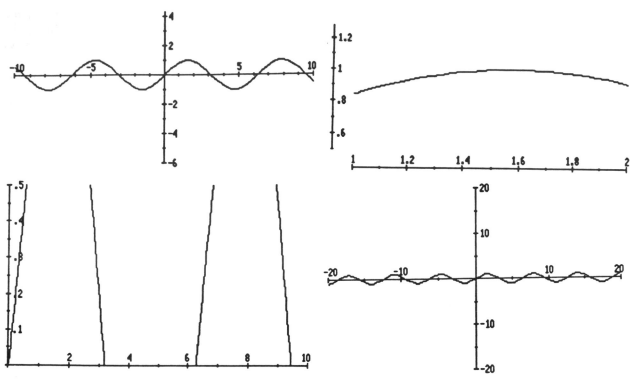

Figure 1

Homework: Solve each of the following problems by adjusting the window in a function graphing program, "zooming in" to get accurate results. [If your software does not have a "zoom" command, you will have to adjust the window yourself.] Your goal is to have the answers to the questions clearly pictured by the graphs you hand in.

1. Estimate the minimum value of $f(x) = 4/\cos x + 4/\sin x$ on the interval $(0, \pi/2)$. Give both x and y coordinates to 2 decimal place accuracy. [This problem is related to the question "Can a 10 foot pole be carried through a right angle turn in a hallway 4 feet wide?"]

2. Maximize the function $f(x) = 4x^7/(2x^{16} + x + 1)$ on the interval $[0, 2]$. Try to get 3 decimal place accuracy in both the x and y coordinates.

3. Let $f(x) = \cos(x)/x$. Find the smallest positive integer N so that $|f(x)| < .1$ whenever $x > N$.

Directions: Hand in two graphs for each of the assigned problems. One graph should show a large portion of the function while the second should be a close-up showing the aspect of the graph that answers the question of the problem. The graphs must be accompanied by a written statement of your conclusions expressed in complete sentences.

Introduction to Limits

Objective: To estimate limits both by examining a graph and by looking at a table of values. To determine how close x must be to its limiting value to produce a value of the function that is within a prescribed distance of its limit.

Prerequisites: Intuitive introduction to limits.

Example: Estimate $\lim_{x \to 0} (1 + x)^{1/x}$ by producing both a graph and a table.

F(X) = (1+X)^(1/X)

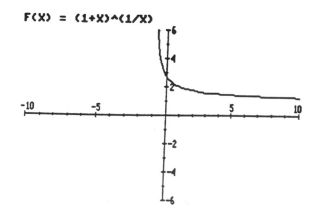

The table suggests that the limiting value is about 2.718, which is consistent with the graph. Notice that if $x = \pm.01$, then $F(x)$ differs from the limit 2.718 by more than .001; if $x = \pm.001$, then the difference is still greater than .001; but if $x = \pm.0001$, then the difference is less than .001. The graph and the table support the conclusion that $|F(x) - L| < .001$ if x is within about .0001 of 0.

TABLE OF VALUES FOR
F(X) = (1+X)^(1/X)

X	F(X)
1	2
.5	2.25
.25	2.4414
.125	2.5658
.1	2.5937
.01	2.7048
.001	2.7169
.0001	2.7181
.00001	2.7183
-.00001	2.7183
-.0001	2.7184
-.001	2.7196
-.01	2.732
-.1	2.868

Homework: Estimate the values of the following limits if they exist. Obtain answers which seem to be accurate to 3 decimal places. Produce both a graph and a table of values that support your answers. In each case where $\lim_{x \to c} f(x) = L$ determine roughly how close x must be to c to get $|f(x) - L| < .001$. If the limit does not exist, explain how you know.

1) $\lim_{x \to 0} \dfrac{\tan x}{2x}$

2) $\lim_{x \to \pi} \dfrac{1 + \cos x}{\sin 2x}$

3) $\lim_{x \to 1} \dfrac{\sin 3x}{x - 1}$

4) $\lim_{x \to 0} \dfrac{\arctan x}{x}$

5) $\lim_{x \to 1} \dfrac{\ln x}{1 - x}$

6) $\lim_{x \to 0} \sin\left(\dfrac{\pi}{x}\right)$ Challenge!

Directions: Write a careful, neat report containing your responses to the assigned problems. Write your report in complete sentences explaining what you did and how you reached your conclusions.

Graphing: Functions & Derivatives

Objective: To learn through the graphs of f, f' and f'' how derivatives indicate characteristics of a function.

Prerequisites: Maximum and minimum points, inflection points, intervals where the function is increasing or decreasing and concavity.

Example:

1. Print out an accurate graph of the function $f(x) = \ln x - \sin x$, $x \in (0, 6]$, $y \in [-3, 3]$. Notice that on this interval, this function has a root, an extreme point and inflection points. Make large dots at these points on the graph. Use the scale on the graph to roughly estimate and label in pencil the coordinates of the root, the extreme point and the inflection points.

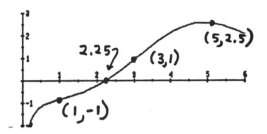

2. Graph f, f' and f'' in one coordinate system and label the intervals where f is increasing/decreasing and the intervals of concavity. (It is allowable for the graph of f' or f'' to go out of the window but select a window that shows all the important features of f.)

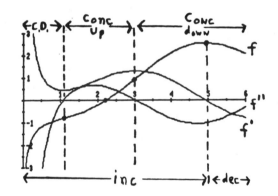

3. By looking at graphs through various windows, obtain better estimates (2 decimal places) of the root, extreme point and inflection points. Use these new values to update the labels on your graph from part 1 above.

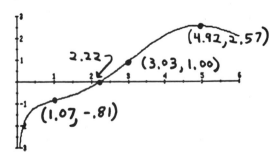

Homework: For each of the functions below, do the same as in 1, 2 and 3 above using appropriate windows.

a. $f(x) = x^3 - 4x^2 + x + 2$, $x \in (-\infty, \infty)$

b. $f(x) = e^{-x}(1 - 3\cos 2x)$, $x \in [-1, 3]$

c. $f(x) = 3x^4 - 8x^3 - 5x^2 + 4x - 15$, $x \in (-\infty, \infty)$

d. $f(x) = \ln x + \cos x$, $x \in (0, 6]$

e. (Personalized cubic; can this curve be your *unique* mathematical signature?) Let $a_1a_2a_3 - b_1b_2 - c_1c_2c_3c_4$ denote your social security number and define $a = (a_1 + a_2 + a_3)/27$, $b = (b_1 + b_2)/9$ and $c = (c_1 + c_2 + c_3 + c_4)/18$. Your personalized cubic is $f(x) = x^3 + ax^2 - bx - c$, $x \in (-\infty, \infty)$

Directions: For each assigned problem you should hand in no more than two pages of computer print-out suitably labelled by hand. At the end of the assignment, write three observations regarding the relationships between the graphs of f, f' and f''.

Computer Assignment

(1) Starting from point A in a lake, 2 miles from the straight shoreline CB, a person will row to shore and then walk along the shore to point B. The person can row at a speed "a" mph and walk at a speed "b" mph where $a < b$. Find the path which requires the minimum time to travel. Use $a = 3$ and $b = 4$.

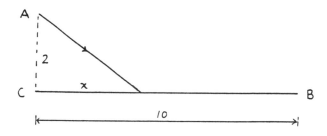

(2) Use the computer to draw a graph of the function $t(x)$ (travel time as a function of x) from problem (1) and zoom in to verify the values obtained in (1) for the minimum-time path.

(3) Generalize the function $t(x)$ by letting the speeds be constants (or parameters) a and b. Use the computer to find the minimum-time path for these cases:

a	b
4	5
9	10
9.9	10

(4) Challenge. Find the smallest ratio a/b which causes the minimum-time path to be the straight line from A to B.

Computer Problems

Due: Dec. 12, 1989

This assignment may be run on any personal computer having "BASIC" capability and a printer. The directions indicate using the program given below, but if you wish you may use the PC MTH 151 diskette available in BAC 106 and the Upham Hall computer lab.

Obtain a printout of the work requested below.

— — — — — — — — — — — — — — — —

1. Enter the program as given and run. (This estimates $\int_0^1 4/(1+x^2)\,dx$ using Simpson's Rule with $n = 10$. Running the program the first time will serve as a test since the exact value of the integral is known to be π). Modify statement 10 to try $n = 50$.

2. Modify lines 10, 40 appropriately and run the program in order to estimate $\int_0^4 e^{-x^2}\,dx$ using $n = 10$. Modify statement 10 to use $n = 100$ and run.

3. Suppose you are faced with the task of estimating $\int_0^4 \sin^2(1000\pi x)dx$ using numerical integration (e.g., Simpson's Rule). (So that we can have an idea as to how well we are doing, let me tell you that the exact value of the integral is 2.) First try Simpson's Rule with $n = 200$ and then answer the two parts below. It may be handy to use 3141.5926536 as a good approximation for 1000π.

(a) Explain what happened. Why is the answer so far off?

(b) Use the fact that this is a periodic function and try to find another approach (still using Simpson's Rule in some way) that will get an accurate answer. Be sure to explain what you did.

_ _ _ _ _ _ _ _ _ _ _ _ _ _ _ _   _ _ _ _ _ _ _ _ _ _ _ _ _ _ _ _ _

Simpson's Rule Program

```
Ok

10 DATA 0,1,10
20 READ A,B,N
30 H=(B-A)/N
40 DEF FNG(X)=4/(1+X^2)
50 S=FNG(A)+FNG(B)+4*FNG(B-H)
60 FOR K=1 TO N-2 STEP 2
70 S=S+4*FNG(A+K*H)+2*FNG(A+(K+1)*H)
80 NEXT K
90 PRINT "The Simpson's Rule estimate is";
   (H/3)*S
Ok

run
The Simpson's Rule estimate is 3.141593
Ok
```

Computer Problem

Due: November 21, 1989

This assignment does not require the use of a diskette and may be run on any personal computer having simple "BASIC" capability and a printer. The attached sheet titled "Directions for using the PC MTH 151 Programs" contains some useful information but directions 1-8 may be essentially ignored. If you wish, you may try the PC MTH 151 diskette to see what is available or to try the directions and explanations for the IBM PC Keyboard.

Obtain a printout of the work requested below. When the printer is ready (shift) PrtSc will command the printer to copy what is on the screen. The keys ctrl PrtSc (together) will command the printer to copy as you enter on the keyboard. Repeating ctrl PrtSc will turn the printer off.

1. Enter the program as given and run. (This estimates a solution of $2\sin(x) = x$ using initial value $x_1 = 2$.) Modify statement 10 to try an initial value of 1 and run. Explain what happened.

2. Modify lines 10, 30 and 40 appropriately in order to approximate the critical point(s) of f given by

$$f(x) = x^4 + 2x^2 - x + 2$$

and then approximate the minimum value of f. (You may wish to just use a calculator for this part.)

Newton's Method Program

```
10 X=2
20 FOR N=1 TO 10
30 Y=2*SIN(X)-X
40 Y1=2*COS(X)-1
50 Z=X-Y/Y1
60 PRINT Z
70 X=Z
80 NEXT N

run
1.900996
1.895512
1.895494
1.895494
1.895494
1.895494
1.895494
1.895494
1.895494
1.895494
Ok
```

Class Experiment

Two sections of MTH 153 used this activity as an experiment performed by student volunteers while their classmates looked on and recorded the observed values of h. The equipment consisted of a two-liter plastic soft drink bottle, whose midsection is essentially cylindrical, with a ruled strip of masking tape attached to the side for depth measurements. Each time the demonstration was performed, the results were in astoundingly close agreement with the Torricelli predictions.

PROBLEM. A small hole is drilled in the side of a cylindrical container and the height of the water level (above the hole) goes from 10 cm down to 3 cm in 68 seconds. Estimate the height at intermediate times.
- The linear model $dh/dt = k$, $h(0) = 10$, $h(68) = 3$ can be seen to have the approximate solution $h(t) = (.103)t + 10$.
- The Torricelli model $dh/dt = k\sqrt{h}$, $h(0) = 10$, $h(68) = 3$ has approximate solution $h(t) = (.00044)t^2 + (-.133)t + 10$.

The table below gives the values of h predicted by each of these models for various intermediate times. We will fill in the column of observed values when we perform the experiment.

Time t	Linear h	Torricelli h	Observed h
0	10.0	10.0	
10	9.0	8.7	
20	7.9	7.5	
30	6.9	6.4	
40	5.9	5.4	
50	4.8	4.4	
60	3.8	3.6	
68	3.0	3.0	

Group Work in Class

We experimented occasionally with having students work together during class time. "Guidelines" was part of the course information sheet handed out on the first day of class. It was necessary to review those points a couple times during the course, especially after a midterm evaluation on which a few people complained that some others were either grossly unprepared for work or were less than helpful. On the whole, though, group work was satisfying enough to warrant continuation. The students were noticeably more lively on those occasions, and they experienced some worthwhile mathematics and interaction.

Guidelines for Work Done in Small Groups

Occasionally we will divide up into small groups for the purpose of discussing problem solutions or even <u>finding</u> problem solutions, as in a quiz. The ability to work cooperatively and efficiently in groups can be a valuable asset, both professionally and nonprofessionally. Consequently I hope you will approach our group opportunities with the thought and energy that they deserve.

Here are some things that can be expected of group members:

a) Prepare yourself in advance as well as possible for the task at hand.

b) Be willing to share the knowledge you have, and offer it when appropriate.

c) Be willing to ask for help and pursue questions until you understand the answers.

d) Be sure that every group member sees work that is done toward a solution. Even members who do not specifically request to see the work should be offered the opportunity. For groups of more than four, I <u>strongly</u> urge that all work be written on the blackboard.

Examples of Group Assignments

10/6/88 When we're ready to start, please join the other members of your group at the blackboard and use your group time to discuss the solution of the Farmer Brown problems (4.1/23, 25). Also think about possible variations, such as 4.1/24. After the group discussions you will each take a quiz on a Farmer Brown variation, and if a majority of your group get it right, then everyone in the group will receive a bonus point.

The class was partitioned into seven groups of four or five members, each group corresponding to one letter of the word "PROBLEM". While students read their group assignments and the instructions above, the instructor wrote those seven letters at various locations on the blackboard.

11/3/88 Find the approximate length (in the same units that would appear on the coordinates axes) of the graph of $y = x^3$ from (0,0) to (2,8). For each correct decimal place in your answer, team members will receive a bonus point. Also, team members of the team that reports the most accurate answer will receive an additional bonus point. The work you submit should show exactly how your result was achieved.

The main intent of this exercise was to have students discover (or <u>re</u>discover, based upon their previous calculus) the idea of approximation by a sum. There were six groups of six or seven members, and they needed about 25 minutes to complete the exercise.

11/11/88 The graph of $x^2 + 4y^2 = 16$ is an ellipse (see diagram). Your job is to find the approximate area of the region inside it, and you may use any reasonable method other than a formula for the area of an ellipse (if you happen to know one). First gather at the board to discuss the problem. You may use no books or other aids except calculators. Have a group secretary record your method and results to hand in. Bonus points will be awarded for accuracy.

12/1/89 *Below are parallel narratives that deal with the idea of Riemann sum approximation. Seven groups of five members worked on the (a) problem (velocity-distance) that day, the prize being 100-peso coins, which bear the mark "$100."*

I (a)

You have just made a 5-hour trip by car and used the cruise control to maintain a constant speed most of the time. Table A shows the time intervals on the left and the mi/hr you maintained during each interval on the right. To figure the total distance traveled use the familiar formula *distance = rate × time* and calculate the Riemann sum in line (B).

I (b)

You have just moved a heavy piece of equipment 5 feet along the floor by means of four steady shoves. Table A shows the interval of distance moved with each shove and the force in pounds that you maintained during each shove. To figure the total amount of work accomplished, use the simple formula *work = force × distance* and calculate the Riemann sum in line (B).

TABLE A:

subinterval	speed or force
$[0, 1]$	50
$[1, 3]$	65
$[3, 4.5]$	55
$[4.5, 5]$	40

(B)   Total $= 50 \times 1 + 65 \times 2 + 55 \times 1.5 + 40 \times 0.5$
       $= 282.5$ (miles or foot-pounds).

II (a)

The purpose of your trip is to observe a particle-acceleration experiment in which the speed of the particle is allowed to vary. The time interval for this experiment is $[0, 5]$, and if x denotes the number of seconds that have elapsed, then the speed is given by the function $f(x)$ in line (C). What is the approximate distance traveled by the particle during this experiment?

II (b)

You moved that object as part of a robotics demonstration in which a robot furniture-mover is going to push the object back to its original place. If x is the distance moved so far, then the force being applied by the robot at that point is given by $f(x)$ in line (C). How much work, approximately, does the robot do?

(C) $f(x) = 9 + 9 \times SQR(30 - x^2), \ 0 \le x \le 5.$

III.

Operating in teams so as to share the work and check each other, answer the question raised above as accurately as you can, given the constraints of time. Make sure everyone in your group knows exactly what you plan to do and who's going to do what. When time is up, be ready to report your result and show work that supports it. The prize for the group with the most accurate answer is $100 apiece (sort of).

12/15/89 *Five groups of seven members were determined by a random drawing of cards as students arrived. The groups worked on Numbers 1,3,4 and 5 at the blackboard.*

1. Here's the report of a test driver after a braking test on a straight track: "I accelerated smoothly from rest and within 15 seconds was cruising at 60 mph. At about 90 seconds into the test I entered the braking area and applied the brakes strongly. The car decelerated but was unable to stop in time. It struck the concrete barrier at a speed of approximately 5 mph."

Your part in this problem is to draw three graphs that might represent the motion of this test drive—one showing the position of the car as a function of time, one showing the velocity as a function of time and one showing the acceleration as a function of time. Put appropriate units on the axes of your graphs.

2. The figure below shows the graph of the function $f(t)$, but the questions in this problem are stated in terms of the function $F(x)$ defined by the formula $F(x) = \int_0^x f(t)dt$.

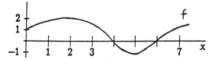

Estimate the answers to these questions on the basis of the graph.

a) What is $F(0)$?
b) For which numbers x does $F(x) = 0$?
c) At which number x does $F(x)$ have its minimum value?
d) At which number x does $F'(x)$ have its minimum value?
e) At which number x does $F''(x)$ have its minimum value?

The remaining problems contain statements about functions f and g, which we assume to have first and second derivatives defined for all real numbers. In each problem, if you believe the statement is true, then make a sketch that illustrates the statement; but if you believe the statement is false, then make a sketch that shows a counterexample to the statement.

3. If $f(x) \le g(x)$ for all x, then $f'(x) \le g'(x)$ for all x.

4. If $f(x) \le g(x)$ for all x, then $f''(x) \le g''(x)$ for all x.

5. If $f(x) \le g(x)$ for all x, then

$$\int_0^1 f(x)\,dx \le \int_0^1 g(x)\,dx$$

Final Exam for MTH 153E

1. If $G(x) = e^{2x}\cos(3x)$, find $G'(1)$ correct to 2 decimal places.

2. Evaluate $\int_{-\pi/2}^0 \cos x\sqrt{1 + \sin x}\,dx$ by means of a substitution.

3. Meterologists claim that in a certain air mass the temperature T and altitude h are related: the rate of change of temperature relative to altitude is proportional to the product of the altitude and the logarithm of the temperature. Express that relationship as a differential equation.

4. This problem deals with the integral

$$\int_1^4 1/(x - 5)\,dx.$$

(A) Evaluate the integral and give a simplified exact answer.

(B) Draw a diagram that clearly shows the graph of $1/(x - 5)$ and the region in the plane that this integral pertains to. Then tell exactly how the value of the integral is related to that region.

5. This problem concerns $\lim\limits_{h \to 0} \dfrac{(4 + h)^{1/2} - 2}{h}$.

(A) This is actually the definition of $f'(c)$ for which function $f(x)$ at which number c?

(B) Using whatever method you like, find the value of this limit.

6. Let f be the function whose graph is shown here. I want you to write down all limit statements, both one-sided and two-sided, that describe the behavior of f as shown in the diagram at $-\infty$, $+\infty$ and each point where f is discontinuous.

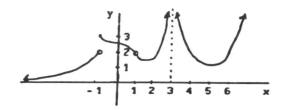

7. Suppose f is a function for which $f''(x) = x^{1/3} + e^{2x}$. Find an expression that shows the entire family of possible formulas for $f(x)$.

8. Use integrals containing "$f(x)$" and "$g(x)$" but no absolute values to set up an expression that represents the total area of the cross-hatched region in the diagram.

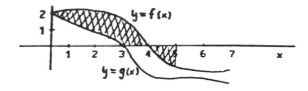

9. When a photo is taken with the aid of a flash-bulb, the rate at which light is produced varies during the time interval of the flash, as shown in the graph and the table below it.

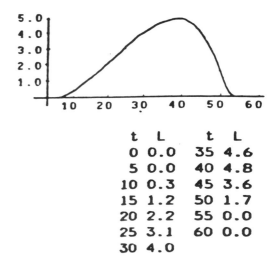

t	L	t	L
0	0.0	35	4.6
5	0.0	40	4.8
10	0.3	45	3.6
15	1.2	50	1.7
20	2.2	55	0.0
25	3.1	60	0.0
30	4.0		

(Light output is measured in millions of lumens and time in milliseconds beginning at the moment of ignition.) The amount of light that reaches the film is measured by the area under the $L(t)$ curve, beginning at the time the shutter opens and ending when it closes. Use this information and calculus ideas we have studied to find the most accurate answers to the questions below.

(A) If the opening and closing of the shutter occur at $t = 20$ and $t = 60$, how much light would reach the film?

(B) What is the average light output of the flash-bulb while the shutter is open?

10. Here is the graph of a function f. Suppose that another function h satisfies the conditions $h(0) = 0$ and $h'(x) = f'(x)$ for all x. Sketch the graph of h on this same coordinate system.

11. A certain object starts at position 0 and moves on a coordinate axis. Let s, v and a be its position, velocity and acceleration, respectively, at each time t, $t \geq 0$. Below is the graph of v as a function of t for $0 \leq t \leq 4$. Your job is to make two careful drawings, one showing a possible graph of s as a function of t and the other showing a possible graph of a as a function of t. In addition to being compatible with the given graph of v, your graphs should be made on separate coordinate systems using the same scale as the given graph. (Notice that the vertical scale has been compressed a bit to save space.)

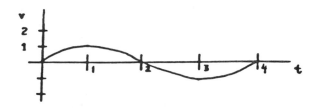

12. Let $f(x) = x^4 - 4x^3 - x$. (A) Using limits, find and briefly explain a reason why the function f clearly cannot possibly have an absolute maximum value. Any claim you make about f should be supported by specific evidence. (No graph sketch or work with f' is necessary!) (B) f has just one critical number, which is near 3. Use Newton's method to find it with 2-decimal-place accuracy. (C) Using information about f', justify the claim that at this critical point of f there is an absolute minimum value of f. (D) Repeat part (C), this time using information about f''.

Grade Distribution:

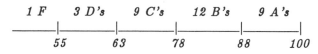

Final Exam for MTH 153C

1. Carefully draw the graph of a function f with domain $(-1, \infty)$ that satisfies all of the conditions listed below.

$$f(0) = 2, \quad f'(0) = 0, \quad f''(0) = 4,$$

$$f(3) = 2, \quad f'(3) = 1, \quad f(4) = 2,$$

$$\lim_{x \to -1^+} f(x) = +\infty, \; \lim_{x \to 2} f(x) = 1, \; \lim_{x \to \infty} f(x) = 2.$$

2. For $y = f(x)$, the derivative can be defined by $f'(x) = \lim_{\Delta x \to 0} \dfrac{\Delta y}{\Delta x}$. Draw a picture that illustrates this definition, showing an interesting function f (just a graph, no formula necessary), a point x, and Δx and Δy and anything else that seems pertinent. Be sure that everything is labeled for identification.

3. The figure below shows the position versus time graph for a certain object moving on a straight line.

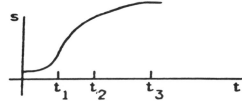

A. Is the object moving faster at time t_1 or at time t_2? How do you know?

B. What is the initial velocity ($t = 0$) of the object? How do you know?

C. During the time interval $[t_2, t_3]$, is the object speeding up or is it slowing down? How do you know?

D. Make a careful sketch of the velocity versus time graph for the object.

4. This problem deals with the differential equation $dy/dx = y(x^{-2})$.

A. Use the method of separation of variables to find the general solution (involving a constant C) for the equation.

B. Now find the particular solution that satisfies the condition $y = 3$ at $x = 1/2$.

5. Let $f(t)$ be the function graphed below. In each part below, be sure that your method and reasoning are made clear to the reader.

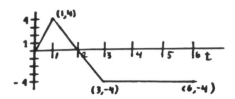

A. Use the method of areas to find $\int_0^6 f(t)dt$.

B. Use the Fundamental Theorem of Calculus to evaluate the integral in A. (Hint: Divide $[0,6]$ into sections and find a simple formula for f on each.)

C. If $F(x) = \int_0^x f(t)dt$ for $f(t)$ shown at the start of this problem, what is $F'(4)$?

6. An artificial lake has the shape shown in the figure below, with six separate length measurements made at intervals of 20 feet. Your job is to obtain as accurate an approximation as you can for the surface area of the lake. Explain in complete sentences how your solution is related to the material of MTH 153.

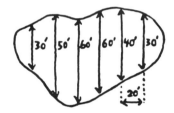

7. A herd of 100 deer is relocated to a small island, and the population begins to grow. Earlier data suggest that its growth will closely fit the formula

$(*)$ $P(t) = -t^4 + 21t^2 + 100$

where P is the size of the population t years after relocation. In solving the problems below, assume formula $(*)$, and give concrete evidence based upon MTH 153 ideas.

A. At what time t will the population cease to grow? (i.e., after time t, there is no more growth.) Justify your answer.

B. What is the maximum population?

C. Sketch a graph of $P(t)$ that includes all inflection points and their t-coordinates.

8. On the coordinate system provided, sketch the graphs of $y = \cos x$ and $y = e^{-x}$ for $0 \le x \le \pi/2$. (Both graphs begin at $(0,1)$, and they cross once between there and $\pi/2$.)

A. Use Newton's method to find the number x (to two decimal places) at which the two curves cross. I suggest that you make 1.3 your initial approximation, and check that your calculator is in radian mode.

B. Find the area of the region that lies between the two curves. (If you didn't find the crossing point in A, then just assume that they cross at 1.23.)

9. If P is the pressure of a gas, V is its volume and T is its temperature, then $PV = kT$, where k is a positive constant whose exact value depends upon the particular gas. Suppose that at a certain instant in time, $T = 20$, $V = 10$, P is decreasing at the rate of 2 pressure units per second and T is increasing at the rate of 3 degrees per second. Is V increasing or decreasing at this moment? Give the best value you can for the rate at which V is changing.

Grade Distribution:

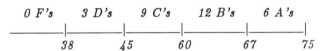

0 F's	3 D's	9 C's	12 B's	6 A's
38	45	60	67	75

Sample Test Questions

These problems are taken from quizzes and exams in various sections of MTH 153 during fall semester, 1989.

● Draw a picture and write a sentence to show that each of these statements is false.

(a) For every function f on $[0,1]$ there is some $c \in (0,1)$ such that

$$f'(c) = \frac{f(1) - f(0)}{1 - 0}.$$

(b) For every function f on $[0,1]$ with $\int_0^1 f(x)\,dx$ > 0 it must be that $f(x) \ge 0$ for all $x \in [0,1]$.

● Define Δy and dy in terms of f, x, Δx and illustrate Δy and dy on this graph:

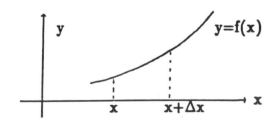

● Both the Mean Value Theorem for Derivatives and the Mean Value Theorem for Integrals guarantee the existence of a pt $c \in [a, b]$ such that, in the case of MVT-D, $f'(c)$ equals the slope of the secant line, and, in the case of MVT-I, $f(c)$ gives an average value for the function on $[a, b]$. For a given function, f, on $[a, b]$ must the $c's$ from these two theorems be the same point in $[a, b]$? (Your answer must include a complete statement of each theorem.)

• If f is differentiable and increasing and if $dx = \Delta x > 0$, then is Δy always greater than dy? Explain using various examples, pictures or counter examples.

• In the proof of the Fundamental Theorem of Calculus where F is the antiderivative of continuous function f on $[a, b]$, we used the Mean Value Theorem for Derivatives to show that for partition $P : x_0 < x_1 < \cdots < x_n$,

$$(*) \; F(x_i) - F(x_{i-1}) = f(\bar{x}_i)\Delta \, x_i$$

where $\bar{x}_i$ is a sample point in (x_{i-1}, x_i). Using the picture below, explain how the Mean Value Theorem for derivatives was used to develop (*).

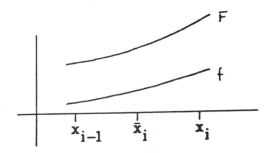

• Here is a graph of $y = f'(x)$. Tell, for which values of x, $f(x)$ is: concave up; decreasing; a local max; at a point of inflection.

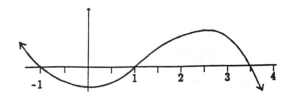

• Find the area of the shaded region shown in the sketch:

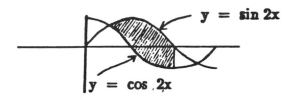

• Indicate the set of x-values with the property that $|f(x) - L| < \epsilon$.

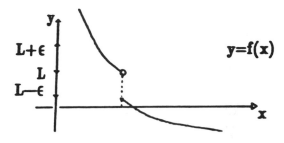

• On the coordinate system below draw the graph of a continuous function f, given that f' has the graph shown below. The domain of f is $[0, \infty)$, and $f(0) = 0$.

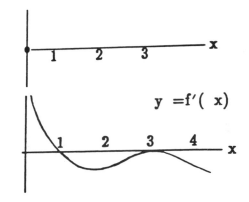

• On the coordinate system below, draw the graph of a differentiable function f that clearly satisfies these conditions: $f(a) \neq f(b)$, and there are exactly four numbers c that satisfy the conclusion of the Mean Value Theorem, for this function.

• Given that

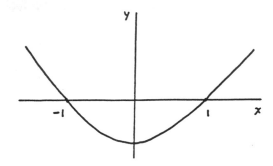

is the graph of the derivative of f, explain why each of the following <u>cannot</u> be the graph of f.

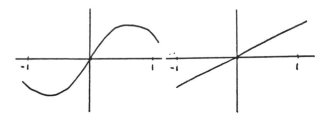

• Find the limits by inspecting the graph of $y = f(x)$ shown below. ($y = -1$ and $x = c$ are asymptotes.)

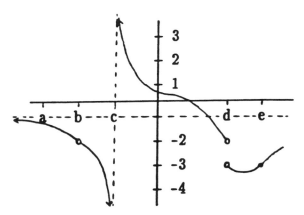

1) $\lim_{x \to b} f(x)$ 2) $\lim_{x \to c+} f(x)$ 3) $\lim_{x \to d+} f(x)$

4) $\lim_{x \to -\infty} f(x)$ 5) $\lim_{x \to d} f(x)$ 6) $\lim_{x \to c-} f(x)$

7) At what points in $[a, e]$, if any, is f discontinuous?

• Sketch a graph of $y = f(x)$ for which

a. The <u>domain</u> is $[-3, 3]$;

b. $\lim_{x \to 1} f(x) = 0$;

c. f is not continuous at 1;

d. $\lim_{x \to -2} f(x)$ does not exist;

e. $\lim_{x \to -2+} f(x) = 3$.

• In the space below, sketch the graph of a function f that satisfies all of the following conditions:

a. Its domain is $(0, 6]$;

b. $\lim_{x \to 0+} f(x) = 2$;

c. f has an inflection point at $x = 3$;

d. The minimum value of f is 1;

e. f is discontinuous at $x = 1$, but

f. $\lim_{x \to 1} f(x)$ exists.

• On the coordinate system below, sketch the graph of a function g that is continuous on $[0, 5]$ and satisfies the values given in the table shown here.

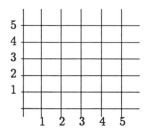

x	$g(x)$	$g'(x)$
1	1	1
2	2	-1
3	3	0
4	4	5

• A certain particle moves along a coordinate line so that its position s is related to the time t by some unknown formula, $s = f(t)$. Below is a graph of s versus t, with s measured in feet and t in seconds.

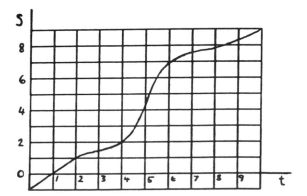

Among other things, it shows that at time $t = 0$, the particle was at position $s = -1$. In answering the questions below, please include appropriate units.

a) What is the particle's position at $t = 6$?

b) What is the average velocity of the particle, during the interval $[0, 6]$?

c) What is the approximate velocity of the particle at $t = 4$? Please explain briefly how you went about answering part (c).

d) At which of the times listed below was the particle's velocity greatest?

Circle one: t = 1   3   5   7

• Recall how the definition of limit mentions numbers ϵ and δ, which are related to each other as well as to values of x and $f(x)$. Below are graphs of two functions that satisfy $\lim_{x \to 4} f(x) = 3$. If $\delta = 2\epsilon$ for one of these functions, which one is it more apt to be?

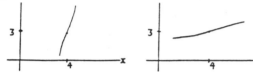

• If a certain calculus problem asks students to find *the rate of change of temperature relative to surface area*, then

a. How could the italicized concept be expressed in the symbolism of calculus?

b. What sort of units might you find given in the answer? (e.g., cu. ft./min.?)

• Sketch a possible graph of f meeting the following conditions where $a < b < c$.

a. $\lim_{x \to a^+} f(x) = f(a)$, but $\lim_{x \to a^-} f(x) = m$ where $m \neq f(a)$.

b. f is discontinuous at b but b is in the domain of f.

c. f is continuous on $(b, c]$.

d. f is continuous on (c, ∞) but $\lim_{x \to c^+} f(x) \neq \lim_{x \to c^-} f(x)$.

University of Michigan–Dearborn

Computer Labs in Calculus

by Margret Höft and David James

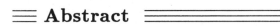
Abstract

Contact:

MARGRET HÖFT and DAVID JAMES, Department of Mathematics, University of Michigan–Dearborn, Dearborn, Michigan 48128. PHONE: 313-593-5414.

Institutional Data:

The University of Michigan–Dearborn is a state supported university located in metropolitan Detroit. It was founded in 1959 as a branch campus of The University of Michigan and has as its primary mission undergraduate education; 90% of the 7,500 students are undergraduates. The College of Arts, Sciences and Letters is the largest academic unit, offering undergraduate degree programs in the departments of behavioral sciences, humanities, mathematics, natural sciences, and social sciences. The second largest unit, the School of Engineering offers both undergraduate and graduate degrees in electrical engineering, industrial and systems engineering, and mechanical engineering. Essentially all students commute to campus from the metropolitan Detroit area, and most are from a working class background. Many are the first in their families to go to college.

The Department of Mathematics and Statistics is an undergraduate department with eighteen full time regular faculty, all holding a Ph.D., four full time lecturers, one visiting professor, and twenty four part time instructors. Approximately 140 students (65 of those are women) are mathematics majors, the third largest major in the college. The department also offers a minor in computational and computer mathematics, and a minor in statistics. Between 20 and 25 students graduate per year with a degree in mathematics. While some go on to full time graduate studies (nine since May 1988), most proceed directly to employment in statistics, mathematics, or computer science related areas, or to a career in mathematics education. A recent survey of mathematics alumni revealed that almost all students stay in the Detroit metropolitan area after they graduate. In addition to its daytime offerings, the department schedules 27% of its 230 classes each year at 4:30 p.m. or later to meet the needs of working adults.

Project Data:

We describe here a one year long pilot project introducing computers and a laboratory component into first semester calculus. In the Fall semester of 1988, three out of twelve sections of Calculus I, and in the Winter semester of 1989, three out of nine sections of Calculus I were taught as a laboratory course. All calculus classes have an upper limit of 32 students per section. In the Fall semester of 1988 there were 86 students, and in the Winter semester there were 82 students enrolled in three laboratory sections. Three faculty members (Professors John

F. Fink, Margret Höft, David James) taught one laboratory section each in both semesters. The three sections were closely coordinated and used the same materials, written by Höft and James.

All departmental sections of Calculus I used the same textbook *Calculus and Analytic Geometry* (7th edition) by G. Thomas and R. Finney (Addison-Wesley), and followed roughly the same syllabus. For the laboratory sections two pieces of software were used, one tutorial, *Exploring Calculus with the IBM PC* by J. Fraleigh and L. Pakula (Addison-Wesley), and one with some symbolic capabilities, *MicroCalc* by H. Flanders. The students used IBM PCs for their computer work, and for demonstrations in the classroom, a faster Zenith computer was available.

We did not try to get external funding for the pilot project. However, we sought and received internal funding from many sources including the Vice Chancellor for Academic Affairs, the Dean of the College of Arts and Sciences, and the Dean of the School of Engineering. We also wrote proposals to request support from Campus Grants and Educational Enhancement Funds, two programs that support research and innovative instruction. All our proposals were funded.

Project Description:

All calculus classes are taught in sections of 32 students per section, and classes meet four times per week for fifty minutes. Students in mathematics, natural sciences, computer science, and engineering enroll in the regular calculus sequence, whereas students in business and management and some social sciences enroll in what we call business calculus. The laboratory calculus sections in the pilot project were part of the regular, not the business, calculus sequence.

The pilot sections were in design similar to a laboratory course in the sciences. For three fifty-minute periods per week the students met with their instructor in a regular classroom setting for lecture and discussion. A computer with an overhead projection system was available in the classroom, and was used by the instructor to demonstrate a calculus concept, or a graph of a function, or it might be used as a tool to solve a computational problem. The fourth period of the week was scheduled for 80 minutes in a classroom containing IBM PCs for student use. In these sessions the instructor spent about twenty minutes in the beginning of the class at the blackboard to introduce the topic of the day. Then the students were given a handout with computer experiments and exercises designed to explore the topic in more detail. The handouts also contained a list of homework problems to be done with the aid of a computer, specific instructions for a laboratory report, and what we called "the discovery problem of the week," a problem selected to challenge the students, including the better ones. This problem usually went somewhat beyond the material covered in the lectures and required the use of the computer, knowledge of the calculus topics, and some ingenuity.

Students worked in teams of two per computer, and discussion, interaction, and comparing of notes between teams was encouraged. The sessions were supervised by the instructor and one student assistant. The instructor's role was to exploit the laboratory setup to teach calculus with the aid of computers, to stimulate experimentation and exploration, and to encourage the students to try out any conjecture that came to their minds. The assistant helped with technical problems. Most students were able to complete the exercises in their handouts during the 60 minute work period in the lab. In addition, they had to meet with their lab partners at least once during the week in either the campus or engineering computer labs to do their homework assignments, to write the lab report, and to work on the week's discovery problem. Each team's lab reports were graded and counted toward the final grade for the course.

We did not give a common final for the three pilot sections, and therefore, we have no comparison of test scores available for students in the pilot sections versus students in the regular sections. At the end of the second semester we solicited student opinions about the pilot sections by distributing a questionnaire (attached in the extended report). Responses to the questionnaire showed that the pilot sections were considered to be beneficial by many of the students, and were generally well received. This semester (Winter 1990) we have expanded the project to include two sections of Calculus II.

☰ Project Report ☰

Getting Started

We had a variety of reasons for starting this pilot project and for changing our approach to first semester calculus teaching. At our university, as at other universities across the nation, the standard approach to first semester calculus teaching, i.e. lecturing and listening, appears to be failing for many students. We have unmotivated students and high drop-out rates, and since a successful calculus course is the gate to all advanced technical, scientific, and engineering studies, it is essential to find more effective teaching strategies, which capture the students' interest, raise their conceptual understanding, and increase their problem solving and analytical skills.

Another motivation for change came from the realization that integration of powerful calculators and computers into the undergraduate mathematics curriculum is overdue, and that the difficult task of implementing this integration must begin in a serious way. There seems to be a sense of agreement emerging that student learning can benefit from the introduction of technological tools, and there are many good arguments in the literature why we mathematicians should start to prepare our students "for their future, and not for our past."

When we were designing the Calculus I course for the pilot project, we were looking for a model that would be appropriate for all sections of calculus, not just for one or two special sections, in a department where many sections of calculus are offered each semester. We were looking for a model lending itself to department-wide implementation not only at our university, but also at the many similar undergraduate institutions where instruction of science and engineering students is a large part of a mathematics department's responsibility. We were not proposing drastic changes in content or curriculum, but tried to work with an existing syllabus that is fairly standard for such institutions. However, we had some changes in pedagogy in mind. We wanted students to be more actively involved in the learning process, we wanted more interaction among the students, and also more interaction between teachers and students. Computers were to be used to encourage exploration and experimentation in mathematics. Students should be expected to solve a few multistep, multilevel problems where

they need more than just one single calculus concept or skill to get the solution. Finally, we wanted the students to do at least some coherent writing using mathematical terminology.

In the spring of 1988 we contacted many mathematics departments in the Midwest by telephone to gather information about computer activities at other institutions. Though we did not find a university or college where department-wide introduction of technology into calculus instruction had been tried, we learned about successful attempts and approaches by individual instructors, and got many helpful suggestions and also recommendations for software.

We came to the conclusion that for our university it would be best to keep the traditional lecture format (so most of the traditional topics can be covered), and supplement it by faculty supervised computer laboratories. It was important to us that the computer lab sessions should be supervised by faculty, thus giving the teacher an opportunity to interact directly with a student sitting at a computer, to make suggestions for "what if" questions, to encourage experimentation on the spot, and generally to foster a spirit of active involvement with the mathematical concepts.

We did not try to get outside funding for the pilot project, but invested a lot of time and energy securing resources from inside the university, thus giving the project plenty of publicity on campus, in particular among higher level administrators such as Deans, Directors, and Provosts who sometimes have funds available for special projects that look promising. Every request we made on behalf of the pilot project was somehow funded. The Department Chair was very supportive but had no money; the Dean of the School of Engineering provided computers for the faculty members involved in the project, money for software, a computer to be used in the classroom for classroom demonstrations. Help from the technical support staff of the School of Engineering (in particular the Supervisor of Data Systems, Ms. Celia Bruce) was invaluable. The Director of the Computing Center lent us two computers and bought some software. The Dean of Arts and Sciences and the Provost for Academic Affairs funded a summer stipend to prepare materials for the project, and provided money for lab assistants. Our proposals to the Campus Grants Fund and to the Educational Enhancement Fund, two programs that provide seed money for projects in research and innovative instruction, were both successful, and a request for released time for D. James was granted. We encountered generosity, good will, and a lot of support across campus.

The Pilot Sections

The Lectures

All calculus classes in the three semester calculus sequence, and most other courses offered in the department, have an upper limit of 32 students. Calculus classes carry four credit hours, and (with a few exceptions) meet four times a week for fifty minute periods. For courses with multiple sections the department issues a syllabus to the instructors of the sections, and it is expected that all sections follow this syllabus. The three Calculus I sections in the pilot project followed this same syllabus.

The pilot sections met for three fifty minute periods per week in a regular classroom for lecture and discussion. A computer with an overhead projection system was available in the classroom, and was used by the instructors during the lectures. Typically, the computer was used in a carefully planned demonstration of a calculus concept, lasting no more than five to ten minutes. Many of the demonstrations illustrated a four step exploratory approach to mathematics: generating examples and conjecturing, and then testing and proving. Students were encouraged to participate actively in the exploration and make conjectures, and they generally responded very well. The computer might be used spontaneously once or twice again during the lecture in response to a student question, or to shorten a lengthy computation, or to display a graph to illustrate the solution of a problem. Some calculus topics obviously benefit more than others from computer displays. While limits of functions, derivatives, Riemann sums, Newton's method, curve sketching, are obvious candidates for computer displays in almost any piece of calculus software, applied problems of some complexity are not so easily accommodated, unless the software has considerable symbolic capabilities. The lectures of the three pilot sections were not closely coordinated, and different instructors used at times different pieces of software for their computer demonstrations.

At the end of a lecture, students were assigned homework problems from the textbook, usually an assortment of drill and conceptual problems, just as in all other calculus sections. Students were expected to solve these problems, but solutions were not collected or graded. Since these problems were from the regular textbook, they did not require a computer to be solved, but students were free to use computers if they wanted to.

The Laboratory Sessions

The fourth period of the week was the laboratory part of the pilot project. This period was scheduled for eighty minutes in a classroom with IBM PC computers. Typically, the first twenty minutes were spent lecturing, and then the students had one full hour to work with the computers. During the lecture the instructor introduced and explained the topic of the day, and provided the theoretical background that was necessary to do the computer exercises. Then the students were given a handout with computer experiments that were designed to explore the topic of the day, and they started their computer work. The lab sessions for the three pilot sections were closely coordinated, and all three sections used the same lab handouts.

The sessions were supervised by the instructor and one student lab assistant. The lab assistant helped with technical and hardware problems, or with routine mathematical and program problems. The instructor's role was to exploit this laboratory setup to teach calculus with the aid of computers. Students worked in teams of two per computer, but discussion, interaction, and comparing of notes between teams was very much encouraged. This gave the students a chance to "talk mathematics" with each other and with the teacher, who circulated around the workstations, answering but also raising questions, and encouraging students to try out any conjecture that came to their minds.

The lab handouts also contained some homework problems that had to be done with the help of a computer, and contained what we called the discovery or challenge problem of the week, a problem selected to challenge the students, including the better ones. These problems usually went somewhat beyond the material covered in the lectures and required the use of the computer, knowledge of the calculus topics, and some ingenuity. Most students were able to complete the assignments in each lab handout during the work session; some would even get a start on the computer homework problems, but for all of them there would be at least something left to be done after the session. The students then would have to make arrangements to meet with their lab partners in the campus or engineering computer labs to complete the assignments, to write the lab report, and to work on the week's challenge problem.

One week after a laboratory session, each team of two students submitted a lab report with solutions of exercises, screen dumps of significant graphs, references to theorems that made a solution of a problem possible, and a solution to the challenge problem. The report also had to include a two-paragraph description of what was accomplished during the lab session, such as which concepts were studied, which assignments were found difficult and why, how long it took to complete the assignments, and where they got stuck and how they got out of it. The rationale for these descriptive reports was twofold. Firstly, the students gained valuable writing experience using mathematical and technical terminology, and were forced to reflect on how they accomplished learning of mathematics in the context of a laboratory. Secondly, the instructors received timely feedback—both positive and negative—from the students, and could make changes and adjustments in subsequent lab assignments, if students had justified complaints. The lab reports were collected and graded and counted towards the final grade for the course. Generally, the lab reports were well written and carefully done and contained correct mathematics. Many students used word processors and laser printers to give their reports a polished look.

The Laboratory Assistants

We were able to get approximately $600 per semester to employ student lab assistants. Three student assistants were involved in the project; two were senior mathematics majors, one was an engineering student. They helped in the supervised lab sessions, but were also available for several hours as tutors in the campus and engineering microcomputer labs to help students finish their computer homework. They also graded the lab reports.

Tests and Grades

Grading and testing were handled individually by the three instructors and varied slightly from section to section. All sections had three one-hour exams, and, though these were scheduled during the same week, the content of the tests was not coordinated. In the weeks with one hour exams we did not schedule a laboratory session. Grading varied from section to section. One instructor counted each of the one hour exams for 15% of the grade, three short quizzes for 5% each, and the lab reports for 15%.

The comprehensive final examination counted for 25% of the course grade. These percentages varied in the other sections, but were quite similar. Only the computer homework from the lab handouts was collected and graded.

Equipment and Software

In 1988, the mathematics department had no instructional computer facilities, but had a variety of microcomputers (mostly IBM family and clones) available for faculty use. The School of Engineering had two microcomputer laboratories, one equipped with Zenith computers (networked), the other with Macintosh computers (networked). The Computing Center of the university operated a large microcomputer lab with IBM PCs and clones (networked), and also one classroom with approximately 30 IBM PCs (not networked). This classroom was used for the supervised calculus laboratories. Since it was heavily scheduled, we had some difficulties getting it reserved for the laboratory times. The room had a blackboard and rows of chairs in the front, so it could be used for lectures. There was also a computer with an overhead projection system in the front of the room, making computer demonstrations during lectures possible. Three rows of tables at the back of the room and a row of tables along the side walls held IBM PC computers and printers. The stations were arranged so that students sitting at the computers faced either the front or the sides of the room. This made it possible for them to see the front of the room at all times. Each printer was connected to two computers, and the printer was located between the two computers, so students from both computers could view and reach their printouts without having to get up. Since we wanted the students to work in pairs, we used only one of the computers in such a computer-printer-computer arrangement. As mentioned before, the computers in this room were not networked, and the students had to insert diskettes to use the software.

The software for the project was chosen with the computer novice in mind. Students were not expected to have any previous experience with computers or, in particular, knowledge of a computer language. We looked for software that required only minimal instruction for use, and would not generate an excessive amount of frustration. We wanted one package with some symbolic capabilities, and one tutorial package, and decided on *MicroCalc* by

H. Flanders as the package with some symbolic capabilities, and *Exploring Calculus on the IBM PC* by J. B. Fraleigh and L. I. Pakula as the tutorial. These were the only two packages used by the students. We bought site licenses for *MicroCalc* (approximately $350), and 24 copies of *Exploring Calculus on the IBM PC* at approximately $22 per copy, where half of these were donated by Addison-Wesley Publishing Co. *Exploring Calculus on the IBM PC* was available for sale in the campus bookstore, but we did not expect students to buy their own copy.

To do their computer homework assignments, all students could use the campus microcomputer lab; only engineering students could use the laboratories of the School of Engineering. The software was installed on the network in the campus lab, and also in the Zenith lab of the School of Engineering, and the students had easy access when they were on campus. Since the supervisors in the general computer labs were "work study" students not necessarily skilled in mathematics, we made sure that for six hours during the week and for five hours on weekends, a senior mathematics student was available in the labs to help the students in the pilot project with mathematical problems.

The three fifty minute lectures per week were held in a classroom in the School of Engineering, and all three pilot sections used the same classroom. For the computer demonstrations during the lectures we used a Zenith computer with a hard drive that had a variety of calculus software installed on it. The classroom demonstrations were not as closely coordinated as the labs, and different instructors used different pieces of calculus software during their lectures, but mostly we used the same software that was used by the students in the labs. The computer belonged to the School of Engineering, and had to be brought into the classroom by the instructor before each class and returned after class. It was on a cart with wheels, and when not in use in a classroom, it was part of the Zenith lab of the School of Engineering. A Telex Magnabyte LCD panel was used to project the computer output to a screen via an overhead projector. The pull down screen that was permanently installed in the room could not be used since it covered the blackboard when pulled down, and therefore a portable screen had to be brought to the classroom and set up on the side of the room. The LCD panel belonged to Audio Visual Department located in the campus library, and was delivered by work study

students before each class. The instructor had to assemble this whole outfit before class and disassemble it after class, a time consuming and at times discouraging task.

Evaluation

In the fall semester of 1988, the students that enrolled in the three pilot sections did not know they were signing up for experimental sections. They were told about the nature of the pilot project on the first day of classes and were offered assistance to change to other sections if they preferred not to be involved. Very few students (three or four) wanted to transfer out, and a few students transferred in when they heard about the project. In the winter semester of 1989, most students were aware of the pilot sections when they signed up. At the end of the winter semester the students were given a questionnaire to solicit their reactions to the new laboratory course. We modeled the questionnaire after a similar one given at Dartmouth, but the questions were changed somewhat to reflect the difference between the course format of the Dartmouth project and our project. The questions and responses are shown in Table 1 below. This was the only evaluation we did. We made no attempt to compare the knowledge level of students in the pilot sections to the knowledge level of those in the traditional sections.

The student responses to the questionnaire substantiated our impressions that the students in the pilot sections were more interested in the material, more responsive, and asked more questions than their counterparts in traditional sections usually do. During the lectures they seemed to be more attentive and seemed to concentrate harder, in particular, when there was something happening on the computer. They had few problems learning how to use the computers, even though 28% described themselves as not being knowledgeable about computers, and only 29% had used computers in a non-computer-science course before. The software turned out to be as user friendly as we had hoped it to be, and the students did not have to overcome an excessive amount of frustration to become comfortable with it.

The teams of two students that were formed in the computer laboratories tended to work and study together on their other homework assignments as well. On a commuter campus, it takes students

Percent Distribution of Responses to Survey

Instructors: Fink / Höft /James
Math 115, Winter 1989
Total Number of Responses : 68

SA= Strongly Agree
A = Agree
N = Neutral/No Opinion
D = Disagree
SD= Strongly Disagree

	SA	A	N	D	SD
1. Using a computer in the classroom contributed to my understanding of the course material	12	74	10	4	0
2. Using a computer in the laboratory contributed to my understanding of the course material	13	65	12	10	0
3. Using a computer in the classroom enhanced my interest in the course material	13	37	40	7	3
4. Using a computer in the laboratory enhanced my interest in the course material	12	34	34	15	5
5. Using computer graphics helped me to understand derivatives	36	52	7	5	0
6. Using computer graphics helped me to understand integration	18	49	18	11	4
7. Using computer graphics helped me to understand Newton's method	22	56	7	15	0
8. Overall, the computer labs were a valuable part of this course	15	54	18	10	3
9. The computer assignments could be completed in a reasonable amount of time	24	32	21	18	5
10. Learning how to use the computers required little help from others	27	54	6	10	3

Table 1

sometimes a long time to make contact with other students, or to form study groups. Computer partnerships seemed to help in that respect. Team work also created problems: When students left the supervised labs with some work still to be done (e.g. the homework assignments and the challenge problem), some tried for the easiest and quickest way to complete the assignments; they either split the work into two parts, and each student worked on just one part, or one student finished the job without any help or input from the other. Even though some successful study groups were formed, for many students it was difficult to arrange a time when they could meet their lab partners outside of scheduled classes. Many, if not most of our students also hold jobs off campus, and they arrange their class schedule around their work schedule, leaving very little flexibility for extra time on campus.

In the Fall semester of 1988 we had (justified) complaints about the length of some of the labs, and the increase in workload for the same number of credit hours. For the Winter semester we shortened the labs and assigned fewer computer homework problems, and we also introduced as a rule that only five of the eight challenge problems had to be completed for full credit. Generally, in the second semester, sailing was a lot smoother than in the first, but students still felt that their workload was larger than for students in the regular sections.

While the lab sessions went smoothly most of the time, the physical setup for the computer demonstrations in the classroom was a different matter. It took too much time to fetch and assemble all the equipment, to connect all the cables, and focus the LCD panel. All that effort sometimes did not produce results; one piece or another simply might not work, and in the few minutes before class started there was not enough time to figure out what the difficulty was, unless it was obvious. Lighting in the classroom also was a problem; when the lights were on, it was impossible to see the projected computer output, and when the lights where out, it was of course impossible to see the blackboard.

To implement this laboratory course for first semester calculus has required a tremendous investment of energy and time. Writing and revising the laboratory materials (and typing and duplicating them for the students) took much faculty time and effort in the beginning, but less time as we became more skilled and experienced, and the second semester did not require as much time as the first. In retrospect, more time had to be spent on securing the necessary resources, learning about hardware as well as software, making sure that the physical setup of the project did not fall apart, than was spent on developing course material. For a smooth operation of three experimental sections, many details needed daily attention. During the second semester, when things should have been easier, we started to give talks and write papers about the project, and we wrote grant proposals to get outside funding for a continuation of the project. A successful pilot project has increased our workload for the foreseeable future; writing this CRAFTY report is only one example of the increase.

Plans for the Future

Towards the end of the pilot project, plans for the future began to take shape. We felt very encouraged by the response and the attitudes of the students and wanted to make computers available also for second semester calculus, and—further down the road—for Calculus III. However it was clear that this would be impossible within existing resources. One already overscheduled classroom with slow IBM PC computers simply would not do. We needed another computer lab, one specifically reserved for teaching calculus. Moreover, it was also clear, that at the end of Calculus I we were outgrowing the software that had served us well for first semester material. For Calculus II and III we wanted one of the powerful computer algebra systems.

During the summer of 1989 we learned that our grant proposal submitted to the Instrumentation and Laboratory Improvement Program of the National Science Foundation would be funded. We have since purchased seventeen Macintosh IIcx computers and set up a computer lab for calculus teaching. Our project was approved for an Educational Grant by Wolfram Research, Inc., and we have *Mathematica* running on the computers in the lab. This semester (Winter 1990) we are teaching two sections of Calculus II (Höft and James, one section each), and one section of Calculus I (Fink) in the new calculus lab. We are following the same format as in the pilot project with three fifty minute lectures per week and one eighty minute lecture-laboratory session. The lectures as well as the lab sessions are in the same room, and the room is reserved for the mathematics department only. One of the Macintosh computers is used for demonstrations during the lectures. It is connected to a Proxima LCD panel, and a stationary overhead projector projects the computer output to a permanently installed screen. This equipment is permanently located there and is available at the flick of one switch. With dimmer light switches we can control the lighting in front and back of the room, and, separately around the projection equipment.

Of the 62 students in the two Calculus II sections, virtually all have previous experience with Macintosh computers; some even have experience with *Mathematica*, which they had seen their friends in the School of Engineering use. This is a drastic change from the previous computer experiences of the students in the pilot project in the fall of 1988, (when *Mathematica* was just being released), and may indicate that student computer sophistication is increasing very rapidly.

Where do we go from here? For the Fall semester of 1990 we anticipate that at least three more faculty members of the department will be teaching sections of Calculus I and II using computers. For the Winter semester of 1991 we expect to include Calculus III in the project.

Sample Materials

Instructions to Students

This is one of three experimental Calculus I classes offered this winter at the University of Michigan–Dearborn. In these three classes, computers will be used to explore the concepts of calculus. The emphasis is not on the use of computers as computational tools, but the computing power of the computer will be used to formulate and test ideas and conjectures that lead to the discovery and understanding of the principles of calculus.

Students must be prepared to use computers all semester, but no knowledge of programming or previous work experience with a computer is required. Software will be provided, and the computer laboratory in the basement of CAB and the engineering computer laboratory will have tutors specifically assigned to these three sections to help with mathematical and technical difficulties. A lab fee of $7.50 per semester has to be paid for use of the lab.

Topics: The three experimental sections will cover the same material as all other sections of Calculus I , and will follow roughly the same syllabus. Only very minor modifications will be made. Students from these sections will be ready for Calculus II at the end of the winter semester.

Laboratory Projects: Students will attend supervised laboratory sessions, where exercises and homework problems will be done using computers. These sessions will be held in 29 CAB. We will meet in 29 CAB every Tuesday during regular class time, on some Tuesdays for lectures , on other Tuesdays for computer sessions, and sometimes we will do a little of both. A more precise schedule will be announced as the semester progresses. In the computer sessions, students will work in teams of two, but discussion, interaction and comparing notes between teams is encouraged. Each team has to submit a **lab report** within a week after the computer session. The lab report and its associated homework problems will be collected and graded.

Tests and Grades: Semester grades will be determined from the total scores you obtain during the term. There will be 3 tests (one class period), each counting for 15% of your total score, 3 quizzes (15 minutes), each counting for 5% of your score, and a comprehensive final will count for 25% of your score. Lab reports will count for 15%. Team members of the same team will get the same score for a lab report.

Syllabus

For courses with multiple sections the mathematics department issues a syllabus to the instructors of the sections. It is expected that all sections roughly follow this syllabus. All calculus sections used the textbook *Calculus and Analytic Geometry* (7th edition) by G. Thomas and R. Finney (Addison-Wesley), and the departmental syllabus for Calculus I is based on this text. We offer twelve sections of Calculus I in the Fall semester and nine in the Winter semester. Three sections in each semester were taught as a laboratory course. Like all the other sections, they used this textbook and roughly followed the same syllabus.

Math 115   Calculus I
Thomas/Finney (7th edition)   (Fall 1988)

Day(s)	Section	Topics
1	1.1, 1.2	Rectangular coordinates, slope
2	1.3	Equations of lines
3, 4	1.4, 1.5	Functions & graphs, absolute value
5, 6	1.6–1.8	Derivatives (informal), velocity
7, 8	1.9	Limits
9	1.10	Asymptotes
10, 11	1.11	Continuity
12, 13	2.1, 2.2	Sum, product, quotient and power rules for derivatives
14	2.3	Implicit differentiation
15	Review	
16	Exam I	
17	2.4	Linear approximation, differentials
18	2.5	Chain rule
19, 20	2.6, 2.7	Trigonometric functions & their derivatives
21	2.87	Parametric equations
22	2.9	Newton's method for approximating roots
23–25	3.1–3.4	Curve sketching, monotonicity, concavity, absolute and local extrema
26, 27	3.5	Max-min problems
28	Review	
29	Exam II	
30	3.6	Related rates problems
31, 32	3.7, 3.8	Rolle's & Mean Value Theorem l'Hôpital's Rule
33–35	4.1–4.4	Indefinite integrals
36–38	4.5–4.7	Definite integrals, Fundamental Theorem
39	4.8	Substitution in integrals
40	4.9	Trapezoid & Simpson's Rules
41	Review	
42	Exam III	
43, 44	6.4, 6.5	Natural log & its properties
45	6.6	Exponential function e^x
46	6.7, 6.8	a^x and $\log_a x$, $a \neq e$
47	6.9	Applications
48, 49	5.1, 5.2	Distance and area between curves
50, 51	5.3, 5.4	Volume by plane cross-sections & by shells
52	Review	
53	Exam IV	

Computer Laboratories

We scheduled eight lab sessions during the Winter semester 1989. There would have been enough time in the semester for nine or ten sessions, but we did not schedule sessions during the weeks when the students had major tests for this class, and also not during the last two weeks of classes. In these weeks the time slot normally reserved for lab sessions was used for regular lecture and discussion. A brief description of four of the labs is given below, and for the other four, complete handouts are attached.

1. Basic Skills

The students learned the basics of using the computer and the software: how to make selections from the menus, how to type in functions, how to edit, how to change the viewing window in a graph, how to decide whether a viewing window is good or bad, how to determine domain and range of a function, and how to get hard copy of their work. They experimented with viewing windows for functions like

$$f(x) = \frac{x^3 - 10x^2 + x + 50}{x - 2}$$

and

$$f(x) = 5\sin 3x - 2\sin\left(\frac{x}{4}\right).$$

2. Graphs, Tables, and Limits

Function tables were introduced. They were used to estimate roots of equations, points of intersection of graphs, and to find limits of functions. The students investigated limits like $\frac{\sin x}{x}$ at zero and $\frac{1.01^x}{x^5}$ at infinity. They also learned how to graph several functions in one picture, and compared for instance the two functions $y = 10^x$ and $y = x^{10}$, looking at them from close up and from far away. They experimented with various values of a, b, and c in $y = x^3 + ax^2 + bx + c$ until they had three functions which looked quite different from each other close up. Then they enlarged the viewing window to get a global look at the same three functions.

3. The Derivative of a Function

A complete handout is attached.

4. Approximations, Scientific Error

Tangent line approximations, e.g. for the function $f(x) = \sqrt{x}\,\frac{\sin(x^2)}{x^2 + 2x + 2}$ for x near $a = 1.1, 1.3, 1.5$ are calculated. Error estimates were done by finding an upper bound M for the values of $|f''(x)|$ near these points graphically. The challenge problem of the week introduced quadratic approximations.

5. Newton's Method

A complete handout is attached.

6. Local Extrema, Points of Inflection

The students found local extrema and points of inflection for some functions, where the derivatives would be very difficult to compute by hand. For the functions $y = x^3 + x^2 + kx$ and $y = x^4 + kx^3 + x^2 + x$, they watched the shape of the curves change for different values of k, and kept track of the number of extrema and inflection points as k changed. By trial and error estimates and graphical feedback from the computer, the students approximated those numerical values of k which produced a major change in the shape of the curve. Then they used their knowledge of derivatives to find the exact value of k by hand.

7. Test Your Calculus Skills

A complete handout is attached.

8. Integration

A complete handout is attached.

Classroom Demonstrations

A short article, *Computer Demonstrations for Calculus I*, describing some of the classroom demonstrations we used in the pilot sections, will appear in the Proceedings of the Second Annual Conference on Technology in Collegiate Mathematics held at Ohio State University in November 1989.

Laboratory 3

The Derivative of a Function

In this session we explore the derivative of a function. We first focus on the definition of the derivative, and then we explore the idea that the derivative of a function is again a function.

Insert the diskette *Exploring Calculus with the IBM PC* into drive A and turn on the computer and the monitor. Move through the screens until you arrive at the main menu.

Part I

Example

Select Program I from the main menu . We want to use Program I to estimate $f'(2)$ for $f(x) = x^3/8$. Since $x^3/8$ is the default function in the program, we type N when asked if we want to change it. We specify 2 as the point where we want the derivative, we supply -6, 6 for xmin, xmax for graphing, and then we use the defaults. For Δx we use the values 2, -2, 1.5, -1.5, 1, -1, .5, $-.5$, .3, $-.3$, .1, $-.1$, .001, $-.001$, .00001, $-.00001$, and .0000001, $-.0000001$. As we watch the values of M-SECANT change for values of Δx closer and closer to zero, we conclude that the value of $f'(2)$ is about 1.5.

Exercises

A. i) Use Program I to estimate $f'(1)$ for $f(x) = \sin(x)$. Start your Δx's at $\Delta x = 1$, then $\Delta x = .5$, and then use smaller and smaller *positive* values for Δx. Watch the screen carefully to see where each new secant line is drawn. Make a screen dump showing the function and several secant lines for positive Δx's. Label (by hand) the values of Δx giving rise to each secant line. What is the approximate value of the limit of the slopes of the secant lines for positive Δx as Δx goes to zero?

For $\Delta x > 0$, we have $m =$_____ .

Repeat the problem for $\Delta x = -1$, then $\Delta x = -.5$, then $-.25$, then $-.1$, etc. for *negative* values of Δx. Again get a screen dump and give the limiting value of the slopes.

For $\Delta x < 0$, we have $m =$____ $f'(1) =$____ .

ii) Approximate $f'(0)$ for $f(x) = \sin(x)$. Repeat the above at $x = 0$. Skip the screen dumps.

$f'(0) =$_____ .

iii) Approximate $f'(1.5708)$ for $f(x) = \sin(x)$; that is, repeat the above at $x = 1.5708$ (which is approximately $\pi/2$). Skip the screendumps.

$f'(1.5708) =$_____ .

B. Repeat exercise A. for the function $f(x) = 4\sin(1/x)$ at $x = 1.5$. Use the default values for the viewing window, then later also try $-2, 2$ for xmin, xmax, and $-10, 10$ for ymin, ymax. To get better resolution, try 320 points. Get two screen dumps as in exercise A. i).

For $\Delta x > 0$, we have $m =$_____ ,

for $\Delta x < 0$, we have $m =$____ , $f'(1.5) =$____ .

C. In this exercise we want to approximate $f'(x)$ for several values of x in the interval $1 \le x \le 2$. We take the function $f(x) = x^3 - 4x^2 + 6x$ and use Program I to find first $f'(1)$. Then we compute $f'(1.2)$, $f'(1.4)$, $f'(1.6)$, $f'(1.8)$ and $f'(2.0)$, and complete the following table:

x	$f'(x)$ (3 digits)
1.0	$f'(x) =$
1.2	$f'(x) =$
1.4	$f'(x) =$
1.6	$f'(x) = .88$
1.8	$f'(x) =$
2.0	$f'(x) =$

$f'(x)$ can be computed this way for any x. This means that $f'(x)$ is itself a function. Use the values in the table above to make a plot on the graph

on the next page. For example, if $x = 1.0$, the corresponding value of $f'(1.0)$ is 1, so we plot the point $(1, 1)$ on the graph. When all six points from the table are plotted, draw a smooth curve passing through those points. You have just constructed the graph of a new function, namely the graph of the function $f'(x)$. Think about it! You will learn a more efficient and faster way to compute $f'(x)$ in Challenge Problem B. on page 4.

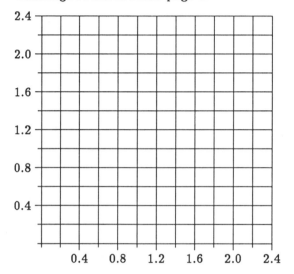

Part II

We will now study the relationships between the shape of the graph for $f(x)$ and that of $f'(x)$ using Program L. Then we will use Program M to test how well you have understood these relationships.

Exercises

A. Select Program L from the main menu and select the f and f' option. Follow directions to graph the function $f(x) = \dfrac{1}{x^2 + 1}$. After the function is graphed, quickly hold down the $<$Ctrl$>$-key and press the $<$Numlock$>$-key. This will freeze the run of the program until any key is pressed. While the action is frozen, decide where the graph of the derivative should be drawn. Then press any key to see the graph of the derivative. Repeat this exercise for the function $f(x) = x^3 - 3x^2$ and the function $f(x) = \sin(2x)$. When you can accurately predict the graph of the derivative, go on to exercise B.

B. Select Program M from the main menu and choose Option 1: Graphing the derivative f'. Select the easy mode (E). Graph derivatives until you are confident of being able to obtain a grade of at least 80%. Make a screen dump where your score is at least 80%.

Lab Report

(1) Write a short account of what you did during your lab session.

(2) Turn in this Lab #3 handout showing the boxes filled in with the answers found during lab. Also turn in screendumps.

(3) Complete Challenge Problems A and B.

Challenge Problems

A. Use Program L to graph $f(x) = 2^x$ and its derivative. Repeat with $f(x) = 3^x$. Then experiment to find as accurately as possible by graphing, a number c such that the graph of c^x and the graph of its derivative coincide. The function c^x is thus unchanged under differentiation. This number c is a very important number in mathematics. Can you guess what c is? Hand in a screen dump for the function that coincides with its derivative and state what value of c was used.

$$c = \text{\underline{\hspace{4cm}}} \ .$$

B. Select Program J from the Main Menu. Read the text introducing the program <u>very carefully</u>. Follow directions to estimate the derivative of the default function already in the program at $x = 1$. Do both columns of numbers lead to the same estimate? What is the difference between the two columns of numbers? How are the two columns generated? Think about the graph that you see on the screen. Do you understand what the straight lines are that you see? Now repeat at $x = -1$.

Next, using Program J, compute $f'(1.1)$ and $f'(1.2)$ again for the function in exercise C of Part I. Write the two values on the lines below. Did you get the same values as you had in exercise C of Part I? Was the method of Program J in this case faster than the method of Program I?

$f'(1.1) = \text{\underline{\hspace{3cm}}}$

$f'(1.2) = \text{\underline{\hspace{3cm}}}$

Now use Program J to estimate the derivative of the function $y = |x| \cdot \cos x$ at $x = 0$. Here you will see that the two methods to compute the derivative disagree. Write a paragraph to explain why you get different answers in this case. Explain which answer is correct and why.

Laboratory 5

Newton's Method for Solving $f(x) = 0$

In this session, we shall work with Newton's method, an iterative procedure producing approx-imate solutions for equations of type $f(x) = 0$, where f is a differentiable function.

Procedure:

(1) Graph $f(x)$, and from the graph make a guess x_1 of where f crosses the x-axis.

(2) Use the first approximation x_1 to get a second x_2, and the second to get a third x_3, etc. To get from the nth approximation x_n to the next approximation x_{n+1}, we use $x_{n+1} = x_n - \dfrac{f(x_n)}{f'(x_n)}$.

Of course this procedure will only work, if $f'(x_n) \neq 0$. See the Appendix at the end of this handout, or your Calculus Textbook for more information.

In Part I, we will use the diskette *Exploring Calculus* and in Part II, we introduce a new diskette *MicroCalc* by H. Flanders.

Part I

Select Program P from the Main Menu of the diskette *Exploring Calculus*.

WARNING: To run Newton's method with Program P, you have to type in a function $f(x)$ and also its derivative, $f'(x)$. Make absolutely sure your derivative is correct! Program P cannot check your derivative for correctness and will use whatever derivative you supply for its computations in step (2) above.

Example

To solve

$$\frac{(x + 1)^3}{12} + \frac{x}{5} - 1 = 0$$

, select Program P from the Main Menu and type Y when asked if you want to change to a new function. Type in the function given above (without the $= 0$ part), calculate the derivative by hand to be

$$\frac{(x + 1)^2}{4} + \frac{1}{5}$$

, and type this in at the appropriate time. Accept the usual defaults for the graphing window. The resulting graph shows a crossing of the x-axis (a root), so when asked if you need a better graph for your rough estimate of the root, answer N(for no). A good rough estimate of the root is $x_1 = 1$. However, for demonstration purposes it will be easier for you to see what is going on, if you start a little farther away, with a rough estimate of, for example $x_1 = 3$. So write 3 in the rough estimate box below, and also type it into the computer. Then press the space bar. The tangent line at $x = 3$ crosses the x-axis at a place that is a better estimate of the true root than 3 was. So we repeat the whole process starting at that point instead of 3, by pressing the space bar. Do this repeatedly, pressing the space bar, until the numbers in the left hand column (the x-values) settle down to 1.1063178682. Notice that the corresponding y-values (the right hand column) get extremely small (-1.388×10^{-17}), but it is the convergence of the left column that is the important thing. Finally, record your root in the box below, accurate to six significant figures.

rough estimate	root
$x_1 = 3$	$x =$

Exercises

Use Program P to find roots in the following problems.

A. Solve $x^3 + 10x^2 - 50 = 0$ (that is, find the roots). You will need to calculate the derivative by hand to type in when needed. The defaults do not give you a good viewing window this time, so make the window bigger by changing the default in the x-direction until you can see all three roots, and changing the default in the y-direction until you can see the entire curve between the roots (Hint: what is $f(0)$ and what is $f(3)$?) Get a screen dump showing three crossings of the x-axis. On the screen dump, point out with arrows the three roots we are trying to approximate, and write down a good guess x_1, y_1, and z_1 for each of the three roots. Then get the three roots as accurately as possible, using Program P. Be sure to continue your iterations for each root until your left hand column repeats its first six significant digits. Get a screen dump showing the tangent lines for one of the roots (your choice).

rough estimates	root
$x_1 =$	$x =$
$y_1 =$	$y =$
$z_1 =$	$z =$

B. Get a nice graph showing all roots of

$$x^2 - \frac{1}{x} - 6 .$$

(Ask yourself what this function does as x approaches 0, and if the graph does not show what you know should be happening, then increase the number of graphing points from its default value of 50.) For the root that is close to zero, try -1 as an initial rough estimate and proceed pressing the

space bar until you settle down to a root; then try -0.5 as your initial rough estimate, then try -0.2. Get screen dumps to show what happens in each of the three cases. On each of the screen dumps you must write a couple of sentences pointing out the surprising things that happen and explaining why. This illustrates how careful you must be in making your initial rough estimate.

C. Use Newton's Method (program P) to determine the square root of two, by solving the equation $x^2 - 2 = 0$. If you start with an initial estimate of 3, how many steps are required to get 15 significant decimal places of accuracy?

$\sqrt{2} =$ _____ .

Number of steps for 15 significant digits: ____ .

Return to the main menu and press *End* to leave *Exploring Calculus*. Remove the diskette.

REMARK: Newton's Method can be used to find the intersection point(s) of two curves, say $g(x) = 3\cos(x)$ and $h(x) = x$. You simply define a new function $f(x) = g(x) - h(x) = 3\cos(x) - x$ and apply Newton's Method to this new function, because the x's which make $f(x)$ zero are precisely the x's for which $g(x) = h(x)$.

Part II

Put the two *MicroCalc* diskettes into drives A and B, and type MC. Experiment with the four arrows (the 2, 4, 6, 8 keys at the right end of the keyboard), which move the selection box around the main menu. Move the box to "Graph y = F(x)" and press the *Enter* key, and read the resulting screen of information.

Example

To get a feeling for how easy it is to use this program, we redo exercise A of Part I where we wanted the real solutions of $x^3 + 10x^2 - 50 = 0$. We first want to graph the function $y = x^3 + 10x^2 - 50$. *MicroCalc* is asking us to type in $F(x)$. We respond by typing $x^3 + 10x^2 - 50$.

You can edit (change and correct) what you typed using the arrow keys (2, 4, 6, 8) and also the delete key on the top row. If things get hopelessly confused, you can always press the *Escape* key (which takes you back to the previous screen) enough times until you arrive at the main menu, where you can start over.

When the equation is as you want it, press the *Enter* key. *MicroCalc* retypes our equation, placing it at the top of the screen. Note that in the retyped version, the * for multiplication has been omitted from $10x^2$. In this program, we may omit the * for multiplication and insert a space instead.

For x0 type in -10, and for x1 type in 10; for y0 type in -100, and for y1 type in 100. Then press *Enter*. You should see a graph, and from this graph you can see that reasonable rough estimates of the three roots are -10, -2, and 2.

Next, press *Escape* to return to the main menu, and then use the arrow keys to move the selection box to *Newton-Raphson* and press *Enter*. Your current function is at the top of the screen, and since it is the one we want, we do not need to type it in again. The next line flashes on for only a few seconds. It says: Computing F'(x). Here we have a major difference between this program and *Exploring Calculus*. This program computes the derivative, whereas in Program P of *Exploring Calculus* we had to supply the derivative.

We are now asked to provide an initial estimate x0 for the root. We choose x0 = 2 by typing "2", and then we press the *Enter* key. Can you interpret the next screen? Press the *Enter* key again to get the second iteration, then *Enter* to get the third iteration, etc., until you see what is happening. We are finding one root of $F(x)$. Keep pressing *Enter* until the xN column stops changing. Write down that root, which should be 2.03801, to six significant digits.

To look at a different root, press the arrow keys until you have highlighted the "Change x0" box, and press *Enter*. When asked for the new x0, type in -2, then press *Enter* repeatedly to see x_n approaching a value near -2.59, while simultaneously $F(x_n)$ goes to zero. Write down this root, and find the third root by starting at x0 = -10.

rough estimates	root
$x_0 = 2$	$x =$
$y_0 = -2$	$y =$
$z_0 = -10$	$z =$

Press *Escape* to get back to the main menu.

Exercises

A. Use the program *Newton-Raphson* to find a root of $f(x) = x^4 + x - 3$ in the interval [1,2]. Such a root exists because $f(1) = -1$ (negative) and $f(2) = 15$ (positive), and hence $f(x)$ must be zero somewhere between $x = 1$ and $x = 2$.

Root between 1 and 2: _____ .

B. Use the program *Newton-Raphson* to solve $\frac{x^2}{100} + \cos\sqrt{8x+1} = 0$. (Note that in this pro-

gram square root is sqrt.) To get initial estimates for the roots, use the program *Graph y = F(x)* from the main menu to graph the function. Experiment with the zoom out and the zoom in feature, until you are sure you see all the roots. Write down an initial estimate for each root, and then use *Newton-Raphson* to determine each root to six significant digits.

rough estimates	root

To end the session, press the *Escape* key several times to get back to the main menu.

Lab Report

(1) Write a two-paragraph account of what you did during your lab session.
(2) Turn in this Lab #5 handout showing the boxes filled with the answers found during the session. Also include screen dumps.
(3) Give a clear explanation of what happened in Part I Exercise B and why this happened. Your explanation should be written on the screen dumps.
(4) What theorem is used in Part II Exercise A, that is, what theorem says $f(x)$ must be zero somewhere between $x = 1$ and $x = 2$? What are the hypotheses for this theorem? Does our function $f(x) = x^4 + x - 3$ satisfy these hypotheses?
(5) Hand in solutions to the homework and the challenge problem.

Homework

Use Program P from *Exploring Calculus* to do the following:

(1) Find two real roots of the equation $x^4 - 2x^3 - x^2 - 2x + 2 = 0$ to six decimal places. First get a screen dump showing there are two roots. Give good initial estimates of each root to use in Program P.

(2) Estimate π to five decimal places by applying Newton's Method to the equation $\sin x = 0$. Does it matter what your initial estimate is? Try an initial estimate of 2, then of 1.9, then of 10. For each of the three initial estimates, choose a good viewing window to see what is going on, and get screen dumps for each one showing the progression of tangent lines.

(3) Let $f(x) = -2x^3 + 3x^2 + x - 1$. The equation $f(x) = 0$ has three roots.

(a) Find the middle root, accurate to six digits.

(b) For some initial estimates, Newton's Method will not converge, but will simply bounce back and forth between two numbers. By looking at the graph of $f(x)$, see if you can guess such an initial estimate x_1. Then check your answer by using your guess x_1 as initial estimate, and watching what happens graphically to your successive estimates.

(c) After you have found your answer to (b), try an initial estimate which is just a little bit bigger than your estimate in (b), and then another which is just a little bit smaller. Explain in words what happens in each of these cases.

Challenge Problem

(Use Program 5 from *Exploring Calculus*)

Newton's Method also works for two equations in two unknowns (and for three equations in three unknowns). Recall that to solve one equation in one unknown, $f(x) = 0$, Newton's Method requires a first estimate x_1 and then it constructs the tangent line to the graph through $(x_1, f(x_1))$. Where this tangent line crosses the x-axis is x_2 which is an improved estimate (usually) of the true root x we are seeking. We repeat the process until the desired accuracy is obtained.

An analogous procedure can be used to solve a system $f(x, y) = 0$, $g(x, y) = 0$ of two equations in two unknowns x and y. We start with an initial approximate solution (x_1, y_1) and find the equations of the tangent planes in 3-space to the graphs of the functions f and g at this point. These two planes can be expected to intersect in a line which can be expected to intersect the xy-plane in some point. This new point (x_2, y_2) is an improved estimate (usually) of the true root (x, y) we are seeking. The process is repeated with this new point (x_2, y_2) in place of the original (x_1, y_1) to find still a better estimate, etc. The main idea here is to replace a hard nonlinear problem with an easier linear one.

Problem: The nonlinear system $x^2 + y^2 = 4$, $xy^2 = 1$ has four solutions. Find all four as accurately as you can. First subtract 4 from both sides of the first equation and 1 from both sides of the second equation to get them in the form $f(x, y) = 0$ and $g(x, y) = 0$. Then get a good first estimate for each of the four solutions. You can do this by graphing both curves (either by hand on graph paper or by using the implicit function option $F(x, y) = 0$ of *MicroCalc*), and estimating the coordinates of the four points of intersection. Use Program 5 of *Exploring Calculus* to complete the problem.

Laboratory 7

Test your Calculus Skills

In this session we use programs on the diskette *Exploring Calculus* that are designed to test your skills in differential Calculus.

Exercises

A. Call up Program N and choose Topic 1 (on the first derivative). Work the quiz repeatedly until you can confidently get a grade of 85%. Make screen dump #1 of a score of 85% or higher. HINT: To estimate the first derivative (=slope) at a point, it helps to hold something with a straight edge (a piece of paper) up to the screen to use as a tangent line to $f(x)$ at the point in question, and then estimate the slope of this tangent line.

B. Choose Topic 2 and work the quiz (on the second derivative) until you can get a grade of 85%. Make screen dump #2 of a score of 85% or higher.

C. Choose Topics 1 and 2 simultaneously and work the quiz until you can get a grade of 85%. Make screen dump #3 of a score of 85% or higher.

D. Call up Program M and choose Topic 1: Graphing the first derivative. Select the easy mode (E). Graph first derivatives until you can get a grade of at least 75%. Make screen dump #4 of a score of 75% or higher. Remember to give your partner a chance too!

E. Call up Program M and choose Topic 2: Graphing the second derivative f''. Select the easy mode (E). Graph second derivatives until you can get a grade of at least 70%. Remember that $f''(x)$ is the rate of change of $f'(x)$, that is, the rate of change of the slope. So if slopes are increasing as you move from left to right (which is quite different from the slopes being positive) then $f''(x)$ is positive, and the more rapid the increase in slopes, the more positive $f''(x)$. Make screen dump #5 of a score of 70% or higher. Remember to give your partner a chance too!

HINT: If you find this hard to do and need practice, call up Program L and graph the function $f(x) = \dfrac{1}{x^2 + 1}$ and its second derivative. After the function is graphed, quickly hold down the control key <Ctrl>, and while holding it down, press the <Numlock> key. This freezes the run of the program until any key is pressed. While the action is frozen, decide where the graph of the second derivative should be drawn. Then press any key to resume the actual graphing of the second derivative. If you need more practice, do the same for the functions $f(x) = x^3 - 3x^2$ and $f(x) = x^4 - 3x^2$ and $f(x) = \sin(2x)$. (If you can't seem to get the hang of graphing $f''(x)$ just by doing examples, you can always resort to finding the slope a half unit to the left and a half unit to the right of each point in question, and then calculating how much the slope has changed over that interval of length 1, which is an estimate of the rate of change of slope.)

Lab Report

(1) Write a short account of what you did during your lab session.

(2) Turn in this Lab #7 handout with screen dumps attached as indicated.

(3) Turn in the homework and challenge problem.

Homework

Call up Program N and choose Topic 7: Motion on a line. Work the quiz on Topic 7 until you can get a grade of 80%. Make screen dump #6 of a score of 80% or higher.

NOTE: *Average velocity over an interval* is approximated by estimating the velocity (slope of f) at several equally spaced points along the interval, and taking their average. (10 points should be plenty.)

Average acceleration is similar to the above but averaging the slopes of f' (can you explain why?)

Resultant distance traveled: the end position minus the beginning position.

Total distance traveled: You must break this down into several separate intervals. If f is increasing to begin with, then the first interval ends where f starts to decrease. What was the distance traveled to that interval? The next interval (f decreasing) continues until f starts to increase. What was the distance traveled on this interval? Etc. Then add up all the distances traveled for the total.

Challenge Problem

Run Program F asking for options 1, 2, 3, 5, 6 (separated by commas), until you can get a grade of at least 70%. Make four screen dumps having grades of 70% or higher.

Laboratory 8

Integration

In this session we use Program S and Program U from the diskette *Exploring Calculus* to estimate

definite integrals. Program S graphically illustrates an approximation by rectangles, trapezoids, and parabolas. Program U gives you a single Simpson's rule (parabolas) estimate for a definite integral.

Exercises

A. Call up Program S and use Options 1, 2, and 3 to estimate the integral $\int_1^3 \left(\frac{1}{x^4} + 1 \right) dx$ by a Riemann midpoint sum, a trapezoidal sum and by Simpson's rule (parabolas). Use $n = 2, 4, 8, 16$ and fill out the table:

	Midpt Rects	Trapezoids	Simpson's
$n = 2$			
$n = 4$			
$n = 8$			
$n = 16$			

Find the exact value of the integral by hand. Answer: _____ .

Which of the three methods is best? _____ .

B. Use Program S to estimate

$$\int_{-1}^8 (4 \sin x + 2.5 \cos 3x) \, dx .$$

Use $n = 4, 16, 32$ and fill out the table below.

	Midpt Rects	Trapezoids	Simpson's
$n = 4$			
$n = 16$			
$n = 32$			

Integrate by hand and express your answer to seven significant digits: _____ .

Which of the three methods is best? _____ .

C. Use Program U to estimate

$$\int_0^4 [12 \sin x - 3 \cos(x/2)] \, dx$$

by Simpson's rule.

If $n = 50$, then the estimate is _____ .

If $n = 100$, then the estimate is_____ .

D. The integrals

$$\int_{-1}^1 2/(1 + x^2) \, dx$$

and

$$2 \int_0^1 2/(1 + x^2) \, dx$$

both have value π. Using Simpson's rule with $n = 10$,

The estimate for the first integral is: _____ .

The estimate for the second integral is: _____ .

Which value is more accurate? Explain why you would expect one to be more accurate than the other.

E. Approximate the value of x for which $\int_0^x \sin\sqrt{t}\,dt = 1$. Graph the integrand $f(x) = \sin\sqrt{x}$ first and get a screen dump. By looking at the graph, try to get a rough estimate of how far to the right you have to go (starting at 0) to accumulate enough area under the curve to be equal to 1 square unit. This will be your first estimate of x in the upper limit of the integral. Then use program U to get better and better estimates for the true value of x.

Approximate value of x: _____ .

Lab Report

(1) Write a short account of what you learned during your lab session.

(2) Turn in this Lab #8 handout with boxes filled in and with screen dumps attached. It has to be clear which screen dump goes with which exercise. Label them!

(3) Answer the question at the end of exercise D.

(4) Turn in the homework and challenge problems.

Homework

(1) Choose Topic 3 of Program N to test your skills in integration. Obtain a screen dump showing a grade of 70% or better.

(2) Choose Option 3 of Program M to draw indefinite integrals (antiderivatives). Use the easy mode (E). Make a screen dump of a grade of 75% or higher.

Challenge Problem

As in Exercise E, use Program U to get an approximate solution of the equation $\int_0^x e^{-t}\,dt = 1$. Again, you should graph the integrand first to get an idea how big x will have to be. Note, that if x gets big, you should also take large values of n for better accuracy. Why? Also do this problem by hand. Something very odd is happening. Explain!

Student Comments

The questionnaire (Table I) that was given to students at the end of the semester, also asked the students to comment on

a. What they liked about the use of computers in the course,

b. What they did not like about the use of computers in the course,

c. What changes they would make.

Following are some typical responses, (grammar, spelling, and punctuation as written by the students):

"I did feel the computer section was beneficial in helping to understand derivatives and integrals. It was also a good experience to get a hands on approach to computers and computer graphics. The one aspect I didn't like about the section was the fact that for several labs a great deal of time had to be spent outside of class to finish the labs."

"What I liked about the computers in this course was that you could actually see the functions and their derivatives. You didn't have to draw them yourself. You could also visualize your own functions and have them graphed to your liking."

"The computers used were a helpful aid in understanding calculus. The only thing that I don't understand is why we didn't get any extra credit for doing these, I mean an extra credit hour because we covered the same material as the other classes plus we had the computer assignments."

"What I liked about the computers was that you could learn how to use them and it helped me with other classes. You could also see the problems happening on the screen and since the visual sense is the strongest it was very helpful."

"I liked using computers because they helped me to better understand the graphing of integrals and derivatives. I also liked learning to integrate computers into mathematics because I think it is becoming very important and will be even more so in the future."

"The computer aspect to this course made the process of learning and perceiving the basic fundamentals, theorems, and processes of calculus a lot easier. The two phased, class and lab, instruction built up the knowledge and usage of calculus better than the plain calc course without the computer. The course is fine enough and at the right instruction pace for most of the students, so as I see it no changes should be required."

"It gave me a better outlook on calculus and made it far more interesting. I would make no changes in the program except to make it available to all calculus I classes."

"The computers helped me understand some of the material easier than just learning it from a book. I would recommend using the computers in the future to help understand the material more."

"There are quite a few computer-math programs on the market and 90% are better than the ones we used. I think the labs should be done by one person, after all, people will work in groups anyway, but everyone should have to turn in their own report."

"The course was interesting and the assignments were well planned. The teacher and aid were willing to help, and the professor gave clear explanations in the beginning of her class periods. I disliked nothing and believe it should remain the same."

"It is easier to deal with a function if you can see it."

"The home work assignments took a little longer than they were supposed to. The only change I would make is to shorten the assignments by a couple of questions."

"I would use the lab time to help students with their problems instead of assigning more homework and projects."

"I did think the computer helped in the understanding of many areas. I hope it is continued. The lab assignments, at times, were difficult to finish in a reasonable period of time."

"Using the computer helped me to understand better what was going on in class. It was a tool that we had to use instead of just sitting in class and having the teacher show us everything. No changes. The labs thoroughly covered the harder material in the chapters. That was good. All in all, they made the class more interesting."

"The only change that I think would help are shorter assignments and faster processing computers."

"The use of computers in this course helped me to gain a better understanding of the course material. This course is very well structured. I learned a great deal from it."

New Mexico State University

STUDENT RESEARCH PROJECTS IN THE CALCULUS CURRICULUM

*by Marcus S. Cohen, Edward D. Gaughan, R. Arthur Knoebel,
Douglas S. Kurtz, and David J. Pengelley*

 Abstract

Contact:

DAVID J. PENGELLEY, Department of Mathematical Sciences, Las Cruces, NM 88003-0001. PHONE: 505-646-3901. E-MAIL: calculus@nmsu.edu.

Institutional Data:

New Mexico State University is a land grant university with an enrollment of about 14,000. It has a graduate program offering doctoral degrees in Agriculture, Education, Engineering, and the Sciences. Most undergraduates are full time students and 28.4% are minority students. Students in the mainstream calculus courses are mainly engineering majors, of which 33.5% are minority students.

The Department of Mathematical Sciences has 36 full time faculty and 24 graduate assistants. There are 54 undergraduate majors, including students from the College of Education with mathematics as a major teaching field. The graduating class of mathematics majors this spring consists of 5 females and 12 males. The department offers the B.S., M.S., and Ph.D. degrees.

Project Data:

The program at New Mexico State University began in the Fall of 1987 and has expanded to include all courses in the calculus sequence. Six faculty and five teaching assistants are teaching half of all our calculus classes with our new materials, known as student research projects. All calculus classes, both project and nonproject, have enrollments of 40-45 students. The text is *Calculus and Analytic Geometry*, Stein, 4th edition. Funding has been received from the National Science Foundation for the period June 1988 to December 1991.

Project Description:

Our program of instruction involves the use of "student research projects" in our mainstream calculus courses. A project is a meaningful and substantial problem, harder and more involved than the usual homework exercise found in the textbook. Students are usually given two weeks to work on the problem. They must analyze the problem, decide on a strategy for the solution, and bring appropriate mathematical ideas to bear on the problem.

Students in the calculus come from mathematics, engineering and the sciences, and those in project classes are self selected in the sense that they may choose between project and nonproject classes. In a typical project class, there will be two two-week projects, quizzes and/or examinations, and a comprehensive final examination. Class size and day-to-day organization of the courses are similar for all classes. During the period when students are working on a project, a "project lab" staffed by teaching assistants and undergraduates is open 10-12 hours a week to offer assistance.

Students submit a written report giving their solution to the projects. The project grade is based on mathematical content and written presentation. For some projects, students work in groups of three or less.

Our department has been quite supportive and we have attracted other faculty to our program. In addition, faculty at nine other schools have become involved.

Assessment of the program is underway. Preliminary data suggests that in the second and third semesters of calculus, pass rates are consistently higher in the project sections than in the nonproject sections. Final examination scores do not appear to be significantly different between project and nonproject sections. In contrast, pass rates in the project sections of the introductory calculus course have shown an inconsistent pattern. Some 30-45% of students in project sections have indicated that projects were the most valuable part of the course. Approximately 50% of project section students evinced a desire to take another projects course. These students frequently cited reasons such as "projects teach how to apply principles," "projects make you think," and preference for projects over exams. Another student quote is "I learned more from doing the project than I did from cramming for an examination."

We will be completing and publishing a book of projects this summer with guidance for instructors on their use.

═══ Project Report ═══

History

The calculus program at New Mexico State University began in the Fall of 1987. Two faculty members, Marcus Cohen and David Pengelley, decided to try something different in their third semester calculus courses. Their aim was to have students discover the excitement, intrigue, and beauty of calculus. Consequently, during that semester, students were assigned three two-week projects in place of the usual hour examinations. Professors Cohen and Pengelley hoped that this experience would fundamentally alter their students' view of

what mathematics is all about, and simultaneously build their self-confidence in what they could achieve through imaginative, theoretical thinking. Students had to use the ideas of calculus, not just cookbook techniques, to solve these problems. To succeed, students needed to discover the calculus as the proper (and usually the only) tool that would help them solve the problem. In a sense, the projects were theoretical—one could not do them without an appreciation of the ideas behind the method.

The results were encouraging. Students were able to experiment with ideas, ask relevant questions, and identify and read sections of the book that were not covered in class. They were able to synthesize and unify ideas by working on a complex problem over a long period of time. As one would expect, good students found projects exciting and viewed them as an individual challenge, while some of the failing students found it necessary to seek intensive help from their professors. In many cases, the professors were able to identify and correct plaguing misconceptions. As a result, these students were able to get back on their feet in calculus.

In the spring of 1988, a faculty member, Douglas Kurtz, and a teaching assistant, David Ruch, also started using student research projects in their second semester calculus courses and the excitement started to spread. That spring, Edward Gaughan and Arthur Knoebel joined the team and a proposal was written to the National Science Foundation. In July 1988, the group was awarded one of the first multiyear Calculus Curriculum Development Grants. Since its modest beginnings, the program has expanded to include all the courses in the calculus sequence, and next fall eight faculty and seven teaching assistants will be using projects in half of the calculus classes in the department. In addition, faculty at a number of other colleges and universities have been using some of our projects in their calculus classes. As a part of our grant funded by the National Science Foundation, four faculty at neighboring institutions have worked closely with us in developing our program. These faculty and their institutions are: David Arterburn, New Mexico Institute of Mining and Technology; Adrienne Dare, Western New Mexico University; Carl Hall, the University of Texas at El Paso; and Richard Metzler, University of New Mexico. Their advice and enthusiasm over the past two years has been very helpful.

The Program

Our program of instruction centers on the use of "student research projects" in our calculus courses. A "project" is a meaningful and substantial problem, harder, more open-ended and more involved than the usual homework exercises found in textbooks. Students are given an extended amount of time to work on the problem, usually two weeks. They must analyze the problem, decide on a strategy for the solution, and bring appropriate mathematical ideas to bear on it. The students need help in the beginning but they are usually surprised at how quickly their problem-solving skills develop. A sample project follows and more are included at the end of this article.

A Greenhouse Extension

Your parents are going to knock out the bottom of the entire length of the south wall of their house and turn it into a greenhouse by replacing some bottom portion of the wall by a huge sloped piece of glass (which is expensive). They have already decided they are going to spend a certain fixed amount. The triangular ends of the greenhouse will be made of various materials they already have lying around.

The floor space in the greenhouse is only considered usable if they can both stand up in it, so part of it will be unusable, but they don't know how much. Of, course this depends on how they configure the greenhouse. They want to choose the dimensions of the greenhouse to get the most usable floor space in it, but they are at a real loss to know what the dimensions should be and how much usable space they will get. Fortunately they know you are taking calculus. Amaze them.

This project is a good example of one of our "minimalist" statements: only the data essential to setting up a well-posed problem are given, and those are given in words, with no clues as to how to turn them into mathematical expressions. Instructors may expect to have to field many questions from students as they first get started with this type of project. On the other hand, some of our project statements tend more toward the "roadmap" approach, in which a student's progress through the project is guided in a step-by-step process. Some such examples are included.

Projects may be used to enhance learning of material that students typically find difficult. One such topic is vector algebra because of its placement in our calculus sequence. Poor student performance on this material prompted us to write projects related to vectors, vector algebra and the relations between vectors and calculus. One such project involved the Starship Enterprise being held captive in an elliptical orbit by the evil Klingons. The students enjoyed this project and many entered into the spirit of it with their own story line. Some students even took the time to learn about the equations of ellipses in polar coordinates (which had not been covered in class) and used polar coordinates to solve the problem.

Projects may also be used to cover important material that is not included in the course. There have been proposals in recent years to eliminate certain techniques of integration from the calculus course. Such topics can be made a part of a project. Two of our projects have involved the use of hyperbolic functions as a technique of integration. Students were asked to read about these functions and then discover and prove the identities necessary to make hyperbolic substitutions useful.

A few of our projects required computers and quite a number required a calculator. However, our goal was to develop a program that could be used to enhance the understanding of calculus without requiring sophisticated equipment and software.

Writing is also a very important part of the process of completing the projects. A written report must be presented to the instructor explaining the solution, using correct English and correct mathematical language and notation. Since the students have two weeks to work on the project, you can expect to see their best work. A student's project grade will be determined by the mathematical content and the written presentation.

The final task of a project is a discretionary ten to fifteen minute private conference between student and instructor. One may choose to interview very good students to encourage them to think about a career in mathematics. A less successful student could be counseled on mistakes and what is needed to improve on subsequent projects. These interviews play the additional role of assuring us that the project solution was the student's own work.

We believe that students who are using projects are learning both critical thinking and scientific writing, using calculus as their laboratory. We are attempting to teach them the tools of mathematical thinking and application. Mathematical thinking is, after all, at the foundation of science and engineering, and the calculus sequence is the best place to learn it.

The Setting

New Mexico State University is the land grant

university for the state of New Mexico and has an enrollment of approximately 14,000. Most of our calculus students come from the College of Engineering and the sciences. The University has a large graduate program offering doctoral degrees in various fields in Agriculture, Education, and Engineering, as well as Biology, Chemistry, Computer Science, Mathematics, Physics and Psychology. Most undergraduates are full-time students and university records show that 28.4% of the undergraduates are minority students.

The Department of Mathematical Sciences has 36 full-time tenure-track faculty and 24 graduate assistants. The department offers the B.S., M.S., and Ph.D. degrees in Mathematics.

Calculus classes at New Mexico State University enroll about 40 students each, and the courses share a common syllabus. Each of the first two semesters of calculus has a common final examination. A MWF class typically meets 45 days a semester for 50 minutes each period. Because both project and nonproject classes are offered to the student, we decided not to attempt any changes in the course content of the three calculus courses. Students may register for a project or for a nonproject class. Typically about half of our calculus sections have been project-based.

How It Works

Individual project classes differ considerably from instructor to instructor. For example, some instructors assign only two projects, while others have assigned up to five projects in a semester. From our experience we know that the project <u>must</u> be a significant part of the the student's grade and that the length of time that students have to work on the project must be specified and limited. These conditions are crucial for creating the student commitment required to successfully complete a challenging project.

What follows is a description of a typical class that assigns projects. In this hypothetical course, there will be two in-class hour examinations, each counting 15% of the course grade; two two-week projects, each counting 15% of the course grade; graded homework worth 10% of the course grade; and a two hour comprehensive final examination worth 30% of the course grade. The first examination will be given sometime during the 4th week and the first project given out immediately after the first examination. The timing of the project and the

examination is adjustable based on what has been covered in class, but we found it was not wise to have students working on a project and preparing for an examination at the same time. The second examination will be given during the 9th week and the second project given out shortly after the examination.

Before the project can be assigned, it must be developed and carefully written. Writing a project, in contrast to writing an examination, may extend over a period of several weeks. A typical project will begin with an idea and both faculty and teaching assistants will have an opportunity to be involved in the process of creation of the project. A large part of our efforts over the past two years has been to create projects. We will publish a collection of projects that includes some guidance for instructors on designing and using them. The final writing and editing should be done in the summer of 1990.

When our program began, we would distribute among the students three to six different projects for each assignment. The idea was to discourage students from copying each other's work. One of the main problems was the need to assign projects of equal difficulty. Since we discovered that cheating was a negligible problem, now we usually assign the same project to the entire class. When the project is handed out, a cover sheet is attached that gives advice and rules of the game. A copy of "How to Work on Your Project: What is Expected" is included in the supplementary material. It is expected that students will need assistance with the work and, in fact, they are encouraged to seek counsel from their instructors, even if they think all is well. As you might imagine, this can consume many hours of faculty time.

To ease the demand on faculty time, we have designed a laboratory where students go and seek assistance on the projects. Initially, this lab was staffed by teaching assistants who were also teaching a section of the same course using projects. Now we have found that we can staff the lab with senior undergraduates. This will make our program more practical at those schools where teaching assistants are not available. Each of the three calculus courses has a lab, open about twelve hours a week when students are working on projects. We try to be open weekend and evening hours to accommodate as many students as possible. In addition, we have some of the lab hours set up for appointment only. This allows students with very tight schedules to avoid long waits in the lab. The lab is closed the two days prior to the day the projects are due. During these two days, students must see their instructor, who has extended office hours, for help. The purpose of this is to avoid unpleasant situations in the lab. Frantic students, who waited until the last minute to begin, must face their own instructor.

The use of graduate teaching assistants in the projects lab and in teaching their own calculus classes with projects has an added benefit: the teaching assistants are trained in the method and contribute their own ideas and energy to the program. One of our graduate students' innovative contribution to the program proved valuable in his obtaining a university position. An unexpected result of our program is the degree of camaraderie that has developed within the group of faculty and teaching assistants involved with projects. We meet regularly as well as talk informally among ourselves about teaching and students' work on the projects. Teaching calculus has become a common topic of coffee room conversation again.

We require students to write a report explaining their solution. Most students are not accustomed to writing mathematics and our experience is that their first efforts may be rather poor. It is a good idea to have students show their instructor a preliminary version of the paper several days before it is due.

For the second project of the semester students are encouraged to work in groups of three or fewer. We allow the students to choose their own groups. The only problem seems to be that some groups find their schedules don't allow them to get together very often. We try to warn them about this problem early enough to give them time to make up their groups. If the group needs assistance, we allow a single representative to go to either the lab or the instructor. In this way, that person must report back to the group, encouraging mathematical dialogue among the students. Each group turns in a single report and each member gives a confidential evaluation of the relative contributions of all members of the group. All members of a group will not necessarily get the same grade on a project.

Grading a project takes about the same amount of time as grading an examination. It is less stressful since you are grading your students' best work. They have thought long and hard about their problem, done background reading, and discarded false starts. They have also eliminated algebraic errors so the problem of determining partial credit is not a big issue. Project grades are letter grades, and

mostly A's and B's. If numbers of students and other time constraints do not allow all students to be interviewed, we interview selected students, usually those who did poorly and need counseling or whose grades are borderline.

Institutional Support

Our department has been quite supportive. Since the beginning of the NSF sponsored part of the program, the number of faculty teaching project classes has grown from five to nine. Our department head has been very encouraging and, with the approval of the college dean, she has given us some released time to develop the program. The College of Engineering, our primary source of calculus students, believes that we are teaching their students valuable problem-solving skills and is supportive of what we are doing. A few critics claim that students are working on projects at the expense of other calculus skills. Results on common examinations seem to refute that claim.

In the Fall of 1990, we will begin to integrate our program into the department. The faculty of our department approved our proposal to give one additional credit hour to students enrolled in calculus classes that assign projects. This is partially in response to students' comments about the amount of time they invest in a project. This will not involve any change in the way project classes are handled.

Is It Working?

Assessment of the program is underway. Preliminary data suggests that pass rates are consistently higher in the project classes than in the nonproject classes, in the second and third semesters of calculus. Scores on final examinations do not appear to be significantly different between project and nonproject sections. In contrast, pass rates in the project classes of the introductory calculus courses are inconclusive.

About 40% of the students from sections which used projects have indicated that projects were the most valuable part of the course. Approximately 50% of project section students evinced a desire to take another projects course. These students frequently cited reasons such as "projects teach you how to apply principles," "projects make you think," and "I learned more from doing the project than I did from cramming all night for an examination."

On the one hand, using projects in a calculus course is more work than teaching in the traditional way. One has to find or design the projects, someone has to arrange and schedule the lab, one spends more time talking to students, and one may be involved in training other faculty to use projects. But the extra time spent is high quality time. Students speak to you about very specific questions that cut to the heart of what they are trying to do. This is in contrast to the typical "I don't understand anything" type of office visit. Good projects, often those with a story line, will excite the students and their papers are a pleasure to read. One learns a lot more about students using projects than using only examinations and we found that our students are capable of much more demanding work than we ever imagined. It is a unique and satisfying experience to compare notes on our students' creative approaches to problems as opposed to bemoaning dismal performances on examinations. Calculus is more fun to teach and, once hooked on projects, it's hard to go back to the old ways.

The Future

As mentioned above, projects will become integrated into the departmental program beginning in the Fall of 1990. Faculty participation in the program is voluntary. So the long range future depends on us, the faculty, to make the program attractive to both students and other faculty. The responses of other faculty so far make us believe this can be done. We hope to export our program to other colleges and universities by giving talks, workshops and minicourses at meetings across the country. We will publish a book containing about one hundred projects including guidance on constructing and using projects.

Presently, we are experimenting with projects in differential equations, and plans are in the works to include projects in the vector calculus course. Projects have been tried in a college trigonometry course and in a high school algebra course with encouraging results.

Beginning in the fall of 1990, we plan to work with several mathematics teachers in the local high schools to incorporate projects into their courses. This program will train teachers to create and use projects. We hope to discover the manner in which projects may best be incorporated into the high school curriculum.

═══ Sample Materials ═══

Instructions to TAs

The purpose of this document is to describe our philosophy about your role and obligations regarding "calculus projects".

You have a dual role in this department. You are both a student and a teacher. It is important for all of us to not lose sight of this duality and to keep the two roles in perspective. It is also important for you to know that we as faculty recognize the importance of both roles. We do not want your teaching to demand so much of your time that your graduate studies suffer. On the other hand, you must be willing to devote the time necessary to meet your responsibilities to your students.

As a teaching assistant involved in using projects, you have duties that differ somewhat from those of other teaching assistants. Some of those are detailed below.

1. You must meet with your course coordinator before the semester begins. Details, such as the number of projects, dates for projects, number of examinations, homework, etc., must be discussed prior to the first class meeting. Your syllabus should be designed in consultation with the course coordinator and should include pertinent information relative to projects and examinations. You should know enough about the use of projects to field related questions on the first day of classes.

2. At least a week before projects are assigned, you will receive copies of the projects for you to solve. In order for you to be able to help students with the project, you must understand it yourself, and the only way to know the project is for you to work out a complete solution. If there are several versions, it is your responsibility to work through each version. You should discuss your solution with your course coordinator before the projects are handed out.

3. During the period that the projects are out, classes should continue in the normal manner. Homework, quizzes, etc., occur as usual. There may be occasions where it is appropriate to spend part of a lecture on a topic directly related to the project.

4. You will be assigned times to work in the lab for project students. This is an important task and you should plan to be prompt and prepared for the lab. You will be expected to keep a head count at the lab. Your course coordinator will have discussed with you the types of hints that you can give, since

you will not actually solve the project for the students. An important part of your job in the lab will be to help students understand what is expected and keep them from heading down blind alleys. Do not agree to proofread a project for someone else's student. This can cause problems.

5. There will be a few days between the last day that the lab will be open and the day the project will be due. During those days, you are expected to have extended office hours for the purpose of helping your students. Be careful that students don't camp in your office and work on their projects. Perhaps you might want to post a sheet and have students sign up for a time (perhaps at 10 minute intervals) to avoid a line in the hall.

6. After the projects are turned in, you should plan to read them and return them promptly. The next class period is ideal; don't let more than one class period pass before you return the papers.

7. Read the project carefully, mark errors, note bad grammar and incorrect use of language and notation. If the errors in language and notation are excessive, make a note on the paper to that effect and don't try to mark all the errors. You may want to correct some of the notation and language errors, but don't correct the mathematics.

8. Discuss with your course coordinator which students you should interview. Return those papers without grades and ask the students to schedule interviews. All other projects should be returned with grades indicated. Grades for students to be interviewed will be assigned after the interview and graded papers should be returned to the student at the first class meeting after the interview if at all possible. Be prepared to discuss the grade with the student if requested. It is not uncommon for students to expect good grades on mediocre papers.

9. After each project cycle is completed, be prepared to give your course coordinator your impressions and comments.

Instructions to Students

Working on your Project: What is Expected

Here are some words of advice and encouragement along with the rules of the game.

First of all, this is a major, lengthy assignment. To do well you should start immediately, and work on it every day. You will probably need all the days you have been given in order to complete your project by the day it is due.

1. Start today. Let your subconscious work for you. It can do amazing things. If you immerse yourself in the project, solutions will come to you at the strangest times.

2. Read the entire project to see what it's all about. Don't worry too much about details the first time through. Do this today.

3. Next, read the project very carefully and make a list of any unfamiliar words or concepts you encounter. If concepts occur that you're not sure about, you must understand those ideas before you can do the project. Even if you understand all the words and terms, don't assume that the project is easy. If you wait until the last few days or so to start, it is doubtful you will be able to finish on time.

4. You may need to do some outside reading. There are lots of books in the library that contain information that might be helpful to you.

5. After you have worked a bit every day on the project, you will find certain parts easy and you will have completed those parts. You will have identified the hard parts and have begun to zero in on the obstacles. You are beginning to become the master of your project. You are so familiar with it that it is easy to sit and work on it if you have a spare minute or two. I recommend you work on the project some every day and keep a journal to record your progress.

6. While I expect you to work independently, I do not expect that you can work through the project without some assistance. I encourage you to come and talk to me about your project, and to get help in the special Projects Lab for your course. Even if you think things are going along smoothly, you should let me see what you are doing. This way I can head you off if you are going in a wrong direction.

7. There will be times when you need help and I'm not available. At those times, you should go to the Projects Lab for assistance. It is open several hours daily for most of the days you will be working on your project. You will receive a separate schedule showing the Lab hours.

8. Don't go to the the Projects Lab expecting someone to work your project for you. The purpose of the lab is to give you guidance and let you know if you are on the right track. Come with specific questions, and be prepared to show clear written work that you have prepared in advance.

9. When you have done the work necessary to complete the project, you need to prepare it in writ-

ten form. The paper you turn in should have a mix of equations, formulas and prose to support your conclusions. Use complete sentences, good grammar and correct punctuation. Spelling is also important. The prose should be written in order to convey to the reader an explanation of what you have done. It should be written in such a way that it can be read and understood by anyone who knows the material in this course. You will be graded on your written presentation as well as the mathematical content.

Sample Projects

In the following sections we provide a selection of sample projects assigned at various times during the term.

Composing Functions

Consider the two functions:

$$f(x) = 1 - x \quad \text{and} \quad g(x) = \frac{1}{x} \; .$$

We can compose them in two ways:

$$f(g(x)) \quad \text{and} \quad g(f(x)) \; .$$

We can go further and compose these two new functions with themselves, and also with the old ones, in a number of ways. Keep composing these functions with new ones as they are generated and figure out simplified formulae for them in terms of the variable x . (Don't forget to compose functions with themselves, like $f(f(x))$.) You might think that more and more new functions will be generated. Surprisingly only a finite number of new ones are generated by composition, even though there may be many different ways of composing f and g to get the same function. Remember that two very different looking formulae may represent the same function.

a) How many distinct functions are there, including f and g themselves?

b) List them.

c) How is each one composed from f and g?

d) How do you know that these are all there are?

e) For what real numbers are all these functions simultaneously defined?

A Continuous Additive Function

In this project you will learn about additive functions and discover that continuous additive functions have a very special form.

Definition: Let f be a function whose domain is the set of all real numbers. Then f is *additive* if $f(x + y) = f(x) + f(y)$ for all real numbers x and y .

a) Give an example of an additive function and show that it is additive. Give an example of a function that is not additive and show that it is not additive.

b) Suppose that f is an additive function and m is any real number. Define a new function g by the formula $g(x) = f(x) - mx$. Show that g is an additive function.

c) Let $m = f(1)$ and show that the function g defined in part (b) above has the property that $g(x + 1) = g(x)$ for all x .

Definition: A function f is *bounded* on a closed interval $[a, b]$ if there is a real number B such that $|f(x)| \leq B$ for all x in $[a, b]$. A function f is *bounded* if there is a real number B such that $|f(x)| \leq B$ for all x.

d) Suppose that f is an additive function that is bounded on $[0,1]$ and that g is defined by $g(x) = f(x) - mx$ where $m = f(1)$. Show that g is bounded.

e) Let g be a bounded additive function; that is, there is a real number B such that $|g(x)| \leq B$ for all x. If there is a real number a such that $g(a)$ is not zero, show that there is a real number t such that $|g(t)| > B$, contradiction. What can you conclude about g? Explain. Hint: what is $g(2a)$, what is $g(3a)$?

At the beginning of the project, you were told that you would discover that continuous additive functions have a very special form. As you might suspect, the preceding parts of this project are designed to help you in that discovery. You will need to use some facts about continuous functions.

f) Show that if f is additive and continuous, there is a real number m such that $f(x) = mx$ for all x.

Preparing for the 1990 Census

The government census bureau is preparing to take the census of the United States in 1990 and you have been hired to help them plan for the counting of the American population. You job is to make a projection of the population so that the bureau will know how many census takers to hire.

Your director, Dr. John Knowitall, wants you to proceed as follows. He wants you to use the actual population data from two consecutive polls, along with the actual rates of population growth at the times of the two polls. He wants you to find a quadratic function which models the population as a function of time and agrees with the actual populations and the rates of growth of the two polls.

a) Your first task in this project is to show Dr. Knowitall that in general it is impossible to find such a function. To do this, let t_0 be the year of the first poll, and let t_1 be the year of the second poll. Also, let p_0 and p_1 denote the population values from the polls and let r_0 and r_1 denote the rates of growth from the two polls. To simplify your computations, write your polynomial in the variable $t - t_0$. (If you do not think this really does simplify the arithmetic, repeat part (a) using a polynomial in the variable t and see what kind of computations are needed.)

Being the resourceful person that you are, you realize that the difficulty was caused by the fact that the quadratic function was a polynomial of too low a degree to allow you to include all the information that Dr. Knowitall wants.

b) Show that by increasing the degree of the polynomial to three, it is always possible to fit the data to a new polynomial (that is, to find a polynomial which agrees with the populations and the rates of growth of any two polls).

c) After completing parts (a) and (b), use the actual census data below along with your results from part (b) to find population functions for each pair of consecutive poll data. Use this to project the population of the United States for 1990, as follows. First use your polynomial modelling function for each two consecutive polls to project the population of the succeeding census and compare the result to the actual census taken. Use all this information to make your projection for 1990. How confident are you of your projection? Discuss why.

Year	Population	Rate of growth (people per year)
1950	151,325,798	2,357,930
1960	179,323,175	2,598,812
1970	203,302,031	2,360,967
1980	226,542,518	2,324,049

Invariant Areas

Consider the hyperbola $xy = 1$, with $x > 0$ and $y > 0$. We are going to show that the lines tangent to this curve have a very special property.

Definition: The line tangent to the curve with equation $y = f(x)$ at a point $(a, f(a))$ on the curve is the line through $(a, f(a))$ with slope $f'(a)$. We call this line a *tangent line*.

The tangent line at any point on the hyperbola forms a right triangle with the coordinate axes. Your first problem is to show that the area of this triangle does not depend on the point chosen. To prove this, you may want to proceed as follows. Expressing y as a function of x, you are led to consider the function $H(x) = 1/x$. What is the equation of the line tangent to the graph of H at $(a, H(a))$? Draw a graph of this function, a typical tangent line, and the triangle in question. What is the area of the triangle? If the area depends on the point $(a, H(a))$ you have made a mistake.

Now comes the interesting part. You are going to find all functions which have this property. Let f be any function such that $f(x) > 0$ whenever $x > 0$, the derivative $f'(x)$ is always negative, and the second derivative $f''(x)$ is never zero. Consider a triangle formed by the coordinate axes and the line tangent to the graph of f at a point $(a, f(a))$. Assume that it has area which is independent of the point $(a, f(a))$.

As in the first part, find an explicit formula for the line tangent to the curve $y = f(x)$ at the point $(a, f(a))$. Then, find the area of the triangle formed by the tangent line and the coordinate axes.

Assume that the area is constant, which gives you an equation in the variable a. Differentiate this equation in a, simplify, factor, and use the conditions on f to arrive at a simple differential equation. Solve this differential equation to find all the solutions to the problem.

Upper and Lower Riemann Sums

Let $f(x)$ be a positive, continuous, increasing function on $[a, b]$. Your goal is to develop a technique to estimate the integral $\int_a^b f(x) \, dx$ within a given margin of error without evaluating the integral.

a) Your first step is to understand two special kinds of Riemann sums. Partition $[a, b]$ into n equal parts. What is the length of each subinterval? For the first kind of Riemann sum, choose the height of each rectangle to be the maximum value of f over the subinterval. How do you know the maximum exists? For what value of x do you get the maximum value? The Riemann sum defined by taking the maximum value of f over each subinterval is called an *upper sum*. Explain why $\int_a^b f(x)\,dx$ is less than or equal to any upper sum. Write explicitly the upper sum when $n = 4$.

b) For the second kind of Riemann sum, repeat part (a), but replace the maximum value of f over each subinterval by the minimum value. This Riemann sum is called a *lower sum*. Explain why $\int_a^b f(x)\,dx$ is greater than or equal to any lower sum. Write explicitly the lower sum when $n = 4$.

c) Let U_4 be the upper sum and L_4 the lower sum for $n = 4$ found above. The difference $U_4 - L_4$ is non-negative. Why? Simplify $U_4 - L_4$ as much as possible. Next, find a very simple expression for $U_n - L_n$. Explain why your expression is correct.

d) Part c) gives you a way to make the upper and lower sums as close to each other as you wish. Suppose ϵ is a given positive real number. How can you use the result of part (c) to estimate $\int_a^b f(x)\,dx$ with a margin of error at most ϵ? Explain.

e) Estimate $\int_{0.5}^2 (1 + x^2)^{1/3}\,dx$ within 10^{-1} using the ideas from part (d).

Extra credit: Can you think of other ways (than the maximum or minimum) to choose the value of f on each subinterval that would give better approximations to the integral than U_n or L_n? Discuss your ideas.

Houdini's Escape

Harry Houdini was a famous escape artist. In this project we relive a trick of his that challenged his mathematical prowess, as well as his skill and bravery. It will challenge these qualities in you as well.

Houdini had his feet shackled to the top of a concrete block which was placed on the bottom of a giant laboratory flask. The cross-sectional radius of the flask, measured in feet, was given as a function of height z from the ground by the formula

$$r(z) = \frac{10}{\sqrt{z}} \, ,$$

with the bottom of the flask at $z = 1$ foot. The flask was then filled with water at a steady rate of 22π cubic feet per minute. Houdini's job was to escape the shackles before he was drowned by the rising water in the flask.

Now Houdini knew it would take him exactly ten minutes to escape the shackles. For dramatic impact, he wanted to time his escape so it was completed precisely at the moment the water level reached the top of his head. Houdini was exactly six feet tall. In the design of the apparatus, he was allowed to specify only one thing: the height of the concrete block he stood on.

a) Your first task is to find out how high this block should be. Express the volume of water in the flask as a function of the height of the liquid above ground level. What is the volume when the water level reaches the top of Houdini's head? (Neglect Houdini's volume and the volume of the block.) What is the height of the block?

b) Let $h(t)$ be the height of the water above ground level at time t. In order to check the progress of his escape moment by moment, Houdini derives the equation for the rate of change dh/dt as a function of $h(t)$ itself. Derive this equation. How fast is the water level changing when the flask first starts to fill? How fast is it changing when the water just reaches the top of his head? Express $h(t)$ as a function of time.

c) Houdini would like to be able to perform this trick with any flask. Help him plan his next trick by generalizing the derivation of part (b). Consider a flask with cross sectional radius $r(z)$ (an arbitrary function of z) and a constant inflow rate $dV(t)/dt = A$. Find dh/dt as a function of $h(t)$.

Extra credit: How would you modify your calculations to take into account Houdini's volume, given Houdini's cross-sectional area as a function of height?

Undetermined Coefficients

There are many integration problems which are straightforward to solve but computationally difficult. Often, however, with a little thought we can guess the form of the answer and then use differentiation to find the antiderivative. In this problem, you are going to learn the technique itself in one special case and then generalize it to other situations.

a) Let A and B be real numbers and consider the integral $\int \cos(Ax)e^{Bx}\,dx$. From the integrals of $\cos(x)$ and e^x, we might guess the antiderivative is a multiple of $\sin(Ax)e^{Bx} + C$. Show this is not correct. By considering the derivative of $\cos(Ax)e^{Bx}$, we might guess that an antiderivative involves a linear combination of trigonometric functions times

e^{Bx}. Use this idea to find the antiderivative of $\cos(Ax)e^{Bx}$ and use this to evaluate

$$\int \cos(3x)e^{4x}\,dx \ .$$

b) Using the technique of part (a), derive and explain a technique for evaluating the integral $\int \cos(Ax)\sin(Bx)\,dx$, where $A^2 \neq B^2$, without performing any integration.

c) Many books have the following formula:

$$\int \cos(Ax)\sin(Bx)\,dx$$
$$= \frac{-\cos(B-A)x}{2(B-A)} + \frac{-\cos(B+A)x}{2(B+A)} + C \ .$$

Show that this is equal to your answer to part (b).

d) Let n be a positive integer and $P(x)$ a polynomial of degree n. Use the technique above to derive a technique for evaluating $\int P(x)e^{Ax}\,dx$. Explain your technique in detail. Evaluate the following integral:

$$\int (4x^6 - 2x^5 + 2x^3 - 10x^2 + 5)e^{4x}\,dx \ .$$

Extra credit: Can you think of other products of classes of functions where we could use these ideas? Discuss the technique in these cases.

Life on the Hyperbolic Curve

We have studied the use of trigonometric substitution in finding the antiderivatives of certain special types of functions. The success of these substitutions depends on certain trigonometric identities and the integrals of special combinations of trigonometric functions. In this project, you will learn about hyperbolic functions and how such functions may be used in place of the trigonometric functions in the substitution method for finding antiderivatives.

First, read about the hyperbolic functions. Any good calculus book can serve as a reference. Then, you will be ready to find the antiderivatives of the functions given below. You are to use substitutions with hyperbolic functions to find the antiderivatives and you must justify any identities or integration formulas that you use. Perhaps you might put all the justifications in an appendix and give references to the results from the appendix as you

use them. Since you are using the hyperbolic functions only as a tool to find the antiderivatives, your answers should involve neither hyperbolic functions nor their inverses.

Find the indefinite integral of each of the functions given below.
a) $(9x^2 + 4)^{-3/2}$
b) $\sqrt{x^2 - 4}$
c) $\dfrac{1}{x^2\sqrt{16 - x^2}}$
d) $\dfrac{1}{\sqrt{1 - x^2}}$

In part (d), you probably expected to get $\arcsin(x) + C$. How do reconcile this with your answer?

Cubics and Coordinate Systems

In this project you will discover some surprising facts about the graphs of cubic polynomials.

a) Carefully graph the polynomial function $f(x) = x^3 - 3x^2$. Be sure to find and identify critical points and inflection points.

b) By now you have discovered that f has a relative maximum at $x = a$ and a relative minimum at $x = b$. Let $P = (a, f(a))$ and $Q = (b, f(b))$. Draw the line through P and Q and let $y = L(x)$ be the equation of that line. Evaluate:

$$\int_a^b [f(x) - L(x)]\,dx \ .$$

c) Repeat parts (a) and (b) above for other cubic polynomials of your choosing until you see a pattern. Make a conjecture based on that pattern.

d) Return now to the function f given in part (a). The chord PQ and the graph of f intersect in three points, P, Q and a third point R. Two regions are thus formed. Find the area of each. What do you notice? Now explain the results observed in parts (b) and (c) in terms of what you have discovered.

e) Read about translations of axes and symmetry. Begin with the graph of $y = x^3 - 3x^2$ and translate the axes in such a way that the new origin is at the point R you found in part (d). What is the equation of the curve in the new coordinate system? Use this equation and symmetry to further explain your discoveries.

f) Make a conjecture concerning the symmetry of the graph of a general cubic polynomial function $f(x) = ax^3 + bx^2 + cx + d$. Now you are to prove that your conjecture is true. You will probably want to use translation of axes as you did in part (e).

Sticking the Tank

You have just been hired by the Environmental Protection Agency (EPA) under the Superfund program to measure the level of toxic wastes in buried tanks across New Mexico. Most of these tanks are cylinders, with their axes horizontal. You are to "stick the tank" by inserting a stick through a hole in the center of the top until it touches the bottom, then pulling it out and reading off the liquid level showing on the stick. They have hired you because you know calculus; they have faith that you can convert the "height on the stick" reading to "filled volume in the tank."

Assuming that the cross-sectional radius of the tank is R and its length is L, calibrate the stick for them. That is, convert height showing on the stick to volume of liquid. Check your results by doing the calculation in two separate ways:

a) Evaluate a definite integral that gives the filled volume in terms of the height h on the stick. Do not use tables! (Suggestion: place the origin of your coordinate system at the center of the circular cross-section. Make a sketch!)

b) Use elementary geometry and trigonometry (no calculus) to obtain the volume.

c) Show that your results for a) and b) are equal.

Since you have been so successful in such endeavors, the EPA sends you out to stick a tank which has the shape of an elliptical cylinder, i.e. whose cross section is an ellipse instead of a circle. The major axis of the ellipse is horizontal. Calibrate the stick for them, using calculus.

Extra credit: Calibrate the stick for tilted tanks of circular and elliptical cross sections.

Space-Capsule Design

You are part of a team of engineers designing the Apollo space-capsule. The capsule is composed of two parts:

1) A cone with a height of 4 meters and a base of radius 3 meters;

2) A re-entry shield in the shape of a parabola revolved about the axis of the cone, which is attached to the cone along the edge of the base of the cone. Its vertex is a distance D below the base of the cone.

Assume the capsule has uniform density ρ. Your project director has specified that the center of mass of the capsule should be below the center of mass of the displaced water because he believes this will give the capsule better stability in heavy seas.

He has given your team the task of finding values of the design parameters D and ρ so that the capsule will float with the vertex of the cone pointing up and with the water-line 2 meters below the top of the cone, in order to keep the exit port 1/3 meter above water.

a) Show your project director that this task is impossible; i.e. there are no values of D and ρ that satisfy the design specifications.

b) Prove that you can solve this dilemma by incorporating a flotation collar in the shape of a torus (doughnut). The collar will be made by taking hollow plastic tubing with a circular cross section of radius 1 meter and wrapping it in a circular ring about the capsule, so that it fits snugly. The collar is designed to float just submerged with its top tangent to the surface of the water. Show that this flotation collar makes the capsule plus collar assembly satisfy the design specifications. Find the density ρ needed to make the capsule float at the 2 meter mark. Assume the weight of the tubing is negligible compared to the weight of the capsule, that the design parameter D is equal to 1 meter, and the density of water is 1.

Your investigations will be guided by a physical principle (I), a formula, which you will derive (II), and a theorem which you will state and prove (III).

I) *Archimedes' Principle:* A body floats in a fluid at the level at which the weight of the displaced fluid equals the weight of the body.

II) Derive a formula for the center of mass of the solid of revolution obtained by revolving the function $x = f(y)$ from $y = c$ to $y = d$ about the y-axis. Explain why the center of mass is on the y-axis so you just have to find $\bar{y}$.

III) Suppose a body is made up of two pieces, which are solids of revolution about the y-axis. Conjecture and prove a result relating the center of mass of the composite body to the centers of mass of the pieces.

HINT: Suppose one piece has center of mass at $\bar{y}_1$ and mass m_1 and the other has center of mass at $\bar{y}_2$ and mass m_2. Calculate the center of mass of the composite object by assuming m_1 is concentrated at $\bar{y}_1$ and all the mass m_2 is concentrated at $\bar{y}_2$. The formula you obtain for $\bar{y}$ should be your conjecture. Proceed with the proof using the definitions of $\bar{y}_1$, $\bar{y}_2$, and $\bar{y}$ in terms of integrals.

Escape from the Klingons

The starship Enterprise has been captured by the evil Klingons and is being held in orbit by a

Klingon tractor beam. The orbit is elliptical with the planet Klingon at one focus of the ellipse. Repeated efforts to escape have been futile and have almost exhausted the fuel supplies. Morale is low and food reserves are dwindling.

In searching the ship's log, Mr. Spock discovers that the Enterprise had been captured long ago by a Klingon tractor beam and had escaped. The key to that escape was to fire the ship's motors at exactly the right position in the orbit. Captain Kirk gives the command to feed the required information into the computer to find that position. But, alas, a Klingon virus has rendered the computer all but useless for this task. Everyone turns to you and asks for your help in solving the problem.

Here is what Mr. Spock has discovered. If F represents the focus of the ellipse and P is the position of the ship on the ellipse, then the vector $\overrightarrow{FP}$ can be written as a sum $\overrightarrow{T} + \overrightarrow{N}$ where $\overrightarrow{T}$ is tangent to the ellipse and $\overrightarrow{N}$ is normal (perpendicular) to the ellipse. The motors must be fired when the ratio $|\overrightarrow{T}|/|\overrightarrow{N}|$ is equal to the eccentricity of the ellipse. Your mission is to save the starship from the evil Klingons.

π in the Sky

You are the "super-programmer" for the world's largest supercomputer. Your boss has taken you aside to give you your next assignment. He explains, confidentially, that for several days now, Project SETI—which is concerned with the search for extraterrestrial intelligent life—has been receiving a string of digits from a powerful point source near Tau Ceti. He suspects that these are the digits in the decimal representation of π. However, he doesn't know when they started transmitting, so they may have gotten quite far out, perhaps to the millionth decimal place or beyond. Your assignment is to calculate π to more decimal places than ever before, in order to compare these computed digits with those from the extragalactic transmission. What you need is an algorithm; that is what you will develop here.

a) Express $\tan^{-1} z$ as the definite integral of some function $f(t)$, with z as the upper limit. Show how, if you could obtain a numerical value for this integral when $z = 1$, then you could get an exact numerical value for π.

b) Now write the integrand $f(t)$ as a geometric series. Use the identity for the sum of the first n

terms of a geometric series to express the integrand as the sum of the first n terms plus an "error term."

c) Integrate this sum term by term to get an expression for $\tan^{-1} z$ consisting of a polynomial plus an error term. Show that the error term has the form of an integral.

d) Give an upper bound on the error incurred when using only the polynomial of part (c) as your approximation to $\tan^{-1} z$. How many terms do you need to give a value for π accurate to a hundred decimal places? To a million? If the computation of each term and its addition to the previously computed terms takes $1\mu s$ on the supercomputer, then how many years will it take to compute π to a hundred decimal places? To a million? To understand why this is a poor series to compute π, evaluate the first six polynomials at $z = 1$.

e) You mention your problem over lunch to Sylvia, the mathematician down the hall. She jots a couple of formulas down for you:

$$(1) \qquad \frac{\pi}{4} = \tan^{-1} \frac{1}{2} + \tan^{-1} \frac{1}{3},$$

$$(2) \qquad \frac{\pi}{4} = 4 \, \tan^{-1} \frac{1}{5} - \tan^{-1} \frac{1}{239}.$$

Verify the correctness of these identities. (It is not sufficient to punch the *arctan* button on your calculator!)

f) How would you convert Sylvia's formulas into algorithms to approximate π? Do her formulas yield better approximations (in terms of time and money) than your original method in parts (a)-(d)? Evaluate the first six polynomials for both parts of (1) and (2) and put these beside the six term approximation in part (d). Compare the number of terms of your original series and the series generated by formulae (1) and (2) needed to approximate π to a million decimal places. How long will these formulae take on the supercomputer to yield approximations to π correct to a million decimal places?

Tribble Trouble

You are the captain of the Starship Enterprise and tribbles have invaded the hold of your ship. You are concerned enough to ask Spock, the Vulcan, for an analysis. As usual, Spock takes a superior attitude. Vulcans are experts with infinite series, although they never developed calculus. Hoping to teach you something, he gives his report in a series of riddles, to which you must supply the answers. Supply answers to questions (a) through (d) below, and then go Spock one better by showing him how Earthlings would use calculus to solve the problem in parts (e) through (h).

Spock Speaks

Let y_0 be the population of tribbles at some initial time $t = 0$, Captain. Then let y_n be the population after n intervals of length Δt have elapsed, i.e. at $t = n\Delta t$. If you allow one more interval to elapse, the population will have increased during the next interval by

$$\Delta y_n = y_{n+1} - y_n .$$

Suppose Δy_n is proportional to both y_n and Δt:

$$\Delta y_n = k \, \Delta t \, y_n ,$$

where k is the growth rate. (You may neglect the death rate in your calculations.)

a) Give an expression for y_1, the total number of tribbles after one interval Δt, in terms of y_0, the initial population at $t = 0$.

b) Give expressions for y_2, y_3, and y_n, the total number after 2, 3, or n intervals respectively, in terms of y_0. Make your expressions as compact as possible and factor out y_0. Rewrite your expression for y_n in terms of the total elapsed time t by setting $\Delta t = t/n$.

c) Now expand the expression for y_n you found in part (b) using the binomial theorem. In order to get better resolution and accuracy, let the time scale of your analysis become finer and finer; that is, let $n \to \infty$ as $\Delta t \to 0$. Assume that you can take this limit term by term; your result is an infinite series. Please write this series.

This series is your answer for the population at time t, Captain. You may calculate the answer to any desired degree of accuracy by including enough terms from the series.

Captain Kirk's Comeback

d) How do you know you can calculate the answer to arbitrary accuracy? For example, how far off would I be if I just added up the first hundred terms? The first N terms? Give me an upper bound for the error after 100 terms, and the error after N terms, Spock, and then prove that the error goes to 0 as N goes to ∞.

Now wipe that superior smile off Spock's face by showing him what can be done with Earthling calculus.

e) Start with the same expression for the increment, $\Delta y = (k\Delta t)y$, as Spock used in part (a), and convert it into a differential equation by taking the limit as $\Delta t \to 0$ and using the definition of the derivative. Solve this equation by separation of variables to find the function $y(t)$ which gives the population at time t in terms of the initial population y_0. It is your turn to smile. (Spock had to run to the ship's library to look up separation of variables.)

Spock's Comeback

Captain, I claim that your solution is no different from mine. You have simply renamed my infinite series as your function because Earthlings cannot handle the concept of infinity. Allow me to demonstrate, Captain. In a book in the ship's library, I came across something you Earthlings call the 'derivative.'

f) Differentiate my series term by term with respect to time and see if the answer isn't k times my series. Therefore, my series does satisfy your differential equation, since its derivative is k times itself. Now compare your function from part (e) with your answer in part (c). Your function and my series must be one and the same!

Kirk's Second Comeback

Not necessarily, Spock. The same differential equation may have many different solutions. Allow me to demonstrate ...

At about this time, McCoy and Scotty burst into your cabin. Scotty says, "You gentlemen can debate theory 'til you're blue in the face, but that does'na help us with our problem. *I'm* prepared to use radiation from the ship's engine just one time to exterminate 99% of the little beggars right now. Just give me the word, Captain!"

McCoy replies, "I cannot condone such a waste of life, Captain. I've got a drug that will cut their growth rate k in half. I think you'll find this a more effective and less brutal method than the one suggested by our engineer."

g) Decide between Scotty's and McCoy's strategies. Which is more effective in the short run? In the long run? Convince them of your answers.

h) In the future, will these two treatments ever result in the same number of tribbles again? When?

Rearranging an Infinite Series

The alternating harmonic series $\sum_{n=1}^{\infty} \frac{(-1)^{n+1}}{n}$ converges to $\ln 2$. Now suppose s is any given number. You are going to prove the amazing result that the alternating harmonic series can be rearranged (that is, its terms can be written in a different order) so that the resulting series converges and has s as its sum.

You will need to understand the precise definitions of the limit of a sequence and the sum of a series.

Hint: Start by trying to reorder the terms to alternately overshoot and undershoot s with various new partial sums.

Tetrahedra

Start with the hyperbola $xy = k$, where k is any arbitrary positive constant. Draw several tangent lines to it at various points in the first quadrant. Calculate the area of each triangle formed by the tangent line and the coordinate axes. What do you notice? Make a conjecture about the areas of these triangles and prove your conjecture.

Now generalize to three dimensions. Let k be a positive constant and consider the tangent plane to the surface $xyz = k$ at a point in the first octant. Consider the volume of the tetrahedron formed by this plane and the coordinate planes. Based on your investigation of hyperbolae, make a conjecture about how this volume changes as the point of tangency varies? Now prove your conjecture.

An Ant in the Sugar Bowl

An ant at the bottom of an almost empty sugar bowl eats the last few remaining grains. It is now too bloated to climb at a vertical angle as ants usually can; the steepest it can climb is at an angle to the horizontal with a tangent equal to 1. The sugar bowl is shaped like a paraboloid,

$$z = x^2 + y^2 \qquad (0 \le z \le 4),$$

where the coordinates are in centimeters.

a) Find the path the ant takes to get to the top of the sugar bowl, assuming it climbs as steeply as

possible. Use polar coordinates (r, θ) in the xy-plane and think of the ant's path as parameterized by r; then find a relation between the differentials $d\theta$ and dr, and integrate this relation to get $\theta(r)$.

b) What is the length of the ant's path from the bottom to the rim? To answer this, first discover a formula for arc length involving dz, dr, and $d\theta$ in three dimensions.

c) Draw a graph of the sugar bowl and the path the ants takes to get out. Hint: you may want to start with the projection $\theta(r)$ of the path in the $r\theta$-plane.

A Leaking Can

A cylindrical can with constant density is filled with your favorite drink and is sitting upright on the xy-plane. Unfortunately, it has a slow leak in the bottom.

a) Where is the center of mass of the can plus drink before the drink starts to trickle out? What do you think will happen to the center of mass as the drink flows out? Will the center of mass continue to sink as the drink empties? Where is the center of mass when the drink is all gone? Use intuitive arguments to support your answers to these questions.

Now we will use calculus to answer these questions.

b) Suppose R is an object in space divided into two pieces, R_1 and R_2. R does not necessarily have uniform density. Let M_1 and M_2 denote the masses of R_1 and R_2 respectively. Suppose $(\bar{x}_1, \bar{y}_1, \bar{z}_1)$ is the center of mass of R_1 and $(\bar{x}_2, \bar{y}_2, \bar{z}_2)$ is the center of mass of R_2. Find formulae for the coordinates $(\bar{x}, \bar{y}, \bar{z})$ of the center of mass of R in terms of $\bar{x}_1$, $\bar{x}_2$, $\bar{y}_1$, $\bar{y}_2$, $\bar{z}_1$, $\bar{z}_2$, M_1, and M_2. Show that $(\bar{x}, \bar{y}, \bar{z})$ lies on the line between the other two centers of mass.

c) Use the result of part (b) to give a formula for the height of the center of mass of the can plus drink when your remaining drink is at height z. Find out how high the drink is when the center of mass of the can plus remaining drink is lowest. Show that at that moment, the center of mass is at the top of the drink. Explain what is going on.

Extra credit: Can you develop an argument not using calculus which shows that the center of mass must reach a minimum, and that the minimum coincides with the top of the drink?

Purdue University

CALCULUS, CONCEPTS, AND COMPUTERS: INNOVATIONS IN LEARNING

by Keith E. Schwingendorf and Ed Dubinsky

Abstract

Contact:

KEITH E. SCHWINGENDORF and ED DUBINSKY, Department of Mathematics, Purdue University, West Lafayette, IN 47907. E-MAIL: ks@math.purdue.edu; bbf@j.cc.purdue.edu.

Institutional Data:

Purdue University is a Midwestern land-grant university with a fall 1989 undergraduate enrollment of 26,994 full time and 2,680 part time students, a graduate enrollment of 3,853 full time and 2,018 part time students. The majority of Purdue students are in engineering, science, management and agriculture.

The Department of Mathematics consists of about 90 full time faculty and 400 undergraduate mathematics majors (over half of which are in mathematics education and actuarial science) with about 100 bachelor's degrees awarded per academic year.

Project Data:

Since the fall of 1988, four teachers (Dubinsky, Daniel Gottlieb, Meyer Jerison and Schwingendorf) have taught in our project. Several experimental sections of Purdue's main line three semester engineering and science calculus sequence have been offered in the past four semesters. Our experimental sections with 25-40 students per section meet in a classroom three days per week and in a TA assisted computer laboratory twice per week. These compare with beginning engineering and science calculus lectures with over four hundred students that meet three times per week and have TA led recitations of about 30 students twice each week.

We have also taught a few experimental sections of our two semester calculus sequence for our management, social science and life science students. These experimental sections had 25-30 students meeting twice per week in a classroom and once each week in our computer laboratory, as compared to regular classes of about 40 students meeting three times per week.

Our required computer laboratory sections use Macintosh computers. We use the Mathematical Programming Language (MPL) *ISETL*, since it involves students in a level of programming that is essential to our theory of learning. We have created interfaces that make it possible for students to utilize the Symbolic Computer System (SCS) *Maple* in conjunction with *ISETL*. Future plans include making our course materials compatible with IBM computers using *ISETL* and *Derive*, and adding a graphics package to *ISETL*.

Although we desired to alter the content of our experimental sections at the beginning of our project, and we are doing so for AY 1990-91, our department has constrained us during the first two years by requiring us to cover the standard course syllabus allowing students to move back and forth between experimental and regular sections. We have used standard texts in our classes together with laboratory materials prepared by us. The distinctive feature of our project is not the technological component, but the fact that technology supports our underlying understanding of how people learn.

We have worked closely with other departments to determine not only what they expect students to learn in calculus, but we have also carefully examined many of their textbooks to see what calculus they actually use.

Our project was funded by an NSF grant for AY 1988-89 ($30,000). The setup of the Department's Macintosh Laboratory was partially funded by the Dean of the School of Science ($20,000), the Dean of Education ($12,000), the Mathematics Department ($18,000), an ILI grant ($47,500) and the Purdue University Computing Center (PUCC) ($13,000 per year). Staffing of the laboratory is partially funded by the Mathematics Department ($14,000 per year). Support from West Educational Publishing for 1989-92 ($45,000) provides funding for our project research and textbook development. In August 1990 the project was awarded a three-year NSF grant of $646,000.

≡ Project Report ≡

Basic Philosophy

The centerpiece of our project philosophy is the development of an emerging theory of how students learn mathematics. According to this theory, students need to construct their own understanding of each mathematical concept. Hence, we believe that the primary role of teaching is not to lecture, explain or otherwise attempt to "transfer" mathematical knowledge, but to create situations for students that will foster their making the necessary mental constructions.

A critical aspect of our approach is a decomposition of each mathematical concept into developmental steps following a Piagetian theory of knowledge based on observations of, and interviews with, students as they attempt to learn a concept. Our laboratory and class activities are designed to induce students to go through these steps. They vary

in nature and include: team projects in the computer laboratory; cooperative problem solving (in small groups) in class; confronting students with inadequacies of their existing mathematical ideas for dealing with a specific phenomenon, insisting they attempt (as a means of focusing their thinking) to discover most of the mathematics which is subsequently introduced in class using a modified Socratic approach; creating, by means of computers, a mathematical environment for students in which non-trivial examples can be constructed *by the students* so that mathematical phenomena become a natural part of their consciousness; and finally, having them perform certain computer tasks that are designed to foster directly their construction of mathematical concepts by passing through the precise steps that our theory suggests.

We believe that our courses increase the possibility of students building a deeper understanding of more sophisticated mathematical concepts.

Use of Computers

The use of a MPL allows students to make mathematical constructions *using a syntax which is very similar to standard mathematical notation* and which treats functions as *first class objects*, i.e, functions are data like numbers and can be operated on, collected together in sets or sequences, be accepted as inputs in other functions and can be the resulting object output by the execution of a procedure. Educational research suggests that one way of helping students construct mathematical objects in their minds is to let them perform operations on what is to become an object. In the MPL we currently use (*ISETL*), functions are constructed in such a way that students learn to think about a function as a *process*, i.e., a procedure which transforms an input into some output. The SCS we currently use (*Maple*) permits students to learn a host of techniques for calculation by performing them quickly and easily on a computer. Finally, graphs are constructed automatically using graphics software, but only after the student has first constructed the function, either as a formula or as a process.

Many of our tasks have students transfer information, sometimes requiring a reconstruction, back and forth between the formula of the SCS, the mathematical constructions in the MPL, and the graphs of the graphics facility. This tends to increase their understanding of the connections between the various representations of mathematical notions.

Curriculum

We are developing new curricula with the help of our advisory board of faculty from the departments and disciplines whose students take calculus. These curricula mix a theoretical development with concrete applications and emphasize ideas as opposed to techniques of calculation. The computer environment permits students to relate to certain relatively advanced notions such as branches of functions, the implicit function theorem, and ideas about curves that approach concepts in algebraic geometry.

Student Population and Dissemination

An important aspect of our project is that our materials are being developed and tested simultaneously in two different student populations including engineering, mathematics and the physical sciences, as well as management, social science and life science. Variations in student attitudes are helping us to synthesize a robust approach to teaching and learning calculus.

Pedagogy and Learning Theory

Our project is guided by the principle that, although the content of the calculus course is very important, a key question is not only *What should we teach students?*, but *How can we help students learn the material that ought to make up the calculus course content?* We assert our belief that learning mathematics must be shifted from what we do for the students to what we can help the students do for themselves. With this in mind, our pedagogical approach is based on our belief in the constructivist (in the epistemological and not mathematical sense) theory of learning. Our pedagogical philosophy involves three different kinds of software tools: a symbolic computer system, a high level mathematical programming language and a graphics facility. Each of these systems, together with their combinations, are used for specific pedagogical purposes, where the primary purpose is to foster students' abilities to make mental constructions of mathematical concepts. The particular software is not critical as long as it satisfies our requirements. For example, *Maple* could be replaced by another SCS like *Mathematica* or *Derive*, *ISETL* by another MPL which treats functions as *first class objects*, and we could use any graphics package that is compatible with our understanding of how students learn.

Although each computer system has value when used by itself, we feel that a very important aspect of our approach relates to using combinations of these systems. Our learning theory interprets mathematical concepts as dynamic processes, as static total entities (objects), and as combinations of processes and objects. It also offers approaches for helping students learn concepts from these points of view. Our use of computer systems is guided by an attempt to foster the learning process described by this theory. Our "team" problem solving approach in the laboratory uses the computer to get students to make mental constructions of mathematical concepts. We then follow up the lab activities with cooperative (small group) problem solving in class, together with class discussion based on a "Socratic lecture" approach—with a bare minimum of traditional lecturing and only after students have had time to struggle with and make the necessary mental constructions of the desired concepts.

Software Tools and Programming

The most important way in which we help students make mental constructions of mathematical concepts is to have them construct mathematical entities (*processes* and *objects*) on the computer in ways that our research suggests will induce them to make corresponding constructions in their minds. What this means, in our context, is that students write programs. We think this is an essential part of our approach.

In our laboratory sections, students work in pairs at one or two machines and often in teams of three to five students using one or more machines. In working with *ISETL* and *Maple*, students make use of a number of software tools that we have produced. These include mathematical tools and "front-end" programs. The mathematical tools include programs that make tables for analyzing functions and procedures for finding zeros and limits of functions. The "front-end" programs improve the interface between the students and the two software systems and also the interface between these two systems.

Constructions of Objects and Processes

We now briefly discuss the learning theory on which we base our instructional design. Roughly speaking, processes are built up out of actions on objects and ultimately converted into new objects which are used for new processes and so on, as a person's mathematical knowledge spirals up to higher and higher levels of sophistication.

For example, numbers are *objects*. On these objects, *actions* can be performed, for example by making arithmetic calculations. When an action such as adding three to a number is repeated with different numbers, there is a tendency to become aware of and *interiorize* this action into a *process*, $x + 3$. This leads to algebra. Processes can also be constructed by composing two processes, say adding three and "squaring" which gives $(x+3)^2$ or by reversing a given process, say adding three, to obtain the process of subtracting three or $x - 3$. A single process such as adding 3 can be *encapsulated* to become an object, in this case, the expression $x + 3$. Now the standard algebraic manipulations with expressions can be seen as actions on these new objects. Figure 1 below displays these different kinds of constructions schematically.

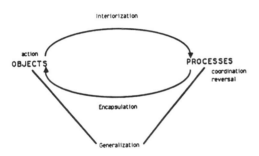

Construction of Objects and Processes

Figure 1

There is an interesting observation that can be made from this analysis. The mathematical notation $(x+3)^2$ can represent two kinds of mental constructions. One is the process obtained by composing the "add 3" process with the "square" process. The other is the object obtained by squaring the object $x + 3$. Although both representations are important, the former carries much greater mathematical content, whereas the latter is more formal and can lead to performing manipulations without much understanding of their content. One of the problems we have with symbolic computer systems is that they emphasize the latter view of expressions like $(x + 3)^2$ and do little to encourage the construction of the processes which carry most of the mathematical content.

There are, of course, much more mathematically sophisticated situations. In calculus, most of the examples in which mathematical concepts are interpreted in terms of objects and processes have to

do with functions. For instance, the input/output point of view treats a function as a process, as does the interpretation that points on a graph come from evaluating the function at the x-coordinate and taking the answer for the y-coordinate. On the other hand, considering differentiation as an operation that transforms one function into another, thinking about iterating the composition of a function with itself, or seeing that the solution of a differential equation must be a whole function (rather than a number or even a structured collection of numbers) requires the interpretation of a function as an object.

Sometimes it is necessary to conceptualize a function simultaneously as a process and as an object. To understand the notation $f'(3)$, for example, requires the idea of transforming a function to its derivative (object) and evaluating at 3 (process). A principle of the theory is that this can only happen if the object conception came as a result of encapsulating the corresponding process conception. In this case, one is able to go back and forth between the two interpretations.

The scheme in Figure 1 illustrates the essential ingredients of the constructions in this theory of learning. It can be used to describe a number of mathematical concepts at all levels of sophistication. Indeed, one of our working assumptions is that the construction of *every* mathematical concept can be analyzed in this way. We point out below how we use these analyses to guide the design of instruction.

One final point regarding this theory is the notion of *reflection*. An important aspect of learning mathematics is to be aware of the mathematical operations that you perform, for example when making calculations, and to reflect on their meaning. This is a critical point for the constructivist theory. It enters explicitly in that both interiorization and encapsulation arise, at least in part, as a result of reflecting on the problem situation and methods of dealing with the problem—both successful and unsuccessful.

Mathematical Concepts

Following are two examples of how mathematical concepts can be interpreted using our theory, along with some implications for teaching. Later in this report we describe how these particular examples are treated using the computer in our course.

Graphs and Functions

There is a general feeling that graphs are extremely important as visual representations of functions and computer graphics systems are a powerful vehicle for getting students to work with graphs in sophisticated situations.

In terms of our learning theory, the conception of function that is important here is the process conception in which the function transforms values in the domain into values in the range. A graph, however is an object. If it does not arise through encapsulating the function process (for example if it arises for the student by looking at pictures drawn on a blackboard or by entering expressions in a Symbolic Computer System and pressing a button) then, we suggest, the connection with a process may not be constructed in the student's mind.

The Fundamental Theorem of Calculus

One can start this circle of ideas with the derivative, beginning with the process of using the difference quotient to estimate the derivative at a point and passing to the limit, encapsulating that process into an object which is the derivative at a point and then varying the point to obtain a function. This entire process is interiorized to obtain the notion of taking a function f and transforming it to another function f'. Conceptually, this requires that the student be able to think of a function as an object to which an action can be applied, resulting in a new object. This is a higher level action which again must be interiorized to obtain a process. In this way, the concept of function is generalized to include functions whose domains and ranges are sets of functions.

The integral is analogous. First, there is the process of partitioning an interval and measuring rectangles to estimate an area. Passing to a limit is more complex here than in the case of derivatives. Nevertheless one can hope that the student will construct a reasonable mental process that, given a function and an interval, produces a number. This process must then be encapsulated so that one endpoint of the interval can be varied so as to obtain a new function whose value at a given point is the encapsulated process (of computing the integral). Thus, again, one interiorizes a process which acts on functions to produce new functions.

The last step in the entire scheme is to coordinate the two processes by composing them. One can apply the integration process and follow it by the differentiation process, and altogether one has again a single process. It seems important for students to construct this process first and then see that, when applied to reasonable functions, it gives back the original function. At this point, one can compose the two processes again, this time in the opposite order to investigate the complication that arises from the fact that the integration process is not one-to-one.

The cognitive difficulties for students in trying to make the mental constructions described here are immense, and the fact that the traditional methods of teaching do little to help them directly in this endeavor goes a long way, in our opinion, to explain why so few of our students emerge (if they do emerge) from the calculus course with any appreciation for, much less understanding of, the incredible intellectual achievement that this theorem represents.

Use of Computers

Now that we have described our theoretical approach and indicated the kinds of analyses of mathematical concepts that we feel must precede the design of specific instructional treatments, we briefly discuss the use of computer systems in implementing our design of instructional materials. We begin with some comments about the issue of having students use prepared software packages versus having them program in some general purpose language. Using a software package that performs some mathematical operations is another way of *showing* the mathematics to the student or getting the student to *use* mathematics. Writing computer code that implements a mathematical process or represents a mathematical object is a *construction* by the programmer and amounts to her or him *doing* mathematics. Moreover, our experience has convinced us that *anytime you construct something on a computer then you are constructing* something *in your mind*. Since, according to our theory, learning mathematics consists of constructing certain processes and objects, it follows for us that the use of computers should emphasize programming and that the programming tasks for students should be so designed as to influence them to make constructions which will contribute to the growth of their mathematical knowledge.

In connection with the above issue of using prepared software packages versus programming, we point to some comments in the March 1990 issue of the *Notices* of the AMS. In Jon Barwise's *Computers and Mathematics* (p. 276), Phil Miles reviews the symbolic computer system *Derive* and raises the concern, echoed by the editor, that "symbolic" mathematics packages may make it even harder for our students to understand the meaning of mathematics." We feel that our comments in this section respond to that concern by showing how a mathematical programming language can make it, if not easier for students, at least more likely that they will succeed to understand mathematics.

There is one major point that we should make about software that is more connected with the computer laboratory than with the content of the course. The software should be as easy as possible for the students to use. This means that it should be "user friendly" and there should be readable, accurate and helpful documentation.

To summarize, there are two important aspects of our approach in using computers as we do. The first is that programming is more important than using software packages, not that the former should be employed to the exclusion of the latter. Indeed, we will show an example of coordinating the two kinds of computer systems and how this can lead to particularly useful tasks for students.

The second is that our entire position collapses on practical grounds unless the programming language and the environment in which it is used are extremely convenient and relatively free of frustrating syntax issues *that are not connected with the mathematical issues.* However, some difficulties *are* connected with the mathematical issues and in these instances the student's frustrations are the right difficulties to struggle with. The examples given below suggest ways in which we keep the above two points in mind as we design our computer materials.

Computer Constructions

We now briefly discuss two examples of using a computer to help students make the necessary mental constructions that, according to our learning theory, will lead them to understand mathematical concepts.

Graphs and Functions

Our instructional approach considers that the connection between the production of data points and a curve on a coordinate system is of paramount importance. Here is what we do to help students construct this connection in their minds.

They are given the task of writing *ISETL* procedures to represent various functions such as the func f given by

```
f := func(x);
    if x < -1 then return -x**3;
    elseif x < 1 then return x**2;
    else return 2*x-1;
    end;
end;
```

Then they can write a line of code such as

```
graph(f,a,b,n,''filename'');
```

We have made it possible for our students to enter *Maple*, perform the following two commands

```
read 'filename';
sketch '';
```

where the result of the first command is a table that appears on the screen and the second results in the points being plotted on a graph.

One might question the use of such a multi-step operation. Our feeling is that not only does it give the student an opportunity (encouraged by the teacher) to reflect on what is going on, but it is closer to the way things work in scientific investigations than the one-shot, push a button approach to which some people are reducing these powerful computer systems.

In any case, the student can study the behavior of the function at the two interesting points, $x = -1$ and 1. The corner at -1 will be obvious, but one does not see the interesting behavior at 1. To study this, the student can write a function which will approximate the derivative of f. A func for this would look like

```
fp := func(x);
    if x < -1 then return -3*x**2;
    elseif x < 1 then return 2*x;
    else return 2;
    end;
end;
```

and the same process can be applied to fp. It is possible to use our scheme to place f and fp on the same graph.

We have students do all this with a number of functions before getting them to write the following *ISETL* program.

```
ada := func(f,e);
    return func(a);
        return (f(a+e) - f(a))/e;
    end;
end;
```

This presents them with some very serious cognitive difficulties related to an object conception of function, but the reward is powerful for the student who eventually learns to write `ada(f,0.00001)` or `ada(f,0.00001)(2)` and understand the meaning of the computer's response.

Learning is further enhanced by the opportunity to *vary the function f* to which ada is applied. This gives an added dimension to the general technique of having students study a picture on which appears both a function and its derivative without indication of which is which. We believe that when these pictures come from functions which the student constructed on the computer the whole experience becomes richer.

The Fundamental Theorem of Calculus

Working with the above func ada helps students take the desired cognitive steps which help them construct a notion of the derivative as an operator that takes a function and transforms it into another function.

The next step is to construct the concept of definite integral. The key computer activity for the students here is to construct a number of *ISETL* funcs which implement various Riemann sums. Here is a func `RiemLeft` which implements a Riemann sum by using left endpoints of the subintervals.

```
RiemLeft := func(f,a,b,n);
    x := [a+((b-a)/n)*(i-1) :
        i in [1..n+1]];
    return %+[f(x(i))*(x(i+1)-x(i)) :
        i in [1..n]];
    end;
```

Notice that, except for using %+ in place of $\sum$, the syntax here is correct mathematical notation. Future enhancements of the language will probably use things like $\sum$. Of course, there are variations corresponding to taking right endpoints, midpoints, minima and maxima. The trapezoid rule is included in the midpoint variation and Simpson's rule is very easy to implement. The students apply these funcs to many functions, draw pictures on graphs produced by *Maple* and see that the mathematical calculation does approximate the area under a curve, or between two curves, for example. The standard properties of the integral (linearity, monotonicity, concatenation of intervals) are more easily understood by the student because they correspond to properties of computer constructs which he or she has made. Even limits can begin to make sense for the student by writing appropriate programs.

The students are helped to encapsulate the process of computing a Riemann sum to obtain an object which is an approximation to the indefinite integral, by writing the following func which corresponds to the step of varying the upper endpoint to obtain a function.

```
Int := func(f,a,n);
    return func(x);
        return RiemLeft(f,a,x,n);
        end;
    end;
```

Students have a hard time writing this. In working with them it seems to us that their struggles are focused on the non-trivial mathematical concepts of using the definite integral with varying endpoint to define a function. A major difficulty occurs when students investigate the *ISETL* expressions:

```
Int(ada(f,e),a,n)(x);
```
and
```
ada(Int(f,a,n),e)(x);
```

Many students overcome the difficulty and all of them seem to get something out of the experience. It is a most satisfying experience, when students produce the three graphs of the functions f, `Int(ada(f,e),a,n)` and `ada(Int(f,a,n),e)`. Two of them are almost identical and the third is displaced by a constant amount. In the end, most students figure out for themselves that the displacement corresponds to the value of $f(a)$ and it is the familiar "constant of integration."

Curriculum

For the past two years we have been working with our advisory board of faculty at Purdue to determine the content needs of client departments and disciplines. Our cooperative effort will produce "leaner" calculus courses that will address the necessary skill areas and promote greater understanding on the part of students as desired by their disciplines.

A second major determinant of our content will be the results of our research into how students learn or why they don't learn calculus.

Our courses will begin with a major study of functions as processes, their source in applications, their graphs, analytical and graphical representations of important transformations of functions, and their role as a framework for understanding all kinds of numerical transformations. This is done in an effort to broaden the students' *concept image* of functions which is usually not well developed prior to the study of calculus.

Standard topics connected with the derivative and integral will be introduced early in connection with the applications that give rise to them. As previously mentioned, computer constructs (by the students) of processes for approximate derivatives and Riemann sums are tools for studying the derivative, integral and the Fundamental Theorem of Calculus simultaneously with the development of an *object* conception of function. In addition, our new curriculum will include:

- Reducing excessive analytic computation by hand and increasing an emphasis on the use of technology to solve or approximately solve calculus problems.

- Increasing the emphasis on problem solving throughout the curriculum, with less emphasis on problems grouped and solved by a particular method. Basic concepts, such as "zeros of an equation or function," "limits of difference quotients" and "limits of Riemann sums" will be used to enhance problem solving skills.

- Increasing the number of problems which are not solvable in closed form and replacing problems which have "neat" solutions by problems which require numerical solutions.

- Emphasizing critical and independent thinking, verbal explanations of the meaning of what students are doing with calculus concepts in problem solving, and less emphasis on working toward an answer based on one particular method of solution. Experimentation and multiple approaches to problem solving will be emphasized in an effort to help students solve non-routine problems.

- Introducing differential equations and their solutions (as functions) soon after the derivative has been defined, including solutions of differential equations as an application of integration, both in closed form and through numerical approximation methods.

- Using the computer to introduce a number of topics not usually accessible at this level, including implicit function theorems, branches of curves, iterations of functions and fixed points.

Although our course is more difficult for the average student, the rewards in deeper understanding of calculus concepts for the students who take our course far outweigh the difficulties they encounter.

Getting Started

Our project began during the spring of 1988 with the writing of our NSF proposal for AY 1988-89. At that time we began to think about how the theory of learning (initially developed by Dubinsky) could be used to design calculus courses for our student populations. With support from School of Science Dean K. L. Kliewer, the Purdue University Computing Center which set up and now maintains our Macintosh Lab, the Departments of Mathematics and Education support of our Mac Lab and an NSF planning grant for AY 1988-89, we made our first approximation in the design of our calculus courses. During the initial design and implementation, we made great strides with unconventional methods in our delivery system using computer activities (including programming), team laboratory work and cooperative (small group) problem solving in class to help our students learn calculus concepts.

The Mathematics Department at Purdue feels that experimental methods should show indications of success with small groups of students and the traditional curriculum before proceeding to larger populations. Therefore our initial efforts were with small groups of 25-40 students using a standard text and lab materials prepared by us. Our success over the past two years has led to an agreement that in fall 1990, we will teach larger classes of 60-70 students together with laboratory sections of 40-45 students. Future scheduling of the lab sections will be done as in traditional laboratory courses, where students will have class assigned independent of the lab section. Consequently, our courses will have the sense of a traditional laboratory course.

Course Operation

Our Macintosh Laboratory is located in the basement of the Mathematical Sciences building which houses the Department of Mathematics. The laboratory has 40 computers and three Laserwriter printers. The Mac Lab is open most of the day and evening hours for use by our calculus classes, other Mathematics Department classes and various other classes (primarily English, Communication and Freshman Engineering computer classes at this time).

Lab and Homework Assignments

The laboratory assignments are mostly computer tasks and the homework assignments are pencil and paper work, including many of the standard problems that students learn to solve in calculus. There is another important difference between the two that relates to our pedagogical approach.

An important part of our theory is the hypothesis that a person constructs new ideas as a response to a perceived problem situation, to straighten out something that is confusing in order to return to a state of mental well-being, to understanding at least the immediate environment.

One of the main purposes of the computer work is to prepare the students in this way. We see the laboratory tasks as setting problem situations before students, getting them to wrap their minds around a situation, getting them ready to think about and develop new mathematical ideas that ultimately lead to the construction of mathematical ideas sufficiently powerful to solve problems that were previously inaccessible.

Therefore it is important that the timing of tasks is such that students have had an opportunity to do appropriate lab work before the relevant topics are considered in class. Classroom discussion has the purpose of bringing out ideas that the students have begun to construct. The homework is used for reinforcement of these emerging concepts. We often ask our students to write brief essays on mathematical ideas in their homework and lab assignments.

One implication of this point of view is that, in the laboratory, students often deal with mathematical ideas before they are explained, use terms before they are defined, and confront problems before they have developed the tools necessary to solve them. Another implication is that the students are not necessarily expected to succeed with the computer tasks in the laboratory. It is only necessary that they struggle with them and think seriously about them.

Typically we have about twelve lab assignments and thirteen or fourteen homework assignments per semester. However, we will probably increase the number of homework assignments due to the lack of maturity on the part of the freshman and sophomores who take our courses. More frequent monitoring of homework seems to be necessary to help our students manage their time more efficiently. Two to three hourly exams plus a comprehensive two hour final are given each semester. The teaching load for one of our classes is equal to one of the two required classes per semester.

Student Assistants

We use both graduate and undergraduate TA's in our classes and labs. In class, the TA's help students with the small group problem solving, mainly by responding to queries and posing questions designed to stimulate the students' thinking. In lab, their role is largely to help students with technical difficulties that arise in connection with operating the hardware and software.

Generally, our TA's know just a little more (and that knowledge is fairly new) about the work than do the students. The situation often becomes one of students and TA's working together to figure something out. We feel that this is an important part of the overall atmosphere of our course. Not only do the assistants help the students learn, but we have found that the assistants, many of whom intend to become high school mathematics teachers, benefit from a deepening of their understanding of mathematics. They add a new component to their understanding of the learning process. A possible offshoot of our experiments could be some new ideas for the preparation of secondary and even college mathematics teachers.

Classroom Work

The classroom, for us, is not so much a place where explanations are made as it is a place where students are confronted with problems we pose that are designed to get them to construct their own mathematical notions that can be used to solve these problems. When we say that students construct their own mathematical ideas, what we are trying to get from the students are autonomous mental constructions. We want them to make mathematical ideas their own, in whatever form makes sense to them. That is what happens internally. Externally, we aim at their being able to solve the same problems, in much the same way (with individual variations) as everybody else.

In class, our students get together in small groups and think about a problem for a few minutes. Then, we bring things together, listen to the various responses, get comments, discussion and even disputes going if possible. At the end, we may or may not give an answer of our own. The important point is the effect this has, subsequently, on their ability to understand and solve problems.

Appraisal and Conclusions

Students

The students in our courses are selected as much as possible to be like the profile of those in the regular classes. Schwingendorf is currently the Undergraduate Mathematics Counseling Coordinator and is in charge, together with the help of counselors in other schools on our campus, of the selection of our students. Prospective students are informed that our course is experimental when they register for it and they are informed that the work load will be greater than the regular calculus class. We should note that each semester a few students request to take our courses not having had our previous course (or courses). We have allowed several of these students to enter our courses and they have had little problem adjusting to the computer work and increased work load. Students are screened like all other students who enter our calculus courses based on their SAT's, high school profile and their performance on an Algebra and Trigonometry Placement Test. During the first year of our EMS course, the SAT profile for our students (Verbal 510 and Math 630) was slightly higher than that of the students in the regular course (Verbal 495 and Math 605). However, during the second year of our project our SAT profile was almost identical to that of the students in the regular course.

As mentioned previously, our students took Departmental common multiple choice final exams during the first two years of our project. Copies of sample final exams, which test primarily mechanical skills, are included in our sample materials. Although we spent significantly less class time in our EMS calculus classes on mechanics than the regular classes, our students averaged 6% and 2% higher on the final exams the two times we taught first semester calculus, and about the same and 7% higher the two times we taught the second semester course. We consider this to be a success and it indicates that de-emphasizing mechanics does not put our students in an inferior position relative to traditional methods of performance.

Exams

During each semester, we give two or three exams to our classes. These exams (see Sample Materials) are intended to indicate the extent to which students have developed conceptual understandings. About half of our exam questions ask the student to solve problems different from any he or she has been assigned in the course, and a few problems could be considered non-routine in nature. The exams are given in the evening and students can take as much time as they need. The students are not allowed to use books or notes, but calculators are permitted. Several of our exam questions are of the essay type.

The exams are graded with partial credit given when understanding is displayed, even if a complete solution is not obtained. The overall class average on these tests are in the 60-70% range. In general, about one-half to two thirds of the students are in the 75% and above range, most of the rest score in the 50-60% range and a small number (usually the same students) score below 50%.

Student Comments

In general, student evaluations of our courses have been favorable. However, a number of students in our EMS classes have complained about the work load which averages about 15 hours per week for our 5 credit hour course. We note that our university emphasizes to students that they should be prepared to work 2 to 3 hours out of class for each hour spent in class, with more time spent per hour in class for courses with a laboratory component. Furthermore, the majority of the student evaluations from our traditional EMS calculus lectures usually indicate that they spend only 5 to 8 hours per week on such courses! The following is a sample of comments from our students:

- "I must admit that having to think so much for a math class frustrated me; I don't believe that truly understanding what I was doing (or why) was a requirement in my previous math classes."
- "... calculus is not a series of problems to solve, but rather a method and thought process involved in solving these problems."
- "In the past three semesters I have explored a new way of thinking and understanding as well as learning the concepts of calculus. Each semester brought new light on what calculus is and how to approach solving a problem."
- "Probably the biggest adjustment I had to make was learning how to solve problems by just thinking instead of using a formula or a pattern. It took me a long time to be able to do this on my own."
- "... I've learned to analyize ideas. Calculus is more than memorizing numerous formulas, it is knowing what the formulas mean, what they do, and where they come from so you don't have to memorize them."
- "An important aspect of this class has been the interaction with other people to solve problems."
- "At the start of the calculus program, we were forced to work in groups, but by the end you would have had to force most of us apart."
- "From the first day when we were asked what

a function is, this class has been a thought provoking one. Getting a better understanding of functions is probably the most important thing I have learned. A process that transforms one thing into another is not what I thought when I came here. The equation was to me what a function was until this class. Realizing a derivative and integral accept and return functions was a big step for me. I don't think of these things as equations anymore, and that has really helped in physics and EE classes."

- "In my opinion Labs are the most beneficial part of this class. Labs indicate to the student what he or she doesn't understand. It helps students to formulate questions."

- "The use of computers enabled me to work with the concepts behind the mathematics, while the homework made sure I could function by hand. The exploration of the concepts gave, and still promises, greater understanding than is possible through hand written solutions in the same length of time. Indeed, the increased versatility given by the computer increases experimentation and depth of understanding exponentially."

- "This particular math class was a lot harder than what I expected it to be, but it taught me concepts of teaching that I would never have learned otherwise."

- "... with a little (well, sometimes a lot) of coaxing, and a few hints, we always managed to figure things out for ourselves, and that made the concepts more solidly comprehended."

- "The true test of knowledge is measured by a person's understanding of why we take certain steps and apply certain formulas and most of all the ability to apply concepts learned to new situations. This is the essence of what I learned in calculus."

Future Plans

Our MPL *ISETL* is currently in the process of being improved. By fall 1990 it will include a graphics facility whose features will go beyond those of the current version of *Maple* we are using. The initial graphics package for *ISETL* will have only crude 3-D graphics; however we are confident that future upgrades will produce a 3-D package superior to most available at this time. Future enhancements of *ISETL* should contribute to its wider use and acceptance in the educational and mathematical community.

We do not believe that our courses will displace the standard departmental courses in the near future. However, we anticipate that our courses will serve an increasingly larger population of students each semester. We hope to show that our teaching methods and instructional materials are not dependent on small classes by experimenting with larger classes (about 60-70 students per class) beginning in fall 1990.

We are currently writing a calculus text for our EMS classes and work has begun on a pre-calculus textbook for such students. A preliminary version of our EMS calculus text will be available by spring 1990. Our text and teaching methods will be used at Purdue, and we expect that more than a dozen other schools around the nation will use our text and teaching methods by fall 1991.

We plan to continue and increase our research studies into how students learn calculus (and mathematics). We do so with the help of both undergraduate and graduate research assistants.

We are planning to conduct training workshops with support of West Educational Publishing. We have also applied for (and recently received) a three-year NSF grant, including funding of summer workshops to train faculty from around the nation in the use of our texts, class materials and teaching methods. We plan to present papers at conferences regarding the research connected with our project, the related theoretical outcomes of our project and the resulting instructional design for our courses.

We have also written two papers that will be included in Carl Leinbach's MAA Notes Series on "Teaching Mathematics as a Laboratory Science." These papers will showcase *ISETL* to the wider mathematics community. Contact the authors for references and details on the availability and cost of class materials.

═══ Sample Materials ═══

The following are examples of Lab and Homework assignments for our EMS calculus classes (MA 161-162-216) which indicate how we use the computer in our classes. We also include some of our tests.

MA 161 Lab Assignment 4

1. For each of the functions described below,

you are given a small positive number which we will call ϵ, a value c in the domain of definition of the function and a number L. Your task is to find another positive number δ having the property that if the independent variable is in the interval from $c - \delta$ to $c + \delta$, then the values of the function lie in the interval from $L - \epsilon$ to $L + \epsilon$. One way to do these is to use **Maple** to sketch the graph and reason from the picture. In each case, explain how you got your value of δ and why it works.

a. The function is given by the expression $\frac{\sin(x-0.5)}{x-0.5}$, $\epsilon = 0.01$, $c = 0.5$, and $L = 1$.

b. The function is given by the expression $\frac{\sqrt{2+x}-\sqrt{2}}{x}$, $\epsilon = 0.01$, $c = 0$, and $L = \frac{\sqrt{2}}{4}$.

2. Write an *ISETL* func ad, i.e., ad(f,a), that will accept a func representing a function f and a point a in the domain of f and return the value of the *difference quotient*,

$$\frac{f(a + 0.001) - f(a)}{0.001}.$$

We call this expression the "approximate derivative of the function at a." Run your func on the following examples.

a. The function is given by the expression, $\sqrt{x}$ and $a = 2$.

b. The function is given by

$$r(u) = \begin{cases} u^3 + 3 & \text{if } u \leq -2 \\ -3u^2 + 7 & \text{if } -2 < u \leq 0 \\ -1 & \text{if } 0 < u \end{cases}$$

and $a = -2$.

c. The same function r with a replaced by $a = 0$.

Change your func by replacing 0.001 by -0.001 (in both the numerator and denominator). Run your new func on the same three examples. Give an explanation of how and why the three answers with the first func differed (or did not differ) from your three answers with the second func.

Use *Maple* to produce a graph of the three functions with x in an interval of length about 0.001 and centered at a. Indicate clearly on each graph the meaning of the expression for the *difference quotient* in each of your two funcs. (Hint: Think about the meaning of the numerator, denominator and the whole difference quotient.)

3. It is possible for an *ISETL* func to return another func, which is a feature having many applications. For example, here is a func that will accept two funcs representing functions and return a func which represents the product of the two functions.

```
mult := func(f,g);
    return func(x);
        return f(x)*g(x);
    end;
end;
```

Using a similar structure, write two funcs plus and comp that will give, respectively, the sum of two functions and the composition of two functions.

Let r be the function represented in problem 2(b) and let s be the function with the following representation

$$s(x) = \begin{cases} 2/x & \text{if } x \leq -1 \\ |\sin(x)| & \text{if } -1 < x \end{cases}$$

Write out by hand as precise as possible an explanation of what would be the meaning of the following two *ISETL* expressions. This explanation should be in terms of the original formulas (expressions) used to define r and s.

```
plus(r,s)(x)  and  comp(r,s)(x)
```

4. *ISETL* tuples can be used to represent finite sequences of numbers, and this can be used to "approximate" a sequence which approaches a number. Consider, for example, the following *ISETL* code where a represents a real number and f is an *ISETL* func representing a function.

```
a := 2;
s := [a + ((-1)**i)/(1000*i) :
    i in [1..60]];
s;
[f(x) :  x in s];
```

Run this code for the following choices of f and a.

a. The function h (from Lab 1 problem No. 4) given by

$$h(x) = \begin{cases} -(x+4)^2 - 2 & \text{if } x < -3 \\ (x+2)^2 - 4 & \text{if } -3 \leq x < 1 \\ x + 4 & \text{if } 1 \leq x < 2 \\ -2x + 5 & \text{if } x \geq 2 \end{cases}$$

with a = -3.

b. The same function h as in part (a) with a = 2.

Now do the following:

a. Give a verbal description of the meaning of the code for s in the second line of code above. Include an explanation of the presence of (-1)**i and of the meaning of s(2).

b. Give a verbal description of the meaning of the fourth line of the code, [f(x) : x in s];. Include a description of the effect of (-1)**i and explain how it would be different if it were not there, and if it were replaced by -1.

c. What do you think could be meant by "other ways of approaching a" and "other rates of approach to a?"

d. Two variations of the *ISETL* code for s (in the second line) that implement your response to the previous question.

e. Based on the result of running the code above for each function, what do you predict should be the value of each function at the indicated point a?

f. How do the three examples differ with respect to the result of evaluating f at a?

MA 161 Lab Assignment 5

1. In this problem we formalize the work you have done in defining a function whose process is obtained from the approximate derivative ad(f,a) with f fixed and a varying. Write a func ada that will accept a func, say f, that represents a function whose domain and range are real numbers and a small positive number e. The func ada is to return a func which represents the function given by the expression

$$\frac{f(a+e) - f(a)}{e}$$

Run your func ada for each of the following functions with an appropriate choice of e in each case. Use *Maple* to sketch a graph which has both a sketch of the original function f and a sketch of the function ada(f,f) which your func ada produces. Choose reasonable intervals to display clearly what is going on. Use $e = 0.00001$.

(a) $g(r) = 5r^3 + 2r^2 - r + 4$ and

(b) $f(z) = |2z - 1| + 1$

2. In the following list of functions, there are some pairs f, g in which g is a reasonable approximation to the function ada(f,e) of the problem 2. Your task is to discover as many of these pairs as you can by using *Maple* to sketch each function f as well as ada(f,e) and then comparing the graph of ada(f,e) with the graphs the other functions in the list. For each pair f, g that you think g is a reasonable approximation to ada(f,e), hand in

a single sheet containing graphs of all three functions f, g and ada(f,e). Connect points on your sketches so that your graphs are solid curves.

a. $a(b) = -\frac{1}{b}$

b. $i(j) = \ln(\frac{1}{j})$, assume $j > 0$

c. $k(m) = \frac{1}{m^2}$

d. $n(p) = p - p\ln(p)$, assume $p > 0$

MA 161 Homework Assignment 5

1. The *ISETL* operation %+ can be used to add up all of the quantities in a tuple, or even quantities depending on the values in a tuple. For example, if P is the partition of the interval from a to b into n equal parts and f is some function whose domain includes this interval then there are many situations in calculus where one would like to consider each subinterval and multiply the value of f at, say, the left endpoint times the length of the subinterval and then add up all of these products. The mathematical notation for this quantity is

$$x_i = a + \frac{(b-a)}{n}(i-1) \ \ i = 1, \ldots, n+1$$

$$\sum_{i=1}^{n} f(x_i)(x_{i+1} - x_i)$$

and, using %+, the *ISETL* code for obtaining it is very similar.

```
x := [a+((b-a)/n)*(i-1) :
        i in [1..n+1]];
%+[f(x(i))*(x(i+1)-x(i)) :
        i in [1..n]];
```

a. Use the following function and a partition of the interval from -3 to 2 into 17 equal parts to compute the above quantity.

$$H(y) = \begin{cases} \dfrac{1.3y^2 + 6y - 4}{y^2 + 2y + 1} & \text{if } y \leq -2.6 \\ y + 1 & \text{if } -2.6 < y \leq 1 \\ \dfrac{1}{3y - 2} & \text{if } 1 < y \end{cases}$$

b. Use *ISETL* and *Maple* to produce a graph of H on the interval from -3 to 2 and indicate on the graph a representation of what is being computed here.

2. *Maple* can be used to calculate the derivative of any function that can be represented by a *Maple* expression. For example if you enter the following in *Maple*

```
> u := (t**2 + 4)/t;
> diff(u,t);
```

then you will get a response that is equivalent to the expression $1 - 4/t^2$ which is the derivative of $(t^2 + 4)/t = t + 4/t$.

In this problem, your task is to use this facility of *Maple* to try to discover the rules for differentiating functions that are given by combining functions. We are interested in six rules: constant multiples, sums, differences, products, quotients, and compositions. Specifically, this means that we want you to guess six formulas for the derivatives of functions given by $cf, f + g, f - g, fg, f/g, f \circ g$ in terms of the derivatives of f and g and the number c.

First find the derivatives formulas for $\sin(x)$ and $\cos(x)$ using *Maple*. Then make up examples of functions by forming combinations of x^n, $\sin(x)$ and $\cos(x)$ as constant multiples, sums, differences, products, quotients and compositions. For example, functions given by expressions like the following:

$$x^2 \sin(x), \quad 3x^2 + \sin(x), \quad \sin(x)/x \quad \text{and} \quad \cos(3x^2)$$

Then use *Maple* to calculate the derivatives of your combinations of functions and see if you can guess the six formulas for the derivatives.

MA 161 Homework Assignment 6

1. In Homework Assignment 5 problem 1, you worked with *ISETL* to compute the quantity

$$\sum_{i=1}^{n} f(x_i)(x_{i+1} - x_i)$$

where f is a function and $\{x_i : i = 1, \ldots n + 1\}$ is a partition of the interval from a to b into n subintervals. In this problem you are to put it all together and write an *ISETL* func which accepts a func, say f which represents a function, the endpoints a and b of an interval contained in the domain of f, and a positive integer n. Your func is to construct the partition of the interval from a to b into n equal subintervals and then return the value of the above expression. Call your func by the name RiemLeft. Apply your RiemLeft in the following situations. Choose your own value of n. Make a sketch of each function on the indicated interval and then interpret your answer for Riemleft "geometrically" on the sketch for each situation.

a. The function is the absolute value function with $a = -1$ and $b = 2$.

b. The function is ada(f,0.0001), where f is the absolute value function and ada(f,0.0001) is the approximate derivative function for f, with $a = -1$ and $b = 2$.

MA 161 Lab Assignment 7

1. The purpose of this problem is to write three funcs that approximately implement each of the following equations in two variables.

$$x^2 + y^2 = 1$$
$$|x|^{2/5} + |y|^{2/3} = 1$$
$$x^3 + y^3 = 4.5xy$$

Each of your funcs is to accept two real numbers x and y and return true or false indicating whether the pair is a reasonable approximation to a solution of the equation. (Hint: First rewrite each equation so that all the terms are on one side of the equation.)

There are two issues which you are to keep in mind at this point.

a. Because most computer operations (like 5th root) are only approximations, you should not have your func base its decision on exact equality, but equality up to a particular tolerance, which we shall call tol. You must decide on how to use tol and choose a value for it.

b. Your func should have exactly two parameters, for the two variables—no more and no less. Use an external constant for tol, i.e., assign a value to tol before you use your func. Hence, tol can be changed externally as needed. An external assignment of tol will be useful in parts 2 and 3 below.

Hand in a copy of the final version of your three funcs and a record of your *ISETL* session in which you apply your funcs for several test values. Make sure that you get some true results for each equation.

2. There is an *ISETL* operation which will draw the graph of an equation such as the ones you worked with in the previous problem. It has the following syntax (similar to that of graph),

```
graphr(eq,I,J,n,''filename'');
```

where eq is a func that approximately implements an equation (like the ones in the previous problem), I is a tuple of two numbers representing an interval on the x-axis (like [1,3], etc.), J is a tuple of two numbers representing an interval on the y-axis, n is positive integer, and filename is a string which represents a file name.

Here is how `graphr` works. First it partitions the intervals `I` and `J` each into n equal subintervals and this gives a partition of the rectangle that `I` and `J` form into n^2 equal subrectangles. Next it runs through all pairs `[x,y]` of points on the plane corresponding to the vertices of the subrectangles. For each point it applies the `func`, `eq` to decide if the point is a reasonable approximation to a solution of the equation. Finally, it places all points which `eq` decided were reasonable approximations into the file, `filename`.

The result of all this is a file with data in a format that *Maple* can use to construct a graph. Do all of this for the *second* of the three equations that you worked with in problem 1. Again there are a number of issues that you will be concerned with.

a. Your choice of `tol` in writing your original equation `func` will affect the graph in a number of interesting ways. For example, the larger the value of `tol`, the more points you will get for your (approximate) graph. You will learn a great deal by trying to figure out what these effects are and then finally choosing reasonable values for `tol` and `n` to obtain an approximate graph of the equation. For example, as you make `tol` smaller, you will want to make *n* larger. For your first try, take `tol = 0.1` and `n = 50`.

b. Your choice of the intervals `I` and `J` is also critical. You must understand enough about the equation so as to make a choice of a rectangle that will include the entire graph, but not be so large that you will only get a small number of points.

3. In this problem you will use *ISETL* to construct and graph an implicitly defined function from an equation. The operation consists of the following steps.

a. Take the *third* equation in part 1 above and use the method of the previous problem to sketch its (approximate) graph. Try $I = [-2, 3]$ and $J = [-2, 3]$, then adjust I and J as desired.

Keep the following in mind: A particular choice of `tol` may have very different effects if x and y are very small, than if they are very large. Try to figure out a way to take care of this automatically by modifying your `func` which approximately implements the equation. At the same time, make sure your `func` does not introduce a potential division by 0.

b. In the *ISETL* folder there is a file containing the `func` `implicit`, which you will have to `!include` to use. The `func` `implicit` is as follows:

```
implicit := func(rel,I0,J0);
  return func(x);
    local f;
    if not ((I0(1) <= x) and
        (x<= I0(2))) then return;
      end;
    f := func(y);
      return rel(x,y);
    end;
    return bis(f,J0(1),J0(2),0.001);
  end;
end;
```

Study the `func` `implicit`. The `func` `implicit` accepts a `func` `rel` which represents a function of two variables, x and y that implements the left hand side of the equation that you took in part (a) (i.e., the expression that you got by setting everything equal to 0), and two `tuples` `I0` and `J0` (like I and J) that are the endpoints of intervals that restrict x and y, respectively. The intervals `I0` and `J0` are assumed to be chosen so that in these intervals, for each x in `I0` there is exactly one pair `[x,y]` which has `rel` return the value 0. The choice of `I0`, `J0` is to be made by inspecting the graph you made in (a). Note that `implicit` uses the `func` `bis`, so you must `!include bis` in *ISETL* to use it.

c. Sketch the graph of the function represented by the `func` which `implicit` produces. Do this for the *third* of the three equations in part 1. Look at its graph to make choices of `I0` and `J0`. Sketch the graph of the resulting implicitly defined function.

MA 161 Lab Assignment 12

In this problem, which is perhaps the most important of all this semester, you will put together several things that you have been using this semester. First, the approximate derivative `func` `ada` from Week 5, then `RiemMid` and finally, the `func` `print1` from Week 6.

The first step in the problem is to write a `func` `Int`, i.e., `Int(f,a,x,n)`, which accepts a `func` representing a function f, a number a and a positive integer n, and returns a `func` whose value, for a number x is the result of applying `RiemMid` to f with the interval from a to x with n subdivisions.

Next adjust the `func` `print1` to obtain a `func` `print3` which will accept only one `func` representing a function f, an interval $[a, b]$, a positive integer n and a filename. The result of `print3` is to place in the file four columns of values. The first gives the numbers x at $n + 1$ evenly spaced points in the interval $[a, b]$; the second gives the value of f at x; the third gives the value obtained by applying the approximate derivative operator ada) to f, applying Int to the resulting func and then evaluating the resulting func at x, i.e., Int(ada(f,e),a,25)(x); and the last column gives the value obtained by applying Int to f, applying ada to the resulting func and then evaluating the resulting func at x, i.e., ada(Int(f,a,25),e)(x);. Use $e = 0.00001$.

Hand in:

1. A hard copy of the result of applying thefunc print3 to the function represented by $x \sin x^2$ on the interval from 0 to $\pi/2$ and your interpretation of the resulting data.

2. A hard copy of the result of applying print3 to the function $\sqrt{1 - x^2}$ on the interval $[0, 1]$ and your interpretation of the resulting data. In particular, is there any way in which this example differs from the previous one? If not, explain why they are the same. If so, explain the cause of the difference.

3. A sketch of the graphs of the functions f, Int(ada(f,e),a,25) & ada(Int(f,a,25),e) for each of the functions in parts (a) and (b) above, on the interval indicated in each part.

MA 162 Lab Assignment 2

1. In this problem, you will use the func RiemMid and a variation of the func Int from last semester. A copy of RiemMid and the variation of Int to be used are attached. The func Int accepts a func representing a function, a number a representing the lower limit of integration in the domain of the function, and a positive integer n. Int returns a func which represents an approximation to the function whose value at x is the definite integral of the original function from a to x.

Run Int on each of the following functions with the given lower limit of integration a, and then use graph in *ISETL* and *Maple* to produce the graphs of the resulting functions on the indicated intervals.

a. $f1(x) = 1/x$, $a = 1$, $[0.01, 2.01]$
b. $f2(x) = 1/\sqrt{1 - x^2}$, $a = 0$, $[0, 1]$
c. $f3(x) = 1/(1 + x^2)$, $a = 0$, $[0, 1]$

2. Consider the following three functions: (a) $\sin x$ (b) $\tan x$ and (c) $e^x = exp(x)$.

Look at the functions that are obtained by composing one of these functions with one of the functions you obtained using Int in the previous problem (remember there are two possible compositions

for each pair of functions). A copy of the func comp from last semester is attached for your use. In some cases, the result is a very interesting and simple function. Try to find these interesting cases by forming the compositions using comp and then graphing the resulting compositions using graph on the corresponding intervals in part 1. List the composition pairs that you find interesting and explain why you think they are interesting. *Hint: The ones that are interesting are all interesting in the same way.*

3. Write a func L which returns a func which applies the func Int from part 1 to the function given by the expression $1/x$ with $a = 1$, so that L has domain all positive real numbers. Make a table of values for x and $L(x)$ using the func printL attached at the end of this assignment. By examining the table of values, experimenting with values of x and $L(x)$, looking at the defining expression for $L(x)$ and thinking about properties of integrals, try to guess when the func L is zero and try to guess formulas that express $L(a \cdot b)$, $L(a/b)$ and $L(a^b)$ in terms of a, b, $L(a)$, and $L(b)$. When making your guesses, remember that the values for $L(x)$ in the table are approximations!

Give a geometric interpretation for the values of $L(x)$ and illustrate your interpretation by drawing appropriately on a sketch (or sketches) of the function represented by the expression $1/x$ for $x > 0$.

4. Use the func bis and your func L from part 3 to approximate the value of x where $L(x) = 1$ to four decimal places.

MA 162 Lab Assignment 6

1. In this problem you will write an *ISETL* func u that will represent an approximate solution of a given differential equation of the form $y' = f(x, y)$, where f is a reasonably well behaved function of *two* variables. Your solution will be a func u that represents a function that is a solution to the differential equation and has for domain the interval $[a, b]$. Thus you will use a number of global constants: the "right hand side" f of the differential equation, the interval $[a, b]$, and in addition, a positive integer n which represents the "mesh size" for your solution. Finally, you will need to "prime the pump" by giving your func the initial value $u(a)$.

Your func will compute u(x) given x by running through the following steps.

a. Construct a partition P of the interval [a,x] into n equal subintervals.

b. Use a for...do...end loop to construct a tuple uP of values u(x), x in P as follows.

   i. The first component of uP is u(a) which is given.

   ii. Assuming that the value $y_i = u(x_i)$ of uP has been determined, the next value is given by

$$y_{i+1} = y_i + f(x_i, y_i)(x_{i+1} - x_i).$$

c. Return the last component of uP, which is the desired u(x).

The above approximation method is often called *Euler's Method*.

Write your func for the following differential equation

$$y' = \frac{x^3 + x - 2xy}{1 + x^2}$$

on the interval $[1, 10]$ with the initial condition $u(1) = 0.375$.

What's do you think is going on in *Euler's Method*? Write a brief essay on what you think is going on in *Euler's Method*. (Hint: What does the equation in part ii of part (b) represent?)

2. Write a general func diffeq that will accept any right hand side and all other required parameters and will return a func which is the the solution to the differential equation corresponding to the right hand side. Run your func on the same example as in the previous problem and sketch a graph using *Maple* and *ISETL* that should look just like the one you got before. Also run diffeq for the differential equation $y' = \cos x + y \tan x$ on the interval $[0, \pi]$ with the initial condition $u(0) = 1$.

MA 162 Lab Assignment 7

1. In this problem you are to write a func ps which is to take a single sequence a (written as a func) and return a sequence S whose n^{th} term is given by

$$S_n = \sum_{i=1}^{n} a_i = a_1 + a_2 + ... + a_n$$

The sequence S is called the sequence of nth *partial sums* of the sequence a. Run your func ps on the following sequences for several values of n:

   (a) $a_n = n$
   (b) $b_n\ 2^n$
   (c) $c_n = \frac{1}{n \ln n}$.

Try to figure out "closed-form" expressions for the first two, i.e., a "simple" expression which represents the nth partial sum. Hand in:

a. A copy of your func ps.

b. A record of an *ISETL* session in which you apply ps to the given sequences and compute some terms of the resulting sequences.

c. Your closed form expressions for the sequences of partial sums for the first two given sequences.

2. For each of the sequences listed below, you are given L, a candidate for its limit and a measure ϵ of closeness to the limit. By playing with various values or inspecting the expression or whatever, make an estimate of from what point on (that means at what index) the terms of the sequence are of distance at most ϵ from the candidate. It may be that there is no such point.

a. The general term is given by

$$(3n - 1)/2n + 1),$$

   where $n = 0, 1, 2, \ldots$ $L = 3/2$, $\epsilon = 0.001$

b. $2, 1, 1/2, 2/3, 3/4, 4/5, 5/6, \ldots$ $L = 1$, $\epsilon = 0.01$

c. $2, 1, 1/2, 0, 2/3, 0, 3/4, 0, 4/5, \ldots$ $L = 1$, $\epsilon = 1$

d. $2, 1, 1/2, 0, 2/3, 0, 3/4, 0, 4/5, \ldots$ $L = 1$, $\epsilon = 0.99$

e. The sequence obtained by applying the func ps you wrote for part 1 to the sequence a, given by $a_n = 2^{-n}$, $n = 0, 1, 2, \ldots$ with $L = 2$, $\epsilon = 0.01$

Hand in your estimate, in each case, of the index from which point on the desired relation holds, and an explanation of the method you used.

MA 161 Lab Assignment 8

1. For this problem you must again modify the func print1 from the last semester. This time the parameters are to be s, n, S, filnm where s is a func that represents a sequence, n is a positive integer, S is a real number which is the sum of the series, and filnm is a filename. In the func, the assignment to P should be changed to make P the tuple of integers from n to n+50. Finally, the print line should have the values of x, the corresponding values of the sequence, the corresponding values of the sequences of partial sums of the alternating series obtained by multiplying every other term of the sequence by -1 (beginning with the second term), and the absolute value of the difference between S and the partial sum. Run your func on each of

the following series and value of its sum, choosing a reasonable value of n.

a. $\dfrac{1}{n}$, $n = 1, 2, 3, \ldots$ $S = \ln 2$

b. $\dfrac{1.5^{2j-1}}{(2j-1)!}$, $j = 1, 2, 3, \ldots$ $S = \sin 1.5$

Hand in your output for each of the two series, and any observations you can make regarding convergence or divergence of the series and the values of the terms of the sequences.

1. For this problem you must again modify the func print1. This time the parameters are to be s, n, filnm where s is a func that represents a sequence, n is a positive integer, and filnm is a filename. In the func, the assignment to P should be changed to make P the tuple of integers from n to n+50. Finally, the print line should have the values of x, the corresponding values of the sequence, the corresponding values of the sequences of partial sums or series (use your func ps from Lab 7), and the ratio of two successive terms of the sequence (that is, the second divided by the first) represented by s. Run your func on each of the following for a reasonable value of n.

a. $\displaystyle\sum_{k=1}^{\infty} 2^k/k!$

b. $\displaystyle\sum_{k=1}^{\infty} 2^k/k^2$

c. $\displaystyle\sum_{k=1}^{\infty} k^3 e^{-k}$

d. $\displaystyle\sum_{k=1}^{\infty} k/(k+1)^2$

Hand in your output for each of the four series, and any observations you can make regarding the convergence or divergence of the four series and the ratios of that you computed.

MA162 Lab Assignment 9

1. In this problem, you are given two lists: a list of series and a list of functions. Based on the work you have been doing you can take each of these series and produce a func that will represent a function which has a certain domain. You may use the 15th partial sum for all series. Your task is to evaluate these functions at various points and do the same for the functions in the second list and make matchups where possible. When there is no matchup, try to guess what *would* match the series or the function. In order to see a list of values to

compare, you might use an appropriate version of one of your print funcs. Here is the list of series.

a. $\displaystyle\sum_{n=0}^{\infty} (-1)^n (x-1)^n$

b. $\displaystyle\sum_{n=0}^{\infty} (-1)^n n (x-1)^{n-1}$

c. $\displaystyle\sum_{n=0}^{\infty} (-1)^n \dfrac{(x-1)^{n+1}}{n+1}$

d. $\displaystyle\sum_{n=0}^{\infty} \dfrac{(x-1)^n}{n!}$

e. $\displaystyle\sum_{n=0}^{\infty} (-1)^n \dfrac{x^{2n}}{(2n)!}$

f. $\displaystyle\sum_{n=0}^{\infty} (-1)^n \dfrac{x^{2n+1}}{(2n+1)!}$

g. $\displaystyle\sum_{n=0}^{\infty} \left(\dfrac{x^n}{n!} + (-1)^n \dfrac{x^{2n}}{(2n)!} \right)$

h. $\displaystyle\sum_{n=0}^{\infty} \left(23.684(-1)^n \dfrac{x^{2n+1}}{(2n+1)!} + 36.7432(-1)^n \dfrac{x^{2n}}{(2n)!} \right)$

i. $\displaystyle\sum_{n=0}^{\infty} n! x^n$

Here is a list of the functions.

a. $r(x) = \cos x$
b. $f(a) = -1/a^2$
c. $h(u) = \ln u$
d. $f(w) = -\cos(w)$
e. $g(t) = 1/t$
f. $f(x) = e^x$
g. $f(x) = 0$
h. $h(x) = e^x + \cos x$
i. $q(x) = \sin x$

Hand in a list of relations of the form $f(x) = \sum_{n=0}^{\infty} a_n (x-c)^n$ for x in {some domain}.

2. In the previous problem, several of the functions given are derivatives or antiderivatives of each other and there are some linear combinations. Suppose that you are given two series representations of functions on some domain $f(x) = \sum_{n=0}^{\infty} a_n (x-c)^n$ and $g(x) = \sum_{n=0}^{\infty} b_n (x-c)^n$ where λ, μ are real numbers. Based on your experiences with the previous problem, what do you think would be series representations for

a. The derivative of f.
b. An antiderivative of g.
c. The function $\lambda f + \mu g$.

MA162 Homework Assignment 9

In this problem you will write a func that encapsulates some of the work you have been doing in your investigations of series. This `func` will accept a `func` which represents a sequence of functions, a second `func` `Rem` representing a function and a small positive number *eps*. It will return a `func` which represents the function defined by the series obtained from the original sequence of functions.

The other two parameters are used to determine the value of n in taking the nth partial sum to evaluate the function you are constructing. What you must do is write the `func` `Rem` which takes a positive integer n and a value of the domain variable and returns an upper bound for the error you would obtain by using the nth partial sum. Then, in your `func` you must include code like

```
n := 1;
while Rem(x,n) > eps do
    n := n+1;
end;
```

and after the `func` runs this code, n will have a good value.

Thus, for each application of your `func` you must not only write a `func` representing the original sequence, but also you must write a different `Rem` for each series.

Another caution: If you allow the `func` which you produce to be given a value at which the original series does not converge, then you could get into an infinite loop.

Run your `func` on the Taylor series for the following functions expanded about the given value of c.

a. $\sin 2x$, $c = 0$
b. $(1 - \cos x)/x^2$, $c = 0$
c. $\ln x$, $c = 1$

Use a variation of `print1` to print a table of three columns for each of the above functions and the Taylor series at the indicated value of c: the value of the independent variable, the value of the function as given at that value of the independent variable, and the corresponding value of the function represented by the `func` you produce. The last two columns should look very similar. Use about 40 values of the independent variable. Hand in a copy of your `func` `Rem` and a record of your output from your variation of `print1` for each function and the indicated Taylor series.

MA 161 Test 1

1. Following is the graph of two functions, f and g on the same coordinate system. One is the derivative of the other. Decide which is which and give your reasons.

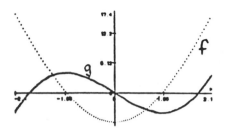

2. Following is the graph of two functions, f and g on the same coordinate system. Sketch a graph of $2f + g$.

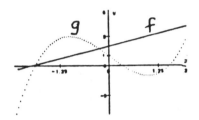

3. Estimating from the graphs of f and g in Problem No. 2, make a rough estimate of the following values. Show your work.
   a. $(f \circ g)(0) =$
   b. $(f \circ g \circ f)(-1.5) =$

4. Following is a graph of a function f. Sketch a graph of the function given by the expression $f(x + \frac{1}{2}) + 1$.

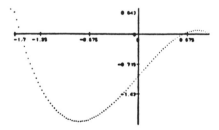

5. Sketch a rough graph of the function given by the expression $\dfrac{3}{(x + 2)^2 (x - 1)}$.

6. Given the following definitions of the functions f and g, find the composition $(f \circ g)(x)$.

$$f(x) = \begin{cases} x^2 & \text{if } x \geq 0 \\ |x|(5 - x) & \text{if } x < 0 \end{cases}$$

$$g(x) = \begin{cases} \dfrac{1}{5 - x} & \text{if } x \neq 5 \\ 3 & \text{if } x = 5 \end{cases}$$

7. A metal bar, which has length 1 when the temperature is 0 degrees centigrade, is expanding as a result of its being subjected to a gradually increasing temperature. The amount of expansion is proportional to the temperature up to 200 degrees centigrade and from thereon it is proportional to the square of the temperature. In both cases, the constant of proportionality is 1. Write a function that represents this situation.

8. Write the derivative of the function given by the following expression: $\cos(\sqrt{1 + \sec(x)})$.

9. For which of the following functions does the relation $f(x^2) = (f(x))^2$ hold? (There could be more than one.) Circle your answer(s).

   (A) $f(x) = \sin(x)$ (B) $f(x) = x^{2/3}$
   (C) $f(x) = x^3 + 5$ (D) $f(x) = x^3$
   (E) $f(x) = \log(x)$

10. Find the indicated limits

   a. Given that $f(x) = \sin\left(\frac{x}{3}\right)$ find $\lim\limits_{x \to \pi} f(x) =$

   b. Given that $f(x) = \begin{cases} \dfrac{x^2 - 4x + 3}{x - 1} & \text{if } x < 1, \\ -2x & \text{if } x > 1 \end{cases}$,

   find $\lim\limits_{x \to 1} f(x) =$

   c. Given that $f(x) = \dfrac{x^2 - 2}{x - \sqrt{2}}$

   find $\lim\limits_{x \to \sqrt{2}} f(x) =$

11. Use the definition to derive the derivative of the function given by the expression $\frac{1}{x+1}$. Show your work.

12. Using any method you can, figure out the derivative of the function given by the expression $|3x - 2|$. Show your work.

13. Write the equation of the line which, at $x = 1$, is tangent to the graph of the function given by the expression $\dfrac{1 + x}{1 + x^2}$.

14. Suppose that $H = F \circ G$. Fill in the blanks in the following table.

x	F(x)	G(x)	H(x)
0	27	1	
1	3	-2	7
2		-3	8
3	16	2	-1
4	-14		27

15. Compute $(f \circ g)'(3)$ using whichever of the following values are needed: $f(1) = -4$, $g(1) = 5$, $f(2) = 0$, $g(2) = 6$, $f(3) = 1$, $g(3) = 2$, and $f'(1) = 2.5$, $g'(1) = 5.3$, $f'(2) = 3$, $g'(2) = 6.7$, $f'(3) = 3.1$, $g'(3) = 2.9$

16. In each of the following, you are given information and you are to state what if any, conclusions can be drawn and why.

   a. You are given that $F(2) = -3$ and $f(3) = 4$. What can you conclude about possible zeros of F in the interval $[2, 3]$? Why?

   b. You are given that F has a zero at 3. What can you be sure of regarding zeros of the function given by $F(x^2)$? Why?

17. Suppose that $f(x) = \begin{cases} x^2 & \text{if } x \leq 3 \\ -x^2 & \text{if } x > 3 \end{cases}$ and you are trying to determine the limit of this function as x approaches 3.000001. Which of the two branches of this function will you be more interested in and why?

18. Given a function g, a point a in its domain, and a small number h, what would be a geometric interpretation of the following quantity? $\sqrt{h^2 + (g(a + h) - g(a))^2}$ Give some indication of how you decided on your answer.

MA 161 Test 2

1. Find the equation of the normal line to the curve defined by the equation $3x^4 + 4y - x^2 \sin y = 3$ at the point $(1, 0)$.

2. For each of the following two curves, give a graphical description of how Newton's method, starting at the indicated x_0, would succeed or fail to approximate the indicated zero. Label each situation as success or failure, and indicate what happens by drawing right on the graph.

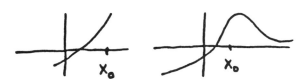

3. Following are the graphs of two equations

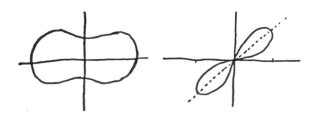

$(x^2 + y^2 + 25)^2 - 100x^2 = 10^8$ $(x^2 + y^2)^2 = 4xy$

Graph 1   **Graph 2**

a. On graph 1, label the intersections of the graph with the coordinate axes and on graph 2, label the points at which the graph is farthest from the origin. Your labels should be the approximate numbers at which these points occur.

b. For each, draw intervals on the coordinate axes to restrict the graph so that it defines a function u for graph1 and v for graph 2, but do this in such a way that the composition $v \circ u$ is defined.

4. Find the third Taylor polynomial for the function given by $x\sqrt{4-x}$ about the point 1.

5. Suppose that you applied the func RiemLeft and the func RiemMax to the same function f on the same interval $[a, b]$. Can you be certain of any relation between the two results? Explain your answer.

6. Make a careful sketch of the graph of the function given by the expression $x^{\frac{2}{3}}(3x + 10)$. Label with values the important points and show the shape of the curve clearly.

7. Boyle's Law for gases states that $pv = c$ where p denotes the pressure, v the volume, and c is a constant. At a certain instant, the volume is 75in^3 and the pressure is decreasing at the rate of $2(\text{lbs/in}^2)/\text{sec}$. At what rate is the volume changing at this instant?

8. How many roots does the equation $x^3 - 6x^2 + 9x + 23 = 0$ have? How did you tell? Approximate one of them as closely as you can.

9. A package can be sent by parcel post only if the sum of its height and girth (the girth is the perimeter of the base) is not more than 96 in. What is the maximum volume that can be sent if the base of the box is a square?

10. The Campbell Soup Company has come to its senses and designed its can so as to minimize the amount of material used. If the radius of the can is increased by 1 percent and the basic design is maintained, what is the corresponding percentage increase in the amount of material?

11. Use the mean value theorem to show that if f is a function whose domain is the interval $[\alpha, \beta]$ and whose derivative has the constant value k on $[\alpha, \beta]$, then f is a linear function.

12. Let f be the function given by

$$f(x) = \begin{cases} 2x & \text{if } x \le 1 \\ 3 + x^2 & \text{if } x > 1 \end{cases}.$$

Find f' and explain the method you used for determining $f'(x)$ for each x.

MA 162 Test 1

1. Use the definition of the natural logarithm function as a definite integral with variable upper limit to show that $\ln(ab) = \ln(a) + \ln(b)$.

2. Let f be the function defined for $x \ge 0$ by $f(x) = \int_0^x \sqrt{1 + (\sin t)^6}\,dt$. Is f an increasing function? Explain why it is or is not.

3. An alloy consists of two elements, one of which is inert and the other decays at a rate which is proportional to the *total* amount of the alloy. Supposing that C_1 is the amount of the inert material and C_2 is the constant of proportionality, find an expression for the function A_1 where $A(t)$ is the amount of the alloy at time t.

4. Do one of the following two (5 pt. bonus if both correct).

a. Describe the method of integration by parts and explain why it works.

b. Describe the method of substitution (for calculating integrals) and explain why it works.

5. Consider the differential equation $y' = f(x, y)$. Give a formula for obtaining an approximate solution of this equation which passes through the point (x_0, y_0). Explain how you got this formula and tell how one might increase the accuracy of the approximation.

6. Suppose that f is the function given by $f(x) = \sin x$. Draw a rough sketch of a function which is an inverse of f and whose range contains the value 20.

7. Given that the function sinh (hyperbolic sine) is given by $\sinh(x) = (e^x - e^{-x})/2$, derive a formula for the derivative of the inverse of this function.

8. Find an expression which gives all solutions of the following differential equation. What can you say about the degree of uniqueness of a solution? $\frac{d^4y}{dx^4} - 5\frac{d^2y}{dx^2} + 4y = 0$.

9. Let (a_n), (b_n) be sequences and form their sum $(a_n + b_n)$. Suppose that you know that the sequence $(a_n + b_n)$ converges. Does it follow that at least one of the sequences (a_n), (b_n) must converge?

If your answer is yes, say as much as you can (in terms of the meaning of convergence of sequences) about why. If your answer is no, give a counterexample.

10. Suppose that you know that the series $\sum_{n=1}^{\infty} 2u_n$ converges. Does it follow that the series $\sum_{n=30}^{\infty} u_n$ must converge?

If your answer is yes, say as much as you can (in terms of the meaning of convergence of series) about why. If your answer is no, give a counterexample.

11. For each of the following series, write C or D according to whether the series converges or diverges:

$$\sum_{n=0}^{\infty} \frac{n}{5n+2} \qquad \sum_{n=1}^{\infty} \frac{1}{\sqrt{n^3+1}}$$

$$\sum_{n=-3}^{\infty} 5\left(-\frac{4}{5}\right)^n \qquad \sum_{n=0}^{\infty} \left(1+\frac{1}{2^n}\right)$$

$$\sum_{n=3}^{\infty} \frac{1}{n(\ln n)^{3.5}}$$

12. Consider the series $\sum_{n=3}^{\infty} \frac{1}{2^n}$. What does it converge to? If S_n is the sum of the first n terms of this series, find N such that $|S_n - L| < 0.001$ for $n \ge N$. (Here L is the sum of the series.)

13. Explain why the series $\sum_{n=2}^{\infty} \frac{1}{n^2-1}$ converges and find its sum.

14. Explain why it is the case that if $q \le 1$ then the series $\sum_{x=1}^{\infty} \frac{1}{x^q}$ converges or diverges as the case may be. (It is not enough to state a rule.)

15. Explain why, if $r > 1$ then the series $\sum_{y=1}^{\infty} \frac{1}{y^r}$ converges or diverges as the case may be. (It is not enough to state a rule.)

MA 162 Test 2

1. Using manipulations, known series and the derivative of $\frac{x}{1+x^2}$, find the Maclaurin series (power series expansion about 0) of $\frac{1-x^2}{(1+x^2)^2}$.

2. Suppose you know that the known function f is equal to its Taylor series expansion on an interval of radius $r > 0$ and centered at c. Write down a formula for the coefficients of this expansion and explain how it is derived.

3. Use power series expansion with remainder term to estimate to within 0.01 the definite integral $\int_0^{1/2} \cos x^2 dx$.

4. Find a power series solution for the differential equation $y'' - 2y' + y = x$ $y(0) = 2$, $y'(0) = 3$.

5. Draw the graph of the following equation $41x^2 - 24xy + 34y^2 - 25 = 0$.

6. Write an equation whose graph is the set of all points p in the plane such that the distance of p from the point $(1, 2)$ is the same as the distance of p from the line $x = 3$. Sketch the graph.

7. Sketch the graph whose equation in polar form is $r = 8\cos 3\theta$.

8. Set up the polar coordinate form of the integral that gives the area inside the curve whose equation is given by $r = 2 + 2\cos\theta$ and above the x-axis. Explain where the formula for the area comes from.

9. Draw a careful picture in 3-dimensions of the line which is determined by the following two equations: $x + 2y + z = 3$ and $3x + y - 2z = 4$.

MA 162 and MA 162 Final Exam

1. Evaluate $\lim_{x \to 0} \frac{1 - \cos 3x}{1 - \cos 2x}$.

(A) 1   (B) $\frac{9}{4}$ (C) $\frac{3}{2}$ (D) 0   (E) $\frac{2}{3}$

2. Find $\tan(\sin^{-1}(2x))$.

(A) $\frac{1}{\sqrt{1-x^2}}$ (B) $\frac{x}{\sqrt{1-x^2}}$ (C) $\frac{2x}{\sqrt{1+4x^2}}$

(D) $\frac{2x}{\sqrt{1-4x^2}}$ (E) $\frac{1}{\sqrt{1-4x^2}}$

3. The graph of $y = xe^{-x}$ looks most like

(A)   (B)

   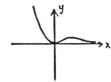

(C)   (D)

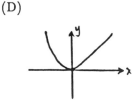

(E)

4. Find the volume of the solid generated by revolving the region bounded by $y = 0$, $y = \frac{1}{1+x^2}$, $x = 0$ and $x = 2$ about the y-axis.

(A) $\pi \ln 3$ (B) $\pi \tan^{-1}(2)$ (C) 2π
(D) $2\pi \tan^{-1}(2)$ (E) $\pi \ln 5$

5. Evaluate $\int_0^1 xe^{x/2}dx$.

(A) $2(2e^{1/2}-1)$ (B) $e^{1/2}-1$ (C) $3e^{1/2}+1$
(D) 1 (E) $2(2-e^{1/2})$

6. $\int \dfrac{x+3}{x^4-4x^3+3x^2}dx$ is of the form

(A) $a\ln|x|+bx^{-1}+c(x-2)^{-1}+C)$
(B) $a\ln|x|+bx^{-1}+c\ln|x+3|+d\ln|x+1|+C)$
(C) $ax^{-1}+b\ln|x+3|+c\ln|x+1|+C)$
(D) $a\ln|x|+bx^{-1}+c\ln|x-1|+d\ln|x-3|+C)$
(E) $ax^{-1}+b\ln|x-3|+c\ln|x-1|+C)$

7. Solve $y'=(1+y^2)x^2, y(0)=1$.

(A) $y=\left(1-\dfrac{x^3}{3}\right)^{-1}$ (B) $y=\tan\left(\dfrac{x^3}{3}+\dfrac{\pi}{4}\right)$

(C) $y=\tan^{-1}\left(\dfrac{x^3}{3}\right)$ (D) $y=e^{x^3/3}$

(E) $y=\sin\left(\dfrac{x^3}{3}+\dfrac{\pi}{2}\right)$

8. $\int_0^{\pi/4}\tan^3 x\,dx=$

(A) $\frac{1}{2}$ (B) $\frac{1}{2}-\ln\frac{1}{\sqrt{2}}$ (C) $\frac{1}{2}+\ln\frac{1}{\sqrt{2}}$
(D) 1 (E) $\ln\sqrt{2}$

9. The substitution $x=2\sin\theta$ transforms $\int_0^2 \dfrac{x^3dx}{\sqrt{4-x^2}}$ into

(A) $8\int_0^{\pi/2}\sin^3\theta\,d\theta$ (B) $16\int_0^1 \sin^3\theta\cos\theta d\theta$

(C) $4\int_0^{\pi/2}\dfrac{\sin^3\theta}{\cos\theta}d\theta$ (D) $16\int_0^{\pi/2}\sin^2\theta d\theta$

(E) $8\int_0^{\pi/2}\cos^3\theta d\theta$

10. The root test shows that $\sum_{n=1}^{\infty}\left(1+\dfrac{1}{n}\right)^{-n^2}$

(A) Diverges. (B) Converges.
(C) The root test gives no information.

11. If $f(x)=xe^x$, then $f^{(15)}(0)=$ (Hint: Expand xe^x in a Maclaurin series.)

(A) $\frac{15!}{14!}$ (B) $15!$ (C) 0 (D) 1 (E) $14!$

12. The Maclaurin expansion of $y=x\sin(-x^2)$ begins with (Hint: start with the Maclaurin series of $y=\sin x$)

(A) $-x^3+\frac{x^6}{3!}-\frac{x^8}{5!}+\dots$ (B) $x-\frac{x^5}{2!}+\frac{x^9}{4!}-\dots$
(C) $1-\frac{x^4}{2!}+\frac{x^8}{4!}-\dots$ (D) $-x^2+\frac{x^6}{3!}-\frac{x^{10}}{5!}+\dots$
(E) $-x^3+\frac{x^7}{3!}-\frac{x^{11}}{5!}+\dots$

13. $\sum_{n=1}^{\infty}(-1)^n\dfrac{2^n}{1+3^n}$

(A) Converges Conditionally)
(B) Converges Absolutely)
(C) Diverges)

14. $1+\frac{1}{2^2}-\frac{1}{3^2}+\frac{1}{4^2}+\frac{1}{5^2}-\frac{1}{6^2}+\frac{1}{7^2}+\frac{1}{8^2}-\frac{1}{9^2}+\dots$
(Every third term is negative.)

(A) Converges Conditionally)
(B) Converges Absolutely)
(C) Diverges)

15. $\int_0^x e^{t^2}dt=$

(A) $\sum_{n=0}^{\infty}\dfrac{x^{2n}}{(2n)!}$ (B) $\sum_{n=0}^{\infty}\dfrac{x^{2n+1}}{n!}$

(C) $\sum_{n=0}^{\infty}\dfrac{x^{2n+1}}{(2n+1)!}$ (D) $\sum_{n=0}^{\infty}\dfrac{x^{1n+1}}{(2n+1)(n!)}$

(E) $\sum_{n=0}^{\infty}\dfrac{x^{n+1}}{(n+1)!}$

16. The ellipse $16(x-3)^2+25(y-7)^2=400$ has one focus at

(A) $(6,7)$ (B) $(7,7)$ (C) $(3,10)$
(D) $(3,11)$ (E) $(3,12)$

17. The graph of $(y-4)^2=12(x-1)$ is a parabola whose directrix is the line $x=-2$. The point $P=(13,16)$ lies on the parabola. The distance from P to the focus is

(A) 12 (B) 15 (C) $\sqrt{306}$ (D) $\sqrt{425}$ (E) $\sqrt{481}$

18. The graph of $r=\sec\theta$ is

(A) a line. (B) a spiral. (C) a hyperbola.
(D) a rose. (E) none of these.

19. The area of the region enclosed by $r=2+\cos\theta$ is

(A) 2π (B) 3π (C) $\frac{5}{2}\pi$ (D) $\frac{9}{2}\pi$ (E) $2\pi+\frac{2}{3}$

20. Which of the following best represents the graph of $r=3\sin 2\theta$, where θ is in the interval $[0,\pi]$?

(A)   (B)

(C)   (D)

(E)

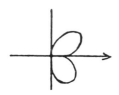

21. Which of the following best represents the graph of $x^2 + y^2 - z^2 + 1 = 0$?

(A)   (B)

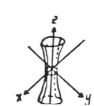

(C)   (D)

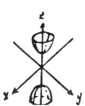   

(E)

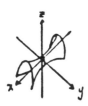

22. The points $(0,0)$, $(1,2)$, $(4,2)$, and $(3,0)$

are vertices of a parallelogram. Find the smaller angle θ of the intersection of the diagonals of this parallelogram.

(A) $\theta = \cos^{-1}\left(\frac{6}{\sqrt{190}}\right)$ (B) $\theta = \cos^{-1}\left(\frac{4}{5}\right)$

(C) $\theta = \cos^{-1}\left(\frac{3}{\sqrt{45}}\right)$ (D) $\theta = \cos^{-1}\left(\frac{4}{\sqrt{160}}\right)$

(E) $\theta = \frac{\pi}{4}$

23. $(3\vec{i} - \vec{j} + \vec{k}) \times (-\vec{i} - 2\vec{j} + \vec{k}) =$

(A) $2\vec{i} - 3\vec{j} + 2\vec{k}$ (B) $-2\vec{i} + 8\vec{j} + 14\vec{k}$
(C) 0 (D) $-3\vec{i} - 2\vec{j} - 5\vec{k}$
(E) $\vec{i} - 4\vec{j} - 7\vec{k}$

24. An equation of the plane that contains $(23, 4, -2)$ and is parallel to $-2x + y + 2z - 7 = 0$ is

(A) $3x + 4y - 2x - 7 = 0$

(B) $3x + 4y - 2z - 29 = 0$

(C) $-2x + y + 2z + 6 = 0$

(D) $-2x + y + 2z = 14 = 0$

(E) $\frac{x-3}{-2} = \frac{y-4}{1} = \frac{z+2}{2}$

25. Symmetric equations of the line that contains $(2, 4, -1)$ and is perpendicular to $-3x + 3y + 5z + 4 = 0$ are

(A) $x = 2, y = 4, z = -1$

(B) $-3x + 3y + 5z + 1 = 0)$

(C) $2x + 4y - z + 4 = 0)$

(D) $\frac{x+3}{2} = \frac{y-3}{4} = \frac{z-5}{-1})$

(E) $\frac{x-2}{-3} = \frac{y-4}{3} = \frac{z+1}{5})$

26. A projectile is launched with an initial speed of v_0 ft/sec from ground level at an angle of 30° from the horizontal. At what time does it reach a point 100 feet horizontally from the launch point?

(A) $20v_0$ sec   (B) $200v_0$ sec   (C) $\frac{200}{\sqrt{3}v_0}$ sec
(D) $\frac{80}{\sqrt{3}v_0}$ sec   (E) $50\sqrt{3}v_0$ sec

St. Olaf College

Viewing Calculus from Graphical and Numerical Perspectives

by Arnold Ostebee and Paul Zorn

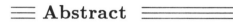 **Abstract**

Contact:

ARNOLD OSTEBEE or PAUL ZORN, Department of Mathematics, St. Olaf College, Northfield, MN 55057. PHONE: 507-663-3420 or 3414. E-MAIL: ostebee@stolaf.edu; zorn@stolaf.edu.

Institutional Data:

St. Olaf is a private, four-year liberal arts college. The student population, of about 3000, is primarily from the Midwest. Nearly all students are full-time; the vast majority live on campus. We offer a B.S. in nursing, but no other graduate or professional programs. We have neither a business nor an engineering school.

The mathematics department is one of the largest on campus, both in number of faculty and in number of majors. Our 27 faculty share 18.25 FTE's. (Some of us have partial appointments in other academic units or grant-supported released time.) Approximately 100 students per year, or about 1 in 7 of *all* St. Olaf graduates, complete mathematics majors.

Project Data:

Since our project began in 1987, eight faculty members have received released time and/or summer support to develop course materials and teach project courses. At least eight more have used project-related materials. During the grant-supported period, about 25% of all Calculus I and II students were enrolled in project courses. At present, at least 50% of all calculus sections make some use of project materials.

Project-related courses have been offered for seven semesters, since Spring 1987. Students in these courses have access to a computer laboratory, but our courses have no formal laboratory component. We have used the symbolic mathematics system *SMP*, running on a network of Sun workstations, but will shift this fall to *Maple* on Macintoshes that are available in many campus locations.

In 1989-90 all sections of calculus used Stein's *Calculus and Analytic Geometry*, fourth edition. For 1990-91, all sections are using the Fifth Edition of Purcell & Varberg. In project-related sections, additional expository and problems materials are used.

External funding for hardware and software was provided by NSF's College Science Instrumentation Program (now called ILI). Support for released time, mainly for materials development, came from the U.S. Department of Education's Fund for the Improvement of Postsecondary Education (FIPSE).

Project Description:

Our project aims to bring graphical and numerical—as well as algebraic—viewpoints to bear on calculus ideas. Doing so entirely "by hand" would be forbiddingly time-consuming and distracting; with computing, these viewpoints become practically accessible.

Our goals have nothing to do with "computerizing" the calculus or with "computer literacy." No prior experience with computing is assumed; no computer *programming* is taught. Formal instruction in using the software is limited to a brief hands-on session at the beginning of each semester. We use computing strictly as a tool to focus our courses more clearly on concepts.

Project versions of Calculus I and II have consistent, distinctive themes. In Calculus I, *graphical representations* are used wherever possible to stress the interplay of algebraic, analytic, and geometric ideas. In Calculus II, the new themes of *approximation and error estimation* are added; the emphasis on graphical viewpoints continues. We make crucial use of the *combination* of graphical, symbolic, and numerical modes of computing in comparing and moving among these disparate but complementary themes.

During our project's three-year funded period (ending in December 1989), we designed, wrote materials for, and taught special versions of Calculus I and II. Project courses have the standard local structure: 3 lecture hours per week, sections of 20–35 students, no formal laboratory component. Project courses are taught in standard classrooms, not in a computer laboratory; an overhead display device is available for computer output. There are no TA's as such, but advanced students are sometimes available for help with homework.

Students in project sections use *SMP*, running on Sun workstations, to complete homework assignments, tests, and other course work. Homework assignments are made each class session; most require both conventional paper-and-pencil work and work with *SMP*. Some tests are given in class, others on a take-home basis. In the latter case, use of *SMP* is generally permitted.

The St. Olaf mathematics department has consistently supported the project. A substantial majority of department faculty have participated directly, with or without project support, in project-related activities.

Materials developed with project support include approximately 500 pages of problems sets, tests, and expository materials. Although faculty from many other institutions have received copies, we have not done formal field-testing. Project materials are available at a cost of $15.00 for printing and postage. Contact Professor Ostebee.

≡ Project Report ≡

Day-by-day Mechanics

All elementary calculus classes at St. Olaf, *SMP*-aided or not, have the same formal structure. Classes meet three times weekly; "hours" are 60 minutes. A typical semester course comprises 39 full class hours. Class sizes range from 20 to 35; we offer, in all, 15–20 sections of Calculus I and Calculus II each semester, of which 3–5 are honors sections. (Honors sections at St. Olaf serve mainly to group students by placement level and ability; they use roughly the same syllabi as other sections.) No scheduled recitation or laboratory sections are attached to any of our calculus courses, but upper division students are available for several hours most evenings for informal help with homework problems.

Students in *SMP*-aided sections have access to a physical computer laboratory. (A description of the laboratory appears below in the Technology section.) They exploit the graphical, numerical, and algebraic powers of *SMP* in completing homework assignments, tests, and other course work.

Homework assignments are normally made at each class meeting. Many assignments involve both conventional paper-and-pencil work and work with *SMP*. Examples of typical homework assignments for both Calculus I and Calculus II are appended.

The logistics of testing and grading are similar in ordinary and *SMP*-aided sections. Our local tradition is to leave such things up to individual instructors (e.g., we don't give common final exams). A typical grading scheme in any calculus section involves three or four tests, a final exam, and regular homework. Take-home tests, written projects (perhaps done in groups), and open-ended assignments may be more common in *SMP*-aided sections, but they are also used in our other calculus classes.

St. Olaf faculty teach 6 courses per year; a calculus section counts as one course. Our FIPSE grant bought a pool of released time and summer support (about three and one-half person-years over a three-year period) with which to compensate faculty (partially!) for the extra time and effort of designing and teaching *SMP*-aided sections. This extra burden on faculty came (and still comes) in several forms: learning to use *SMP* itself, meeting with other project faculty, course planning, and, most of all, writing and rewriting new material. We shared the pool of released time broadly among faculty in the department. In a typical semester, three faculty members (two veterans and a neophyte) received one course release each. Veterans spent their time mainly in developing material, neophytes in learning the ropes.

Getting Started

Our project had its genesis in 1983 when our department, along with five others, received a grant from the Sloan Foundation to study the roles of discrete and continuous mathematics in the first two undergraduate years. (Reports on the Sloan-supported projects were published in *Discrete Mathematics in the First Two Years*, MAA Notes Number 15.) The Sloan project mainly had to do with discrete mathematics. A secondary effort was to seek an appropriate balance between "continuous" and "discrete" points of view, in the curriculum generally and in our calculus courses particularly. We came to two conclusions: first, that students should see *both* continuous and discrete methods and ideas in calculus; and second, that the supposed dichotomy between discrete and continuous mathematics is artificial. Calculus is really about the *interplay* of discrete and continuous ideas: difference quotient and derivative, Riemann sum and Riemann integral, infinite series and improper integral, estimate and exact value. Seeing the subject from both points of view adds depth and shading.

From here it was a short step to beginning efforts with symbolic computing. With it, we hoped to show the numerical and graphical, as well as algebraic, sides of topics like convergence of Riemann sums to an integral. We bought the computer algebra system *SMP* in 1984 because, at that time, it offered the best combination of numerical, graphical, and algebraic computing power. In 1985 we used *SMP* in a January term course on applications of calculus.

These early experiences led us, in 1985–86, to submit two complementary grant proposals: the first, for around $35,000, to NSF's ILI (*nee* CSIP) program, for matching funds for equipment purchase; the second, to FIPSE, for around $200,000 over three years, mainly to buy released time. With support from ILI we were able, by spring 1987, to equip a lab with enough Sun workstations to support several *SMP*-aided calculus sections. The FIPSE grant supported development of course materials and other activities related to the use of *SMP* in calculus courses.

Our project changed substantially in scope and emphasis over its lifetime. At first we saw things rather mechanically: we would provide hardware, software, and manuals, and then get out of the way. Things turned out differently: both the problems and opportunities the tools raised were much greater than we foresaw. We had imagined that we could, with minor adjustments, *append* graphical and numerical computing to an essentially standard course. We were wrong: the special viewpoints and emphases we aimed for turned out to fit poorly in a conventional course. We had planned to write only terse exercise sets that would use *SMP*, but we found we needed a considerable amount of expository material as well. Indeed, we found ourselves redesigning our courses from the ground up.

Local Support

The climate of support for the project, both within the mathematics department and from the College, has always been favorable. To be sure, mathematics faculty vary in their interest in and commitment to the project, but no more than one would expect in a large, diverse, and busy department. An explicit goal from the start has been to spread "ownership" of the project around the department, rather than to make it the fiefdom, and burden, of a few zealots. We took various steps to foster that goal.

• The pool of released time was spread thinly: eight faculty had some of it.

• We organized a 3–5 day workshop for department faculty in each of the three summers of the funded period. Nominal stipends were paid. The workshops served partly to train faculty who were new to the project, but *everyone* was welcomed; over the years, almost all tenured and tenure-track faculty participated. The workshops aimed to demystify the hardware and software; to inform faculty about the project's goals, strategies, and logistics; to discuss pedagogy and curriculum; to give faculty experience in designing their own course materials; and to *get* advice and counsel. Each workshop concluded with an extended presentation to the entire department, covering project history, goals, and progress to date.

• Materials developed for the project courses were provided to *all* faculty teaching calculus. Little of this material is inextricably tied to *SMP*; more characteristic are graphical, numerical, and certain types of word problems. Many faculty now use these materials—some without also using *SMP*.

We should emphasize that this level of departmental support did not stem from any deep or general dissatisfaction with the *status quo*: calculus enrollments, success rates, and continuation through the mathematics major have long been at impressive levels at St. Olaf. The project was seen as an effort to improve an already good thing, not as a defense against disaster.

Other Departments

Since changes in calculus courses affect "client" departments, we have tried to apprise other departments of our efforts, and to get their advice. In summer 1989 we held a one-day workshop for members of the physics, chemistry, biology, and economics departments. The participants supported our project's goals. For example, they were far less concerned with details of course coverage than with students' general ability and willingness to use the calculus—especially in its graphical, numerical, and qualitative aspects—as a tool in their disciplines.

Obstacles, Problems, and Strategies

We encountered few institutional obstacles. Both the College and the department have been helpful in finding funds, arranging schedules, accommodating released times, and providing space.

A partial exception to this happy rule resides in the rigid credit system at St. Olaf: students receive the same credit for nearly all academic courses. (That credit system is now under review; it may change.) It would have been difficult, for example, to give extra academic credit for a formal laboratory component in *SMP*-aided calculus courses, had we chosen to require one. In practice, we decided *against* a formal laboratory approach, but with more flexibility we might have decided otherwise.

Student attitudes and preparation posed —and still pose—larger and more difficult problems. We cite a few:

• Shiny machines and powerful programs notwithstanding, our project's main focus is the *concepts* of calculus. Such a focus clashes with many students' (well-founded) expectations. These students expect a calculus course focused primarily on routine manipulations. Even able students may cling tenaciously to this expectation;

exercises that take more than a few moments' thought are sometimes viewed as beyond the pale. Coping with more substantial problems requires surprising amounts of help, reassurance, and class time.

• Poor algebra is one verse in the conventional litany of complaints about student preparation. The problem is real and our students suffer from it. But just as important, is students' tendency to see *too much* of mathematics as algebra. For example, many students *identify* functions (transcendental functions, too!) with algebraic formulas. If functions are nothing but formulas, then the calculus is nothing but a compendium of rules (some of them quite arcane) for formula manipulation.

We try to counter this false identification of functions with algebraic formulas by stressing *other* representations: graphical, numerical, tabular, verbal, etc. It is difficult, but we think important, to help students master and move among these various styles of representation.

• Students have many problems with the idea of function. One that haunts us particularly is the notion of *bounds*. Our stress (cf. section on Content) on approximation and error estimation entails frequent use of upper and lower bounds for functions over closed intervals. Students find bounding problems, important and basic as they are, surprisingly elusive; they evoke the worst sort of wishful thinking.

We attribute this problem—and others —to students' weak and static (and often downright wrong—"What's the 'formula' for $\sin(x)$?") identification of "function" with "formula." Studying graphical and numerical representations of functions helps, but this takes time and effort.

Content and Philosophy

A Diagnosis

The worst fault of standard calculus courses is that they overemphasize routine techniques— formal differentiation, antidifferentiation, convergence testing, etc.—while slighting the conceptual foundations of the subject. Too often, analytic objects (integral, derivative, convergence, etc.) are represented and manipulated only algebraically (i.e., *via* symbolic manipulation of explicit elementary functions). Overfed on "techniques of integration," for example, many students fail to distinguish between integration and antidifferentiation. They are helpless when they encounter problems that require numerical or graphical methods.

Viewing calculus either as an introduction to pure mathematics or as a foundation for applications leads to the same conclusion—that concepts, not techniques, are truly fundamental to the course. Whatever use they make of the calculus, students need more than a compendium of manipulative techniques. The *sine qua non* for a useful command of the calculus is a conceptual understanding that is deep and flexible enough to accommodate diverse applications.

A Prescription

Bringing graphical and numerical, as well as algebraic, viewpoints to bear on calculus ideas is the philosophical foundation of our project. Combining, comparing, and moving among graphical, numerical, and algebraic "representations" of central concepts is our key strategy for improving conceptual understanding; it pervades and unifies our courses. In treating integration, for example, we augment the usual algebraic antidifferentiation methods with both graphical and numerical methods, the latter with careful error estimates.

The Role of Computing

We use computers for one main purpose: *to facilitate combining graphical, numerical, and symbolic viewpoints on the calculus.* Incorporating these viewpoints into a calculus course "by hand" is forbiddingly time-consuming and distracting; with computing they are practically accessible.

Numerical computing has been used in calculus for decades, without dramatic effect. Why should things be different now? The answer, we think, has to do with the availability of *symbolic computer systems.* (We use the acronym SCS for such systems; examples include *Mathematica, Maple, Derive, Reduce,* and *SMP.*) Their crucial new feature, for our purposes, is that they make graphical, numerical, and symbolic computing power available in one package, without the distraction of programming.

With only rudimentary knowledge of a modern SCS students can draw on a powerful kit of graphical, numerical, and symbolic tools. To plot a function, to estimate its integral numerically, or to expand it in Taylor series using *SMP* requires only a one-line command:

```
Graph[Sin[x^2],x,0,6 ]
Midpoint[Sin[x^2],x,0,1,10]
Taylor[Sin[x^2],x,0,9]
```

None of the *SMP* commands our students use are more complicated than these.

Distinctive Features

Our main goals are conceptual: to help students understand calculus ideas more deeply and apply them more effectively. Such goals are unexceptionable and almost universally shared; more interesting is *how*, in practical terms, we try to realize them. Full syllabi for our project courses are appended. Here we summarize a few distinctive features.

Several themes pervade both semesters of our courses:

* *Combining graphical, numerical, and algebraic viewpoints and representations.* We insist throughout the course that students manipulate and compare graphical, numerical, and algebraic representations of mathematical objects.

* *The calculus as language.* We present the calculus both as a vocabulary for *describing* the phenomena of change and as a mechanism for *deducing* their consequences. We try to emphasize both aspects from the beginning. Thus, for example, we introduce differential equations, particularly in the descriptive sense, early in Calculus I.

* *Concepts vs. rigor.* Our course is quite conceptual but not especially rigorous. We think that stating and proving general forms of theorems is less valuable (at this level) than helping students understand concretely what theorems say, why they are reasonable, and why they matter.

(What theorems say might seem to be no problem: they say what they say. But students, we find, have surprising difficulties with the *statements*—let alone the proofs—of theorems. Logical formalities (hypothesis, conclusion, contrapositive, etc.) are often misunderstood. Sometimes the trouble goes deeper. The fundamental theorem, for example, is often taken to *define* the definite integral.)

Thus, we do not expect our students to produce more polished formal proofs than other calculus students, but we do expect them to solve more varied problems and demonstrate a better command of calculus ideas. Computing helps illuminate concepts in various ways: by quickly working out computational examples, by illustrating results graphically and numerically, and by revealing patterns.

- *A course mainly in one variable.* Our treatment of multivariable topics is limited mainly to partial derivatives and multiple integrals. This stems partly from an effort to keep things "lean" and partly from local conditions. Since most St. Olaf students take Calculus III *after* a semester of Linear Algebra, we can delay multivariable topics and also teach a slightly more ambitious Calculus III.

- *Programming is not taught.* Our students only begin to tap the enormous power of symbolic computer systems. For example, *SMP* is a high-level programming language, but few of our students have any inkling of this. We have chosen to emphasize the results, rather than the processes, of computation. (That we avoid programming in our courses is not meant to deny the value of having students create their own mental structures and implement them concretely.)

Calculus I

The most distinctive feature of our first semester course is its *emphasis on graphs.* From the very start, where the important standard functions are introduced, graphs—most of them drawn by machine—are everywhere. We regard qualitative, geometric properties of the standard functions as no less important than their algebraic properties. Later, as the ideas of derivative, concavity, and integral are introduced, our students freely use computer graphics to illustrate and emphasize connections between analytic properties of a function and geometric properties of its graph. (We first introduce the definite integral, for example, in terms of area.) Graph-*sketching* is, by comparison, de-emphasized: students do some of it, but mainly to fix ideas. The graphical theme appears regularly in exercises and tests as well as in exposition: many problems involve graphically- or tabularly-presented functions.

Another feature of this course is the lower-than-usual proportion of student activity devoted to routine symbol manipulation. We assign some drill problems, but we de-emphasize pathology. Instead we emphasize other sorts of problems: problems that combine algebraic, graphical, and numerical viewpoints; multi-step problems; "translation" problems (e.g., rephrasing geometric properties of a curve in terms of derivatives); problems that build

qualitative intuition (e.g., relative rates of growth). Examples of most of these genres are included in the appendices.

Symbol manipulation skills are necessary, but hardly sufficient, for real understanding. For example, most students quickly master the *algebraic* mechanics of the chain rule, and so might *seem* to "understand" the rule. But few students, we find, can answer even the simplest questions about derivatives of compositions of *graphically-presented* functions. Representing functions in varying styles—graphical, tabular, "black box"—can help bring students (albeit sometimes reluctantly) to the heart of the matter.

Calculus II

The graphical emphasis of Calculus I continues undiminished in Calculus II. Added to it is another recurring theme: *approximation and error estimation.* It appears in numerical integration, in applications of the definite integral, in improper integrals, and in our study of series.

Emphasizing approximation and error estimation leads to our greatest departure from standard practice. It shapes both our syllabus and our use of computing. The combination of numerical, graphical, and symbolic computing is often important in our course, but here it is essential.

Consider, for example, our treatment of infinite series. Standard practice is to emphasize almost exclusively the problem of testing series for convergence or divergence. Students learn to label series either "convergent" or "divergent," but often have little understanding of what either label means. With computing help, on the other hand, we can emphasize the more interesting and instructive problem of *estimating the limit* of a series to specified accuracy.

Our approach to series involves several nonstandard ingredients: tail analysis, error estimates, comparing rates of convergence, etc. Little of this appears in standard texts; computing is almost essential for most of it. (See the appended tests and problem sets for more details.) This treatment, we think, is not only more "applicable" than the standard one, but also promotes a deeper understanding of the phenomenon of convergence itself.

A word about the general role of numerical techniques may be in order: although we emphasize them far more than is usual in calculus courses, we make no effort to "cover" numerical analysis as a topic in its own right. Our experience suggests that students benefit *conceptually* from numerical approaches to analytic ideas. Thus, students study numerical methods that illustrate and clarify underlying analytic ideas. Traditional numerical considerations—stability, robustness, etc.—are not discussed.

Pedagogy

Our courses are not, in terms of pedagogy, greatly exotic. For example:

- We teach in ordinary classrooms, not in a computer laboratory. (No large enough lab exists on campus, in any case.) An overhead display device is used to show *SMP* computations and their results. Instructors' styles vary, but most of us, most of the time, give blackboard-and-chalk lectures with occasional machine computations and graphical displays.
- We encourage students to work on homework assignments in groups of two or three; all members of a group get the same grade on each assignment. Since the exercises we assign can be quite non-routine, we often collect homework at the *second* following class period and expect to spend some extra class time on it.
- Tests can be either in-class or take-home. Most of us give some of both, depending on the topic. Take-home tests are usually "open book, open *SMP*." In Calculus II, for example, topics—and hence also tests—seem to divide about equally into "pencil" and "machine" categories. Concerns about cheating on take-home tests occasionally arise, but no serious problems have developed. St. Olaf's honor system (students sign a pledge after each exam) works to our advantage here.
- Although mathematical writing is not, as such, a consistent theme in our courses, for most of us it is at least a *leitmotif*. We regularly assign essay questions on homework and tests.
- We spend very little class time on the mechanics of symbolic computation. Instead, we hold brief, intensive workshops at the beginning of each semester and employ advanced students to help *SMP* beginners with technical questions.

Technology

A network of Sun-3/50 workstations is the technical base of our project. Students have access to eight of these machines, housed in a laboratory within the mathematics department. Several additional workstations are scattered around the department. The network is accessible (by remote login) from elsewhere on campus. Our system can handle, at most, 16 simultaneous *SMP* users. This capacity has so far proved sufficient for about 100 students per semester. On the other hand, we have not been able to provide continued access to *SMP* for all of our "alumni," desirable as that would be.

Mathematics faculty use the Sun network both as a general computing resource and as an interface to other campus computing systems. Most important for project purposes is technical text processing: the system supports a full—and fully used—implementation of TeX.

We chose our particular hardware and software several years ago. They have served us well, although today we would not—indeed, could not—make exactly the same choices. *Any* suitable system would offer, in convenient form, a powerful combination of graphical, numerical, and symbolic capabilities. Choices depend on price, convenience, and compatibility with other local equipment. Given these considerations, in order to expand our project to all calculus sections we are switching in 1990-91 to *Maple* which runs on a widely available campus Macintosh network.

Appraisal and Conclusions

Our three-year project has generally achieved its original goal: to bring high-level computing to bear on the teaching and learning of calculus. On the other hand, our project comprises many goals and activities; no *single* verdict of success or failure applies equally well to all of them. Thus we discuss four aspects of the project—products, faculty, students, and assessment—separately below.

Products

The most important tangible products of our project are syllabi and supporting materials for two significantly redesigned calculus courses. Qualitatively, this is what we envisioned three years ago. Quantitatively, we accomplished considerably more than we expected.

Our plans for redesigning our courses became more ambitious with time. Originally we intended to "append" symbolic, numerical, and graphical computing to our existing calculus courses. We would write some supporting material (problem sets, mainly) for the purpose, but the courses themselves would remain little changed.

In the event, more and different material was needed. For example, we found that standard textbooks, which strongly emphasize symbol manipulation activities, are often at odds with the special goals of our courses. Students (and faculty!) need pertinent, readable expository material that supports numerical and graphical points of view. During the last year of the project we began, in earnest, to write expository material. By now we have produced several hundred pages of materials of all types: examples, tests, exposition, etc.

We also found unexpectedly high demand for written materials from other faculty, both here and elsewhere. Last fall, for example, we provided a collection of supplemental problems to all Calculus I instructors at St. Olaf. Most of them used our problems, especially those of a graphical nature. We have also sent copies of our materials to scores of faculty at other institutions. In addition, we expect some of our materials to appear in the calculus resource collection being compiled by the ACM/GLCA consortium.

Faculty

From the beginning, we have tried to involve as many mathematics faculty as possible in our project. By now, over half the department has participated in some formal way.

Not all department faculty would wish, at least at present, to teach SCS-aided sections of calculus. Reasons would range from logistics (too busy, unfamiliar with the software, etc.) to differing priorities or philosophy (e.g., concern for coverage of particular topics). Although SCS-aided sections of calculus have always been a minority, they have never been "ghettoized" or regarded as marginal within the department's instructional effort. On the contrary, department faculty, as a group, have reliably supported the project and its goals, and helped in various informal ways: with student advising, by accommodating the project's special scheduling and space needs, and by offering constructive suggestions. Some faculty are quite enthusiastic: one veteran teacher remarked that for the first time in 20 years, she felt that her students really understood what it means to say that a series converges.

Our project has been, to put it mildly, labor-intensive. One should distinguish, though, between the work of developing courses and materials on the one hand, and teaching project courses on the other. For *development* work, our FIPSE grant provided a significant amount of released time. This released time never seemed quite enough, but it was absolutely vital to our project. Without it we could never have written and produced the materials we did. Some (but much less) released time was made available to compensate faculty for the extra effort of *teaching SMP*-aided courses. This extra work was significant, especially for faculty entirely new to the project, but it was not overwhelming, and it tended to decrease as more and better materials appeared, and as a routine developed. By now, several faculty use *SMP* entirely on their own time. As the syllabi and course materials have stabilized, more and more faculty have begun to use our materials. This trend seems likely to continue.

Everything else aside, our project has led to an enormous amount of valuable discussion about calculus and *SMP* in particular, and about computing and undergraduate mathematics in general. One of our faculty—not a participant in the project—put it something like this:

> This project, with its emphasis on concepts, has really forced those of us teaching the traditional course to take a hard look at what we are doing. How, we must ask ourselves, would our students fare on the final exams being given in the experimental sections? Is it true that an insistence on mastery of techniques obscures a student's grasp of central ideas? The chance to see the homework assignments and exams being used in the experimental courses has certainly caused us to sharpen the focus in our courses. The presence of a project like this has a salutary effect on all of our courses.

Students

We think that calculus students benefit from our *SMP*-aided courses in several important ways: by encountering the subject at a more conceptual level; by consistently handling graphical, numerical, and algebraic representations of the main ideas; and by harnessing some of the combined power of two great engines—calculus and the computer.

We are not, on the other hand, near Utopia. The problems we alluded to above (poor understanding of functions and bounds, difficulties in "translating" among various representations of calculus ideas, etc.) remain. Indeed, they may be *more* serious for our courses than for standard courses, which depend less on such conceptual skills.

Students' difficulties with calculus ideas are surprisingly great. In hindsight, we may have been naïve. First-year students' earlier training emphasizes symbol manipulations; stressing concepts goes against expectations. Calculus textbooks usually only exacerbate the problem.

Whatever the reasons, we have found it harder than we expected to help students achieve the conceptual understanding—and even the *attitude* toward conceptual understanding—for which we aim. For example, our hopes that *SMP* could help students *discover* nontrivial ideas in calculus, in open-ended settings, have seldom been realized. Designing problems and activities that are neither too vague to make sense nor too routine to be worthwhile seems to be very difficult. A formal laboratory situation, in which faculty could offer help when needed, might be an answer. For reasons discussed above, we have not yet tried this approach.

Students are able, in principle, to select themselves into or out of *SMP*-aided calculus sections. The qualifier is necessary because, in practice, few students choose calculus sections based *solely* on the presence or absence of *SMP*. Many other factors intrude: time of day, grapevine information, instructor's reputation, what is available in a given semester, etc. Since the department offers both honors and non-honors versions of both Calculus I and Calculus II every semester, some students have little or even no choice between *SMP*-aided and conventional sections. With so much "noise" in the system, clear enrollment patterns are hard to detect. For example, one year we experienced significant "defection" from *SMP*-aided to ordinary sections between semesters; in another year just the opposite occurred.

One thing *is* clear: stronger students, such as those in our honors sections, thrive better than weaker students in our *SMP*-aided sections and seem to benefit more. We attribute this less to the extra burden (in reality, not great) of learning to use *SMP* than to the sharper conceptual focus of our courses. For average students, seeing the principles beyond computation—whether done by hand or by machine—remains difficult. For the very weakest students, the mechanics of *SMP* itself may indeed impose an additional burden.

Assessment

Measuring the success or failure of projects like ours in *quantitative* terms, especially in the short run, is difficult both in principle and in practice. Common final exams, for example, address only what is *common* to the courses under comparison. (Our limited experience with such testing revealed no consistent differences.) We are more interested in what is *uncommon* in our courses: do students really benefit, as we expect, from seeing things from graphical and numerical as well as algebraic viewpoints? How do our courses affect students' algebraic facility, their ability to *use* calculus ideas effectively, and their attitudes to mathematics? Do students really achieve better conceptual understanding? How will they fare in later mathematics courses?

For various reasons, we have done relatively little quantitative assessment of our project. Time was one factor: we always saw course development itself as our first priority. Just as important were philosophical questions: what to measure, how and when to measure it, with what instruments, and against what standard. These questions are particularly difficult to answer when mathematical standards themselves are changing, partly under the influence of computing.

Nevertheless, we believe that assessment, difficult as it is, is important both to help shape local efforts and for external credibility. For example, we hope to compare performance of students from different calculus sections in our elementary real analysis course. This course, generally taken late in the sophomore or early in the junior year, seems to us to be most closely related to the goals of our special calculus courses.

We have also, all along, surveyed student attitudes in our *SMP*-aided courses. The results are quite mixed. For example, some students find using *SMP* easy, enjoyable, and valuable; others seem to find it unfriendly and distracting. On one point most students agree: the *SMP*-aided courses are harder and require more work than standard courses. (We regard this perception as accurate, but it poses a potential problem. If students are offered two ostensibly parallel tracks through the calculus sequence, each earning equal credit, why should they choose the harder alternative? On the other hand, the project courses do not seem *unreasonably* difficult. We hope and expect that in the long run the current two-track system will tend to merge into one.) The extra labors, happily, are generally seen as requited: most students describe the *SMP*-aided courses as "more worthwhile."

Now that a substantial cadre of *SMP* "alumni" exists, we have begun to compile "demographic" information: comparative attrition rates between semesters of calculus, rates of enrolling in subsequent mathematics courses and of majoring in mathematics, GPA, etc. Complete results are not yet available, but some patterns have emerged.

One category of special interest is gender. Our

evidence suggests, troublingly, that women students sometimes avoid computer-aided courses. Over the test period, 54% of the students in non-*SMP* courses were women; in *SMP* courses the comparable figure was 36%. Contrary to some expectations, the women who did enroll in *SMP* courses were *not*, as a group, mathematically stronger (as measured by SAT and ACT scores) than women who enrolled in standard courses. The data suggest, if anything, the opposite. Although women may avoid *SMP*-aided courses, none of our data suggest that taking such courses does women (or men, for that matter) any harm. Both men and women students who took *SMP*-aided courses fared at least as well in later courses in mathematics, economics, and the natural sciences, and were at least as likely as other students to major in mathematics.

Future Plans

Our funding period ended in late 1989, but local prospects look good for continuing and, perhaps, expanding most of our project activities. For example, we are planning to make *Maple* available on our campus-wide Macintosh network. This would allow us to offer a larger number of SCS-aided calculus sections each semester. It would also give our students *continuing* access, throughout their undergraduate careers, to mathematical computing. (At present, limited capacity obviates such a policy.)

Specific plans for higher-level mathematical computing are also afoot. In 1989 the department submitted a successful proposal to the ILI program of NSF to establish a second laboratory for mathematical computing. The laboratory will support a planned Advanced Mathematical Computing Requirement for all St. Olaf mathematics majors.

Among the clearest lessons of our experience is the need—expressed by both students and faculty—for an appropriate textbook. Such a book would thoroughly integrate, not merely append, the special viewpoints of graphical, numerical, and symbolic computing. It would also help solve some of the myriad practical problems that always accompany curricular changes. With this in mind, we sought and received support from the NSF USEME-Calculus program for development of such a book.

Significant calculus reform at the national level is not a foregone conclusion, but prospects look promising, at least in the long run. Changes, if they do occur, will surely be driven largely by developments in mathematical computing. Projects like

ours, we hope, will permit the mathematics community to anticipate and build upon, rather than merely react to, this changing environment.

≡ Sample Materials ≡

Contents

The following sections contain examples of homework assignments and examinations from Calculus I and II. The sample homework assignments were selected from those used during the fall 1989 semester. Approximately 30 homework assignments were made in each course during the semester. Selected examinations from a single Calculus I section and a single Calculus II section have also been included.

Syllabus for Calculus I

Course Overview. (1 day)
- Exam dates, homework policies, etc. Preview of course contents.

Functions and Their Graphs. (7 days)
- Polynomials and rational functions. Interpreting graphs of functions. The domain and range of a function. The absolute value function. Inequalities: graphical interpretation and solution techniques.
- New functions from old: graphical interpretation of addition, scalar multiplication, translation of the origin, change of scale, composition, etc. Piecewise-defined functions. Numerically presented functions (e.g., from a table). CAS commands for defining, evaluating, and graphing functions.
- Using functions to model relationships (word problems including mixing problems, radioactive decay, etc.).
- The trigonometric functions: graphs, periodic behavior, and basic identities. Relationship between radian and degree measure.
- x^r (r any rational number). b^x and $\log_b x$. Properties of logarithms.
- Limits at infinity and infinite limits. Horizontal and vertical asymptotes. Intuitive notion of the limit of a function at a point. Properties of limits.

- Continuity. The Intermediate Value Theorem. Max-Min Existence Theorem. Upper and lower bounds.

The Derivative. (17 days)

- Average and instantaneous rates of change. Define derivative using difference quotients. Geometrical/graphical interpretation in terms of secant and tangent lines. Show that many functions look like lines when viewed "up close".
- The derivative function. Derivatives of sums. Rules for differentiating x^r (r any rational number). Idea of an antiderivative. Guess-and-check antidifferentiation. Idea of a differential equation. Initial value problems.
- Relationship between the sign of f' and intervals where f is increasing/decreasing; critical points. Repeated derivatives. Relationship between the sign of f'' and the concavity of f; points of inflection.
- The geometry of derivatives (continued): Relative extrema. Extrema on a closed interval.
- Root finding: Bisection method and Newton's method.
- Motion problems: solution by antidifferentiation, etc.
- Derivatives of $\sin x$ and $\cos x$. Why radian measure?
- Derivatives of products and quotients. Guess-and-check antidifferentiation.
- Related rates and max-min problems.
- Derivatives of b^x, and $\log_b x$. The number e. Guess-and-check antidifferentiation (use oracle for antiderivative of $\ln x$).
- Exponential growth and decay (population growth, radioactive decay, Newton's Law of Cooling).
- The chain rule. More guess-and-check antidifferentiation.
- The chain rule (continued). Introduction to inverse trigonometric functions. Derivatives of $\tan^{-1} x$ and $\sin^{-1} x$.
- Separable and first-order linear differential equations.
- The substitution technique for finding antiderivatives.
- Continuity versus differentiability. Rolle's Theorem, the Mean Value Theorem, and their corollaries. Geometric interpretation of these results.
- Linear approximation (with quadratic error bounds).

The Riemann Integral. (7 days)

- Definition of $\int_a^b f(x)\,dx$ as area. Geometric evaluation of definite integrals. Basic properties of the integral (linearity, etc.).
- Sigma notation. Definition of the definite integral as the limit of Riemann sums. Left, right, midpoint, upper, lower, and "random" Riemann sums. CAS computation of Riemann sums.
- Estimating definite integrals using Riemann sums. Error bounds for left and right Riemann sum approximations when the integrand is monotonic over the interval of integration.
- Computing areas and volumes by "cutting and slicing" techniques. Non-geometric integrals (e.g., compound interest, total mass of an object).
- The area function $A(x) = \int_a^x f(t)\,dt$. Examine graphs of f and A on the same axes and observe how properties of f are reflected in behavior of A.
- The Fundamental Theorem: If $f(x)$ is continuous, then $F(x) = \int_a^x f(t)\,dt$ is differentiable and $F'(x) = f(x)$. Furthermore, $\int_a^b f(x)\,dx = F(b) - F(a)$.
- The Fundamental Theorem (continued). Implications for the solution of initial value problems.

End-of-semester Review. (1 day)

Examinations. (3 days)

Available for Optional Topics, etc. (2 days)

Syllabus for Calculus II

Course Overview and Review. (3 days)

- Exam dates, homework policies, etc. Preview course contents. Review geometric interpretation of the derivative, differentiation techniques, etc. Review definition of the definite integral in terms of Riemann sums. Basic properties of the definite integral.
- The Fundamental Theorem: If $f(x)$ is continuous, then $G(x) = \int_a^x f(t)\,dt$ is differentiable and $G'(x) = f(x)$. The notations $\int_a^x f(x)\,dx$ and $\int f(x)\,dx$. Functions defined by integration (e.g., $\ln x$).
- The Fundamental Theorem: If $F'(x) = f(x)$, then $\int_a^b f(x)\,dx = F(b) - F(a)$. Geometrical significance of the constant of integration. Notion

of *elementary* antiderivative. Simple antidifferentiation problems: polynomials, sin $2x$, etc., using a guess-and-check approach.

Numerical Integration. (5 days)

- Error bounds for Left and Right Riemann sum approximations to $\int_a^b f(x)\,dx$ when $f(x)$ is monotonic on $[a, b]$. Use of these error bounds to achieve numerical estimates guaranteed to have a specified accuracy. Distinction between theoretical error bound and actual approximation error.
- Error bounds for Left and Right Riemann sum approximations to $\int_a^b f(x)\,dx$ when f is *not* monotonic on $[a, b]$. Using a graph of f' to compute bounds on the approximation error.
- Midpoint and Trapezoid Rules. Qualitative discussion of errors: When are these rules exact? When do they underestimate or overestimate the exact result? Error bound formulae. Estimating error bounds using graphs of the appropriate derivative functions.
- Trapezoid Rule as an average of Left and Right Riemann sum estimates (exact for linear functions). Simpson's Rule as a weighted average of Midpoint and Trapezoid Rule estimates.
- Review/summary day.

Optional Topics: Numerical integration of tabulated data. Computational strategies for numerical estimation of definite integrals—subdividing the interval of integration, etc. Monte Carlo integration.

Antidifferentiation Techniques. (6 days)

- Antidifferentiation using substitution. Relationship with the chain rule for differentiation.
- Antidifferentiation by parts. Relationship with the product rule for differentiation. Reduction formulae.
- Antidifferentiation of $(ax + b)^{-n}$ when $a \neq 0$ and $(ax^2 + bx + c)^{-n}$ when $b^2 - 4ac < 0$.
- Antidifferentiation of rational functions using partial fractions. (CAS is used to compute the coefficients in the partial fraction decompositions.)
- Using tables of integrals and/or a CAS.
- Review/summary day.

Improper Integrals. (3 days)

- Two types of improper integrals: integrals over an unbounded interval and integrals with an unbounded integrand. Definitions of convergence and divergence. Evaluation of improper integrals using antidifferentiation techniques.
- Comparison and absolute comparison tests for convergence.
- Numerical techniques for estimating improper integrals: making the tail small (i.e., treating it as zero) and estimating the value of the tail.

Optional Topics: Limit comparison test. Applications of improper integrals in probability and statistics.

Sequences and Series. (11 days)

- Taylor's Theorem (derive using repeated antidifferentiation by parts). Definition and geometric interpretation of Taylor polynomials.
- Ideas of pointwise and functional convergence (motivated by graphical examples of Taylor polynomials). Using the remainder term to compute error bounds. Qualitative behavior of the remainder term as n increases.
- Intuitive notion of the limit of a sequence of real numbers. Definition of convergence of a sequence. Examples (r^n, $k^n/n!$, $n/3^n$, etc.) and theorems.
- L'Hôpital's Rule. Emphasis is on the asymptotic behavior of functions.
- Taylor series. Formal power series as a generalization of Taylor series. Definitions of series convergence and divergence. Geometric and harmonic series.
- The integral test. Use of the integral test to estimate the sum of an infinite series of positive terms.
- Comparison and n^{th}-term tests. Absolute convergence implies convergence.
- Finding the radius of convergence of a power series. Power series representation of a function.
- Alternating series test. Numerical estimation of convergent alternating series.
- Algebraic operations on power series: addition, subtraction, substitution for x, integration, and differentiation.
- Review/summary day.

Optional Topics: Ratio test. Rearrangements of conditionally convergent series. Limit comparison test.

Polar Coordinates (2 days)

- Definitions and graphs.
- Areas.

Optional Topics: Parametric equations.

Multivariable Topics. (3 days)

• Partial derivatives.
• The definite integral of a function over a region in the plane. Double integrals over rectangular regions.
• Double integrals over non-rectangular regions.
  Optional Topics: Level curves.

Available for Optional Topics, etc. (2 days)

End-of-semester Review. (1 day)

Examinations. (3 days)

Calculus I Assignments

The Geometry of Derivatives

1. The graph of a function f appears below. Answer each of following questions using this graph.

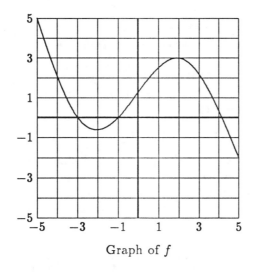

Graph of f

a. For what values of x, if any, is $f'(x)$ negative?
b. For what values of x, if any, is $f'(x)$ positive?
c. For what values of x, if any, is $f'(x) = 0$?
d. For what values of x, if any, is $f''(x)$ negative?
e. For what values of x, if any, is $f''(x)$ positive?

2. Find a quadratic polynomial (i.e., a polynomial of the form $ax^2 + bx + c$) that is zero at $x = 1$, decreasing when $x < 2$, and increasing when $x > 2$.

3. Each of the graphs in the second column of figures below is the graph of the derivative of one of the functions graphed in the first column. Match the corresponding functions and derivatives.

(a) 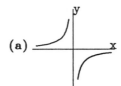   **(i)**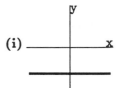

(b) 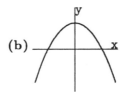   **(ii)**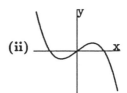

(c) 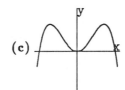   **(iii)**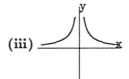

(d) 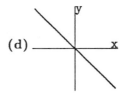   **(iv)**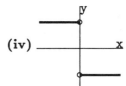

(e) 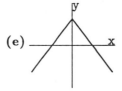   **(v)**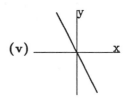

4. Each of the graphs in the second column of figures below is the graph of the *second* derivative of one of the functions graphed in the first column. Match the corresponding functions and second derivatives.

(a) 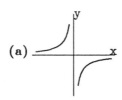   **(i)**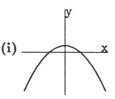

(b) 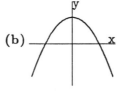   **(ii)**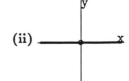

(c) 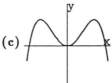   **(iii)**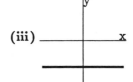

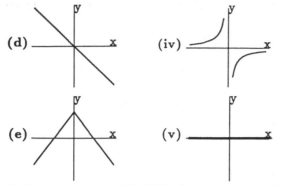

5. Interpret each of the following sentences using derivatives. That is, explain what each sentence is saying about the behavior of a function. In each case, be sure to indicate what the function is and the meaning of any variables you use.

  a. The child's temperature is still rising, but the penicillin seems to be taking effect.

  b. The cost of a new car continues to increase at an ever increasing rate.

  c. The work force is growing more slowly now than it was five years ago.

  d. During the past two years, the United States has continued to cut its consumption of imported oil.

6. (Problem from Stein.)

7. An oil tank is to be drained for cleaning. Suppose that there are $V(t) = 100,000 - 4,000t + 40t^2$ gallons of oil left in the tank t minutes after the drain valve is opened.

  a. What is the average rate at which oil drains out of the tank during the first twenty minutes?

  b. What is the rate at which oil is draining out of the tank twenty minutes after the drain valve is opened?

  c. Explain what $V''(t)$ tells you about the rate at which oil is draining from the tank.

8. In each case below, decide if there is a function that has the given properties. If so, sketch the graph of such a function. If not, explain why not.

  a. $\lim_{x \to \infty} f(x) = 5$ and $\lim_{x \to \infty} f'(x) = 1$

  b. For all x, $f(x) < 0$ and $f'(x) < 0$.

  c. For all x, $f(x) > 0$ and $f''(x) > 0$.

  d. $f(x)$ has a local maximum at $x = 3$ and $f''(3) = 0$.

The Chain Rule

1. Let $h = (f \circ g)$. Use the information about f and g given in the table below to fill in the missing information about h and h'.

x	$f(x)$	$f'(x)$	$g(x)$	$g'(x)$	$h(x)$	$h'(x)$
1	1	2	4	3	?	?
2	2	1	3	4	?	?
3	4	3	1	2	?	?
4	3	4	2	1	?	?

2. Assume g is an unknown function that is differentiable everywhere (i.e., $g'(x)$ exists for all x). Find the first derivative of each of the following functions.

  a. $e^{g(x)}$

  b. $\sin\big(g(x)\big)$

  c. $\big(g(x)\big)^5 + 4\big(g(x)\big)^3 - 2g(x) + \frac{\pi}{2}$

  d. $\ln\big(g(x)\big)$

  e. $g(e^x)$

  f. $g(\sin x)$

  g. $g\big(x^5 + 4x^3 - 2x + \frac{\pi}{2}\big)$

  h. $g(\ln x)$

3–9. (Various problems from Stein.)

10. Let $h(x) = f\left(\dfrac{g(x)}{x^2 + 1}\right)$. If $f(1) = 3$, $f(2) = 5$, $f'(1) = 7$, $f'(2) = 11$, $g(1) = 2$, and $g'(1) = 4$, what is $h'(1)$?

11. Use the identity $\dfrac{f(x)}{g(x)} = f(x) \cdot \big(g(x)\big)^{-1}$ together with the product and chain rules to derive the quotient rule.

12. Let $h = g \circ f$ where $f(x) = x\sqrt{x^2 - 1}$ and g is a differentiable function defined on $[-10, 10]$ such that g' is the function graphed below.

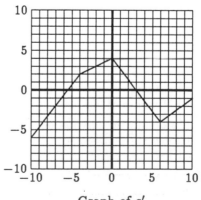

Graph of g'

  a. Evaluate $f'(2)$, $g'(2)$, and $h'(2)$.

b. On which subintervals of $[-5, 5]$, if any, is f concave down?

c. Does g have any local maxima in the interval $[-10, 10]$? If so, where?

d. Does g have any inflection points in the interval $[-10, 10]$? If so, where?

e. On which subintervals of $[-3, 3]$, if any, is h increasing?

Fundamental Theorem of Calculus

1. Let $g(x) = \int_0^x f(t)\, dt$, where $f(t)$ is the function graphed below:

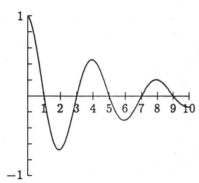

Graph of f

a. Does $g(x)$ have any local maxima within the interval $[0, 10]$? If so, where are they located?

b. Does $g(x)$ have any local minima within the interval $[0, 10]$? If so, where are they located?

c. At what value of x does $g(x)$ attain its maximum value on the interval $[0, 10]$?

d. At what value of x does $g(x)$ attain its minimum value on the interval $[0, 10]$?

e. On which subinterval(s) of $[0, 10]$, if any, is the graph of $g(x)$ concave up?

f. Sketch a plausible graph of g.

2. Let $F(x) = \int_0^x f(t)\, dt$ where f is the function graphed below. (The graph of f is made up of straight lines and a semicircle.)

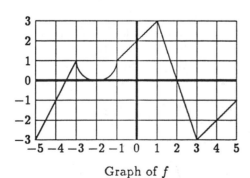

Graph of f

a. Evaluate $F(-1)$, $F(0)$, and $F(2)$.

b. Identify all the critical points of F in the interval $[-5, 5]$.

c. Identify all the inflection points of F in the interval $[-5, 5]$.

d. Let $G(x) = \int_0^x F(t)\, dt$. On which subintervals of $[-5, 5]$, if any, is G concave upward?

3. The function $\ln x$ is sometimes *defined* by integration:

$$\ln x = \int_1^x \frac{1}{t}\, dt$$

a. What theorem can be used to show that $\frac{d}{dx} \ln x = \frac{1}{x}$?

b. Derive the identity $\ln(ax) = \ln a + \ln x$ by carrying out and justifying each of the following steps:

i. Write $\ln(ax) = \int_1^{ax} \frac{1}{t}\, dt$.

ii. Differentiate the right side of the equation from step (i) with respect to x and deduce that $\ln(ax) = \ln x + C$.

iii. By choosing an appropriate x, show that $C = \ln a$.

4. Let $F(x) = \int_1^x f(t)\, dt$ where f is the function graphed below.

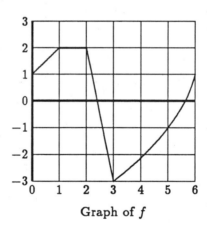

Graph of f

a. Suppose $\int_0^5 f(t)\, dt = -\frac{17}{16}$. What is $F(5)$?

b. Justify the assertion that F has **exactly** one root between 2 and 4.

c. If Newton's method is used to locate this root of F and the initial estimate used is $x_0 = 3$, what is x_1? (HINT: What is $F'(3)$?)

5. A company is planning to phase out a product because demand for it is declining. Demand for the product is currently 800 units/month and dropping by 10 units/month each month. To maintain customer and employee relations, the company has announced that it will continue to produce the product for one more year. At the present time, the company is producing the product at a rate of 900 units/month and 1680 units of the product are in inventory.

a. Let $D(t)$ be the demand rate for the product. Explain why $\int_0^t D(s)\,ds$ is the total demand for the product in t months' time.

b. Give an expression for the inventory at time t. (HINT: The amount of the product in inventory is the difference between supply and demand at the end of t months.)

c. The company would like to reduce production at a constant rate of R units/month, with R chosen so as to reduce its inventory to zero at the end of 12 months. What should R be to reduce the inventory to zero at $t = 12$?

Calculus I Tests and Examinations

All tests and examinations in both Calculus I and II consisted of a take-home part and an in-class part; only the take-home parts are provided in this sample; (The in-class components, being relatively conventional, are of less general interest.) Instructions similar to those provided below for Test No. 1 appear on all course examinations, but are not repeated in each sample test provided below. All tests and examinations at St. Olaf are taken under an honor system.

Test No. 1

The following sources are legal while working on this test: (i) Your textbook (ii) Anything handed out in class (iii) Your own notes (iv) *SMP*. Please don't use anything else; don't discuss the test with anyone except your instructor.

It's important—especially on a test like this—that you present your solutions legibly, fully, and in organized fashion. Show *all* work (within reason, of course—don't show arithmetic calculations) for each problem; explain your reasoning carefully, in complete sentences. Full credit will not be given for mysterious or unsupported answers.

1. (15 pts) The number of deer in a forest at time t years after the beginning of a conservation study is shown on the graph below.

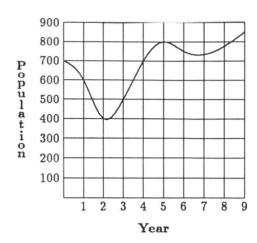

a. Over which of the following periods did the population of deer decline at the rate of 50 deer per year?
(i) 1–2 (ii) 1–3 (iii) 1–4 (iv) 2–3 (v) 5–6

b. When was the population of deer increasing most rapidly?

c. Approximately how fast was the deer population increasing (or decreasing) at time $t = 1.5$? (Your answer should include appropriate units.)

2. (15 pts) Find the derivative and an antiderivative of each of the following functions. (Be sure to label which of your answers is the derivative and which is an antiderivative.)

a. $2 - 3\sqrt{x}$

b. $2x^{-1/3} + \frac{1}{7}x^{14}$

c. $2x^{-2} - 3/x^2$

3. (5 pts) Suppose $f'(x) > 0$ for all $x \in (2, 7)$. Is $f(4) < f(5)$? Explain.

4. (15 pts) Sketch the graph of a function f that has *all* of the following properties:

a. The domain of $f = \{\, x \mid x \in \mathbf{R},\ x \neq -3,$ and $x \neq 2\,\}$.

b. The range of f is the interval $(-3, \infty)$.

c. $\lim_{t \to 2} f(t) = -2$

d. $\lim_{t \to \infty} f(t) = -3$

e. The line $y = 3$ is a horizontal asymptote of f.

f. f is continuous on its domain.

Be sure to label and indicate units on each axis.

5. (15 pts) Let f be a continuous function. Some values of f are given below:

x	0	0.9	0.99	1	1.01	1.1	1.5
$f(x)$	3	-7	-8	-9	-6	-5	0

x	1.9	1.99	1.999	2.0	2.001	2.01	2.1
$f(x)$	25.34	33.97	34.896	35	35.104	36.05	46.18

a. What is $\lim\limits_{x\to 1} f(x)$? Justify your answer.

b. Estimate $f'(2)$. (Be sure to show your work!)

c. One root of f is at $x = 1.5$. Can you determine if f has any other roots? If so, specify an interval in which another root can be found.

6. (10 pts) Is there an a so that

$$\lim_{x\to 3} \frac{2x^2 - 3ax + x - a - 1}{x^2 - 2x - 3}$$

exists? Explain your answer.

7. (10 pts) The graphs of three functions appear below. Which is f, which is f', and which is f''? Explain your reasoning.

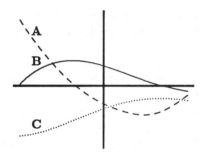

8. (15 pts) Suppose $U(t)$ is the number of people unemployed in a country t months after the election of a fiscally conservative president. Translate each of the following facts about the graph of $U(t)$ into statements about the unemployment situation. (Don't forget to specify units in your answers.)

a. The y-intercept of $U(t)$ is 2,000,000.

b. $U(2) = 3,000,000$.

c. $U'(20) = 10,000$.

d. $U''(36) = 800$ and $U'(36) = 0$.

BONUS PROBLEM

1. (10 pts) Find values of a and b so that the line $2x + 3y = a$ is tangent to the graph of $f(x) = bx^2$ at the point where $x = 3$.

Test No. 2

1. (10 pts) For each of the following differential equations, find a solution that satisfies the given condition.

a. $y' = (2 - y)^2$, $y(0) = 1$

b. $y' - 3y = xe^{3x}$, $y(0) = 4$

2. (10 pts) Sketch the graph of a continuous function g that has *all* of the following properties.

a. The domain of g is the interval $[-3, 5]$.

b. $g''(x) < 0$ for all $x \in (-2, 0)$

c. $g''(x) > 0$ for all $x \in (0, 2)$

d. $g'(-2) < 0$

e. $g'(4) > 0$

f. $g'(1) = 0$ but $g(1)$ is neither the maximum nor the minimum value of g over the interval $[-2, 4]$

3. (10 pts) Right circular cylindrical cans are to be manufactured to contain a given volume V. There is no waste involved in cutting the material for the vertical sides, but the top and bottom (circles of radius r) are cut from squares that measure $2r$ units on a side. Thus, the total amount of material consumed for each can is $A = 2\pi rh + 8r^2$ where h is the height of the can. Find the ratio of height to diameter for the most economical can (i.e., the can requiring the least material).

4. (10 pts) The police guard gave Sara a cold look, but his voice was polite as he he directed her to the room she sought. "Don't touch anything, please, Ms. Abrams. The Chief said I had to let you in, but he said to tell you to mind your fingers." "Thank you," Sara replied cooly. "The Chief knows he can trust me." The guard opened his mouth as if to speak, but he merely shook his head and withdrew.

Sara was standing in what appeared to be a combination bedroom and laboratory. A relative had found Dr. Howell's body on the floor of this room that morning. By 9:00 a.m., the coroner had completed his examination; he stated that death was due to a severe blow to the head, and that Dr. Howell had been dead between 36 and 40 hours. It seemed critical to Sara to know exactly when Dr. Howell had died so that she could eliminate certain suspects. But how could she possibly discover exactly when he was killed? Puzzled, she wandered around the small, cluttered room, being careful not to touch anything. The old doctor apparently was conducting an experiment when he was killed. Sara absent-mindedly read from the notebook which was lying open on the bench.

The fungus grows at a rate proportional to its current weight.

"Great," she thought, "I'm here to investigate a murder and instead I'm getting a biology lesson." At a loss for what else to do, she continued reading.

To exemplify this biological truth, I place the fungus on a scale and record its weight at various times:

10 g	5:30 p.m.
12 g	6:15 p.m.
13	

"Hmm," Sara mused, "the poor guy didn't even get to finish the last entry." Sara suddenly frowned in concentration. She searched her pockets until she found a pencil stub and an old receipt. When the guard entered the room a few minutes later, Sara had just finished scribbling on the receipt. She smiled as she shoved the receipt and the pencil stub back into her pocket.

"Don't worry," Sara said cheerfully, "I'm leaving. I now know exactly when Dr. Howell was killed." The guard looked sourly at Sara's back as she left the room.

When was Dr. Howell killed? How did Sara figure this out?

5. (15 pts) The minute and hour hands on the face of a clock are 7 feet and 5 feet long, respectively. How rapidly is the distance between the tips of the hands increasing when the clock reads 9:00?

6. (20 pts) A disgruntled calculus student is waiting by an open window 100 feet above the ground with a left-over Halloween pumpkin. The student is waiting for a certain 6-foot-tall mathematics professor to walk by.

a. If the student throws the pumpkin straight down with a velocity of 5 feet/second, how long before the professor arrives should the student launch the pumpkin in order to score a direct hit on the professor's head? (Assume that acceleration due to gravity is 32 feet/second^2.)

b. Fortunately for the professor, the student neglected to take the effect of air resistance into account. In this situation, the effect of air resistance can be modeled by an acceleration proportional to the pumpkin's velocity. If the constant of proportionality for the pumpkin is 0.01 sec^{-1}, how far above the professor's head is the pumpkin at the instant in time when the professor is immediately below the pumpkin?

Test No. 3

1. (15 pts) Evaluate each of the following expressions by hand. (Show your work!)

a. $\dfrac{d}{dx} \tan^{-1} x e^x$

b. $\dfrac{d}{dt} \sin^{-1}\left(1 - t^2\right)$

c. $\dfrac{\partial^2}{\partial u\, \partial v}\, uv^2 \cos u$

2. (20 pts) Find an antiderivative of each of the following functions.

a. $x^3 \left(x^4 - 1\right)^2$

b. $\dfrac{x+1}{\sqrt[3]{3x^2 + 6x + 5}}$

c. $\dfrac{e^x}{1 + e^{2x}}$

d. $x \cos\left(1 - x^2\right)$

3. (15 pts) Let f be a continuous function that is positive on the interval $[0, 4]$ and differentiable on $(0, 4)$. Answer "yes" or "no" to each of the following questions about f. Then, if you answer "yes," explain how you know that the statement is true. (Your answer should include references to appropriate theorems in Stein and/or relevant homework problems.) If you answer "no," give an example of a function that shows that the statement is not always true.

a. Is it *always* possible to find a $c \in [0, 4]$ such that the rectangle with height $f(c)$ and base $[0, 4]$ has the same area as the region bounded above by the graph of f, below by the x-axis, on the left by the line $x = 0$, and on the right by the line $x = 4$?

b. Is it *always* possible to find a $c \in [0, 4]$ such that $f(x) \le f(c)$ for all $x \in [0, 4]$?

c. Is it *always* possible to find a $c \in [0, 4]$ such that the instantaneous rate of change of f at c equals the average rate of change of f over the interval $[0, 4]$?

d. Is it *always* possible to find a $c \in [0, 4]$ such that the line tangent to the graph of f at $(c, f(c))$ is horizontal?

4. (10 pts) Let $f(x) = \begin{cases} ax, & x \le 1 \\ bx^2 + x + 1, & x > 1. \end{cases}$

a. Find values of a and b so that f is continuous but not differentiable at $x = 1$.

b. Find values of a and b so that f is differentiable at $x = 1$.

5. (10 pts) Assume the function f is differentiable everywhere and that the line $y = 3x + 4$ is tangent to the graph of f at $x = 2$. If Newton's method is used to find a root of f and the initial estimate used is $x_0 = 2$, what is x_1?

6. (10 pts) Find quadratic (i.e., degree two) polynomials $p(x)$ and $q(x)$ such that for every $x \in [0, 1]$

$$p(x) - q(x) \le e^x \le p(x) + q(x).$$

That is, for each $x \in [0, 1]$, $e^x = p(x) \pm q(x)$.

7. (10 pts) Let f be the function graphed below.

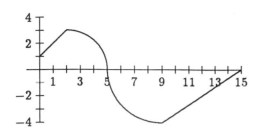

NOTE: The graph of f consists of two straight line segments and two quarter-circles.

a. Evaluate $\displaystyle\int_0^{15} f(x)\,dx$.

b. Evaluate $\displaystyle\int_9^{12} f(x)\,dx$.

c. Evaluate $\displaystyle\int_0^{15} |f(x)|\,dx$.

8. (10 pts) Let g be the function graphed below. Estimate the value of $\displaystyle\int_{-5}^{5} g(x)\,dx$ by evaluating a Riemann sum with 5 equal length subintervals, then draw a sketch that illustrates this Riemann sum geometrically.

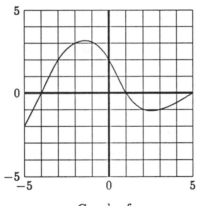

Graph of g

Final Examination

1. (10 pts) Suppose $\displaystyle\lim_{x\to 2} f(x) = 3$ and $f(2) = 4$. State whether each of the following statements is true or false. Then, explain why you gave the answer you did.

a. The number 2 is in the domain of f.

b. f is continuous at $x = 2$.

c. $f'(2)$ exists.

2. (5 pts) Evaluate

$$\lim_{x\to\pi} \frac{5^{\sin x} - 1}{x - \pi}$$

exactly (i.e., as the value of an elementary function).

3. (15 pts) For each of the following differential equations, find a solution $y(x)$ that satisfies the given condition.

a. $y' = \dfrac{2x}{y + x^2 y}$, $\quad y(2) = 3$

b. $xy' + 3y = 4x$, $\quad y(1) = 2$

4. (10 pts) The profit earned from the manufacture of x widgets is $0.0001x^3 - 0.21x^2 + 135x - 20{,}000$ dollars. Because of other constraints, at least 600 widgets must be manufactured; however, it is impossible to produce more than 1000 widgets. How many widgets should be produced if the manufacturer wishes to earn the maximum profit?

5. (50 pts) Let g be a differentiable function defined on the interval $[0, 10]$ such that the graph of g passes through the point $(5, 2)$ and the derivative of g is the function sketched below.

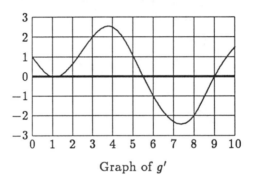

Graph of g'

a. At which values of x in the interval $[0, 10]$ does g have local extrema? Which correspond to local maxima and which to local minima?

b. The graph of g' has a local maximum at $x = 3.8$ and a local minimum at $x = 7.4$. What do these facts imply about the graph of g?

c. If Newton's method is used to locate a root of g and $x_0 = 5$, what is x_1?

d. Find real numbers A and B such that $g(0) = A \pm B$.

e. How many solutions of the equation $g(x) = 0$ exist in the interval $[0, 10]$?

For the remainder of this problem let $G(x) = \int_0^x g'(t)\,dt$.

f. Use a left Riemann sum with 3 equal subintervals to compute an estimate of $G(9)$.

g. How are the graphs of g and G related?

6. (15 pts) Sand is pouring from a pipe at a constant rate. The falling sand forms a conical pile whose height is always one-fourth the diameter of the base.

a. Let $V(t)$ be the volume of the sand in the pile at time t. Give a physical interpretation of each of the following expressions and indicate whether each is positive, negative, or zero when the pile is 4 feet high.

$$\text{(i) } \frac{dV}{dt} \qquad \text{(ii) } \frac{d^2V}{dt^2}$$

b. If the sand is pouring from the pipe at a rate of 16 cubic feet per minute, how fast is the height of the pile increasing when the pile is 4 feet high?

7. (5 pts) Help relieve the tedium of grading exams by giving me something fun to read: write down your favorite joke.

BONUS PROBLEMS

1. (10 pts) The figure below shows a graph of the function $f(x) = \sqrt{x}$ and the line tangent to the graph of f at $x = a$. Find the value of a.

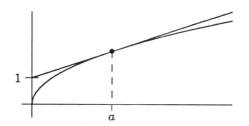

2. (10 pts) Show that $\dfrac{x}{1+x} < \ln(1+x)$ for all positive real numbers x.

3. (10 pts) A cylindrical tank 50 feet long is standing on one end. Liquid from the tank drains from a hole near the bottom at a rate proportional to the depth of the liquid in the tank. The capacity of the tank is 3000 gallons. When the tank is full, the liquid drains at a rate of 1 gallon/minute. How long after the tank is filled will the tank become half empty?

Calculus II Assignnments

Midpoint, Trapezoid, & Left Sums

These exercises are about M_n, T_n, and L_n, the Midpoint, Trapezoid, and Left approximating sums for $I = \int_a^b f(x)\,dx$. The subscript n is always refers to the number of (equal) subdivisions of $[a, b]$. *SMP* commands like these will be handy:

```
NLeft[Sin[x],x,0,1,6]
NMidpoint[Cos[x],x,0,1,10]
NTrapezoid[x^2,x,0,1,12]
LeftGraph[Sin[x],x,0,1,6]
MidGraph[Cos[x],x,0,1,10]
TrapGraph[x^2,x,0,1,12]
```

So will these error bound formulas (see the handout for more details):

$$|I - M_n| \le \frac{K_2(b-a)^3}{24n^2}$$

$$|I - T_n| \le \frac{K_2(b-a)^3}{12n^2}$$

$$|I - L_n| \le \frac{K_1(b-a)^2}{2n}$$

As you know, K_1 and K_2 are upper bounds for $|f'|$ and $|f''|$ respectively. Estimate them graphically—using *SMP*'s D[f[x],x] and Graph commands.

Problems

1. Let $I = \int_0^1 e^{x^2}\,dx$.

a. Compute M_2 and T_2 by hand and calculator, not with *SMP*.

b. Find M_{10}, T_{10}, and L_{10}. What does the error bound formula say in each case?

c. How many subdivisions are needed with each method to ensure accuracy within 0.001? Use any method to compute I that accurately.

2. Let $I = \int_0^1 x^3\,dx$.

a. Find M_{10}, T_{10}, L_{10}, and I itself (by antidifferentiation.)

b. For M_{10}, T_{10}, and L_{10}, compute the *estimated* error (from the error formula) and the *actual* error committed.

3. Compare the error bound for M_n to the error bound for M_{2n}. (In other words, what's the effect of doubling n?) Do the same for L_n and L_{2n}.

4. Here are three integrals that are inconvenient or impossible in closed form. Approximate each using M_{20}, and discuss the maximum possible error for each. Note: *SMP*'s notations for e^x and $\ln x$ are Exp[x] and Log[x].

a. $\int_{\pi/4}^{\pi} \frac{\cos x}{x}\,dx$

b. $\int_0^3 \sin x^3\,dx$

c. $\int_1^3 e^x \ln x\,dx$

5. Write a sentence or two explaining the following cryptic remark: *The Trapezoid rule is the average of the Left and Right rules.*

More Series, Some Alternating

1. (Three problems from Stein.)

2. The limit comparison test is convenient in cases where no *direct* comparison suggests itself. On the other hand, whenever the limit comparison test works, so does the ordinary comparison test. For example, given $\sum 1/(k+1)$, we can apply a *limit* comparison with $\sum 1/k$, but not a *direct* comparison. But $\sum 1/(k+1)$ can be *directly* compared with $\sum 1/(2k)$. For each of the following series, apply first a limit comparison and then a direct comparison.

a. $\displaystyle\sum_{k=2}^{\infty} \frac{2}{k-1}$

b. $\displaystyle\sum_{k=2}^{\infty} \frac{k}{k^3-3}$

c. $\displaystyle\sum_{k=1}^{\infty} \frac{\sqrt{k}}{k^2+k-1}$

3. Give an example of a divergent infinite series $\sum a_k$ such that $a_k > 0$ and $a_{k+1}/a_k < 1$ for all $k \geq 1$.

4. Consider $\displaystyle\sum_{k=1}^{\infty} \frac{(-1)^{k+1}}{k^3}$.

a. Explain why this series converges.

b. Compute S_{10}. How close does the alternating series theorem say S_{10} must be to the limit?

c. Another way to see how well S_{10} approximates the limit is to estimate the absolute value of the tail:

$$|R_{10}| = \left|\sum_{k=11}^{\infty} \frac{(-1)^{k+1}}{k^3}\right| \leq \sum_{k=11}^{\infty} \frac{1}{k^3}$$

Estimate the last quantity using an appropriate *integral*.

5. The alternating harmonic series $\displaystyle\sum_{k=1}^{\infty} \frac{(-1)^{k+1}}{k}$ converges. It's a fact—take it on faith for now—that the limit is $\ln 2$.

a. Compute S_{100}. How far is it *actually* off from $\ln 2$? How much error would the theorem permit? Does S_{100} undershoot or overshoot? Could you have guessed this without knowing the limit?

b. According to the theorem, how large would n have to be to guarantee that S_n approximates $\ln 2$ to within .000001. (Don't ask *SMP* to compute this S_n!)

c. Since we know that every partial sum over-shoots or undershoots, we might try to estimate the limit using only *half* the last term: try $S_{20} \pm a_{21}/2$ as an estimate. Decide whether to use $+$ or $-$. Then compute this new estimate. How far off $\ln 2$ are we now?

More Power Series; Taylor Series

1. Use the Maclaurin series

$$\frac{1}{1+x} = \sum_{j=0}^{\infty} (-1)^j x^j = 1 - x + x^2 - x^3 + \ldots$$

to find a power series representation of each of the following functions. No proofs necessary—just write the answers.

a. $f(x) = 1/1-x$

b. $f(x) = 1/1+x^2$

c. $f(x) = x^2/1+x$

d. $f(x) = x/1-x^4$

e. $f(x) = (1+x)^{-2}$

f. $f(x) = \tan^{-1} x$

g. $f(x) = \ln \left|\dfrac{1+x}{1-x}\right|$

h. $f(x) = \int_0^x \ln(1+t)\,dt$

2. Find the function represented by each of the following power series by recognizing how each is related to a more familiar power series.

a. $\displaystyle\sum_{k=1}^{\infty} kx^k$

b. $\displaystyle\sum_{k=1}^{\infty} (-1)^{k+1} x^k$

c. $\displaystyle\sum_{k=0}^{\infty} \frac{x^k}{(k+1)!}$

d. $\displaystyle\sum_{k=1}^{\infty} \frac{2^k x^k}{k}$

3. For each of the functions below, find the Maclaurin polynomials of degree $n = 2, 4,$ and 8. (Use the `Taylor` command.) Then *graphically* estimate the maximum error committed by approximating the function by each of these Maclaurin polynomials on the interval $[-2, 2]$. (Just graph the function and the three Taylor polynomials on the same axes; approximate answers are OK.)

a. e^x

b. $1/(x+3)$

c. $\sqrt{x+4}$

4. Find a Taylor polynomial that approximates the value of $h(x) = \frac{1}{2}(e^x - e^{-x})$ with error less than 0.01 for any x such that $|x| \le 1$. (Experiment with various Taylor polynomials, using the Graph command. No proof needed.)

Calculus II Tests and Examinations

Test No. 1

1. (20 pts) (Do this problem by hand and calculator, not with *SMP* .) Show all work in painful detail. Let $I = \int_1^3 \frac{1}{x}\,dx$.

a. Compute I exactly (by antidifferentiation.)

b. Compute M_4 and S_4, the Midpoint and Simpson approximations to I, each with 4 subdivisions. Show work that indicates you know what you're doing—a numerical answer alone is not sufficient.

c. What do the error formulas say about the maximal possible error committed by M_4 and S_4? Again, show all work.

d. What is the *least* n for which the Midpoint error formula guarantees accuracy to within 0.001?

e. Same question as (d), but for Simpson.

2. (10 pts) (*SMP* is OK on this problem.) Compute $\int_1^2 [\sin(x)/x]\,dx$ with error less than 0.0001. Use any method and number of subdivisions you like, but explain carefully and convincingly why your answer is really good to within 0.0001.

3. (10 pts) Let T_2, M_2, and S_2 denote, as always, the Trapezoid, Midpoint, and Simpson approximations to an integral, each with two subdivisions. Explain and defend each of the following statements:

a. On $I = \int_0^1 x^2\,dx$, T_2 and M_2 both commit the *maximum possible error.*

b. On $I = \int_0^1 x^4\,dx$, S_2 commits the *maximum possible error.*

4. (25 pts) Compute the following antiderivatives. Show all work—answers alone aren't sufficient. (You may try *SMP* if you like, but an *SMP* output is not an acceptable answer.)

$$\int \frac{x}{\sqrt{2-x}}\,dx$$

$$\int \frac{x}{\sqrt{2-x^2}}\,dx$$

$$\int \frac{dx}{x^2 + ax + a^2} \quad \text{(for } a > 0\text{)}$$

$$\int \frac{7x+3}{x(x+1)}\,dx$$

$$\int x^4 \tan^{-1} x\,dx$$

5. (12 pts)

a. Show, using integration by parts, why antiderivative formula No. 59 (in the orange frontpapers of Stein) is true.

b. *Use* formula No. 59 to find $\int e^{\sqrt[3]{x}}\,dx$. Hint: first substitute $u = \sqrt[3]{x}$.

6. (10 pts) Decide whether each of the following integration formulas is true or false. Justify your conclusions, perhaps by differentiation. (Hint: Be careful with trigonometric functions—they can be written in more than one way.)

a. $\int \frac{1}{\cos x}\,dx = \frac{1}{\sin x} + C$

b. $\int \frac{1}{\sec x}\,dx = \frac{1}{\csc x} + C$

7. (15 pts) Let $f(x) = \int_{-5}^x k(t)\,dt$ where $k(t)$ is the function graphed below. *Note:* Be careful—the graph shows the function k, not f.

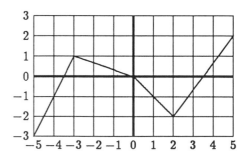

a. What does the Second Fundamental Theorem say about k and f? (Don't *prove* anything; just *state* what the theorem says in this setting.)

b. At what values of x does $f(x)$ have local minima on the interval $[-5,5]$? (No explanation needed.)

c. At what value of x does $f(x)$ achieve its minimum value on the interval $[-5,5]$? (No explanation needed.)

d. On which subinterval(s) of $[-5, 5]$, if any, is the graph of $f(x)$ concave up? (Explain very briefly the basis for your answer.)

Test No. 3

1. (30 pts) Tell whether each of the following series converges absolutely, converges conditionally (i.e. converges with plus and minus signs but not without), or simply diverges. (Don't worry about the limits themselves.) *Show* which tests you used, and how you used them. Don't forget the word *lim* when it is needed. Example of unacceptable answer: *Converges. Limit comparison.* (It's OK to use *SMP* to investigate these series, but *SMP* is not a reason for convergence or divergence.)

a. $\sum_{k=1}^{\infty} \frac{4k+1}{k^3 + \frac{k}{10}}$

b. $\sum_{n=0}^{\infty} \frac{3^n - 4^n}{5^n}$

c. $\sum_{k=1}^{\infty} \frac{-1)^k}{4k - 3}$

d. $\sum_{n=0}^{\infty} \frac{1}{2 + \frac{1}{n^2}}$

2. (10 pts) Find each of the following limits *exactly* , showing work.

a. $137 + \pi - 42 + 1 + \frac{1}{3} + \frac{1}{9} + \frac{1}{27} + \frac{1}{81} + \ldots + \frac{1}{3^n} + \ldots$

b. $\sum_{k=0}^{\infty} \left(\frac{-e}{\pi} \right)^k$

3. (20 pts) For each of the following convergent series: (i) explain briefly *why* it converges; and (ii) Find the limit, correct to within 0.001. Justify your claims, perhaps referring either to the alternating series theorem or to "upper tails".

a. $\sum_{k=1}^{\infty} \frac{1}{k^3 + 2}$

b. $\sum_{k=1}^{\infty} \frac{-1)^{k+1}}{k^3 + 2}$

4. (10 pts) A certain series $\sum_{k=1}^{\infty} a_k$ has partial sums s_n given by the formula $s_n = 5 - 3/n$. Note: Be careful—a_k is not s_k. Answer the questions below. Justify each answer with one brief English sentence.

a. Does $\sum_{k=1}^{\infty} a_k$ converge? To what?

b. Find $\lim_{k \to \infty} a_k$.

c. Find $\sum_{k=1}^{100} a_k$.

5. (10 pts) Theorem 1 on p. 524 says, in effect, that the set of x for which a power series $\sum a_k x^k$ converges is an interval centered at the origin. This problem is in the same spirit (but much easier!)

Suppose that a power series $\sum a_k x^k$ converges absolutely for $x = 3$. Show that it also converges absolutely for $x = -2$. Hint: Comparison.

6. (20 pts) Consider the function

$$f(x) = 1 + x + \frac{x^2}{2!} + \frac{x^3}{3!} + \ldots = \sum_{k=0}^{\infty} \frac{x^k}{k!}$$

a. What's the domain of f? In other words, for which x does the series converge? Why?

b. Write $f'(x)$ and $\int f(x)\, dx$ as power series. Where does each converge?

c. Estimate $f(-1)$ with error less than 0.001. Explain your answer. Hint: $f(-1)$ is the sum of an *alternating* series.

d. By definition,

$$f(1) = \sum_{k=0}^{\infty} \frac{1}{k!} = 1 + 1 + \frac{1}{2!} + \frac{1}{3!} + \ldots$$

Let's approximate $f(1)$ with a partial sum: According to *SMP*,

$$1 + 1 + \frac{1}{2!} + \frac{1}{3!} + \ldots + \frac{1}{7!} = 2.71825$$

How close to the *exact* value of $f(1)$ might this be? Hints: Use an integral to bound the upper tail $R_7 = \sum_{k=8}^{\infty} \frac{1}{k!}$. The inequality $\frac{1}{k!} < \frac{1}{2^{k-1}}$ may be handy.

e. *(Optional, for extra credit)* Actually—this should not be surprising, in light of the previous problem—$f(x) = e^x$. It follows that

$$e^{-x^2} = 1 - x^2 + \frac{x^4}{2!} - \frac{x^6}{3!} + \ldots$$

and hence that

$$\int_0^1 e^{-x^2}\, dx = \int_0^1 \left(1 - x^2 + \frac{x^4}{2!} - \frac{x^6}{3!} + \ldots \right) dx$$
$$= \int_0^1 1\, dx - \int_0^1 x^2\, dx + \int_0^1 \frac{x^4}{2!}\, dx - \int_0^1 \frac{x^6}{3!}\, dx + \ldots$$

Use this to estimate $\int_0^1 e^{-x^2}\, dx$ with error less than 0.0001.

Final Examination

1. (63 pts) NOTE: Do any 9, but *only* 9, of the 10 parts of this problem. In each case, either evaluate the quantity *exactly*, or approximate it with error less than .001. Justify all your claims; e.g., by referring to an appropriate error estimate formula. **Example of *insufficient* justification: "*SMP* says so."**

a. $\displaystyle\int_1^\infty \frac{dx}{x^5+1}$

b. $\displaystyle\sum_{k=3}^\infty (2/3)^k$

c. $\displaystyle\int_3^\infty (2/3)^x \, dx$

d. $\displaystyle\sum_{k=0}^\infty \frac{1}{3^k+1}$

e. $\displaystyle\int_0^1 \sin x^3 \, dx$

f. $\displaystyle\int_0^1 x^2 \sin x^3 \, dx$

g. $\displaystyle\sum_{k=1}^\infty \frac{-1)^{k+1}}{k^6}$

h. $\displaystyle\sum_{k=1}^\infty \frac{1}{k^6}$

i. the length of the curve $y = \dfrac{x^2}{2} - \dfrac{\ln x}{4}$, from $x = 2$ to $x = 3$

j. the shaded area shown:

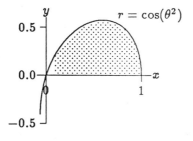

2. (15 pts) Antidifferentiate; show all work:

a. $\displaystyle\int \frac{7\sin x + 3)}{\sin x)(\sin x + 1)} \cos x \, dx$

b. $\displaystyle\int x^2 \sin(2x) \, dx$

c. $\displaystyle\int \frac{\sqrt{1+x^2}}{x} \, dx$

3. (22 pts) All parts of this problem refer to the function $f(x) = \dfrac{1}{2+x}$.

a. Write the Maclaurin series for $f(x)$. Where does it converge?

b. Find the third Maclaurin polynomial $P_3(x)$.

c. Use Taylor's theorem to estimate how much $P_3(1)$ can differ from $f(1)$—i.e., use the theorem to estimate $|f(1) - P_3(1)|$.

d. Compute (by calculator or by hand) the *actual* error $|f(1) - P_3(1)|$.

Abstracts

This section contain brief contributed reports of calculus projects across the nation, ranging from quite modest one-person efforts involving only a few students to significant efforts involving several faculty and hundreds of students on more than one campus. No judgments of quality are implied by the selection of projects reported on in this section, or by those not mentioned. Our goal in this section is to provide a clear picture of every innovative effort we could identify, including especially information about available materials and persons to contact.

Allegheny College

Contact:

RICHARD F. MCDERMOT, Department of Mathematics, Allegheny College, Meadville, PA 16335. PHONE: 814-332-3393. FAX: 814-337-0988.

Institutional Data:

Allegheny College is a four-year private liberal arts college with approximately 1900 full-time undergraduate students, most of whom are residential. Over thirty percent major in mathematics or one of the natural sciences.

The Mathematics Department, which is separate from the Computer Science Department, has ten full-time faculty members. A typical graduating class of 360 students will include about 18 mathematics majors. The Computer Science Department, with four full-time faculty, averages about 13 majors per year.

Project Data:

The 1989-90 academic year is the first year of the project. During this year three faculty members are teaching a total of six sections of Calculus I and Calculus II, using Mathematica in an adjacent classroom and laboratory containing a network of thirteen NeXT and two Macintosh computers which are used both for classroom demonstrations and for investigative calculus laboratories. The project courses have had about ten to fifteen students per section. Regular sections of these courses average about twenty-five students. During the first year of the project we have continued to use the same textbook, *Calculus with Analytic Geometry* by Ellis and Gulick, as is used in the other sections. In the future we expect to use our own combination of text and Mathematica notebook materials. External funding from the Vira Heinz Endowments is being used to support this project as well as broader-based exploration of the uses of symbolic computing in the curriculum.

Project Description:

The graphic, symbolic, and numerical capabilities of Mathematica are used to motivate, enliven, and investigate concepts of the calculus. We are developing text materials and Mathematica notebooks which involve carefully chosen classroom demonstrations and investigative laboratory exercises, tied together through the general theme of approximation. Attention is paid throughout to translating verbal descriptions of problems into mathematical formulations, and to encouraging students to write mathematics in good English.

Investigative laboratory exercises are used to explore calculus concepts more deeply than in a traditional approach. We give two examples. In one laboratory, done long before the students know the Fundamental Theorem, students solve realistic integration problems involving piecewise monotone functions by expressing upper and lower sums in terms of left and right Riemann sums, then generating a sequence of approximations with known error bounds by evaluating their upper and lower sums. In another laboratory, using Mathematica's animated graphics facility, students are led through an investigation of the way in which a cubic polynomial depends on its linear coefficient. In the exercises they explore the dependencies of a cubic on its quadratic coefficient, the behavior of other one-parameter families of functions, and some one-parameter families of their own.

The course is intended for all students in the regular calculus sequence. We expect to extend it for less-well-prepared students into a new integrated precalculus and calculus course, and to extend the use of Mathematica into other mathematics and science courses. Other faculty are generally supportive, but there is some skepticism. The administration is very supportive. In the first year we are experimenting with a variety of classroom approaches, but will probably settle on three 50-minute class periods and one longer laboratory period per week. Laboratory notebooks and demonstrations should be available for field testing in 1990-91.

Babson College

Contact:

ALICE P. MILLER, Babson College, Wellesley, MA 02157. PHONE: 617-239-4476.

Institutional Data:

Babson College is a private business specialty school with approximately 1500 full-time undergraduates and an MBA program of comparable size. One hundred percent of its majors are in business and management. The quantitative methods division consists of 20 full-time faculty members. One semester of calculus is among the requirements for graduation.

Project Data:

The honors calculus program, geared toward seasoned veterans of high school calculus, has been in operation for five consecutive fall semesters. The program is small; there are two sections of honors calculus with 25-30 students in each, as compared to eight or nine sections of regular calculus with up to 38 students in each. To date only one faculty member has taught the course, although the program has the full backing of the department. A second faculty member will teach the course in the fall of 1990.

Computer laboratories, scheduled three to four times per semester, are incorporated into the course to reinforce concepts and engage students in mathematical experimentation and discovery. Students work alone or in pairs at IBM PC's (with 256K and a graphics adapter) using the software package True Basic CALCULUS. The current textbook for the course is *Applied Calculus: An Intuitive Approach* by Faber, Freedman, and Kaplan. In 1989 sections of the field-test version of *Integrated Calculus* by Cozzens and Porter were used as supplementary reading in the course.

Project Description:

The National Research Council's 1989 publication "Everybody Counts: A Report to the Nation on the Future of Mathematics" maintains that "rarely is anything learned well until it is revisited from a more advanced perspective." The honors calculus program at Babson College offers such a perspective to successful veterans of high school calculus who otherwise would be required to "repeat" a standard introductory course. The distinguishing feature of the program is its increased emphasis on understanding concepts and applying them to a broad range of problem situations. The course encourages an attitude of thinking and questioning through classroom dialogue, thought-provoking computational and essay questions on examinations, journal writing, and exploratory computer assignments.

While MSAT and algebra pretest scores confirm the relative strength of the honors students (in comparison to the regular students) in algebra and general mathematics skills, insufficient testing has been done to date to compare the two groups in terms of growth in conceptual understanding. Analyses of overall calculus grades and responses to student opinion surveys reveal that when MSAT scores are taken into account, the stratified calculus program has not contributed significantly to grade inflation, but it has been associated with a small but significant increase in student satisfaction with lectures and class discussions. Overall, the program has proven successful in identifying and retaining students who are both qualified and willing to tackle the study of calculus at a more difficult level than is required by the curriculum, and thereby to build on, rather than to duplicate, their high school calculus experience.

Butler University

Contact:

JUDITH H. MORREL, Department of Mathematical Sciences, Butler University, 4600 Sunset Avenue, Indianapolis, IN 46208. PHONE: 317-283-9723.

Project Data:

Butler University is a small, private university with a curriculum grounded in the traditional liberal arts. Butler attracts better-than-average students who, with some exceptions, are relatively well-prepared in mathematics. The Department of Mathematical Sciences has 11 full-time faculty members and offers major programs in mathematics, actuarial science, and computer science. In a typical year approximately 20 seniors graduate with degrees in the mathematical sciences. Typical class size for calculus is 30-35 students.

Project Data:

As of Spring 1990 only the project director has taught in the project, although one more will do so in the Fall of 1990. We only offer two sections of first semester calculus! In the initial course (Fall 1988), 35 students participated and there were 33 in the regular section. This semester (Spring 1990) there is one section of the project course and there are no other sections. Thus far, the new course has only been taught for the first semester; the second semester of the new course will be taught for the first time in the Fall of 1990. Although there is no computer laboratory requirement, we have used True Basic's Calculus 3.0 on eight departmental microcomputers (Zenith 240 series). In addition, the students are required to purchase Casio graphing calculators. The textbook is *Calculus with Analytic Geometry* by George Simmons.

Initial funding for the project came from the Lilly Endowment, Inc. as a summer stipend for new course development in the Summer of 1988. Since September of 1989, the National Science Foundation has supported the project with a USEME Calculus grant.

Project Description:

Intended for majors in mathematics and the sciences, the project assumes the most important objectives for students to achieve in calculus are intuitive and conceptual understandings of the basic notions of change vs. stasis, instantaneous behavior vs. behavior on the average, and approximation techniques vs. exact ones. The course also assumes the most important skill to be learned in elementary calculus is that of problem solving (i.e., the ability to solve problems which involve the basic concepts listed above). Thus, the course is designed with an increased emphasis on problem-solving; two days a week are used for in-class problem solving. Sometimes the class works in small groups on problems. On other occasions, the class and the instructor work on (usually harder) problems as a "committee of the whole." There are also outside Problem Sets, not tied to a particular section of the text, which include multi-step problems, open-ended ones, some requiring library or journal research, and a few requiring computer use. The students have at least three weeks to work on each set and are encouraged to work together.

The presentation of the course is based on reducing lecturing and actively involving students with each other, with the instructor, and with the calculus. Lecturing, when it occurs (about 2 times a week in a 5-hour course), often uses applications to introduce concepts. Theory is demonstrated with explanations, not rigorous proofs. Sometimes new concepts are introduced in brainstorming sessions, using hand-outs and give-and-take between the students and the instructor to work through definitions and/or applications of new ideas.

Sample problem sets, syllabi, and in-class hand-outs are currently available free of charge from the project director. A database of problems suitable for use on outside problem sets will be available in the Summer of 1990. If requested on paper, there will be a charge for duplicating. The database, stored as TeX files on a Macintosh, can also be distributed on a provided disk.

Chatham College

Contact:

WILLIAM A. BECK, Department of Mathematics, Chatham College, Woodland Road, Pittsburgh, PA 15232. PHONE: 412-365-1221.

Institutional Data:

Chatham College is a four-year private liberal arts college for women. The College enrollment is between 550 and 600 full-time equivalents. Approximately one-third of the student body consists of women over twenty-five years of age. The Department of Mathematics consists of two full-time faculty members, a part-time lecturer in physics, and a part-time director of the Mathematical Skills Program (computation, geometry, introductory algebra).

The number of seniors who select majors in mathematics typically ranges from two to five. Interdepartmental majors have combined mathematics with chemistry, economics, management, music, or information science. One or two seniors each year have elected to seek certification for teaching elementary or secondary school mathematics.

Project Data:

The two-term course entitled Calculus, Physics, Applications I, II has been evolving at Chatham College since 1962. Initially the project was a joint effort between a member of the Department of Chemistry (who was responsible for the physics offerings of the College) and a member of the Department of Mathematics. Since 1966, however, the course has been presented by a member of the Department of Mathematics, which at that time assumed responsibility for the offerings in physics. Course enrollments have ranged from ten to twenty-five per term. The course is a prerequisite for Physics I, II, and Physical Chemistry, as well as an elective for the major in mathematics. Each year two or three computer-based projects are among the laboratory projects which are integral to the course. This spring students will use Derive in conjunction with homework assignments. This year, 1989-90, the only text material for the course is *Introduction to Mathematics-Physics* by Beck, Richey, and Trammell. Support for the course from the beginning has been the regular operating budget of the Department with no external fund sources.

Project Description:

Calculus, Physics, Applications I, II is an introduction to calculus for science and mathematics students which demonstrates the relationship and interaction of mathematics with the physics of motion. It is presented over two terms, each consisting of thirty-eight class meetings. The typical student is a freshman or sophomore with a major in biology, chemistry, or mathematics. Weekly laboratory exercises are self-scheduled and are integral to the course, providing motivation or reinforcement for the classroom work. Laboratory journals are kept and are submitted for periodic review, as well

as consulted during examinations. Classroom work is principally guided discussion with regular homework exercises for practice with skills and problem solving.

The course begins with a survey of principles of measurement, handling of numerical approximations, and the rudiments of statistical analysis. Construction of coordinate systems in one, two, and three-dimensional space is followed by a review of the concept of function and the construction of functions. Description of motion by means of vector-valued functions in three-dimensional space leads to the formulation of measurement of velocity and acceleration. Out of this is abstracted the concept of derivative. Application of the derivative to analysis of real-valued functions, related rates, and discovery of optimum values of a function completes the first term. In the second term, a study of circular motion motivates the introduction of the circular functions and leads to the formulation of the Newtonian theory of motion. Antidifferentiation, integration, and an introduction to power series provide applications to the discovery of solutions for differential equations, measurement of areas and volumes, and the properties of the exponential and logarithmic functions. During the last six weeks of the course, students pursue concurrently an independent laboratory project on some aspect of physical motion.

Information and copies of course materials may be obtained from William A. Beck. The cost of the text materials is approximately nine dollars plus postage.

City College of CUNY

Contact:

JACOB BARSHAY AND EDWARD GROSSMAN, Department of Mathematics, City College of CUNY, New York, NY 10031. PHONE: 212-650-5173 and 212-650-5147.

Institutional Data:

The City College of CUNY is a public, non-residential college offering Baccalaureate and Master's Degrees, and some on-campus Doctoral programs; it has a School of Engineering. There are 7200 full-time and 3600 part-time undergraduates enrolled, with 300 full-time and 2500 part-time students in the graduate program. Minority enrollment of Black and Hispanics consist of 67% with

25% of the science and engineering degrees going to minorities. There are 49 full-time faculty in the Department. In 1989, 50 students declared mathematics majors and 12 Bachelors degrees were granted.

Project Data:

This project has been under way since Fall 1989 when two sections were taught. The normal section size is 36; we have had 11, 12, and 20 students in the three project sections. Five members of the faculty are participating in the project.

We use *Calculus with Analytic Geometry* by Purcell and Varberg along with problem worksheets. The developmental work for the project was funded under a College Title III grant from the US Department of Education.

Project Description:

The calculus labs at CUNY are the vehicle for a pedagogical restructuring of introductory calculus. The goals are to develop a better conceptual understanding of calculus and improve students' ability to solve problems.

The laboratory sections of the course meet every day for 2 hours instead of the usual 4 hours per week. Six hours are conducted by the faculty member, and 4 hours are problem-based laboratories conducted by undergraduate mentors. Class size is limited to 24.

Special problem sets have been developed for the labs to reinforce the mathematical ideas introduced in class. In general, lab problems are more challenging than those assigned for homework, and students must devote considerable effort to solving them. This is facilitated by having the students work together in small groups under the guidance of the mentors.

Each week students are expected to write up and submit several of the laboratory problems. These are graded by the instructor and mentors together, with the mentors gradually assuming more responsibility. Copies of the best solutions are distributed, and oral presentations are sometimes given.

The two additional contact hours with the instructor are also devoted to problem solving. This might take the form of mini-labs at the end of a two-hour class, or time devoted to a writing exercise with peer evaluation and discussion.

We follow the customary syllabus for the course, although we have experimented with some topics not usually covered (e.g., Newton's method and the midpoint approximation to the definite integral).

Students in the special sections take the same departmental final as regular students. The results of our first semester were striking: mean course grade of 3.2 for lab students compared with 1.8 for all others.

The problem collection for the first semester is available from Grossman for a nominal charge to cover postage and handling.

Colby College

Contact:

DON SMALL, Department of Mathematics, Colby College, Waterville, ME 04901. PHONE: 207-872-3255. E-MAIL: dbsmall@colby.uucp.

Institutional Data:

Colby College is a four-year, independent, liberal arts, co-educational, residential college. Approximately 40% of the student body of 1650-1700 are drawn from outside New England. The combined median SAT scores are approximately 1200. There are no graduate programs. The Mathematics Department has 9 full-time faculty, with 8-10 graduating majors.

Project Data:

The development of this honors calculus course began under a Sloan grant in 1983 to develop a freshman-sophomore program giving equal weight to both discrete and continuous mathematics. Six Colby faculty members have taught the course during the past 13 semesters. The course has been taught with and without a computer laboratory. The Maple computer algebra system has been used when a laboratory was required. The textbook used is *Calculus, An Integrated Approach* by D.B. Small and J.M. Hosack. This text was class-tested in an honors course at the University of Houston.

Project Description:

This is a two-semester calculus honors course for students who have earned a grade of B or higher in a two-semester high school calculus course. The course integrates the treatment of one and several variables, focusing on concepts rather than dimension. Students are provided a thorough review of one-variable calculus in the process of extending concepts to higher dimensions. The emphasis is on development of concepts and comprehension rather than computation. Consistent use of the basic approximation process and error-bound analysis provide a unifying approach. Non-routine exercises (e.g., exploratory projects, graphical analysis, making up examples) are used to engage students in the development of concepts.

An objective of the course is to build on the high school calculus experience of the students rather than "starting over again." An important ingredient in the success of the course is that students view the course as being very different from their high school course. This is partially a result of introducing multivariable material early in the course.

The course has been taught in four fifty-minute lecture periods per week for two thirteen-week semesters. For the past two years, it has also been taught in three fifty-minute lecture periods and a ninety-minute laboratory per week for two thirteen-week semesters. The instructor and a student assistant conduct the laboratory. Students use Mac IIs or Mac IIcxs with the Maple computer algebra system. The use of calculators or computers is strongly encouraged. Laboratory worksheets and homework assignments are taken from *Explorations in Calculus with Computer Algebra Systems,* by D.B. Small and J.M. Hosack.

The College of Wooster

Contact:

JOHN RAMSAY, Department of Mathematics, The College of Wooster, Wooster, OH 44691. PHONE: 216-263-2579. E-MAIL: jramsay@wooster.bitnet.

Institutional Data:

The College of Wooster is a private-four-year liberal arts institution with approximately 1800 full-time students. Most students live on campus or in program houses owned by the College. The Department of Mathematical Sciences provides a major in mathematics and in computer science. Of the seven full-time faculty members in the Department, five are mathematicians and two are computer scientists. A typical graduating class will include 8-10 mathematics majors and 6-8 computer science majors.

Project Data:

Two of the five mathematics faculty have taught in the project, two more are preparing to teach it in the fall 1990, and the final member will initiate it in the spring of 1991. In the first semester, the CAS-enhanced section (first-semester calculus) had

23 students, and the two standard sections had 19 and 22 students. In the second semester both sections were CAS-enhanced and had 26 and 31 students. (In the fall, after having been told of the plans to use the computer, students were given the opportunity to switch sections. No students chose to do so.) The new course has been taught for two semesters now.

A computer laboratory is not required but students must have easy access, given the amount of their work which requires computer use. We used our computer classroom for two software orientation labs at the beginning of the semester. In addition to this, one instructor gave all exams in the computer classroom, and the other had students in the computer classroom for three calculus "lab" sessions. We are using the computer algebra system Maple. The software is available to students from any Macintosh attached to our campus-wide Macintosh network. This includes privately-owned machines in student residence hall rooms, most faculty and staff machines in campus buildings, as well as approximately 50 public access machines.

We are using Berkey's *Calculus,* but reliance on the text has diminished considerably. In particular, limits, curve sketching, antiderivatives, and introduction to integration are done very differently than the text, if not independently.

No external funding was used in this first year. External funding is being sought to incorporate Maple into all introductory-level mathematics courses over the next two years.

Project Description:

The Wooster Project began in the summer of 1989 with CAS research and material preparation. The CAS Maple was used in a pilot course in the fall and in all sections of Calculus I in the spring. Maple is used in the classroom (overhead projection unit), is a part of homework assignments, and is available on portions of examinations. We are attempting to teach students to use Maple properly as a tool, and have thus strived not to separate the Maple component from the rest of the course. Labs are not computer labs, but calculus labs (which may or may not utilize Maple on each portion), and are only done when they are specifically relevant to material being covered in lecture. Also, homework assignments only specify Maple use when experience with commands and syntax are needed.

Having received positive responses from faculty and students involved in the CAS-enhanced courses in 1989-90, the Department plans to incorporate Maple into the entire calculus sequence in 1990-91. In addition to the unanimous commitment within the Department, the College has been very supportive of the project. Faculty research, equipment, training, and travel support have been provided. In addition, the College has supported student assistants at each stage. A student assisted in the preliminary summer research and material development, two students (one each semester) have worked on developing tutorial manuals, and two summer research students are assisting in the development of materials to be used in 1990-91.

We are happy to share our labs and assignment sheets with anyone interested. A $5 "donation" to offset photocopy and mail expenses would be appreciated.

Colorado School of Mines

Contact:

JOAN R. HUNDHAUSEN and F. RICHARD YEATTS, Department of Mathematics and Department of Physics, Colorado School of Mines, Golden, CO 80401. PHONE: 303-273-3867; 303-273-3846.

Institutional Data:

The Colorado School of Mines is a public institution specializing in engineering and science. Current enrollment is 2400, including approximately 1700 undergraduates. Women comprise 20% of the student population, while the percentage of minority students is about 7%. In recent years, the school has enrolled increasing numbers of international students, and they presently comprise 13% of the student body.

The Mathematics Department includes 20 full-time faculty, and approximately 27 students graduate with majors in mathematics each year.

Project Data:

Supported by a three-year NSF grant in Curriculum Development (Calculus), this experimental course is being taught by the project directors for the second time during the 1989-90 academic year. This is a full-year course integrating the subject matter of Calculus I and II and Physics I (Mechanics) split into five credits per semester. During the first semester students earn 4 credits in Calculus I and 1 credit in Physics I; the second semester includes 3 credits of Calculus II and the remaining 2 credits of Physics I.

Students volunteer to participate in the course, and attempts are made to select those with adequate precalculus backgrounds. In the fall of 1988, 23 students enrolled, with 17 continuing for the second semester. The comparable figures for the current year are 22 and 15, respectively.

The class meets for six hours each week; two of these are devoted to a well-planned, though largely informal, Laboratory/Workshop session. Moderate use is made of the HP28S throughout the course; these calculators are available on loan to any student who needs one. Textbooks for 1989-90 are *Calculus with Analytic Geometry* by Johnson and Riess (field-test edition), and *Physics for Students of Science and Engineering* by Giancoli.

Project Description:

The course is designed to foster both a deeper understanding of the concepts of calculus and to promote transferability, i.e., the students should have the facility to recognize and apply the mathematics they have learned in calculus to other contexts in advanced physics and engineering courses. Thus, we place much emphasis upon geometry, modeling, versatility in the use of symbolism and parameters, and problem solving. Since some routine calculation can be relegated to the calculator, more time can be devoted to problem definition, interpretation, and analysis of solutions.

Students learn to program the HP28S calculator so that its symbolic, numerical, and graphical power is available to them. The Euler Algorithm is used to introduce students to numerical processes. This rather simple application of the tangent line then serves as an introduction to algorithmic thinking, the solution of differential equations, and, via the Riemann Sum, the Fundamental Theorem of Calculus.

Four hours of the six weekly meetings are conducted in a fairly standard lecture mode; the remaining two hours are devoted to the Laboratory/Workshop. These latter sessions often feature hands-on physics experiments for which students gather data which are then correlated with some analytic verification. Other sessions may involve numerical exploration using the HP28S or the discovery or application of calculus concepts, with guidance provided by specially-designed worksheets.

The Workshops are intended to provide opportunity for integration of the calculus and physics. Both instructors are present during Workshop sessions and there is usually a great deal of interaction among students and with instructors. A student assistant who completed the course the previous year is also on hand to offer guidance and encouragement and to perform some of the grading of homework, quizzes, and workshop assignments.

Student work is examined and assessed regularly. Four exams are given each semester; each covers both calculus and physics topics with integration of the subject matter as appropriate.

Evaluation of the program is an on-going process which includes tracking students through their subsequent courses. A comparable group of students serves as a control. Thus far, similar performances of the two groups on some common examination material has been observed. Both national and local advisory committees provide guidance and assist in the evaluation process.

Workshop materials are being designed and extensively revised during this second year of implementation. Materials will be discussed and shared with participants in a summer faculty workshop to be held in July 1990 and should be available after that time.

Comm. College of Philadelphia

Contact:

ALAIN SCHREMMER, JOSE MASON, Community College of Philadelphia, 1700 Spring Garden Street, Philadelphia, PA 19130. PHONE: 215-751-8413. FRANCESCA SCHREMMER, West Chester University, West Chester, PA 19380. PHONE: 215-436-2641.

Institutional Data:

Community College of Philadelphia is an urban two-year college with, in Spring 1989, out of a full-time equivalent total of 6872 students, 46.1% Black, 42.3% White, 5.1% Hispanic, and 6.5% Asian. Also, 37.5% were men and 62.5% were women.

In the Mathematics Department there are, on the average, 135 sections of 36 students each semester, only about 60 of which are taught by 17 full-time faculty. The Engineering Department is a separate entity which teaches its own precalculus-calculus sequence.

Project Data:

So far 10 faculty members have taught in our Lagrange Calculus Project. Out of 7216 students who enrolled in Basic Algebra between 1980-1985, 31% passed with an A meaning that they were ready for Precalculus I, but of these students only 1.42% completed Calculus I and 0.35% completed Calculus II. The corresponding data for the Lagrange project are: 34% completed the equivalent of Calculus I and 10% Calculus II.

There are currently two sections of Calculus I offered each semester by the Mathematics Department (in addition to one offered by the Engineering Department), and two sections of Differential Calculus II.

Three sections of Differential Calculus I and two sections of Differential Calculus II have been taught each semester since Fall 1989, plus one of each in the Summer. At this time we do not have a computer laboratory.

Project Description:

As much as being an alternative to the standard course, our "Lagrange Calculus Project" is designed to bring calculus literacy and competency within practical reach of an often over-looked population, the large number of women, minorities, and returning adults who "choose not to continue their study of mathematics through calculus, thereby closing career options in mathematics, engineering, and the sciences."

While the conventional approach builds on limits (a concept difficult and completely new to the students), our approach is based on an idea due to Lagrange. It builds on just basic algebra because the Taylor polynomial approximations can "always" be easily computed from the outset. It is rigorous and provides a more logical, geometrically intuitive, conceptual organization. It enables the students to comprehend calculus, to consider problems mathematically, to document their investigation, and to prove their assertions in written assignments.

During the first semester, the usual Precalculus is replaced by the Differential Calculus of low-degree polynomials which, in the second semester, are generalized to the Taylor expansions of the full Differential Calculus. In the third semester, the study of Dynamical Systems will be a natural continuation of Lagrange's approach and will be an alternative to the Integral Calculus, which is much better suited to the needs of those students who intend to pursue a career in the sciences but not, *a priori*, in mathematics, physics, or engineering.

Fifteen faculty from nine institutions in five states have joined seven faculty at Community College of Philadelphia to apply for support to field test the materials already developed for the first two semesters at CCP; revise and adapt these materials, or write new materials suited to the particular circumstances of each institution; develop special problem sets to further develop the problem-solving abilities of the students; and develop, field test, revise, adapt, or write new materials for the Dynamical Systems course.

Reprints and other materials (exams, assignments, etc.) are available from the PI. Workbooks for Differential Calculus I and II are available from MetaMath for $20 each: Mattei and Schremmer's *Differential Calculus, A Lagrangian Approach, Volumes 1 and 2.* A newsletter entitled "Lagrange Calculus: Unlimited, for Just Plain Folks as Well as Math Majors!" is published each semester and is available free upon request.

Cuyahoga Community College

Contact:

SAMUEL W. SPERO, Department of Mathematics, Cuyahoga Community College, Metro Campus, 2900 Community College Avenue, Cleveland, OH 44115. PHONE: 216-987-4561.

Institutional Data:

Cuyahoga Community College is a three-campus, public, two-year college enrolling about 25,000 students a year. The Metro Campus, located in downtown Cleveland, is the second largest of the campuses. Most of its students, whose average age is 28, attend part-time and come from a broad spectrum of ethnic and minority backgrounds.

The Mathematics Department at the Metro Campus has 11 full-time faculty members. While our largest enrollment is in our developmental math courses, we do teach 3 different calculus courses: a one-quarter (12 weeks) course for business majors; a one-quarter course for technicians; and a four-quarter university-level calculus sequence.

Project Data:

Since 1980, the author, with no external funding except for release time, has been experimenting with the use of electronic spreadsheets in the various calculus courses taught in the department. Initially the author used Supercalc 3, then Supercalc 4, and is presently using VP Planner Plus which will be the spreadsheet used in the published workbook. The students use the various PC laboratories available on campus, but no formal laboratory

period is required. The author is the only instructor involved in this particular approach, although some instructors on the other campuses also require computer projects. Now that the testing of the approach has been completed and the published version of the materials is about to appear, a proposal will be submitted to the department to adopt formally this approach.

The textbooks used in the calculus courses are Purcell for the traditional calculus sequence; Washington for the technical calculus; and Mizrahi and Sullivan for the business calculus.

Project Description:

The purpose of this project has been to introduce the use of the electronic spreadsheet as a computational and graphing tool in the study of the calculus. Working within the existing curricula and using the traditional textbooks, we have been using the electronic spreadsheet to facilitate the completion of the many exercises in the traditional textbooks which could benefit from computational and graphing help.

The result of this project is a workbook entitled *The Electronic Spreadsheet in Calculus: Graphing and Numerical Methods,* which will be published in August 1990. The workbook comes bundled with the Lotus compatible spreadsheet VP Planner Plus (Educational Version) and will sell for under $15. The workbook is meant to be self-instructional and provides the student with the skills required to use the electronic spreadsheet for graphing functions, as well as computing the various numerical algorithms studied in the traditional calculus courses (Newton's Method, Trapezoidal Rule, Method of Least Squares, and Euler's Method).

The workbook has been tested in preliminary versions in all of the different calculus courses we teach. While no formal evaluation has been conducted, student response has been excellent in that they not only complete the assignments in the calculus course, but subsequently use the electronic spreadsheet in their other courses, as well as at work.

In using the workbook in class, we suggest that about two hours be invested at the outset to train the students in the use of the electronic spreadsheet (a self-instructional approach is included as Part I of the workbook). No additional classroom time is required. In a typical calculus course we assign 3-5 projects from the workbook to be handed in. Students use the electronic spreadsheet on their own for other assignments.

DeAnza College

Contact:

CHRIS AVERY, DeAnza College, 21250 Stevens Creek Boulevard, Cupertino, CA 95014. PHONE: 408-265-5659.

Institutional Data:

DeAnza College is a two-year community college with 27138 students of whom 71.1% are part-time students. There are 21 full-time instructors in the Mathematics Department.

Project Data:

Three faculty members have taught the Computer-Aided Business Calculus course. The course enrolls a maximum of 40 students in both the regular section as well as the computer-aided section. This section has been taught for three years. The principal software is Interactive Applied Calculus and is available for Apple IIe and IBM or compatibles. The textbook used is *Interactive Applied Calculus* by Avery and Soler. The textbook was written to support the computer pedagogical model. It requires intensive use of the computer on the part of the student as well as the instructor.

Project Description:

The DeAnza College Business Calculus course is designed for transfer students with degree objectives in business or economics. These students require an overview of calculus including differentiation, integration, exponential and logarithmic functions, differential equations, and topics in multivariate calculus including an introduction to Lagrange multiplier techniques. The prerequisite for this course is the completion of intermediate algebra. The computer project was started in 1986 in order to allow students with a limited mathematics background access to the power and utility of the above concepts. With the computer, there is a heavy emphasis on real rather than contrived applications. The students work through course material using the numerical algorithms and graphing capabilities of the computer.

The standard course is offered on the quarter system five days per week for 12 weeks. The computer-based course is the same with four days of lectures and one day in the lab. A 65-station Apple IIe lab is available for lab work.

Quizzes, two mid-terms, and a final are used to evaluate students. In addition there are 8-12 computer lab assignments given. There is an emphasis on cooperative learning. Each mid-term is in two parts. An in-class test is taken on an individual basis. A take-home component that requires the computer is completed in groups. This support system is an important component in developing the student's sense of confidence.

All materials have been developed at DeAnza College. They include the text *Interactive Applied Calculus* by Avery and Soler ($15) with accompanying software ($15), and a lab manual *42 Applied Calculus Lab Assignments* ($10). These materials are available from the Math Lab.

Dowling College

Contact:

RUSSELL JAY HENDEL, Department of Mathematics, Dowling College, Oakdale, NY 11769. PHONE: 516-244-3339.

Institutional Data:

Dowling is a four-year private college that enrolls 2500 students. There are 6 full-time mathematics faculty, 2 computer scientists, and 10-20 math majors.

Project Data:

Four faculty have been involved in the project for 5 semesters. All calculus students take the project courses. Stein's *Calculus* is the required text, and scientific-graphics calculators are mandatory.

Project Description:

Our classes meet for either 2 two-hour periods per week, or for 3 one-and-a-quarter-hour periods per week during a 15-week semester. Our goal was that every calculus concept, or problem, should be seen both formally and numerically. We wished to show students that one accomplishment of calculus was the discovery of closed formulae for many numerical estimation problems. Sophisticated software was not available but students did use scientific calculators for numerical calculations. This formal-numerical approach was emphasized in six areas:

Differentiation: There was continual emphasis on obtaining both exact answers and numerical approximations to derivative problems.

Differentials: Numerous problems of the form "Estimate $\sqrt{65}$" were presented. There was an emphasis on asking students to supply the values of $f(x)$, x, and Δx.

Series: Calculation of the exact value of infinite sums were always coupled with calculation of large finite sums.

Riemann Sums: There was emphasis on calculating both exact values of integrals as well as large finite Riemann sums. We frequently used the discrete formulae for the sums of consecutive integers, squares, and cubes.

Taylor Series: Initial terms of Taylor series were used to reformulate problems involving transcendental functions in terms of problems involving algebraic functions. There was also emphasis on using Taylor series for calculating numerical approximations of values of transcendental functions.

Numerical Methods: We emphasized simultaneous solution of differential equations in closed form and by numerical methods.

The development of this project has been informal. A collection of problem types is available at no cost (other than possibly mailing and xeroxing). Interested people should contact the author.

Drexel University

Contact:

LOREN ARGABRIGHT AND CHRIS RORRES, Department of Mathematics and Computer Science, Drexel University, Philadelphia, PA 19104. PHONE: 215-895-2668.

Institutional Data:

Drexel University is a private, nonsectarian, urban university. Total enrollment is approximately 12,000 students, including 7,000 full-time day undergraduates. The largest enrollments are in engineering and business.

In 1983, Drexel became the nation's first university to require every entering freshman to have personal access to a microcomputer. The current student package consists of an Apple Macintosh SE and a software bundle including Apple's Hyper-Card, a word-processor, a spreadsheet, and other programs.

The Department of Mathematics and Computer Science offers undergraduate and graduate degree programs in both Mathematics and Computer Science. The Department has 37 full-time faculty members, approximately 300 undergraduate majors, and 100 graduate majors.

Project Data:

Work on a preliminary version of HyperCalculus was initiated in the spring of 1989 and is continuing throughout the 1989-90 academic year. This initial phase of the project, which has been supported by a Drexel software development grant, consists of creating a series of experimental HyperCard stacks that are being used and tested in the freshman calculus sequence during the 1989-90 year. All course sections are involved, and all of the experimental stacks are being made available to students for installation on their personal computers. Enrollment in the calculus sequence for science and engineering majors is approximately 600 students per term. There are four lecture sections, plus recitations, and all of the faculty lecturers are using the experimental stacks for some of the lecture presentations. During the spring term a preliminary evaluation will be conducted to ascertain initial student and faculty reaction to the experimental stacks.

Project Description:

This project, expected to extend over a two-year period, will undertake the development of a prototype collection of hypertext materials for calculus instruction. Throughout the development, special efforts will be made to ensure that the design concepts used are generic and portable to other hypertext environments.

The guiding philosophy of the project will be to use the computer and display system to provide faculty with tools that enable them to present material more efficiently, or in a better way, than would otherwise be possible; and to provide students with learning materials that aid in the development of conceptual understanding, that guide and encourage exploratory learning, and that serve to expand horizons by exposing students to ideas and information for which there is no room in the typical course syllabus or text.

There will be two basic categories of stacks, or stack segments, within HyperCalculus: those designed to be used for classroom presentations; and those intended for use as a self-study tool by students. The basic idea of the lecture segments will be to provide concise graphic displays and animations that serve to establish the conceptual framework within which a topic will subsequently be developed. These segments will be primarily visual in nature, with little in the way of text or explanatory detail, except for key items of terminology and notation. These segments will also be built into the self-study stacks that will be made available to students and, in this regard, the lecture demonstrations will serve to introduce students to the use of HyperCard at the browsing level. Starting from the lecture segments as a base, the self-study stack on a particular topic will include additional materials designed specifically for student use. Typically, this will consist of cards that, when appropriately grouped and linked, will constitute a conceptual outline of the subject matter, including all of the central concepts, theorems, and formulas. At this level of detail, the function of the stack will be to provide a framework within which students can quickly access and interact with chunks of material, exploring ideas, gaining an overview, and beginning to recognize linkages between conceptually related chunks. It should be emphasized that this treatment will fall far short of duplicating the text in electronic form. Only the most essential topics will be included. There will be no proofs or derivations of formulas. Examples will be few in number and moderate in complexity, and computational details will be kept to a minimum. The idea is to provide only the "bare bones," linked together in such a way as to form a conceptual skeleton of the subject matter.

Although Drexel will apply for copyright protection of HyperCalculus materials, this is not intended to limit dissemination or replicability elsewhere. Information as to availability of materials, including ordering procedures and cost, will be forthcoming.

Eastern Michigan University

Contact:

BETTE WARREN, Department of Mathematics, Eastern Michigan University, Ypsilanti, MI 48197. PHONE: 313-487-0121 and 313-487-1444.

Institutional Data:

Eastern Michigan University is a publicly supported comprehensive university offering bachelor's and master's degree programs. Total enrollment is 24,000 students of which about 17,500 are undergraduates. While we have a sizable population of traditional dorm and apartment dwelling undergraduates, the College also attracts many non-traditional students through extensive evening course offerings and a second bachelor's degree program.

There are 30 full-time faculty members in the Mathematics Department. Over one-third of our classes (almost all of the intermediate and college algebra, and about half of the sections of business calculus and finite mathematics) are taught by part-time instructors. In a typical year between 100 and 150 students graduate with mathematics majors or minors (both regular and secondary education). About 25 elementary education majors graduate with concentrations in mathematics; these elementary education students must take calculus.

Project Data:

We are currently in Phase II of this project. In Phase I, Bette Warren developed nine pencil-and-paper labs for an honors section of Calculus I. Tim Carroll revised four of the labs and used them in his honors Calculus I the following semester.

This winter and next fall, 1990, the Department is running a two-credit-hour lab course which is open to all students concurrently enrolled in Calculus I. Ten students are enrolled in the course this semester (Winter 1990). We are running Micro-Calc on 8 PC-compatibles (mainly Zeniths) in our Department's microcomputer lab. The only textbook is Thomas and Finney. By Fall 1990 a lab manual will be available. So far this has been a very low cost operation, sustained entirely by the Department.

Project Description:

This pilot project is intended to prepare the way for us to institute a weekly microcomputer lab requirement in calculus. The class will meet for two hours once a week in the microcomputer lab. We usually spend the first 30-45 minutes on interactive instructions for the lab. The instructor then circulates, asking and answering questions for the remainder of the lab period. Each student is expected to write a report incorporating data (tables and occasionally sketches), verbal descriptions of observations, calculations and explanations. Students are encouraged to work together in the lab and to discuss the report with friends and classmates. Each student is responsible for writing his or her own report.

We want all students to be successful and yet show each of them something surprising. Our lab exercises lead students through a sequence of examples designed to help them generalize their findings. We ask only that they observe, describe what they see, and explain how it relates to other exercises. A good write-up of the core lab earns a B. Students are encouraged to answer more probing extra-credit questions for a higher grade. In all cases, students are allowed to resubmit revised lab reports to raise their grades.

We have found that in order to get students to visualize and make connections, we must usually proceed from pictures to numerical calculations to algebraic solutions. The initial visualization should be as free from explicit calculations as possible. Students should retain essential control over the process. At times this requires low-tech improvisation like putting Saran Wrap on the computer screen and letting students draw on it with overhead projector pens. We try to give our students usable skills, but our primary goal is conceptual clarity.

A lab manual will be available by September 1990 for the cost of duplication and postage. Contact Bette Warren for further details.

Elmhurst College

Contact:

JAMES KULICH, Department of Mathematics, Elmhurst College, 190 Prospect, Elmhurst, IL 60126. PHONE: 708-617-3570. E-MAIL: elmhcx9!jimk@antares.mcs.anl.gov.

Institutional Data:

Elmhurst College is a private four-year comprehensive undergraduate college. We have approximately 1700 full-time students with a wide variety of majors. Roughly 60% of our full-time students are commuters. The Mathematics Department has 5 full-time faculty members and is separate from the Computer Science Department. We graduate roughly 6 majors per year.

Project Data:

Three faculty members have been involved in our project, two in the teaching of calculus, and one in pre-calculus. Three sections of Calculus I have been taught to date, two with 22 students each, and one with 9 students. One section of Calculus II is currently being taught with 24 students. The Precalculus class had 5 students. All day sections of Calculus I and II now use technology. This is the second semester in which we are fully implementing technology in our calculus courses. Less organized efforts date back another two semesters.

Computer laboratory assignments are required and count significantly. At the present time they are done outside of the normal class time. An in-class lab period will be added in the Fall of 1990. Our principal software packages are Derive and Epic. Students work on Zenith PC clones which are available at most times on two campus networks. Classroom demonstrations are done using an overhead projection device connected to a Zenith PC. Some 286 class machines are now available to the students.

We are currently using Purcell and Varberg's *Calculus*. We've had no external funding at the beginning of the project. However, we recently received funding from the Culpeper Foundation to work with two other area colleges in expanding this program. We've also received funding from the State of Illinois to run a program on this technology for area high school teachers.

Project Description:

Our project has involved designing a computer-oriented calculus course for our entering students who plan to major primarily in mathematics, computer science, or another science. We currently meet for three 65-minute sessions each week. This will change to four 50-minute sessions next fall in order to introduce an in-class lab component.

Students are assigned laboratory projects every two weeks. These projects are of two types: one type of project involves having the students examine a traditional topic such as limits or integration by substitution in a way which uses the graphical, symbolic, and numerical capabilities of the computer; the other type of project involves having the students model an application of recent class topics. Situations involving parameters are often studied with the students being asked to observe patterns and draw conjectures. Questions asked of the students in the labs become increasingly open-ended as the semester progresses. Calculus I students complete 6 labs and Calculus II students complete 5 labs. Students are encouraged to work in groups but each student must turn in a written report on their own work.

We have not yet engaged in any formal assessment of the project. Most student comments have been favorable, but not all. Most students indicated an interest in continuing with the present methods. We find that students' writing skills are generally weak and that much guidance is necessary in preparing the lab reports. Also, students need considerable aid (especially at the beginning of the course) when being asked to formulate conjectures and answer open-ended questions. Exams are taken in the classroom without computers. However, a considerable amount of graphical and numerical information that would be provided by the computer is usually attached to the exam. Test questions which assume this information are written. Pure mechanical manipulation questions have been de-emphasized on exams.

We have no TA's at Elmhurst College. All grading is done by faculty. We have the full support of the Department and the Dean for the project. Materials used beyond the textbook and software are written by us.

Franklin and Marshall College

Contact:

ALAN LEVINE, GEORGE ROSENSTEIN, Department of Mathematics and Astronomy, Franklin and Marshall College, Box 3003, Lancaster, PA 17604. PHONE: 717-291-4238. E-MAIL: a_levine@-fandm.bitnet or g_rosenstein@fandm.bitnet.

Institutional Data:

Franklin and Marshall College is a highly selective, residential liberal arts college offering a Bachelor of Arts degree only. There are about 1850 full-time students currently enrolled. The Department of Mathematics and Astronomy includes nine full-time faculty in mathematics. Typically there are about 15 mathematics majors each year, a small percentage of whom go to graduate school in mathematics.

Project Data:

In 1989-90, two faculty members taught three sections of the project course during the Fall semester and two sections during the Spring semester. Three faculty members are scheduled to teach four sections in the Fall 1990 semester.

The three sections taught in the Fall 1989 semester enrolled eighty-three students who were assigned to the course randomly. There were four sections of the standard course with 107 students enrolled. The two sections taught in the Spring 1990 semester enrolled thirty-one students; three sections of the standard course enrolled forty-three students. Students were (generally) not permitted

to switch between project sections and standard sections between semesters.

There are occasional laboratories, some of which require the use of a computer, primarily to graph functions. We have used home-made programs exclusively. The software we are using is also used in the standard course and has been used in the department for about five years. This is a Macintosh campus. Students are encouraged to buy Macs and all faculty have Macs on their desks. We have used text materials that Professors Levine and Rosenstein have generated. This project has been funded in part by the Howard Hughes Medical Institute.

Project Description:

This course sequence is designed for Franklin and Marshall students who wish to study calculus. It meets four times each week in sections of about 25 students each. Computers and calculators are used where they are appropriate, somewhat more intensively than the standard sections. Approximately once each chapter laboratory exercises, which may require the use of a computer, are inserted. Sometimes lab reports are required.

The course originated in discussions of what might constitute a "lean and lively" calculus sequence. The project uses new text and teaching materials to address problems of student passivity, a perceived emphasis on mechanical procedures, and rigor at the expense of intuition. Two text-and-workbook volumes which encourage the active participation of the student have been written.

Students complete examples in the text and respond to "challenges" which encourage them to consider implications and extensions of the concepts presented. Each chapter contains a few routine exercises to reinforce the ideas of the section, but artificially complex calculations have been curtailed. In their place, more conceptual problems have been inserted to integrate and extend the ideas introduced. These problems require students to justify their conclusions and to express themselves mathematically.

The topics in the course were chosen using two criteria: students who take only one semester of calculus should see both differential and integral calculus, and students should see calculus as a useful and interesting tool for solving problems. For example, an application leading to a separable differential equation is used as the starting point for a discussion of the differential equation $y' = ky$ which leads to the development of exponential functions.

This project has been supported by the members of the mathematics staff with varying degrees of enthusiasm. It will be expanded slightly next year; revised materials will be used.

Frostburg State University

Contact:

MARCELLE BESSMAN, Department of Mathematics, Frostburg State University, Frostburg, MD 21532. PHONE: 301-689-4453. E-MAIL: r2nkbes@fre.towson.edu.

Institutional Data:

Frostburg State University, a member of the University of Maryland System, is a small, comprehensive, largely undergraduate institution located in Western Maryland. The number of mathematics majors is 80. Approximately 8-10 students graduate each year with a major in mathematics. The Department of Mathematics has 15 full-time faculty.

Project Data:

This project was started in Fall 1989 with one faculty member teaching three calculus sections: two Calculus I and one Calculus III. There were two other Calculus I sections, and another Calculus III section. Students self-registered in all sections using their usual criteria. The average class size for all mathematics classes is 25.

All Calculus classes meet three times a week, 50-minutes per session. The project classes used the same tests as the other sections. They differed from the regular classes in that the computer was regularly used in the classroom, and computer hardware plus software formed the "mathematics laboratory."

This Spring one teacher is teaching a Calculus I class and Calculus for Business Applications II using essentially the same methodology and an expanded set of materials. In Fall 1990 the teacher will be teaching a Calculus I class and two sections of Calculus for Business Applications I using this method. In the next two semesters the teacher will teach Calculus II and III, thus giving students the opportunity to continue in this project through the full calculus sequence. Since the enrollment in all sections is by student self-registration, there could be a broad range of student computer skills and level of familiarity with MicroCalc each

semester. It is intended that in Calculus II and III students familiar with the software will assist their less-familiar classmates.

Project Description:

At the beginning of the Fall semester the computing experience of the students in the three classes ranged from novice to computer science major. In order to enable this diverse group to use the software, a two-hour evening instructional session was held in the computer laboratory during the first week of class. After that session students were given weekly laboratory assignments designed to guide them in using MicroCalc, not only to carry out computations and symbolic manipulations, but also to investigate or explore mathematical concepts and processes. The design and content of these labs was based on the laboratory assignments used by David Smith at Duke.

By manipulating graphs and tabulated data, students investigated, for example, the meaning of function, domain and range, global and point-wise behavior of functions, the concept of limit of a function, the secant-tangent-derivative relationship, polynomial approximation of functions, error estimates, sequences and series, and graphing in polar coordinates. The students are guided in not only using the strengths but also in recognizing the limitations of this computational tool.

The laboratory assignments required the student to answer questions or describe observations in a written report. A student laboratory assistant was available 6 hours a week to assist them at the computer. As the semester progressed, students became more comfortable with the computer and better able to use the software as a research tool.

Other members of the mathematics faculty informally monitored the progress of this project, reviewing laboratory assignments, and on occasion, using this or other software in their classrooms.

Sample laboratory exercises are available from Marcelle Bessman.

Furman University

Contact:

DAN SLOUGHTER, Department of Mathematics, Furman University, Greenville, SC 29613. PHONE: 803-294-3233. E-MAIL: sloughter@frmnvax1.

Institutional Data:

Furman University is a private liberal arts college. The institution is devoted almost entirely to undergraduate education. The undergraduate enrollment is usually slightly more than 2500 students. The Department of Mathematics currently has 10 full-time faculty. There is no graduate program in mathematics. In recent years, 10-15 students graduate each year with majors in mathematics and a similar number of graduates with joint computer science-mathematics majors.

Project Data:

Three faculty members have taught in the project. For Fall term 1988 there were 2 sections of the first term of the new course, and 4 sections of the standard course. All sections had approximately 30 students. For Winter term 1989 there was one section of the second term of the new course, with 12 students, and two sections of the second term of the standard course, with approximately 20 students in each. For Fall term 1989 there were three sections of the first term of the new course, and three sections of the standard course. All sections had approximately 30 students. In Winter term 1990 there was 1 section of the second term of the new course, with 22 students, and 2 sections of the second term of the standard course, with approximately 20 students in each.

Two terms of the new course were taught in the 1988-89 academic year; there was no multivariable section of the new course. Three terms of the new course will be taught in the 1989-90 academic year; a multivariable section is being added for Spring term.

Students are required to use the computer laboratories to do some of their homework as well as take-home quizzes. There are no formal computer laboratory classes. The principal software used in the one-variable calculus sections has been Math-CAD. We have also made limited use of True Basic and Reduce. We now have a small laboratory of 5 Sun Sparcstations, and we will be using them primarily with Mathematica.

We have been using three computer laboratories; each of these has approximately 20 Hewlett Packard Vectras, AT compatible machines with 640K of RAM and EGA graphics. For classroom demonstrations we have a Sharp QA-50 computer projection panel.

This year we used *Calculus and Analytic Geometry* by Thomas and Finney for the one-variable

course, and will use Hurley's *Multivariable Calculus* for the multivariable course. However, the text for the one-variable course was used only as additional reading and as a source of problems. No calculus text currently on the market covers all the material treated in this course, nor do any treat the topics in the same manner that this course does. This project has been funded by NSF grants.

Project Description:

The goal of this project is a complete restructuring of one-variable calculus. I have tried to start afresh, seeking to create a one-variable calculus course that fits nicely into two terms, yet covers the important ideas and procedures that the students will need in their later mathematics and science courses. The course I have developed breaks down into 8 units:

1. The fundamental notion of a limiting process: This idea is treated, at the beginning, from the viewpoint of limits of sequences, making use of examples from area computations, geometric series, and difference equations.
2. Functions: Functions are treated with an emphasis on the dependence between quantities and the notion of cause and effect. The sine function is introduced as a model of oscillatory motion, with computer graphing aiding the discussion of sound waves.
3. Approximation of a nonlinear function by a first-degree polynomial: The idea of approximation is emphasized throughout the course as a cornerstone of calculus. The derivative is introduced through the notion of the best linear approximation.
4. Quadrature: Again, the notion of approximation is fundamental as numerical techniques of integration are presented before the Fundamental Theorem of Calculus. The basic integration techniques of substitution and parts are presented immediately after the Fundamental Theorem.
5. Taylor polynomials and Taylor series: The second term of the course begins with an extension of the notion of approximating functions by polynomials, which is then extended further to the notion of a Taylor series representation of a function.
6. More transcendental functions: The exponential function is defined using its power series representation and is then shown to be the only function satisfying the basic differential equation $y' = y$. The logarithm function is introduced as

the inverse of the exponential function. The inverse and hyperbolic trigonometric functions are also introduced at this time.

7. Complex numbers: This is an introduction to the algebra of complex numbers, as well as basic complex valued functions, polar notation, and conic sections (through the solution of the two-body problem).

8. Differential equations: As time permits, a discussion of techniques, exact and numerical, for solving differential equations. Hence the course proceeds from discrete modelling with difference equations at the beginning to continuous modelling with differential equations at the end.

I will send a more detailed description of the course, along with outlines, syllabi, and sample handouts used in class to anyone who requests more information.

Grand Valley State University

Contact:

CHARLENE E. BECKMANN, Department of Mathematics and Computer Science, Grand Valley State University, Allendale, MI 49401. PHONE: 616-895-2066. E-MAIL: 21874ceb@msu.bitnet.

Institutional Data:

Grand Valley State University is a public comprehensive institution offering a liberal arts and professional education with a small but growing graduate program. Enrollment was over 11,000 students in 1990. Most students are commuters with full-time jobs, or students living at home to keep costs low. The Department of Mathematics and Computer Science has 24 full-time tenure-track faculty, 16 of whom teach mathematics. Approximately 20 of 1200 graduating seniors graduate with majors in mathematics.

Project Data:

All of the faculty teaching Calculus I are participating in the project at some level. Four sections of Calculus I are offered in the Fall semester, and two or three sections in the Winter. Classes range in size from 25-35 students with an occasional small section of 10-15 students. All students in the mainstream calculus sequence are participating in the project at some level (depending on the faculty member teaching the section).

The author began teaching this course in 1986. The project is currently in its seventh semester for Calculus I and in its second semester for Calculus II.

Students are required to purchase Casio graphing calculators. A computer laboratory is used by students for project work (students complete 4 projects per semester). While providing valuable learning experiences for students, the computer lab is not essential to this project. Several programs have been written for the Casio which accompany the materials mentioned above. Computer software that is used by students on projects (but not generally for daily work) includes MicroCalc, Master Grapher, and Calculus Workshop Programs (homemade by Beckmann and P.K. Wong of Michigan State University).

Hardware used most often by students are Casio graphing calculators. Also available is a microcomputer lab with 70 stations connected by a token-ring network running a Novelle operating system. The lab contains several IBM PS-2s and IBM-PC compatible Zenith 286 machines. Each station can also be used as a stand-alone.

The current textbook used is *Calculus* by Berkey. The discovery lessons and homework assignments provide some textual material for students.

Project Description:

The course is based on theoretical and empirical literature concerning understanding of mathematics and how such understanding is enhanced through the use of technology. In the course, students are actively engaged in creating their understanding of new ideas through making connections with already familiar ideas. The availability of a graphics tool for all students at all times allows a strong graphic library of functions to be built for use throughout the course. Materials are written to surprise students, to encourage them to wonder if what they see works for all functions, or if they can find exceptions to the investigations they are undertaking.

A discovery/investigative atmosphere is fostered in the class. Classes are interactive. Depending on the topic and the instructor, materials are used for in-class laboratory investigations or project work outside of class. Labs meet within the scheduled class hours and are not scheduled on a separate basis (some weeks might be all lab activities, other weeks no lab activities, depending on topic).

Assessments include 3 course exams, 4 projects, 5 short quizzes, and a comprehensive final exam.

Projects are completed by small groups of 2-4 students, with each student in the group being responsible for the material covered. Students take turns writing up the project report (all students must have written a project report by the end of the semester).

The project has full backing of the Department. Calculus is one of the most sought-after courses to teach in the Department. Course materials will be available by the beginning of August 1990. These are expected to sell for approximately $16. Also available are home-made computer programs for the IBM, Apple, and Macintosh which illustrate some of the ideas covered in the course. Interested parties should send a computer disk indicating the type of machine to the author.

Gustavus Adolphus College

Contact:

MICHAEL D. HVIDSTEN, Department of Mathematics, Gustavus Adolphus College, St. Peter, MN 56082. PHONE: 507-931-7480. E-MAIL: hvidsten@gacvax1.bitnet.

Institutional Data:

Gustavus Adolphus College is a private four-year undergraduate liberal arts college located in St. Peter, Minnesota. Approximately 2300 students are enrolled with about 4% of these students of minority background. The largest number of graduates, about 130 each year, comes from the Economics/Management Department. Almost all of the students live on campus and are full-time. The Department of Mathematics/Computer Science has 13 full-time faculty members, 9 of whom teach mathematics. We will graduate 23 majors this year, 13 in computer science and 10 in mathematics.

Project Data:

Two members of the Department, Mike Hvidsten and John Holte, have taught in this project. The number of students in the new course equaled the number in a standard course, which is usually about thirty. The new course was taught during 1988-89 (two semesters) and 89-90 (one semester).

Computer labs are an integral component of the course. True Basic Calculus was used as the lab software. We use IBM PC's in our micro-lab.

The text used was Larson and Hostetler's *Calculus, Third Edition*. Funding for development of this course was provided through a Bush Foundation grant.

Project Description:

The main change we have made in our calculus course is the addition of nine laboratory projects to the regular calculus syllabus. The new course meets M-W-F for lecture and on Tuesdays for a lab or for a help session. All but two of the labs take place in the micro-computer lab on campus and last one hour. Two of the labs are done in the classroom. Students work in pairs on the labs, and at the next class period jointly hand in a lab report. The lab report details the activities carried out and results found in the lab. The instructor is present during the lab hour for help.

The nine labs either reinforce or introduce a calculus concept by means of a mathematical "experiment." The projects are functions and graphing, limits and continuity, derivatives and slope, error analysis, maxima and minima, Newton's method, areas and sums, the natural logarithm, and growth and decay. The labs in error analysis and areas are done in the classroom and are not computer-based.

We have equipped our classrooms with computer projection equipment so that instructors can use the lab software during class time. One difficulty we have had in implementing the labs has been the numerical imbalance of students to computers at our campus. This imbalance should be alleviated soon with the addition of two new micro-labs.

Comparison of test scores from the old Calculus I and new Calculus has shown no statistically significant improvement. However, students seem to understand the material covered in the lab more readily than material covered in lecture. Overall, the new course has been successful, and our Department will continue offering such a course.

Printed materials, including a sample syllabus and copies of the nine labs, are available at no cost from Mike Hvidsten.

Harvard Consortium

Contact:

ANDREW M. GLEASON, DEBORAH HUGHES HALLETT, and CARL BRETTSCHNEIDER, Department of Mathematics, Harvard University, Cambridge, MA 02138.

Institutional Data:

The Core Calculus Consortium includes the University of Arizona, Chelmsford High School, Colgate University, Harvard University, Haverford College, University of Southern Mississippi, Stanford University, and Suffolk County Community College.

The University of Arizona is a four-year state university with 28,000 students, including many Hispanics. The mathematics department consists of 60 FTE faculty with 50 graduating majors annually. The calculus text is Anton; other courses are available for Business and Engineering students.

Chelmsford High School is a four-year public high school with 1700 students with a mathematics department consisting of 15 FTE.

Colgate University is a four-year liberal arts university with 2700 students. There are 10 FTE faculty in the mathematics department with 20 math majors graduating annually. The calculus text is Stewart.

Harvard University is a four-year private university with 17,000 students. There are 35 FTE faculty with 27 math majors graduating annually. The calculus text is Thomas and Finney.

Haverford College is a four-year liberal arts college with 1000 students, 5 FTE faculty, and 10 math majors graduating annually. The calculus text is Shenk.

University of Southern Mississippi is a four-year state university with a student body of 13,000. The mathematics department faculty consists of 27 FTE with 30 graduating math majors annually. The calculus text is Shenk.

Stanford University is a four-year private university with a student body of 17,000. There are 30 FTE faculty with 25 math majors graduating annually. The calculus text is Edwards and Penney.

Suffolk County Community College is a two-year public community college with a student body of 14,000. The math department consists of 25 FTE faculty with 3 math majors graduating annually. The calculus text is Larson and Hostetler.

Project Data:

So far seven faculty in six of the eight institutions have taught in the project. The total number of students affected is 250-300.

Project Description:

We believe that the calculus curriculum needs to be completely re-thought. We have designed a new syllabus, and we are currently writing and testing the materials to support it. In designing the new syllabus we followed three guidelines:

- Start from scratch. Do not look at the old syllabus and try to decide which topics can be left out; decide which topics are so central that they must be included.
- Show students what calculus can do, not what it can't do. In a freshman-level course we should be showing students the power of calculus, not the special cases in which it fails.
- Be realistic about students' abilities and the amount of time they will spend on calculus. It is far better to teach a few topics well than many topics superficially.

Our project is based on our belief that three aspects of calculus—graphical, numerical, and analytical—should be emphasized throughout. We call this approach "The Rule of Three," and are working to design a core curriculum for a 2-quarter or 2-semester course based on this principle.

For example, consider the traditional introduction of the exponential function as the inverse of $\ln(x)$, which is itself defined as an integral. This approach is lost on students who don't like logs and are not yet comfortable with integrals. Worse still, it obscures the most important fact about the exponential function—that its derivative is equal to itself. However, using a graphing program or a spreadsheet, we show students that e is the one value of a making the function a^x and its derivative equal.

The ideas of integration and of solving differential equations come into our course early. We find that direction fields provide a way of introducing differential equations which illuminate both the graphical and numerical viewpoints.

Our assignments also reflect the Rule of Three. We regularly include some non-routine problems designed to re-establish the idea that mathematical techniques begin with common sense.

Humboldt State University

Contact:

MARTIN E. FLASHMAN, Department of Mathematics, Humboldt State University, Arcata, CA 95521. PHONE: 707-826-4950. E-MAIL: gmflash@calstate.bitnet.

Institutional Data:

Humboldt State University is a four-year public university, part of the California State University, with approximately 7000 students, and with programs in traditional liberal arts, sciences, business, and natural resources. The Mathematics Department has 18 full-time faculty and approximately 30 mathematics majors in a typical graduating class.

Project Data:

Currently this is a project of a single faculty member teaching one section of 25-30 students out of a total of three or four sections each semester. This has been going on for about 8 years at HSU, and prior to that for two years at Bard College. A computer lab has not been required, and access to computers has varied from some in-class use at times to practically none except for classroom demonstrations. The project has been using Arb-Plot on an Apple IIe for the last few years, but is in most aspects software independent. I have used notes to supplement standard texts and continue to replace materials with drafts from the author's own text. This project does not have any external funding at this time.

Project Description:

The author is working on a thematic revision of the calculus of one-variable for both traditional and business calculus courses. The object is to make sense of calculus by focusing on the themes of differential equations and estimation. A modelling approach assists in distinguishing the various interpretations of the calculus from the calculus itself. Visual and computational elements reinforce the sensibility of the approach. This approach is used in a variety of class formats, and the approach has been kept sufficiently robust to work with most technology—hand-held calculators to Mathematica. Student assignments are designed to involve the students actively in using the concepts as well as the techniques developed in the course.

The author has written drafts of several sections of a book using this approach, and is currently working on the chapter related to Taylor theory which precedes material on infinite sequences and series. Course materials (outline, preliminary chapters, notes, and problem sets) are available from Flashman for review and classroom use at a nominal reproduction and mailing charge.

Iowa State University

Contact:

ELGIN JOHNSTON, Department of Mathematics, Iowa State University, Ames, IA 50011. PHONE: 515-294-7294. E-MAIL: s1mth@isumvs.bitnet.

Institutional Data:

Iowa State University is a large public land-grant institution with 21,000 undergraduates and 4,000 graduate students. Most students live on campus. The Department of Mathematics has 60 full-time faculty members, 70 graduate students, and 150 undergraduate majors.

Project Data:

Our NSF multi-year project "Revitalizing an Engineering/Physical Science Calculus" ($63,000 during the first year) began in January 1989. We taught two experimental sections of calculus during Spring Semester, 1989. During the 1989-90 academic year, eight faculty members have taught experimental sections of calculus. These sections have averaged about 35 students each. During a typical semester of the three-semester sequence classes there are 50 sections with an average of 40 students per section. The project sections are taught with a required computer laboratory. The laboratory, built with NSF and matching institutional funds, contains 44 Macintosh SE/30s, 2 SE/30 servers for two 22 micro networks, 4 ImageWriters, and 2 LaserWriters. There are also 10 Macintosh IIcx machines in faculty offices. Each Macintosh is equipped with a copy of Mathematica. We use Swokowski's *Calculus with Analytic Geometry, Second Alternate Edition*.

Project Description:

Our regular calculus course is a service course (each class meets 4 times per week; 4 semester credits). The majority of the students are in engineering. We use departmental mid-term and final exams based on routine, algorithmic calculus questions. The experimental sections also take these exams and so far they have done better than average. We follow roughly the same syllabus as the non-experimental sections. We do decrease the time spent on all techniques of integration (except integration by parts and partial fractions); our approach to infinite series and sequences is to strongly emphasize power series and de-emphasize all topics not essential to power series; we decrease the amount of hand-graphing of curves and surfaces; and we decrease the number of multiple integrals worked out by hand (except for "setting-up" problems).

These changes have given us some of the time needed by students to solve physically-based, multistep problems using Mathematica as necessary. We assign approximately six of these problems per semester, and expect a "professional" write-up of the results. These projects constitute about 25% of each student's course grade. Students work in teams of two; each team turns in one report; and each member receives the same grade.

We characterize our project as problem-centered. Most of our energy has gone into finding, writing, and refining problems which engage the students. We hope students will talk to each other about these problems and about the mathematics contained in them. (This has, in fact, been happening!) We hope these students will be better problem solvers and expositors in their remaining university courses and subsequent careers.

We have written materials for easing the students into the Macintosh environment. The problems contain most of the necessary Mathematica commands and syntax. Students learn from each other, from the instructor (acting as a consultant), and the undergraduate lab monitor who is always present in the laboratory. The labs are open at least 50 hours per week, including evenings and weekends. Our materials are available at cost.

At this point perhaps one-fourth of the mathematics faculty are somewhat supportive. Most of the rest are more or less neutral or not much interested, and a few are opposed. We have formed a liaison committee with the engineering and physical science departments. They are supportive of the project, but have a wait-and-see attitude. We will have written problem sets for our three-semester calculus sequence by September 1990. These materials are available at cost from Professor Johnston.

Ithaca College

Contact:

STEVE HILBERT, Department of Mathematics and Computer Science, Ithaca College, Ithaca, NY 14850. PHONE: 607-274-3107. E-MAIL: hilbert@ithaca.bitnet.

Institutional Data:

Ithaca College is the largest residential private undergraduate college in New York State with a current enrollment of approximately 6200 students. It is a comprehensive college comprised of a School of Humanities and Sciences surrounded by four professional schools: Business; Communications; Health Sciences and Human Performance; and Music. The Department of Mathematics and Computer Science at Ithaca College houses five major programs: Mathematics, Mathematics-Physics, Mathematics-Economics, Computer Science-Mathematics, and Computer Science. An additional program culminating in provisional New York State teacher certification may be added to each of the first four majors listed above. Over the last five years approximately two-thirds of our mathematics graduates—approximately 15 per year—have been women.

Project Data:

The courses we are developing are our mainstream calculus courses taken by our own majors, science students, and other students who want to take more than one semester of calculus. In both courses the instructor meets the class four times a week. Our project involves five senior faculty members, four of whom have taught (or are currently teaching) Calculus I or II using our new curriculum. The first three sections of Calculus I and the first section of Calculus II were experimental with 12-15 students. Subsequent sections have had enrollments around 25, which is typical for all of our courses. Our new Calculus I is in its third semester, Calculus II is in its second. There is no computer laboratory in either course, but instructors typically make some use of computer demonstrations, and from time to time students may use the computer in their projects. Software and hardware have varied from instructor to instructor: True Basic (IBM PC), Analyzer* (Macintosh), and Mathematica (SUN SparcStation) have all been utilized. We continue to use the text that is used in all calculus courses—Purcell and Varberg's *Calculus with Analytic Geometry, Fifth Edition.*

During 1988-89 we were funded through an NSF Planning Grant. At that time we interviewed groups of faculty from a variety of disciplines on our campus. On the basis of these interviews we identified problems and began to develop classroom materials. The projects deal with the central ideas of calculus, but many of them are concerned with topics that are not often covered in a calculus course. Some traditional material covered included optimizing profit, exponential growth, relationship between distance and velocity, and infinite series. The non-traditional material included projects on the design of a navigation buoy, logistic growth, difference equations, and the design of a security system.

Project Description:

We use large open-ended problems to drive the curriculum; students work on these problems in groups of about three, spending 2-3 weeks on each project. In some cases projects require the students to gather, compute, and analyze data, and some projects involve using functions presented as graphs or tables of data as opposed to algebraic formulas. Our courses emphasize graphs in the first semester (a need expressed repeatedly by the colleagues we interviewed), and gradually introduce an additional focus on numerical methods during the second semester.

Our broad goals in this curriculum are increased student understanding of concepts and the unity of topics in calculus, deeper geometric understanding, increased independence of thought, improved ability to solve problems, and in the long run lower attrition in the calculus sequence and in the subsequent study of mathematics. Preliminary evidence, based on three experimental sections of Calculus I and one of Calculus II already completed, indicates that the courses are meeting the immediate goals very well. Students in the projects-based sections performed significantly better than students in the traditional course on a common final examination; indeed, the average in the projects-based sections was 10% higher. Furthermore, the proportion of students successfully completing the projects-based sections was significantly higher than for our traditional sections. Finally, the students' independence increased dramatically throughout the term; as the semester went on the students worked on the projects almost completely on their own.

Some of our projects and exams are available from the author at no cost.

Kean College of New Jersey

Contact:

DOROTHY GOLDBERG and CARLON KRANTZ, Department of Mathematics and Computer Science, Kean College of New Jersey, Union, NJ 07083. PHONE: 201-527-2105; 201-527-2494.

Institutional Data:

Kean College is a public institution located in Northern New Jersey within the New York metropolitan area. Enrollment of 12,000 is divided almost equally between full-time and part-time students. Kean is committed to the recruitment and retention of minorities who currently represent approximately 24% of the full-time students at Kean. The majority of students commute to school.

There are 23 full-time faculty in the Department who teach either mathematics or computer science courses, or both. There are eight mathematics majors and 100 computer science majors in a typical graduating class.

Project Data:

The project began in Spring 1989 with a pilot study in which two faculty members each taught one Calculus I class integrating the computer. Enrollment in computer-based classes was limited to 24 students as opposed to 30 in conventional classes.

In Fall 1989 the use of the computer was mandated for all Calculus I classes. The eight classes scheduled were taught by eight different instructors. A computer laboratory-classroom serves as the locale for class meetings and at all other times is an open lab staffed by undergraduate mathematics majors who serve as lab assistants and tutors. The instructors use the computer for teaching in an interactive mode. Students must complete a series of 9 computer lab assignments which are graded.

The principal software now consists of True Basic Calculus and Microcalc; Derive is to be incorporated next. The hardware configuration consists of 15 IBM PS/2 Model 50Z microcomputers equipped with math coprocessors and color displays, networked via a Model 80 file server. A Sharp QA-50 LCD overhead display panel facilitates instructional use of the computer.

Following a Fall 1989 pilot study, all three Calculus II classes offered in Spring 1990 were taught using the computer. The project was originally funded by the Department of Higher Education of the State of New Jersey.

Project Description:

The major objective of the project was to develop a computer-based sequence of 4 calculus courses in which the computer is used to teach basic concepts and as a tool for solving calculus problems. It was our intention to provide a learning environment in which students, actively engaged in using the computer, would develop a greater understanding of calculus and would be motivated to study more mathematics. Course content is essentially what it was in the traditional treatment.

To realize our objectives, a laboratory-classroom was established. At this time all Calculus I and II classes meet here twice a week for 75 minutes. The instructor is free to use an overhead display panel; the students may work in pairs at their microcomputers. At other times the lab is staffed by tutors; calculus students may freely use the facility which is open approximately 28 hours per week.

Maximum class size is 24. At this time the majority of Calculus I instructors give 9 common lab assignments, which are graded. Calculus II instructors give 6 such assignments. Later on, as the instructors become more experienced in using the software available, they may develop their own lab assignments. Calculus III and IV have been targeted for pilot studies in the 1990-91 academic year.

Both the Department and the Chair fully support the program. Most classes are taught by resident faculty, a few by adjuncts. To make certain that a large pool of faculty is capable of teaching in the program, a 4-day workshop was given Spring 1989; a 2-day workshop is scheduled for 1990. Both resident and adjunct faculty members participate and receive stipends.

Copies of lab assignments are available from the Department.

Knox College

Contact:

DENNIS M. SCHNEIDER, Department of Mathematics and Computer Science, Knox College, Galesburg, IL 61401. PHONE: 309-343-0112, x 420.

Institutional Data:

Knox College is an independent, highly selective, four-year liberal arts college enrolling approximately 1000 men and women from 42 states and 31 foreign countries. The College offers the Bachelor of Arts degree in 28 academic majors, as well as three interdisciplinary concentrations. The Mathematics Department has seven faculty (one of whom holds a half-time appointment as Director of the Computer Center), and has an average of 15 majors in a typical graduating class.

Project Data:

During the academic year 1989-90, the course was offered twice at Knox (with an average of 20 students), and was taught by two members of the department. Computer laboratory work is required. The computer laboratory is open 16 hours a day during the week, and about 12 hours a day on the weekends. The courseware is being developed using Mathematica on the Macintosh (SE/30's and Mac II's). Our current textbook for both single and multivariable calculus is *Calculus* by Finney and Thomas. We received external funding from NSF and Pew Charitable Trusts.

Project Description:

The Department is currently engaged in a project to exploit thoroughly the graphical, numerical, and symbolic computing power offered by new technology in the teaching of calculus. Our initial efforts resulted in a set of Mathematica Notebooks for multivariable calculus. The Notebooks are designed for use in a standard multivariable calculus course. The advantage of this approach is that it allows us to move into this new technological arena at a gradual pace. Existing textbooks can be used and our Mathematica Notebooks can supplement or replace sections of the textbook. Instructors at other institutions can introduce these Notebooks into their courses at a pace they find comfortable, rather than being faced with mounting a totally new "electronic" course.

Our multivariable calculus class meets four times per week for ten weeks. Classroom meetings are multi-media events, involving both the computer and the blackboard. Instructors lead students in discussions of (rather than lectures about) mathematical ideas thereby providing them with a model of how to actively explore a mathematical idea. The laboratory provides the students with a powerful environment in which to continue the explorations begun in class and work on problems that require further exploration (as opposed to working in small groups filling in a lab manual). TA's serve as laboratory assistants and graders of the computer homework. The next stage of our project will be to develop Notebooks for single-variable calculus. The final goal of the project is to produce a textbook that is thoroughly integrated with these Notebooks. Since we do not believe that technology should totally replace pencil and paper, our textbook will incorporate a good dose of pencil-and-paper assignments. There are currently eight Notebooks for multivariable calculus near completion. Instructors interested in class testing these Notebooks should contact Dennis Schneider.

La Trobe University (Australia)

Contact:

ROBERT S. SMITH, Department of Mathematics and Statistics, Miami University, Oxford, OH 45056. PHONE: 513-529-3556. E-MAIL: rssmith-@miavx1.bitnet.

From January to June 1989 I was a Visiting Fellow at La Trobe University in Bundoora, Victoria, Australia, where I studied and assisted with a course in numerical methods taught through applications and using a spreadsheet program.

Institutional Data:

La Trobe is one of the three major universities in the state of Victoria. It is a public institution of about 13,000 students offering degrees at all levels. There are no professional schools and there is no engineering. As is the case in all Australian universities, the undergraduate degree program is a three-year program and there is no liberal education at the university level.

Of the 25 full-time members of the Mathematics Department, about 45% are in applied mathematics and about 55% are in pure mathematics. The Department instructs about 1100 first-year students, 200 second-year students, and 90 third-year students, and it graduates 70-80 majors per year.

Project Data:

The numerical methods course was in its second offering when I was at La Trobe. Each offering was given by Peter Forrester, a theoretical physicist turned mathematics lecturer. Since the course neither replaced another nor was like any other, there was no control for a comparative evaluation. However, the students at La Trobe have reacted quite favorably to the course, and the faculty has been very enthusiastic as well. Based on these subjective responses, students and faculty alike have hailed

the course as a success and it is now a part of the curriculum.

There is no textbook for the course. The projects, like most of the course materials in mathematics at La Trobe, are authored by the faculty, typed in TEX, printed on a laser printer, and offset in the departmental printery.

A weekly computer lab is required. The computer lab is equipped with 16 IBM PCjx's (a version of the IBM PCjr) on a local area network served by an IBM PC. The software used by the students and by Forrester during the lectures is VP-Planner, a Lotus clone.

Project Description:

The numerical methods course at La Trobe is taught to second-year mathematics students who are studying both pure and applied mathematics. Unlike most American mathematics courses which draw their content from one area, the numerical methods course at La Trobe contains topics from calculus, linear algebra, numerical analysis, difference equations, and differential equations.

The project-oriented course is offered over a 13-week term. In each 2-week period there are three 50-minute lectures delivered to the class of 45 students. Once a week, Tutors (the equivalent American rank is Instructor) conduct computer labs for a period of 110 minutes with about 15 students per lab.

In the first week, students learn the basics of VP-Planner, practice using the spreadsheet with simple mathematical algorithms, and complete an ungraded project. In the remaining 12 weeks, students learn numerical methods and complete 6 two-week graded projects, each based on applications drawn from chemistry, mathematics, medicine, and physics. The appropriate numerical methods from applied mathematics are taught and used to complete the projects. Grades are based on the student's performance on the projects and a comprehensive final examination.

Students are required to analyze the projects and lay out the spreadsheets for the projects before coming to the computer lab. All computer work for the projects must be completed during the supervised weekly labs. Between the weekly labs the Tutors keep the students' computer disks. The Lecturer and the Tutors all hold office hours to assist students and share equally in the assessment of projects.

The electronic spreadsheet is ideal for implementing algorithms which rely upon iterative procedures. Analyses which can be adapted to a tabular format are particularly well-suited to the spreadsheet since it can yield instant updates in response to parameter changes. Thus the spreadsheet is an excellent tool for teaching the numerical methods of applied mathematics—methods which rely heavily upon iteration, approximation, and estimate variations based on parameter changes. But perhaps the best pedagogical reason for teaching mathematics on a spreadsheet is that it is easy and fun for both students and instructor.

The concept of a numerical methods course taught with applications using spreadsheets has received such a positive response here at Miami that we intend to construct a course based on La Trobe's and offer it to our second-year mathematics students as a calculus capstone.

Sample course materials and further information about the La Trobe University course in numerical methods can be obtained from the author at Miami University.

Meridian Community College

Contact:

WANDA DIXON, Mathematics and Science Division, Meridian Community College, Meridian, MS 39307. PHONE: 601-483-8241.

Institutional Data:

Meridian Community College is located in Meridian, Mississippi. It is a public two-year school with a student body of around 3000 students (60% full-time, 25% Black, and 66% women). There are six full-time faculty members in the mathematics department.

Project Data:

In August 1989 MCC received a planning grant from the National Science Foundation to plan a new one-semester business calculus course. We plan to submit a full proposal to NSF next winter to actually develop the materials needed to teach this new course. As we see it now, this course should take the following directions.

It will take advantage of technology without being hardware or software specific. We will assume that the student has access to graphing calculators or computers; thus, the thrust will be interpretation of graphs, rather than drawing them. We will give pictures and ask for interpretations rather than formulas and ask for graphs.

Business students need the concepts of functionally related variables in numerical (data presented), graphical, and analytic forms. They need concepts of limiting values and processes, including instantaneous rates and steady state levels. They need a thorough grounding in exponential growth and decay, and some knowledge of logarithmic change. In addition they need an understanding of the modelling and representation processes involving numerical, graphical, and analytic models of economic and business behavior, with clear concepts of the potential for and limitations of the analytic (formula) models.

In the traditional mathematics courses, which students take prior to business calculus, there is little emphasis on the reasonableness of answers, on knowing whether the number on the print-out or display screen makes sense for the problem being solved. While school mathematics will change toward correcting this serious lack, there is currently a need for all business calculus courses to help students develop the number sense necessary to avoid wrong or absurd answers. We propose a continuing strand in our business calculus to address this issue.

The business calculus student has had at least two years of high school algebra and usually either college algebra or placement out of it. The extra algebraic practice or techniques this student gets in a traditional business calculus course are of very little long-term value. In the world of work, she is a passive, not active, user or creator of mathematics; thus she must understand how to think and interpret numerically and graphically with only occasional use of analytic data. However, the mathematical concepts of calculus, maxima and minima, rates of change, and even increasing and decreasing rates of change are important as the framework of ideas involving numerical or graphical integration. We are committed to significantly de-emphasizing the algebraic (formula) aspects of calculus.

Montgomery College

Contact:

ELIZABETH TELES and MARY KAY ABBEY, Department of Mathematics, Montgomery College, Takoma Park Campus, Takoma Park, MD 20912. PHONE: 301-650-1439; 301-650-1435. E-MAIL: ci89999@montcolc.bitnet.

Institutional Data:

Montgomery College is a multi-campus, urban, public, open-admission, two-year college. The Takoma Park Campus has approximately 4500 students about 30% of whom take mathematics each semester. There are many minority students as well as students from 86 different countries. The students are evenly divided between transfer programs and career curricula. The Mathematics Department has eight full-time faculty, one staff person in charge of the Math Tutoring Center, and 20 part-time instructors. Each semester there are 4 sections of Calculus I, 2 or 3 sections of Calculus II, and 1 section of Multivariable Calculus. Normal class size is 20-30 students. Full-time faculty teach most of the calculus sections.

Project Data:

Two faculty members are in charge of our project. The project materials were tested on one section of Calculus I in the Fall of 1989 before being used in the entire Calculus sequence in the Spring of 1990. Of the 147 students enrolled in Calculus this Spring, 47 participated directly in this project.

A non-computer science teaching room houses 20 IBM-PC's equipped with co-processors and VGA graphics, 10 printers, and a Mitsubishi large screen monitor. Another large screen monitor is available for demonstration purposes in a regular classroom.

Calculus I uses Microcalc. The Calculus II students use both Microcalc and Derive while the Multivariable Calculus students use Microcalc and Mathematica. Mathematica is principally utilized for relative extrema and Lagrange multipliers. There are plans to incorporate Acrospin into the Multivariable Calculus syllabus for the rotation of surfaces.

Calculus with Analytic Geometry, Alternate Edition by Swokowski is used by all calculus sections. All laboratory materials were written by our participating faculty.

While no external money was used for this project, the College has lent support in the form of the computer teaching room and staff, software, and funds for the faculty to attend workshops and seminars dealing with the implementation of the computer.

Project Description:

Hands-on opportunities provided by the computer make calculus a conceptual course rather than a formula-driven one. Specifically, the computer laboratory exercises were written to help the students develop a thorough understanding of calculus and the uses of its underlying ideas and to encourage students to explore the concepts of calculus by changing hypotheses, formalizing conjectures, and then verbalizing the results.

The calculus classes meet for five 50-minute periods per week during a 15-week semester. The lab exercises constitute 10% of each student's final grade. The Calculus I students are in the lab for one 50-minute period each week. Students of Calculus II and Multivariable Calculus are not on a fixed schedule, but they do spend a minimum of 10 hours with computer instruction.

The lab assignments consist of very specific exercises with detailed steps chosen to demonstrate effects of changes, e.g., in domain, number of iterations, etc. Additional problems are given in which the students need to determine the steps necessary for examining the principles of calculus in this setting. They are free to design their own problems or choose problems from the text. Finally the students submit a lab report in which they explain their results.

Student retention has improved. Historically, passing rates have been between 40%-60%. Since the project began, retention rates for Calculus I, II, and III have been 75%, 85%, and 88%, respectively.

The following positive indications from this preliminary effort encourage us to develop a true experimental study. Non-computer-oriented faculty are asking to participate in the use of the materials. As an outgrowth of the lecture-lab format, students have formed study groups. Student misconceptions are revealed more readily in the lab setting. Student-generated examples lead to lively discussions. Student surveys and faculty observations reveal that students are more genuinely interested in class. Assignments are completed more thoroughly and neatly. Students do not pack up early. Absence and tardiness have almost disappeared.

Anyone interested in participating in this study or in using our materials, please contact the authors.

Morehouse College

Contact:

HENRY GORE, Department of Mathematics, Morehouse College, Atlanta, GA 30314.

Institutional Data:

Morehouse College is an independent, fully accredited, predominantly Black liberal arts college for men. Morehouse College has an enrollment of 2600 students. More than 96% of the students are full-time, and about half are housed in college residence halls. The curriculum of Morehouse includes not only traditional liberal arts majors in the humanities, the social sciences, mathematics, and the natural sciences, but also concentrations in business administration, computer science, teacher education, and engineering as well.

The Department of Mathematics has a faculty of thirteen full-time members and six part-time instructors. The Department has an average of sixty majors per year.

Project Data:

The project "Effective Strategies for Teaching Calculus at the College Level" is sponsored by the Exxon Education Foundation. The purpose of this project is to collect accurate and comprehensive information about the effective strategies being used today in the teaching of calculus at the college level. This information is being collected by means of three activities: questionnaires to be completed by mathematics departments across the country; a two-day conference on the campus of Morehouse College; and on-site visitations to observe calculus classes in session on the campuses of all undergraduate institutions in the Atlanta University Center, and also on the campuses of other participating colleges and universities in the Atlanta area.

This project is designed to aid in the improvement of mathematics instructions at the college level by disseminating the collected information to colleges, universities, and professional organizations throughout the country. The *Conference Proceedings* are now available upon request from Professor Henry Gore.

Nazareth College of Rochester

Contact:

NELSON G. RICH, Department of Mathematics and Computer Science, Nazareth College of Rochester, 4245 East Avenue, Rochester, NY 14610. PHONE: 716-586-2525. FAX: 716-586-2452.

Institutional Data:

Nazareth College is an independent, coeducational, liberal arts college located in a suburb of Rochester. The college's current full-time undergraduate enrollment is just over 1400 students, with nearly as many part-time students enrolled in both undergraduate or graduate study.

Nazareth, which began as a college for women, has been coeducational since 1973, and approximately 75% of the current full-time students are female. Approximately two-thirds of the majors in mathematics are female, and about 80% of all mathematics majors enter the teaching profession.

The Department of Mathematics and Computer Science has 5 full-time and 4 part-time faculty, and has graduated approximately 10 mathematics majors and 5 computer and information science majors per year for the past 5 years.

Project Data:

We began in Fall 1989 by piloting Calculus I, taught by one of the PI's. The initial enrollment was 40. During the current semester (Spring 1990) the principal investigators are teaching Calculus I, fully integrating a CAS into the course, while piloting Calculus II. Initial enrollments were 16 in each course. By the Spring 1991 semester, Calculus I, II, and III will be project sections taught by full-time members of the Mathematics faculty.

We have elected to take a conservative approach to integrating a Computer Algebra System (CAS) into the calculus curriculum by teaching the "standard" topics in classical calculus, but using classroom demonstrations and homework assignments in Maple to augment the traditional course. We do not yet have a formal laboratory segment to the course, but instead use classroom demonstrations and follow these with homework assignments for which students must submit hard copy listings of their solutions. Our current textbook is *Calculus with Analytic Geometry, Fifth Edition* by Purcell and Varberg.

We have been using Maple running on a Digital VAX-11/750 to produce demonstrations and problem sets. A new Sun 4/390 Sparc server multi-user system is being installed running Maple Version 4.3. Students typically access Maple through IBM PS/2's emulating VT 241 color graphics terminals in order to take advantage of high resolution graphics capability and Maple's plot procedure.

Our work has been generously supported by Nazareth College and a National Science Foundation grant.

Project Description:

The major features of the calculus project at Nazareth are the use of a CAS (Maple), journal writing, and teaching teachers. It is our hope that with the use of a tool like Maple, we can significantly reduce the computational burden of students, help them to discover patterns that often lead to deep understanding, and in general pursue the deeper conceptual issues of the course. At the present time we have developed approximately six demonstrations illustrating one or more features of Maple and their use in Calculus I, and corresponding problem sets for students. We have developed approximately four such demonstrations, and problem sets for Calculus II. We hope to have developed at least 6-7 such demonstration/problem set pairs for use in three semesters of calculus.

The writing component of our project is an outgrowth of the recent major revision in the college's core curriculum which combines specialization in a major with a multi-disciplinary approach to learning. We require students to maintain a journal throughout Calculus I which significantly contributed to a student's final grade. The journal is composed of the following four main sections: daily entries, topic summaries, important definitions and theorems, and historical perspectives.

Because of the college's long tradition of offering area high schools in-service workshops for teachers and programs to encourage their students to pursue college-level studies in mathematics, we believe we are in a particularly strong position to bring this approach to the teaching and learning of mathematics to this often forgotten audience. We plan to provide an in-service experience or a summer institute on our campus for area high school teachers of calculus.

We have made available the following materials: project summary sheet, Calculus I course information sheet, Maple demonstrations for Calculus, and problem sets for calculus using Maple.

As long as grant funds for dissemination remain available, we are happy to send these materials to any interested parties.

Northeastern University

Contact:

RICHARD PORTER, Department of Mathematics, Northeastern University, Boston, MA 02115.
PHONE: 617-437-5641.   E-MAIL: nucalculus@northeastern.edu.

Institutional Data:

The Mathematics Department has 54 full-time faculty, 35 teaching assistants, and 100 part-time teachers. There are 25-30 undergraduate majors in a typical graduating class. Among private institutions in the country, Northeastern ranks first in total degrees in electrical engineering, first in Black engineering students enrolled (exclusive of predominantly Black institutions), and third in women enrolled in engineering.

Project Data:

Class testing of a new calculus curriculum for all entering students began in the Fall of 1989. 24 teachers including faculty, teaching assistants, and part-time teachers have used the calculus materials written by members of the Department Undergraduate Curriculum Committee. This material is being tested in all first-year calculus courses for majors in mathematics, chemistry, physics, and engineering. The total enrollment in Fall and Winter quarters of these courses is 1032. Graphics assignments are supported by the software Curves and Diffs, developed by Mark Bridger of Northeastern. In addition to the handout material and graphics software, students use Anton's *Calculus, Third Edition* text.

Project Description:

The goal is for students to understand basic concepts and their applications. The curriculum is designed to engage students actively in solving problems that require them to think. Differential equations are introduced early in the first quarter through the geometry of flow fields and graphical solutions. In the first two weeks students solve problems using exponentials, logarithms, and trigonometric functions. The problems on these topics increase in depth as the course progresses. Students explore the mathematics of the catenary using a series of small problems for class discussion and student reports. The average grade on the common final exam in Fall 1989 was 10% higher than the previous Fall. Further changes will be made.

The project started in the Fall of 1988 when the Department Chair Margaret Cozzens reconstituted the Undergraduate Curriculum Committee with people who had previously been working independently on curriculum reform. The strong support of the Department Chair and other members of the Department, combined with the willing co-operation of the administration and faculty in the College of Engineering, have been and continue to be essential.

There are three class meetings per week of 65 minutes each. Typical class size is 30-35 students. There are no recitation sections. Graphics software for the course is installed in five computer labs on campus. The teachers guide to the catenary assignments, together with a Diffs introduction with exercises, are available at cost plus epsilon. Contact us for an up-to-date price on the catenary and Diffs material, and on the availability of a calculus manuscript based on the goals of the project. The software Diffs and Curves is inexpensive and available for IBM's and compatibles from Bridge Software.

North. Virginia Comm. College

Contact:

JONATHAN WILKIN, Department of Mathematics, Northern Virginia Community College, Alexandria Campus, 3001 North Beauregard Street, Alexandria, VA 22311. PHONE: 703-845-6341.

Institutional Data:

Northern Virginia Community College is a public two-year college with five campuses. At the Alexandria Campus there are approximately 10,000 full-time equivalent students, with student head count being about 16,000. Over half the student population is part-time, with perhaps 50% of our course offerings beginning after 4:30 p.m. The Mathematics Department at the Alexandria Campus has 11 full-time faculty; there are virtually no mathematics majors at the College. Almost all of our calculus students aspire to an engineering or computer science major.

Project Data:

This calculus project was a "seat-of-the-pants" project I conducted during the Spring Semester 1990. It involved only me, though one of my colleagues ran her calculus section using computers virtually as much as I. I began the term with two sections having a combined enrollment of about 45. The day section lost half its enrollment the first week, while the night section maintained its enrollment throughout the term. I required work in a computer laboratory that counted 40% of the grade.

The laboratory, having both Microcalc and Derive on 286 based computers with CGA graphics and NEC 24 pin printers, was also available to the students for the mid-term and half of the final. Our current textbook is Grossman's *Calculus, Fourth Edition*. There was no external funding.

Project Description:

Our Calculus I course is a rather standard five-credit course. The sections involved in this project were simply those assigned to me. They were not advertised as different, and comparisons are difficult as we do not have common finals. That's a good thing, too, since using Microcalc and Derive from day one of the course changes the emphasis given to various topics as compared with a traditionally taught lecture section. I think that my students would have performed very poorly on a traditional Calculus I final.

A primary change in emphasis for me was the treatment of graphing. Rather than spending weeks on techniques of graphing, we began with graphing functions on the computer and using these graphs to illustrate and supplement the calculus notions. The spreadsheet-like "table of values" in Microcalc allowed approximations of derivatives using the definition of the derivative; the "calculus-sum" routine in Derive allowed calculation of Riemann sums of some complexity.

The laboratory exercises consisted of five rather lengthy assignments. The first assignment was to acquaint the students with Microcalc and the use of graphing to see relationships. Later assignments were designed to have the students use Microcalc to see relationships, to combine them as necessary, and occasionally to recognize where the computer processes simply yielded incorrect results. A couple of questions succeeded in those regards almost too well.

On all computer assignments and the computer-allowed portion of the final, student interaction and discussion was allowed and encouraged. This was both the best and the worst part of the project. It was great in that students shared ideas, successes, and failures. It was the worst in that some students did virtually nothing other than use another's printout. I did require written reports, but often even they were much too similar.

In the fall, when I plan both to continue into the second semester and begin again with another first semester section, I hope to have similar but not identical problems for each student in the section.

Also, I feel that shorter, more frequent computer assignments are needed, as well as weekly in-class, non-computer-assisted quizzes to maintain what I feel is an appropriate balance.

Oklahoma State University

Contact:

BENNY EVANS and JERRY JOHNSON, Department of Mathematics, Oklahoma State University, Stillwater, OK 74078. PHONE: 405-744-5688. E-MAIL: johnson%nemo.math.okstate.edu@relay.cs.-net.

Institutional Data:

Oklahoma State University is a comprehensive land grant institution serving 20,000 students on its main campus. The university offers in excess of 60 undergraduate degree programs and numerous graduate programs at the masters and doctoral levels.

The Department of Mathematics has 30 continuing faculty members and offers BS, BA, MS, Ed.D., and Ph.D. degrees. We have approximately 75 undergraduate majors and 50 graduate students, most of whom are teaching assistants who cover the bulk of our precalculus courses.

Project Description:

Working under the National Science Foundation grant during the Spring and Summer of 1989, we produced a manual *Microcomputer Resource Manual for College Mathematics Teachers*. This manual contains, among other things, laboratory projects for the whole range of undergraduate mathematics including about 20 calculus units. (Many of these are adaptable for classroom demonstrations as well.) Each one consists of an assignment sheet for the student and an instructor's sheet that explains the salient points of the problem and contains a complete solution. It will be available for distribution to interested parties in the late Spring of 1990 for the cost of postage only.

Many of the exercises in the manual are designed to be done with a computer algebra system, but we tried to make them generic so that they are not tied to any particular software. There are two principal pieces of commercial software that we use: the popular computer algebra system Derive and the new animated 3D graphing program GyroGraphics.

Using these materials as a supplement, Johnson taught Honors Calculus I in the Fall 1989, and Evans is currently teaching Honors Calculus II (Spring, 1990) both with 25 students—nearly all freshmen (class size for regular Calculus is about 35-40 students). These two courses are being taught as part of field-testing activities for the Resource Manual. Although we receive no release time for this, it is a part of our NSF grant proposal. Both courses are taught from the text *Calculus* by Stewart. They meet five 50-minute periods per week in a traditional lecture format, with two major exceptions: first, several outside laboratory assignments from the Resource Manual are required, and second, the class meets one day each week in the computer laboratory for demonstrations or in-class work on the micros.

Because only 11 of the 51 machines in our lab are IBM compatible, two or three students must team up on a computer, but we have actually found this to be helpful. In Johnson's Calculus I class, the students were often assigned computer problems to be done in class as teams, and all members of the team received the same grade on the assignment. This had the effect of involving the weaker students in the effort and stimulating discussion of the exercise among the team members.

No scientific comparison of performance was undertaken, but Johnson particularly noticed that his students' understanding of Riemann sums was better than in previous classes. Many of his students, in both written and oral evaluations, stated that they believed their understanding of concepts was helped by the computer work. Some even said that they felt their performance on exams was better because of it.

Oregon State University

Contact:

THOMAS DICK, Department of Mathematics, Oregon State University, Corvallis, OR 97331.

Institutional Data:

Oregon State University is Oregon's land and sea grant university, with a student population of approximately 15,000 undergraduate and graduate students representing 90 countries and every state. The Mathematics Department has 37 faculty members, 50 full-time graduate students, and approximately 34 mathematics majors graduate each year.

Project Data:

Charles Patton and Thomas Dick are the materials authors. Two faculty members and two graduate students have taught in the experimental classes (approximately 30 students each) at Oregon State University, and experimental classes have been taught each term (quarter system) since Fall 1988. Total enrollment in the traditional sections ranges from 200-700 depending on the term. The pilot classes to date have utilized the HP-28S calculators on a daily basis, with each student in the experimental classes having constant access to the machines for classwork, homework, and testing.

In the early stages of the project, we have used the same text as the traditional sections (Edward and Penney). PWS-Kent Publishers has contracted to publish the materials in a 2-volume soft cover preliminary edition. *Volume I* (covering the first year of calculus) is scheduled to be available by early 1991, and *Volume II* (covering the third semester of calculus material) is scheduled to be ready by early 1992. PWS-Kent has also provided funds for materials development and for in-service workshops on the use of the materials. Hewlett-Packard has provided support for the project with equipment grants. The National Science Foundation has funded the project with both a planning grant (1988-89) and a multi-year grant (1989-93).

Project Description:

This project has as its goal the development and implementation of a calculus curriculum which would make integral use of symbolic, graphical, and numerical tools made widely accessible by computer software and calculators. The major activities of the project are curriculum materials development; class-testing of these materials in a wide variety of settings, representing all sizes and types of institutions with calculus instructional responsibilities, including high schools, two-year and four-year colleges, and research universities; and a provision of instructional workshops and continuing instructional support for those teachers using the curriculum materials and technology.

The curriculum materials will include

- A calculus text covering material corresponding to a three-semester sequence in introductory calculus, including multivariate and vector calculus. The text is being written under the assumption that each student will have access to the numerical, symbolic, and graphical tools provided by either a calculator (such as the HP-28S) or a personal computer with appropriate software. However, the text itself will be "technology-neutral" in the sense that no reference to a particular machine or software is made.
- Student laboratory supplements containing basic reference material and enhancements to the use of a particular technology, and a collection of laboratory activities.
- An instructor's manual providing hints and examples for effective classroom demonstrations of calculus concepts using the available technology, as well as commentary on the laboratory activities.

Initial class-testing of materials at Oregon State University has been with the HP-28S calculator, and the laboratory supplement is being prepared to address the use of the HP-28S and the new HP-48SX. In later stages of the project, additional supplements are planned for the use of appropriate software packages such as Maple or Derive. Assessment procedures for the project include comparisons of student scores on common exams and standardized test measures.

Penn State Consortium

Contact:

MARY McCAMMON, Department of Mathematics, Penn State University, University Park, PA 16802. PHONE: 814-865-7527.

Institutional Data:

This Consortium includes Penn State University, Indiana University of Pennsylvania, Duquesne University, Slippery Rock University, and State College Area High School.

All of the universities have graduate programs. Duquesne is a private university, while the rest are public. Student populations range from about 6500 at Duquesne to about 35,000 at Penn State's University Park Campus. All have a mix of residential and commuting students. None of the institutions, including the high school, has a large minority student population. Penn State also has 17 regional two-year campuses, with student populations ranging from about 500-3000. Four of these are involved in the project.

Penn State University has 50 FTE mathematics faculty at the University Park Campus and 100 on other campuses; IUP has 35 FTE faculty and 40 majors annually; Duquesne has 12 FTE faculty with 10 majors; and Slippery Rock has 11 FTE faculty with 15 majors annually.

Project Data:

Each university offers a freshman calculus sequence of two 4-credit courses for majors in the sciences. At Penn State and Slippery Rock, this sequence is also populated by engineering students. At IUP there is a separate sequence for mathematics majors.

Each university also offers business calculus. At IUP, this is a sequence of two 4-credit courses. At Penn State it is a 4-credit course followed by a 2-credit course. At Duquesne and Slippery Rock it is a single 3-credit course.

Calculus is taught by full-time faculty in small sections of approximately 35 students at all institutions, with the exception of Penn State's University Park Campus. Calculus is taught there in lectures of 300-400 students, with recitations run by TA's.

The project involves 15 mathematics faculty distributed as follows: one at Penn State (University Park Campus), eight at Penn State's regional campuses, three at Slippery Rock University, and one each at Duquesne, IUP, and State College Area High School.

The project is focusing on the calculus sequence for science majors at each university. To this date it has not involved the large lecture sections at University Park, nor the calculus classes at the high school. Each of these is scheduled for involvement by the 1991-92 academic year. The project currently affects about 1000 students per year.

The unifying feature of the project across all schools is a core curriculum, based on a set of goals and objectives defined by participating faculty. The curriculum is compatible with several different technological configurations which we have used in our classes. When fully implemented, it will assume the availability of a graphing utility at all times, and of a computer algebra system in a separate lab. It will also feature an extensive writing component.

Project Description:

A team including mathematicians from four Pennsylvania universities and one high school have developed, and are currently class-testing, a flexible core curriculum for a freshman science and engineering calculus sequence. This curriculum, based on a well-defined set of goals and objectives, emphasizes problem solving and communication skills, and

is compatible with a variety of technological configurations which we have used in our classes. Teaching materials currently being developed will gradually evolve into a textbook with accompanying software. We will arrange to have preliminary drafts of the book field-tested in several institutions, including our own.

By concurrently developing a training program for both graduate and undergraduate assistants, we intend to demonstrate that our curriculum can be taught in both large and small classes. To promote an adequate level of preparation among students we will initiate a program of cooperative activities between high schools and universities, including the design and implementation of both courses and placement tests. These activities will build on a pilot program which we began two years ago. Both our training program for teaching assistants and our network with high schools will be transportable to other institutions across the country.

Polytechnic School

Contact:

RICHARD SISLEY, Polytechnic School, 1030 East California Boulevard, Pasadena, CA 91106. PHONE: 818-792-2147.

Institutional Data:

Polytechnic School is an independent pre-K through college preparatory school. The school enrolls 320 students in grades 9-12. Admission is selective. The mean SAT mathematics score of graduating classes is between 620 and 670. The Mathematics Department employs 5 teachers.

Project Data:

Two members of the mathematics faculty have taught one or both of the two courses in the project. 20-30 juniors in a class of 80 take the first project course. As seniors most of the same 20-30 take the second project course.

The two courses of the project were first taught to a section of the graduating class of 1988. The project courses have been revised and taught to new sections since September 1988. Classroom computer demonstrations are frequent. Students in the first project course do computer lab work on data analysis.

Software used in the first course includes "Hands on Statistics," a simple home-made program to calculate trapezoid estimates; "Master Grapher" from

Ohio State; "Surfs" from Bridge Software; "Analysis," a data analysis package from the North Carolina School of Science and Mathematics; and "The Mathematics Exploration Toolkit" by IBM.

Software used in the second course includes "The Mathematics Exploration Toolkit" for showing cobweb diagrams for difference equations; home-made software animating convergence for dynamic systems modelled by systems of linear difference equations; "The Omnifarious Plotter" for graphing in polar coordinates; and Derive for doing complicated symbol manipulation.

The calculus text for both courses is *Calculus* by Best and Penner. In the first course the students use teacher-prepared handouts on genetics and data analysis. In the second course we use *Discrete Mathematics with Finite Difference Equations* by Sandefur. Development work was funded by Polytechnic School.

Project Description:

The two-year-long courses of our project are intended to fuse the standard calculus of the BC Level Advanced Placement Syllabus with the study of contemporary applications whose mathematical treatment draws both on calculus and the computational capabilities of computers. The first of the courses, Mathematical Modelling I with Calculus AB, opens with a unit on the mathematics of genetics and leads to using areas of regions under normal curves to analyze data from Mendel's experiments. Calculus then begins with an introduction to the integral concept. After the development of elementary calculus has reached the study of the derivative, the techniques learned are applied to finding the slope and y-intercept of least squares best fit lines. Partial differentiation is introduced in seeking an alternative method. Nonlinear curve fitting is accomplished by fitting least square lines to appropriately mapped data sets and then converting linear models into nonlinear models by reversing the mapping. After the AP exam in early May, correlation analysis is added to the tools used in doing least squares line modelling.

The second course of the Sequence, Mathematical Modelling II with Calculus BC, integrates the study of finite difference equations with standard calculus. Calculus is used to analyze stability at equilibrium points. Differential equation models and difference equation models are compared and contrasted. Similarities in techniques are noted, and so is the power of difference equations to model

phenomena exhibiting periodicities and chaos. After the AP exam in May, the students study systems of difference equations and are introduced to the significance of eigenvalues and eigenvectors.

Sections of 15-25 students meet for four 50-minute periods per week. Homework assignments are supposed to occupy an equal amount of time outside of class. The class sessions are conducted in the lecture-discussion format with accompanying computer demonstrations. Computer lab sessions are conducted in a lab with Zenith brand IBM clones with "286" chips.

Handouts (including exercises) on the mathematics of genetics and curve fitting are available from Richard Sisley. Professor Sandefur should be contacted to make inquiries about using the materials he has written on finite difference equations.

Queen's University

Contact:

PETER D. TAYLOR, Department of Mathematics and Statistics, Kingston, Ontario, Canada K7L 3N6. PHONE: 613-545-2434.

Institutional Data:

Queen's is a major Canadian university with 11,000 undergraduates in Arts and Science, Engineering and Commerce. The Department of Mathematics and Statistics has 42 full-time staff members, and 20 students graduate per year with some sort of mathematics concentration in Arts and Science, and another 20 students graduate in a unique Engineering Mathematics program.

Project Data:

I am the only faculty member involved in the project, and over the past three years have taught one class of 60 students each year to develop the materials. There is very little emphasis on the use of computers. I use my own course notes as the text, and I have no external funding.

Project Description:

My objective is to develop a calculus curriculum for the Ontario OAC course (the final year of high school), but what I come up with should be suitable as the first course in any technical or liberal arts college, and in most applications-oriented programs in university. There are three ways in which my approach differs from the traditional technique-oriented first course in Calculus:

1. It attempts to be a process-based curriculum. The material and problems are chosen with an eye on classroom activity, small group work, and student project work.
2. It emphasizes applications to the physical and behavioral sciences, particularly the ideas of mathematical modelling. I try to get the class talking about the world: how does an oven work, how does a car use fuel, how fast should a bird fly, what size of an investment to make, etc.
3. It emphasizes geometric thinking and motivation. The student is often asked to argue from pictures. A lot of what we do concerns the relationship between algebra and geometry.

I cover new material in 2 one-hour classes per week, and then have a two-hour informal lab with small group work and individual counseling. One of my biggest and most problematic objectives is to learn how to test and examine properly a process-based curriculum. I hope to have a reasonable draft of a book available in Spring 1991.

Rensselaer Polytechnic Institute

Contact:

WILLIAM E. BOYCE, Mathematical Sciences Department, Rensselaer Polytechnic Institute, Troy, NY 12180. PHONE: 518-276-6898. E-MAIL: w._e._boyce@mts.rpi.edu.

Institutional Data:

Rensselaer is a private technological university with a large graduate program. Approximately 4000 full-time undergraduates attend with more than 80% majoring in engineering or science. A typical graduating class has 25 mathematics majors. There are 28 full-time mathematics faculty.

Project Data:

Three faculty have taught in the project. In 1988-89, about 60 students were enrolled in the project course. In 1989-90, the number increased to 85, and in 1990-91 about 300 students will be involved. Approximately 1000 students take calculus each semester.

Computer homework has been required on a regular basis. Supervised laboratory sessions have been used only for orientation and tutorial purposes at the beginning of the Fall term. Maple is the principal software used. Current hardware is an IBM 3090 running the MTS operating system.

Students access the system either via public terminals or through their own microcomputers used as graphics terminals. Maple is also available on some public workstations on the campus. *Calculus* by Boyce and DiPrima is the required text. This project has been supported by NSF grants, plus internal funding.

Project Description:

The basic goal of the project is to use contemporary computer hardware and software to revitalize instruction in our mainstream calculus course. More specifically, we want to promote a deeper conceptual understanding of the fundamental ideas of calculus and the ways in which they can be used to solve problems arising in other scientific and technical fields; and to equip students with problem solving tools that are powerful, versatile, and up-to-date, and to promote confidence and good judgment in their use.

Computer use is a regular and routine part of the course—for symbolic manipulations, for graphical displays, and for numerical computations. There has been a shift away from hand calculation toward problem formulation and conceptual understanding.

Our primary audience consists of engineering and science majors, the majority of whom have had a calculus course in secondary school. During 1988-89 and 1989-90 the pilot sections of the project course were small (20-30 students), but beginning in 1990-91 we will begin to scale up to the larger class size typical of our traditional course. Eventually the project course will become the standard course.

A typical section meets for four 50-minute periods per week. Classrooms have a graphics terminal connected to our mainframe computer, where Maple resides. The contents of the monitor screen are displayed on an overhead projector. The instructor has a second overhead projector for writing and for prepared transparencies. Homework is assigned regularly, including both standard textbook problems and more substantial, computer-oriented problems. One to three problems must be handed in each week. A typical problem solution includes computer printout, one or more computer-generated plots, and written explanations and conclusions.

Special laboratory sessions are arranged to help students learn the rudiments of the software, our computer system. Many students rapidly attain an adequate level of skill in computer operation, but some find this to be a significant challenge. This year we have one student assistant who helps with the computer laboratory and grading homework.

Although computers are not available during tests and examinations, the nature of our tests is gradually evolving. The emphasis on hand-executed algorithms is being reduced. In its place there are questions that require the formulation of problems, drawing conclusions from given data, or listing a sequence of Maple commands that will accomplish a given objective.

An important aspect of this project is the creation of a set of computer-oriented examples and problems for a calculus course. A preliminary version of this document has been made available to individuals at a number of institutions for comments and suggestions. While the supply lasts, copies will be furnished on request to other interested people.

Rollins College

Contact:

DOUGLAS CHILD, Rollins College, Campus Box 2743, Winter Park, FL 32789. PHONE: 407-646-2667. E-MAIL: child@rollins.bitnet.

Institutional Data:

Rollins, a small, independent liberal arts college is the oldest four-year institution of higher education in the state of Florida. The college has a full-time enrollment of 1450 students; the curriculum is based on the 4-1-4 academic calendar.

The Department of Mathematical Sciences consists of 10 faculty members and offers a major in mathematics and computer science. Counting both, the Department typically graduates 10-12 majors per year.

Project Data:

Four different faculty members have taught at least one course in the project. In 1989-90, all of our Calculus I and II classes are being taught in the project, a total of about 80 students. Over a four-year period, 11 courses have been taught of which 6 were taught in this current year (1989-90). A computing laboratory is required once or twice a week, as seems appropriate for the subject matter. We use Calculus T/L, a user-friendly interface to Maple written by Child on Macintosh II's in the lab. We are currently using Stein's *Calculus and Analytic Geometry*. The National Science Foundation is funding our project.

Project Description:

The Rollins project focuses on the first two terms of mainstream calculus, Calculus I and II, with most of the work thus far being done on Calculus I. We are particularly concerned with the understanding of the fundamental concepts of calculus.

Assessment of the project is on-going. In the Fall term 1989 a pre-test was administered in the first week of the course, and an almost identical post-test was administered in the last week. Two other methods used are daily class journals kept by the students, and informal student interviewing done weekly.

The courses meet 4 days a week, two 50-minute periods and two 75-minute periods. The computing laboratories are held during the longer periods.

Our teaching philosophy requires engaging the students in doing mathematics. Hence, they are usually actively doing something mathematical either in class or in the laboratory, and they come to expect it. Both the lab sets and work sheets (in class) involve a degree of open-ended thinking, followed by a written response in paragraph form. Frequently we lead the students in a loose way to uncover some specific result, and ask for their conclusions and conjectures with supporting evidence or proof. The students are urged to consult with each other, with the text, and with the instructor, and of course they use the computer in the labs.

The project software, Calculus T/L, has been widely tested at numerous sites. The curriculum approach has not been tested elsewhere.

Calculus T/L with a User's Guide is currently available from Brooks/Cole; we are planning a 300-page single-variable calculus text which will assume the software capabilities of Calculus T/L without assuming Calculus T/L itself.

Rose-Hulman Institute of Tech.

Contact:

ROBERT LOPEZ and ELTON GRAVES, Department of Mathematics, Rose-Hulman Institute of Technology, Terre Haute, IN 47803. PHONE: 812-877-1511. E-MAIL: lopez@rosevc.rose-hulman.edu and graves@rosevc.rose-hulman.edu.

Institutional Data:

Rose-Hulman Institute of Technology (RHIT) is a small, private, select undergraduate college of science and engineering. The normal undergraduate enrollment is about 1300. In addition there are approximately 50 graduate students in the engineering disciplines.

The Mathematics Department, which presently contains 15 full-time members, graduates about 12 majors each year. The preponderance of the department's teaching efforts is in the area of service courses for the engineering programs.

Project Data:

There are presently three faculty members who have taught in the project. In the Fall of 1990, two additional faculty will be involved in the project courses, for a total of five faculty involved.

In the Fall of 1988, two sections of 30 students each (out of a freshman class of 380 students, all of whom begin calculus as entering freshmen) started the Maple calculus course. In the Fall of 1989, four sections of 28 students each started the Maple calculus course. In 1989-90, the sophomores from the first year did two quarters of differential equations using Maple. The Maple calculus course, consisting of three quarters of calculus, has been taught twice. The Maple differential equations course, a two-quarter sequence, has been taught once.

A computer lab is essential in this project. The course is taught with each student sitting at a workstation. The workstations are used in class, for assignments, and for exams; thus every learning activity is a lab activity. To accomplish this we have two rooms devoted to the project. Symlab (short for symbolic, algebraic, and numerical computations laboratory) is a teaching classroom with marker board, 29 workstations, and an overhead projection system. An adjoining room is a workroom equipped with 13 workstations, 13 other terminals, and a printer.

Maple (Version 4.3 running under VMS) is the chief piece of software being used. In the Spring of 1990 Matlab was used to provide 3-dimensional graphics. The project is run on a network of 45 diskless DEC VAXstation 2000 workstations, served by a VAXstation 3200 with hard disks.

The textbook acquired by the students is *Calculus* by Boyce and DiPrima. The text is rarely used, except as a reading reference by those students so inclined. The project was funded by NSF via an ILI grant of $100,000 awarded in 1988. This was a matching award and the total expenditures for hardware, software, and room reconfiguration was in excess of $240,000.

Project Description:

At RHIT, computer algebra is to become the tool of first recourse in teaching, learning, and doing engineering analysis. Thus, calculus and differential equations are being taught with hands-on use of Maple. By allowing students to have access to Maple during lectures, homework, and exams, we have found that there is a great potential for making meaningful changes in college calculus and engineering curricula. Each course is a combination of in-class experimentation and practice, regular homework assignments, and special projects. The class periods are an amalgam of lecture, experimentation, and supervised activity.

It is these learning activities that most radically change the expectations and performance of the students. The students who have completed the five-course sequence have demonstrated, both in their mathematics classes and in their engineering classes, a unique mastery of engineering analysis. The course materials are being generated by the instructors involved. There are presently materials for complete coverage of all 5 quarters of mathematics. At the moment, these materials (class notes, homework assignments, special projects, and exams) are not in a form suitable for dissemination, but that condition is being remedied.

The project has the strong support of the Dean of the Faculty and the Chairman of the Mathematics Department. There is a growing involvement by the members of the Mathematics Department. There is also support in the Engineering Departments where symbolic computing is being assigned in a number of upper-level engineering courses. Grant proposals for classrooms equipped similar to Symlab have been written to support upper-level engineering courses.

We believe that this project has, and will continue to provide the impetus for meaningful calculus and engineering course revisions at RHIT.

San Jose State University

Contact:

MICHAEL J. BEESON, Department of Mathematics and Computer Science, San Jose State University, San Jose, CA 95192. E-MAIL: beeson@ucscc.ucsc.edu.

Institutional Data:

San Jose State University is a four-year public university, part of the 19-campus California State

University system, with more than 20,000 students. There are more than 40 full-time faculty and as many temporary and part-time faculty.

Project Data:

The project has so far mainly consisted of the design and development of MathPert (as in "Math Expert"), a computerized environment explicitly designed to support the learning of algebra, trigonometry, and first-semester calculus. So far, it has been used in the classroom only by its author, and that in only two classes, one in calculus and one in trigonometry. Funding is being sought for a more extensive experiment including a computer laboratory for student use of MathPert, classroom use by other faculty, and preparation of a teacher's guide.

MathPert should be ready for distribution sometime in 1990. It will be distributed on three disks: one for algebra, one for trigonometry, and one for first-semester calculus. The first commercially available versions will run on an IBM PC/AT. Funding for the initial stages of this project was provided by a National Science Foundation grant.

Project Description:

MathPert is intended to replace paper and pencil for homework, and chalk and blackboard in the classroom, but not to replace textbooks or teachers. Students type in their homework problems (or MathPert can generate them), and proceed to direct the step-by-step solution of the problem. At each step the student tells the computer what to do next by choosing an operator from a menu. The successive steps of the solution are displayed on the screen along with their justifications, and can be printed out and turned in. MathPert is designed to be useful even in connection with a traditional course, although it should be even more beneficial when the instructor is taking advantage of it too.

If a student needs help, MathPert can generate the complete solution by itself (it includes a sufficiently powerful expert system). In this respect MathPert is similar to Mathematica or MACSYMA, but it differs in that it is cognitively faithful. This means that it generates solutions using student-style methods, with each successive step comprehensible to the student so that the student should be able to generate a similar solution. It also differs from these other systems in that it is glass box instead of black box, meaning that you get a step-by-step solution, not just a one-line answer with no hint of how it was obtained.

MathPert also includes an internal model of the student's current state of knowledge, which it uses to customize the output to the individual. Thus, an advanced calculus student will generate solutions in which much more is done at each step than in a solution generated by a beginning algebra student. A common-denominator step, for example, would be done all at once, but for someone still learning common denominators, it might take five steps.

Seaver College

Contact:

DON HANCOCK, JOHN JACOB, and DON THOMPSON, Department of Mathematics, Seaver College, Pepperdine University, Malibu, CA 90263. PHONE: 213-456-4321. E-MAIL: hancock@pepvax.bitnet; jacob@pepvax.bitnet; or thompson@pepvax.bitnet.

Institutional Data:

Presently enrolling nearly 2400 full-time students, Seaver College is the undergraduate arm of Pepperdine University. It is a selective, residential, private liberal arts college. The Mathematics Department has 6 full-time faculty, and the number of mathematics and mathematics-computer science majors in a typical graduating class ranges from 10-15. There are two calculus tracks: a two-semester applied calculus with linear algebra sequence required of only business and economics majors, and a three-semester calculus sequence whose clientele is 80% science or mathematics majors. Calculus classes have an enrollment cap of 35 students, but generally average about 20-25 students. Student assistants are available for homework grading and limited tutoring.

Project Data:

Our project is designed to replace all sections of the three-semester calculus sequence taken by science majors, but is being phased-in one course per semester. Two faculty have taught in the pilot courses, which began with two sections of Calculus I in Fall 1989 and spread to both sections of Calculus I and II for Winter 1990. By Fall 1990 the transition will be complete, with each project section organized into 4 hours of lecture and 1-1.5 hours of instructor-directed laboratory per week. During the lab, students work in pairs on Macintoshes using the computer algebra system Maple (and occasionally other home-made software). The primary materials we use, referred to below as Notes, are distributed to the students free of charge and come from a draft of a textbook we are writing.

Student and faculty enthusiasm in our project has been very positive. Test results seem to indicate that our students (especially the top half) are learning significantly more; however, comparisons are not easily quantified since our project courses are more conceptual, and thus more difficult than before.

Project Description:

From a pedagogical viewpoint, our project attempts to separate the learning and understanding of fundamental concepts from the mastering of mechanical computational skills. By using a computer algebra system, students learn to recognize and appreciate the difference between conceptually difficult problems and computationally difficult ones. This permits us to emphasize the theoretical, conceptual, and creative aspects of calculus in the lectures and the Notes. We believe it is of utmost importance to force students to read, write, and verbalize mathematics correctly and precisely.

Some goals of the laboratory component of the project are to help instill in students the excitement of mathematical inquiry and experimentation; to reinforce understanding of concepts discussed in class by actively exploring them from a graphical, analytical, and numerical perspective; to gain experience writing, verifying, and analyzing algorithms using Maple's programming and symbolic capabilities; and to learn that even powerful computer systems are remarkably "dumb" without intelligent human intervention. A recurring theme is that difficult and practical problems are multistep and must be tackled by first breaking them into more manageable subproblems.

The content of our calculus sequence is distinctive in several key ways and reflects our belief that the natural, fundamental interplay between continuous and discrete processes should be emphasized. We begin with a detailed consideration of sequences and their limits, motivating the students with applications to various approximation procedures. The more difficult topic of limits of functions is then able to proceed easily and smoothly. To reinforce the fundamental perspective of discrete as an approximation to continuous, there is early and continued use of numerical methods for approximating functions, roots, derivatives, integrals, solutions to differential equations, and much more. This also permits us to work with a wide range of functions (not

just those given in closed form), and to investigate more intriguing problems. The more subtle idea of continuous as the limiting case of discrete is also examined, such as when students study continuous probability distributions and derive certain differential equations from the corresponding difference equations. The time to cover such non-traditional topics is freed up by removing many seldomly used or messy symbolic techniques from the curriculum.

To encourage our students to think more and write less during lectures, we routinely use an overhead to project transparencies corresponding to the distributed Notes. Lectures are enriched and enlivened by classroom demonstrations on a computer running Maple. A substantial and wide variety of written homework problems, with a heavy emphasis on ideas and writing as opposed to plugging-and-chugging, are assigned twice a week. During each lab session, students are guided by a 5-7 page handout which outlines the concepts to be explored, introduces new features of Maple or other software they will be using, and contains 3-7 computer problems to be completed by the following week.

Siena College

Contact:

EMELIE KENNEY, THOMAS ROUSSEAU, and SUSAN HURLEY, Department of Mathematics, Siena College, Loudonville, NY 12211. PHONE: 518-783-2440.

Institutional Data:

Siena College is a private four-year college. Of the approximately 2600 undergraduates, about 8% are commuters.

The Department of Mathematics has 6 full-time faculty, and graduates about 20 mathematics majors annually. The majority of students taught in the first two years of mathematics are those majoring in the sciences (biology, chemistry, computer science, and physics). A small number of social scientists also take introductory courses.

Project Data:

All faculty members have participated in the project since its inception two years ago. The existing program involves a formal laboratory period for Calculus I, Calculus II, and Mathematics II, a second calculus course for those students who terminate their mathematics instruction after two semesters. Other courses—Linear Algebra, Differential Equations, and Applied Mathematics—utilize the laboratory facility, but have no formal laboratory component at this time. The Department is committed to the extension of the laboratory concept to these and other appropriate courses.

Project Description:

The project model has three aspects: a formal laboratory period, collaboration, and writing. Classes meet 5 hours weekly for 3 hours of lecture and a laboratory. The laboratory period is separately scheduled and conducted in a mathematics laboratory which is equipped with Macintosh IIcx computers using the software Mathematica.

The objective is critical examination of mathematical reasoning, mathematical discourse, the structure of calculus, mathematical modelling, and the nature of mathematical truth. This objective is accomplished through carefully designed laboratory exercises, peer group discussion, and the submission of a laboratory report. Students use the computer as a tool to aid their mathematical analysis. The laboratory is used to introduce some ideas that will be formally presented in lecture, to reinforce concepts that are particularly difficult, and to provide interesting applications of calculus that are not typically found in introductory courses.

The formal laboratory material comprises exercises that are independent of the textbook used, written to improve students' ability to think quantitatively, and designed to be solved in groups. Students choose study partners with whom they discuss their work at all stages. Laboratory assistants, upper-class students, and a laboratory instructor encourage group interaction. Laboratory reports have, during different semesters, been required from each individual and from each group. Examinations routinely include "non-standard" problems which exercise the skills developed in the laboratory.

Two faculty members are developing formal computer exercises to supplement the laboratory exercises. One is pursuing the extension of the laboratory concept to the next set of courses, and one plans to complete *Writing Intensive Laboratory Courses in Calculus and Differential Equations*.

Simmons College

Contact:

DONNA BEERS, Department of Mathematics, Simmons College, 300 The Fenway, Boston, MA 02115. PHONE: 617-738-2166.

Institutional Data:

Simmons College is a small comprehensive urban institution serving 1625 undergraduate women and 1255 women and men in graduate and related studies. The number of majors in a typical graduating class ranges from 7-10.

Project Data:

The calculus project at Simmons College was initiated in Fall 1988. Its focus is first-year calculus. All sections of Calculus I and II have participated with an average of 16 students per section. Four faculty members have taught in the project thus far. Students attend three 50-minute lectures and one 80-minute lab per week. Half the labs are computer labs using Flanders MicroCalc software on Compaqs, the other half are paper-and-pencil labs organized as cooperative group learning experiences. Here students work in groups of four on problem sets that reinforce topics covered in lecture. The textbook for Calculus I and II is Berkey's *Calculus of One Variable*. The calculus project receives no outside funding.

Project Description:

In 1988-89, the Mathematics Department introduced a weekly 1 hour 20 minute laboratory into all Calculus I and II sections. Contributing factors were departmental dissatisfaction over the rate and difficulty with which students mastered the material, our positive experiences with computer-based statistics laboratories and with paper-and-pencil laboratories in intermediate computer science courses; and the interest of several faculty in exploring cooperative group learning techniques. While individual grading policies differ, labs amount to roughly 20% of a student's final grade.

Half the labs require personal computers, while the rest are conducted in cooperative group learning fashion. For each session we prepare a handout that states the purpose of the session and the MicroCalc modules that will be featured. Handouts are designed to lead students to draw conclusions or make conjectures through experimental investigation. These must be reported at appropriate places on the hand-out which is turned in for a letter grade. By introducing computer labs into our first-year calculus course, we hope to reinforce and make vivid precalculus and calculus concepts through computer graphics and to foster inquiry and use of the computer as a tool for investigation.

We call the cooperative group learning labs paper-and-pencil because the computer is not used in these sessions. Instead, a class is divided into heterogeneous groups of four students to work on problem sets. Every member of a team must be able to do every problem. Solutions are reached through consensus, and each group turns in one problem set, every member receiving the grade earned by the team paper. The point of these labs is to get students to talk calculus in order to increase their competence and self-confidence.

Copies of laboratory assignments may be obtained from the author. They are available free of charge or for the cost of xeroxing, depending on the size of the request.

Spelman College

Contact:

NAGAMBAL SHAH, Department of Mathematics, Spelman College, 350 Spelman Lane, SW, Atlanta, GA 30314. PHONE: 404-223-7625.

Institutional Data:

Spelman College is a private, predominantly black, four-year undergraduate liberal arts women's college. Approximately 67% of the students are residential, and the remainder of the students commute. There are approximately 1760 students enrolled at Spelman with 37% of the student population majoring in the natural sciences and engineering. The dual-degree engineering students obtain a natural science degree from Spelman College and an engineering degree from another institution offering a degree in engineering. There are approximately 115 full-time faculty members at the College.

Project Description:

A planning grant was provided by the National Science Foundation to develop a program that investigates the problems encountered by minorities and women while taking calculus that impede their success. Because levels of success are being studied, both the negative and positive factors that contribute to success will be addressed. An understanding of these factors and their relationship with other aspects of student life may suggest ways to enhance the techniques currently employed in the teaching of calculus. These enhancements may

negate the negative factors encountered by minorities and women in calculus and thereby increase the pool of people available to pursue careers in mathematics and science.

The aspects of the program plan completed to date include a delineation of the positive and negative factors that may influence the success of minorities and women enrolled in calculus; an outline of the strategy developed to determine the legitimacy of these factors for minorities and women who are or were enrolled in calculus courses; development of an instrument to solicit data from students regarding these factors; surveying a population of students of calculus deemed representative of minorities and women; a preliminary analysis of the survey data; a comparison between surveyed student attitudes toward computer-aided teaching tools and individual student assessments of these tools; and recommendations for future analyses of the student survey data.

A preliminary assessment of the data highlights reasons for a lack of success in calculus. These reasons include the student's degree of preparation in mathematics prior to calculus enrollment, the student's motivation to excel and their degree of confidence in their ability to succeed, the influence of extracurricular activities on levels of student success, the student's level of mathematics maturity as determined from study habits, and the student's perception of computer-aided teaching tools. The results were not compared with students outside the targeted groups; however, this is recommended for future studies. Initial survey results suggest ways to modify calculus teaching approaches; however, further analysis of data is needed to confirm the suggested teaching and curriculum changes.

Tennessee Technological Univ.

Contact:

JOHN SELDEN AND ANNIE SELDEN, Department of Mathematics, Tennessee Technological University, Box 5054, Cookeville, TN 38505. PHONE: 615-372-3441. E-MAIL: js9484@tntech.

Institutional Data:

Tennessee Technological University is a public, non-metropolitan, comprehensive university with an engineering emphasis. Many students are the first in their families to attend college and are well-motivated. TTU has about 6000 full-time undergraduates. In 1988-89, TTU granted 1090 Bachelor's Degrees, of which 10 were in mathematics. The Mathematics Department has 25 full-time faculty and 8 graduate TA's.

Project Data:

Only two faculty have taught in the project. The project and standard courses are the same size, about 30-35 students. We have just completed our second 2-semester sequence. Students attend class 7 hours per week instead of the normal 5 hours per week, but no separate lab is required. HP 28S calculators are required, and an AT running Derive and Mathematica is available in the classroom. We teach using our own notes instead of a book. Previously there was no external funding, however, we have recently received an NSF ILI grant to up-grade to 30 Mac IIs.

Project Description:

This two-semester sequence called Enhanced Calculus is meant to be taught in parallel with the traditional course and has the same syllabus. Its main feature is its problem-solving character. Students solve problems for which they have not been given algorithms or sample solutions. They work from notes containing definitions, theorems, and problems, but no solutions. Students present their work in class and receive advice and criticism, but there are no formal lectures. This teaching style greatly increases student-teacher and student-student interaction, and facilitates the discovery and correction of misconceptions.

The HP 28S calculators are useful and popular with the students, but not essential. No programming is required and calculator instruction is minimal—perhaps three hours. The additional two hours per week of class time is also not essential; during this time the students work in groups of five or six.

Each semester there are four tests and an exam. Tests last two hours and usually consist of four problems, which are not similar to problems in the notes. To encourage review, the exam does contain problems similar to those in the notes. Students use their calculators and notes on the exam and tests. Occasionally we also assign special projects or ask for problem solutions to be written up carefully and handed in.

We recruit volunteers from in-coming freshmen and have no special prerequisites. In very preliminary comparisons, our students performed without notes or calculator about as well as, if not a

little better than, traditional students. In addition, we think their self-confidence, reasoning ability, grasp of fundamental concepts, and attitudes toward mathematics are improved. The students and our Chairperson like the course and the administration has been supportive, but some of our more conservative faculty remain skeptical.

The notes are available at no cost. The AMS-TEX disk version is easy to edit.

Trinity University

Contact:

PHOEBE T. JUDSON, Department of Mathematics, Trinity University, 715 Stadium Drive, San Antonio, TX 78212. PHONE: 512-736-8243. E-MAIL: pjudson@trinity.bitnet.

Institutional Data:

Trinity University is a private four-year liberal arts institution. Full-time undergraduate enrollment is approximately 2500. First and second year students are required to live on campus. Profile of the 1989-90 first year student class: 1206 mean SAT, 51/49 male-female ratio; 18% minority. The Mathematics Department has 10 full-time faculty. 15-20 mathematics majors graduate each year, and about the same number of minors graduate. Each faculty member has a computer in his or her office.

Project Data:

Four faculty are currently involved in the project. We expect the entire department to be involved by the time the three-year project is completed. There is an upper limit of 20 students in each project course; 30 in standard courses. We have been using Maple in the calculus for four semesters. The formal laboratory is currently in its first offering.

20 Macintosh SE computers are networked with an SE30 (4 MB memory, 80 MB hard drive) as file server. The file server is to be upgraded to a Macintosh IIx. Ellis and Lodi's *Maple for the Calculus Student* was a recommended supplement. An NSF ILI grant provided matching funds to equip the laboratory.

Project Description:

The purpose of our project is to improve the instruction of the calculus by incorporating the use of a computer algebra system. The project is directed at first-year calculus students and is implemented in two distinct ways: several sections of Calculus I and II assign homework which requires students to use Maple; and a formal, one-credit-hour laboratory session with weekly meetings is offered for students in Calculus I or II.

The project, in either form, consists of 10-12 calculus assignments which require the use of a computer algebra system. The assignments are exploratory in nature, and most require written analyses. The assignments are made weekly, and students are given one week to complete each task. The assignments are intended to reinforce the basic concepts of the calculus. Students are asked to explain, illustrate, conjecture, and generalize. Maple is used to generate examples, test conjectures, and provide graphs.

The formal laboratory was offered either at a fixed time or as independent study due to scheduling difficulty. The instructor conducted a formal lab at a fixed time each week. Eight of the sixteen students took the lab at the fixed time, and eight arranged independent study. Each lab session required approximately two hours of computer time. A student assistant was available during the first few lab sessions. Students learned to use Maple easily, and the assistant was no longer needed after the third lab session.

Offering the lab on an independent study basis did not noticeably increase the work load for the instructor. A sign-up sheet was placed on the instructor's door, and students were generally able to pick up and complete the lab assignments on a weekly basis without private conferences with the instructor.

Lab materials are available on disk from Judson. Send an unformatted 3.5 disk (2 MB if possible) and $2 to cover postage and handling. Materials are written in the latest version of Word and can easily be modified to suit individual needs.

University of Arizona

Contact:

DAVID LOMEN AND DAVID LOVELOCK, Department of Mathematics, University of Arizona, Tucson, AZ 85721. PHONE: 602-621-6868. E-MAIL: lomen@ariz.rvax.bitnet.

Institutional Data:

The University of Arizona is a state university with an enrollment of about 35,000 students, 7,000 of whom are graduate students. The largest minority group is Hispanic. The mathematics department has 60 full-time professors and about 40 adjunct faculty. It graduates about 50 majors each year.

Project Data:

Eleven faculty members are involved with our calculus reform project. Most of our experimental efforts require the students to have access to a computer. We use a standard calculus book (Anton) and develop supplements as needed. External funds were received from the National Science Foundation (an ILI grant of $75,000 and a Calculus Initiative grant of $104,806). In addition, computer equipment was donated by Apple, Everex, Intel, Silicon Graphics, and Sun.

Project Description:

Most of our efforts have gone into the development of four different, but complementary, forms of supplements, viz. computer *software*, a *laboratory* manual for first-semester calculus, a 150-page discussion of sequences and series, a collection of *projects* for first-semester calculus, and an *experimental course* taken concurrently with a second-semester calculus course. None of these supplements depend upon a particular calculus text, but most require the use of technology.

All our *software* is designed so that neither the instructor nor the student need be "computer literate." The software was developed to aid students in solving problems; to encourage students to treat mathematics as an experimental subject; to capture the spirit and excitement of current developments in mathematics; and to help the instructors and students during lectures and examinations. To date we have developed 28 pieces of software for MS-DOS machines. The software ranges from exploring one-dimensional iterative maps (period doubling, chaos) to the graphing of Fourier partial sums, from the curve-fitting of data by eye, to the plotting of direction fields and integral curves. This software may be freely copied and distributed, but not sold for profit. A complete list of our software and what each package does is available.

We have developed a series of *laboratory* exercises to supplement a first-semester calculus course that can be used with any calculus book and appropriate software. Topics covered in the labs are interpolation, equation solving, least squares, numerical integration, and Fourier analysis. One class period per week of our five-hour, first-semester calculus course was replaced by the laboratory requirement. The assignments are meant to be independent of class lectures. The idea is to expose students to the topics by examples and illustrations.

The 150-page supplement for Calculus II presents sequences and series in a manner different from the traditional method. It starts by considering polynomials and introduces power series as a generalization of polynomials. Infinite sums come about as a natural consequence of using power series to define functions.

The calculus *projects* that have been developed are also independent of any particular calculus book or computer software (although we have developed software specifically to ease the completion of some projects). These projects were developed with many goals in mind, some of which are to have a student make a conjecture and attempt to prove it; to give non-"template" type problems that require some reasoning; to give a student more practice at translating between words, diagrams, and mathematical symbols; to have problems that require more than one step; to encourage the "rule of three," analytical, numerical, graphical; to give the student experience when to use technology and when to use pencil and paper; and to present projects whose solutions require techniques from various parts of the course.

A two-unit *experimental course,* using MathCad, has been developed that complements the second semester of calculus. The entire course is held in a completely remodeled classroom containing 30 MS-DOS 386 machines, one for each student, and a dedicated computer for the instructor to project computer images onto whiteboards. Some of our other experimental sections of calculus also use this classroom.

The software, lab manual, projects, and experimental course handouts are available from the Mathematics Department for the cost of the materials and mailing.

Univ. of California, San Diego

Contact:

AL SHENK, Department of Mathematics, University of California at San Diego, La Jolla, CA 92093. PHONE: 619-534-2654. E-MAIL: ashenk@ucsd.bitnet.

Institutional Data:

UCSD is a four-year, state-funded university with 14,000 undergraduates and a large graduate program. The Mathematics Department has 50 members and approximately 100 graduating majors each year.

Project Data:

Our calculus project has been running since the fall quarter of 1988 in most of our first-year calculus classes for mathematics, science, and engineering majors. This has involved 34 lectures per term and over 5000 students. A Macintosh Plus lab has been used since the Fall of 1989 by all second-quarter students. Their computer assignments take from 10-15 hours per quarter and use True Basic programs. A grant from Apple, Inc. was used to obtain the hardware for the lab. Other costs were covered by a UCSD Instructional Improvement Grant.

Project Description:

The classroom aspect of the project has been in three phases. Starting in the Fall quarter of 1988, the courses were based on a "Menu of Core and Elective Topics" from *Calculus and Analytic Geometry* by Al Shenk. The "Menu" was distributed to students, and instructors used it to design their individual "leaner and livelier" classes. The Core topics covered the basic ideas of calculus and what we felt students needed to be prepared for their subsequent courses. Instructors were to cover all the Core topics and a number of Elective topics chosen according to their taste and interest and what they felt were the needs of their students. The core material took 50%-80% of class time. All the instructors liked the freedom from a crowded, rigid syllabus, and they used their elective time in a variety of ways.

In the second phase, additional modifications are being made to follow more of the recommendations from recent calculus reform conferences. The new courses require graphing calculators in the homework and have a greater emphasis on problem solving. They are based on Core material from the text that has been streamlined even further and on new graphical problems, discussion questions, and graphing calculator exercises that are being prepared by Al Shenk. Students receive a *Students' Guide* which contains a new "Menu," the new exercises, and Outlines and Overviews of groups of chapters. The *Instructor's Guide* also contains listings of all of the recommended Core exercises from the text, grouped by topic, so instructors can plan their assignments without referring to the book. Finally, this fall we will run several large classes as workshops in which students will work in groups of two or three at their desks on word problems and discussion questions, with periods for the professors to give instructions and answer questions. The main goals are to have students see calculus as more than just a set of problem-solving algorithms, and to have them be active rather than passive in class. Also they will have to study the text more, since its contents will not be dictated to them.

University of Connecticut

Contact:

JAMES F. HURLEY, Department of Mathematics, U-9, University of Connecticut, 196 Auditorium Road, Room 111, Storrs, CT 06269. PHONE: 203-486-1292; 486-6452; 486-3923. E-MAIL: hurley@uconnvm.bitnet.

Institutional Data:

The University of Connecticut is a medium-sized public university with a large graduate program. Besides the main campus at Storrs, there are five small regional campuses. The students are concentrated mainly in the College of Liberal Arts and Sciences and the School of Engineering. The student body at Storrs lives almost exclusively on campus or in private off-campus apartments.

The Department of Mathematics has 33 full-time faculty at Storrs, and 16 full-time faculty at the regional campuses. Courses taught at all campuses are identically numbered, and use the same texts and outlines. All mathematics students must come to Storrs to finish undergraduate degrees. The average number of students receiving baccalaureate degrees with majors in the Department has been 78 per year.

Project Data:

Since 1983-84, a total of 22 faculty have taught in the Department's high-track, four-semester, 18-credit, enhanced calculus sequence, Math 120-121-220-221; all but three of those are Storrs faculty. Through the Spring of 1990, Math 120-121 has been offered during 16 semesters, and Math 220-221 for the past six semesters. In 1989-90, an experiment began to apply the general approach of the first three-fourths of the enhanced sequence to the standard, three-semester, 12-credit, main-track calculus

sequence, Math 110-111-210. By the end of 1990-91, three full-time faculty and two or three teaching assistants will have taught the lecture portion of that project.

The enhanced first-year course has averaged approximately 100 students over the past three years, while the second-year portion has averaged about 50. Corresponding figures for the main-track courses are roughly 500 in the first-year and 300 in the second. In 1989-90, enrollment in the experimental main-track sections was limited to 25 in the Fall and 35 in the Spring, compared to more than 200 in the corresponding conventional sections of each course. For 1990-91, enrollment in the experimental sections will increase to 50-60 students each semester. Both the enhanced and experimental main-track courses require a one-hour weekly computer laboratory. The enhanced second-year course makes software available that students are encouraged to use in completing assignments, and the experimental third-semester main-track course will require the use of similar software to complete assigned homework. Both projects use True Basic as the principal software with locally produced programs written in that language by the faculty and studied and used by the students in the laboratory work. In addition, True Basic's *Calculus* package is used for some symbolic work in the enhanced sequence, and both it and the computer algebra package *Theorist* are being incorporated into the experimental project.

The hardware used in the enhanced sequence consists of 25 IBM Personal Computers and AT&T 6300's in the Department's PC Lab. For assigned homework in the enhanced sequence, students can also access 25 Zenith 157's and 25 IBM PS II's in laboratories operated by the University's Computer Center. A Hewlett-Packard 7475 plotter is available in the PC Lab, along with dot-matrix printers that are used for submitting laboratory reports. The experimental main-track project uses the Department's new Macintosh Laboratory, which has 27 Macintosh SE/30's and three Macintosh II's. The University's MacLab has another 14 Macintosh SE's and II's available for student use in working on assigned laboratory projects. Through 1989, the text for Math 120-121 was *Single Variable Calculus* by Robert A. Adams. The current text is *Calculus* by Boyce and DiPrima. The computer laboratory works through *Computer Laboratory Manual for Calculus* by James F. Hurley and Charles W. Paskewitz. The second-year text is Hurley's *Intermediate Calculus*. The experimental main-track calculus course is using the same text as the other main-track sections, *Calculus* by Hurley. (Faculty and laboratory assistants guide student experimentation during the lab, and students work on laboratory project sheets that are based on the laboratory activity.)

The Department received two large internal grants from the Dean of the College of Liberal Arts and Sciences to purchase equipment and convert a classroom to the PC Lab. To integrate computing into the main-track calculus course we have received one grant from the National Science Foundation Instrumentation and Laboratory Improvement Program (1988-90), and another from the State of Connecticut's High Technology Program (1989-90).

Project Description:

Besides the laboratory period, both courses have one problem-discussion hour per week. Math 120-121-220-221, which is aimed at honor-level students, has three lectures per week. Math 110-111-210 meets four hours per week and is aimed at the average main-track students. The traditional sections of Math 110-111 have three hours of lecture per week; the experimental sections have two—one hour of lecture is replaced by the laboratory activity. Math 120-121 is taught by regular faculty in small sections, with either a graduate or undergraduate teaching assistant in charge of the laboratory. Both faculty and teaching assistants give the experimental Math 110-111 lectures and labs. Other departments will be involved in perfecting the revised outlines being developed for the experimental course. The University of Massachusetts, Amherst, has been testing the main-track approach since the Fall of 1988, using Hurley-Paskewitz and locally produced handouts for the laboratory part of the course. The Department strongly supports the enhanced program, and the Head has endorsed the experimental main-track project in external grant proposals.

A common feature of both projects is the attention devoted to the algorithms the computer uses to perform numerical calculations and to plot graphs of functions and equations. Supplied code is studied as simple, explicit sets of instructions that implement the underlying algorithms and illustrate basic concepts. Weekly assignments require modification of the programs to fit a variety of situations, and interpretation of results that at times are affected by round-off or overflow error. The simplicity of the code is illustrated by the following definition of one function that is examined in terms of limits,

continuity, differentiability, and graphical behavior during the first semester. Its ordinary mathematical definition

$$f(x) = \begin{cases} x \sin \frac{1}{x} & \text{for } x \neq 0 \\ 0 & \text{for } x = 0 \end{cases}$$

corresponds to the following True Basic code.

```
def f(x)
    if x <> 0 then let f = x*sin(1/x) else
            let f = 0
end def
```

Such code combines easily with similar commands to allow in-depth examination of this and other interesting and important functions that formerly weren't considered in introductory calculus. The results often help illuminate theory that otherwise receives little attention.

Giving the students complete working programs with visible code allows emphasis to be placed on the computational significance of the associated mathematical concepts, and seems to promote improved conceptual understanding. For example, the definite integral is clearly understood as a limit of Riemann sums, which leads naturally to numerical methods of evaluation as an alternative of equal value to the fundamental theorem. Numerical computation of limits, derivatives, and integrals also fosters an intuitive understanding not only of those individual notions but also of the very nature of the phenomenon of convergence. When students encounter the latter concept formally for infinite sequences and series, they bring unusually sharp intuition to that study.

Students do not use computers on examinations. Graphics calculators are permitted. In Fall 1989, common final examinations were administered to all sections of Math 110, both computer-integrated and traditional. The mean scores were 68.2 in the former, and 60.0 in the latter. That difference corresponds to a full letter grade (C versus D).

Besides the computer laboratory manual, there are instructional guides and program disks for both types of lab. The four guides (Math 110, 111, 120, and 121) are available for $2.50 each, and the program disks are free to those willing to provide feedback to the author and supply blank disks (one 5.25 inch for Math 120-121, two 3.5 inch for Math 110-111). Make checks payable to the University of Connecticut Mathematics Department. Address requests and other correspondance to the Contact individual.

University of Hartford

Contact:

JOHN WILLIAMS, Department of Mathematics, University of Hartford, West Hartford, CT 06117. PHONE: 203-243-4825. E-MAIL: jwilliams-@hartford.bitnet.

Institutional Data:

The University of Hartford is a private four-year institution with 6 colleges including a college of engineering. We have over 4000 full-time undergraduates and about 3000 part-time students. The Mathematics, Physics, and Computer Science Department has 21 full-time faculty and over 30 adjunct instructors. In a typical graduating class we have 15 mathematics majors, 25 computer science majors, 10 other science majors, and 110 engineering majors.

Project Data:

Three faculty members have taught the computer lab sections of calculus; 4 more members are scheduled to teach sections in the Fall of 1990. In the past two years, 2 out of 7 sections of calculus have been taught with a weekly computer lab; in the Fall of 1990 all scheduled sections will include a computer lab. The principal software we use is Epic and Derive. Our hardware is 12 networked AT&T 6300's in a computer classroom and other IBM clones outside of class. In addition, we require that the students purchase Casio fx7000g graphing calculators. The textbook we have used for the past two years is *Calculus and Analytical Geometry* by Edwards and Penney, but it has no influence on the project. We were able to use some funds left over from an Exxon Education Grant for purchasing software and paying student lab assistants.

Project Description:

Three types of technology are used in the project: a graphing calculator (the Casio fx7000g), calculus software for graphical and numerical work (Epic), and a computer algebra system (Derive). The majority of the labs can be done with any one of the three; however, each has its strengths and weaknesses, and we continue to find it convenient to use all three. We supply the students with 8 programs which make the calculator essentially equivalent to a piece of calculus software, combining some features found in no one piece of software.

The computers are available during regular class time as well as for the once-a-week lab. During the labs the students work in pairs with one pair at each computer. If the students don't have time to finish the assignment during class, they must either go to one of the computer labs on campus, or finish it with their graphing calculators. The following week each student must turn in a lab report. These must be written in correct English and addressed to someone who is not familiar with the topic (i.e., not just answers for the teacher, but a real explanation of the topic). The reports are evaluated for mathematical content and for clarity. Since each student must submit a report, both partners must understand the lab.

Another component of the project is the use of graphing calculators for in-class use and for homework problems. The calculators can be used during a lecture to encourage class participation; for instance, the students can graph a function being used in an example. For homework problems the calculators are used as an alternate means of solution; e.g., using a graph and root finder to find max's, min's, and inflection points, or to evaluate numerically a definite integral. Despite the lack of specific calculator homework problems in most textbooks, it is interesting to note that the students in their lab evaluations overwhelmingly rated the calculators the most helpful of the three technological components. One student wrote that the Casio is "like having Epic at your finger tips."

In Lab 1 of Calculus I the students collect data on a real pendulum, and then adjust the parameters of the equation $y = ae^{-kt}\cos(\omega t)$ to get the best possible fit with the data. Here students get to see realistic applications, and through exploration get to discover how parameters affect the shape of a graph. This lab is used as a basis for several other labs, where students approximate velocities (before derivatives are introduced), approximate the time it takes the pendulum to stop using a graph, and approximate the total distance traveled using Riemann sums. In other labs students discover the product rule using computer algebra, discover the Fundamental Theorem of Calculus, investigate the relationship between the amplitude of a vibrating string and its length, and discover and explain the relationship between the graph of a function and the graph of its derivative. In Calculus II, two labs are based on real data collected for a falling balloon. Another lab is based on the actual distances of the earth from the sun, and still another investigates a population model.

The topics offered in the labs correspond to the standard topics covered in a calculus course, with the addition of a differential equations lab. Thus the lab manual we are preparing can be used with a variety of different calculus courses. By the end of Spring 1990, all the labs will be rewritten so that they are not software specific, and we will write a teacher supplement with suggestions on how to make the labs a learning vehicle.

By using both graphing calculators and computer software we have gained insights into the strengths and weaknesses of each. In particular, we have found that the Casio can be set up so as to be almost as easy to use and as powerful as calculus software (at least the non-symbolic type). We have developed our labs so that they can be used with any piece of software or graphing calculator; the labs focus on the important concepts of calculus rather than on the technology.

By using commercially available software and calculators our students are learning to use a real mathematical tool rather than something specifically designed for just one or two courses. Our students can use what they learn in our courses in other courses and on the job. Many of our labs use real data, some from experiments which are performed in class. Students see how real mathematical models work and what their limitations are. The technology we use is easy to learn so that the majority of the time in lab is spent using the device. Consequently, our approach is transferrable to other schools. Finally, our labs follow the standard calculus syllabus; they can be used as out-of-class projects or, as we do, in a weekly lab format. Since new material is introduced in the labs, lecture time is cut down, leaving enough time for the weekly lab.

After teaching the experimental sections in both the Fall of 1988 and the Fall of 1989, short-answer questions were incorporated into the final exams of the experimental sections and of standard sections. The short-answer part consisted of 4 questions designed by the investigators to emphasize concepts and graphical understanding, and 4 questions designed to emphasize standard by-hand calculations. No calculators or computers were allowed for this part of the final. The students from the lab sections performed much better than the other students on the four conceptual questions, and about the same on the standard problems. This was true both for 1988 and 1989. Thus, it appears that the lab students are gaining conceptual and graphical under-

standing without sacrificing standard manipulation skills.

We also studied how the students who took the experimental sections in 1988-89 did in advanced courses in the Fall of 1989. Keeping in mind that the sample size is quite small, we found no statistically significant differences between the performances of computer vs. non-computer students in either Calculus III or Differential Equations. Thus, we tentatively conclude that our lab section students are getting more bang-for-the-buck. In addition to learning to use technology effectively, they can still solve standard by-hand problems.

According to student evaluations, the calculator-computer format is helpful and enjoyable. Most students rate the graphing calculator to be more helpful than the software. We found this to be somewhat of a surprise; it seems that the ability to take the device home is crucial to the students.

Preliminary copies of the labs are available by writing for copies.

University of Iowa

Contact:

KEITH D. STROYAN, Department of Mathematics, University of Iowa, Iowa City, IA 52242. PHONE: 319-335-0789. E-MAIL: stroyan@math.uiowa.edu.

Institutional Data:

The University of Iowa is a large public research university with a total enrollment of almost 30,000 students. About 20,000 of these students are undergraduates. We have about 150 undergraduate and 120 graduate (M.S. and Ph.D.) students majoring in mathematics. There are also separate undergraduate programs in computer science, statistics, and actuarial science. The Mathematics Department has a faculty of about 40 individuals.

Project Data:

We have offered Accelerated Calculus for about seven years with five different faculty instructors. This year we expanded the size of the course and scope of the curriculum revision. One of those faculty members (with cooperation and advice of several others) oversees the program.

Approximately 10% of our non-business calculus students took Accelerated Calculus in the Fall 1989. Accelerated Calculus has a very diverse group of students whose only common asset is good preparation in high school mathematics. The course is not restricted to mathematics majors. Another 1500 students took business calculus during 1989-90.

Regular use of Mathematica Notebooks running on a network of NeXT computers is required. Most assignments have students modify and use Notebooks which we have written for them. A few assignments require them to write their own. In addition to the Mathematica materials we have written, we also have written all the text materials (in cooperation with several other schools).

We received funding for equipment from the NSF's ILI program and support for materials development from the NSF Calculus program.

Project Description:

The Accelerated Calculus program at the University of Iowa is developing materials that present calculus as the language of science. We motivate topics in calculus with scientific problems of clear contemporary importance. Because our audience is diverse, ranging from engineering to the arts, we have a wide variety of areas of application including physics, population biology, and economics. Solution of difficult scientific problems is what leads our curriculum.

Part of our approach involves the use of a laboratory. The University of Iowa has a long history of successfully using computers in a laboratory with calculus, but we are expanding the use of computing in two directions. First, we are now using the up-to-date NeXT/Mathematica environment and giving our students a start on their education in scientific computing in the weekly "electronic homework" assignments. Second, we assign "projects" or term papers to teams of students several times per year. These projects have the students follow one of the scientific applications in depth and write a technical report on their work. The papers usually involve some computing.

We are also developing a cost-effective method of presenting calculus at a large university where students work on open-ended projects with assistance from graduate students, upper-class undergraduates, and faculty. The computer lab serves as a focal point for this interaction, adding a personal touch to the somewhat impersonal lecture-discussion format without prohibitive amounts of labor. We have two lectures, two discussions with graduate students, and 50 open hours in our dedicated NeXT lab. Students can get help with regular class work as well as work on computing in the lab since it is staffed by faculty, graduate T.A.'s, and well-prepared upper-class undergraduates.

We are sharing materials with the projects at Iowa State University, Duke University, and the Five College Consortium in Massachusetts. Our goal is to develop materials and methods that are transportable to public universities across the nation.

University of Rhode Island

Contact:

EDMUND A. LAMAGNA, Department of Computer Science and Statistics, The University of Rhode Island, Tyler Hall, Kingston, RI 02881. PHONE: 401-792-2701.

Institutional Data:

The University of Rhode Island is a medium-sized state university that enrolls approximately 12,000 undergraduates and 3000 graduate students on its main campus in Kingston, and has a full-time teaching faculty of about 750. The student body is diverse in its interests, ranging from engineering and physics to the health sciences and the humanities. The Department of Mathematics has 23 permanent faculty members, and a typical graduating class contains approximately 75 majors.

Project Data:

At the present time only Professor Diane L. Johnson has taught calculus using the Calculus Companion. She teaches courses of the standard size, about 35 students. The system was first classroom-tested in the Fall semester of 1989 and is presently being used. The application of our software requires Maple running on a Macintosh computer. The system runs on Macintosh SE's and Macintosh II's. We are following the textbook used by the Mathematics Department at the University, specifically Fraleigh's *Calculus with Analytic Geometry*. This project is being funded by grants from the National Science Foundation under its calculus initiative program.

Project Description:

We are currently developing a unique computational environment in which students use the computer as both a tutoring device and a computational aid. Our system consists of a user-friendly interface to the symbolic mathematics package, Maple, together with numerical computation and graphical display routines. Equations are easily constructed and modified, appearing exactly as they do in text-books. Moreover, multiple windows allow the student to see and work with several formulas at once. We are also developing specialized graphics which are designed to illustrate important concepts of calculus.

An important aspect of the project is to devise effective means for incorporating the computer into the two-semester introductory calculus sequence. We are restructuring our curriculum based on the model of the "lean and lively calculus." Applications from science and engineering are used throughout to motivate topics and to provide examples. Our approach provides for rapid computation using symbolic manipulation, extensive graphical support, simplified use of numerical methods, and the incorporation of more interesting applications.

The Calculus Companion enables students to concentrate on learning the important concepts of calculus, guiding them through complex multi-step problem solving, and freeing them from boring and error-prone calculations. Interaction with the computer also encourages clarity of ideas and precision of expression. Our goal is to produce a general purpose tool that can be applied to any calculus problem, not a closed system that can only step through pre-programmed exercises. The computer therefore encourages creativity through rapid feedback and experimentation, making the exploration of mathematics more exciting and enjoyable.

Univ. of Southern California

Contact:

W. PROSKUROWSKI, Department of Mathematics, University of Southern California, Los Angeles, CA 90089. PHONE: 213-743-2567. E-MAIL: proskuro@castor.usc.edu.

Institutional Data:

The University of Southern California is a large private university with about 20,000 full-time undergraduates. In the Department of Mathematics there are about 35 full-time faculty, 90 graduate students, and a small body of undergraduate majors. The majority of undergraduate students comes from the Engineering Departments.

Project Data:

Calculus I and II are taught each semester in about 10-15 parallel classes with 60 students each.

About half of the instructors have used the micro-computer enhancement which was introduced (in an earlier version) in the Fall of 1983. The Department has a dedicated calculus microcomputer laboratory with 30 networked IBM PC-XT's with math co-processors and 4 Okidata printers. Software was developed with no external funding; students pay lab fees.

Project Description:

The University of Southern California offers a novel two-semester calculus sequence utilizing a multi-user, networked microcomputer system with graphics terminals. The students run user-friendly, menu-driven software which illustrates calculus concepts and helps to solve calculus problems. The mathematical content of our courses is virtually the same as that of traditional calculus courses. The number of hours of instruction is also unchanged, even with the enhancement of the micro-computer system. The software is a cohesive set of numerical (and one symbolic) modules. Each module is designed to be highly interactive. This keeps the student from becoming a passive observer. The student defines the function, specifies the input parameters, and guides the module through its calculations. The computer responds with an integrated text and graphics display that attracts the student's attention and stimulates his imagination. Available materials are *Comp-U-Calc*, User's manual/workbook and program diskettes by Proskurowski and Hickernell for about $25.

Univ. of So. Carolina at Aiken

Contact:

ROBERT PHILLIPS, Department of Mathematics, University of South Carolina at Aiken, 171 University Parkway, Aiken, SC 29801. PHONE: 803-648-6851, x 3471.

Institutional Data:

USCA is a four-year institution, one of nine campuses in the University of South Carolina system. The campus has approximately 2500 undergraduate students; most are commuters. The Department has 11 members. On average, eight students per year graduate with a joint BA degree in mathematics and computer science.

Project Data:

The calculus laboratory was designed by three faculty, and has been taught by two of them. 15-25 students are enrolled in each section, and all sections are taught in the same way. Full implementation begins in the Fall 1990.

True Basic Calculus will be the principal software used, and HP 28S's and IBM PC's will comprise the hardware. Anton's *Calculus* is the required textbook. The South Carolina Commission of Higher Education has granted $75,000 to this project, but this funding is still subject to further state budgetary cuts.

Project Description:

The project consists of adding a weekly two-hour calculus laboratory to our existing four lecture hours per week Calculus I and II courses. The laboratory hours are required of all calculus students. The purpose of the laboratory is to collect both computer and physically generated data in support of the calculus curriculum. Each lab project requires a written report detailing data collection and conclusions reached. Faculty will be assisted in the lab by an advanced undergraduate student.

This project is strongly supported by the Department. The natural scientists on campus will assist in designing physical experiments and in the generation of empirical data. Computer-generated data is directed at numerical and graphical evidence for limits, derivatives, differentials, proper and improper integrals, sequences, and series. Close attention will be paid to the use of numerical methods for optimization and integration problems, especially where symbolic methods fail or are too difficult to apply.

A primary goal is to extend the application of calculus to "real" problems; hence, the need for physical experiments. Over the next few years, faculty teaching in the project are expected to assist in the creation of a laboratory manual based upon those laboratory experiments which prove to be pedagogically sound.

University of Vermont

Contact:

JAMES W. BURGMEIER, Department of Mathematics and Statistics, University of Vermont, Burlington, VT 05405. PHONE: 802-656-2940.

Institutional Data:

The University of Vermont is a public institution with a modest but expanding graduate program in mathematics. Approximately half of the undergraduate student body is from Vermont. The University enrollment is about 8500 undergraduates and 1500 graduate students.

The Department of Mathematics and Statistics has about 25 full-time faculty. Two separate degrees are offered in mathematics: a B.A. degree for Arts and Science majors, and a B.S. degree for students majoring in Mathematics from the engineering college. A typical graduating class confers about 40 B.A. degrees and about 35 B.S. degrees. The Mathematics Department has the largest enrollment of any department on campus. More than 4800 students take mathematics courses each year.

Project Data:

At this time the author is the only faculty member involved in this project. However, newly acquired equipment for several classrooms will make it possible for other faculty to join the project. At least ten faculty members will participate in the project in the Fall semester of 1990.

This project has involved about 90 students in 3 sections of a first-semester calculus course that has total enrollment of about 900. The course is taken by majors in the life sciences, social sciences, and managerial sciences. The project has been active for 2 semesters. The software used is Epic running on AT&T 6300 microcomputers. Each student has their own computer at all times during each class meeting. The textbook has been *Calculus and Its Applications* by Goldstein, Schneider, and Lay but will be changing to *Calculus with Applications* by Burgmeier, Boisen, and Larsen for the Fall semester of 1990. Finally, an NSF ILI grant has been awarded to the University, and will be used to outfit several classrooms with a computer and screen-projection equipment.

Project Description:

The central theme of this project is an experiment to use microcomputers intensively as a tool to teach differential calculus to non-mathematics majors. Students use the computers during class, do homework with the computers, and are allowed (encouraged) to use the computers during examinations. Although this course has been taught for many years, the syllabus has several drawbacks: too much time is spent on constructing graphs by hand;

more time should be devoted to modelling; and examples and exercises typically are artificially constructed to provide an easy, pleasant solution, misleading the student into thinking that most similar problems can be solved by hand.

There are several reasons why a computer will help: computers provide speed, accuracy, and impartiality; computers can do animation; many calculation-oriented problems can be done easily with a computer, but are too tedious to do even with a calculator (e.g., Riemann sums); computers enable the class to assimilate some topics more quickly; and the computer makes students active rather than passive learners.

Although the daily format of the class is traditional, many traditional topics are covered in non-traditional ways. For example, roots, critical points, inflection points, and asymptotes are introduced graphically first, then algebraically. Questions ranging from roots to max/mins to tangent lines to Riemann sums are handled by the software. The emphasis of the course is now how to phrase a problem (mathematically) so that the software is useful in answering the question. Understanding the problem now occupies more time than the mechanics of working the problem.

There have been some surprises in this experiment. The most startling surprise was the fact that the students ask more mathematical questions with this approach than with the traditional approach. The software provides insight, confidence, and problem-solving power to the students. However, effective use of computers by the students requires careful planning of the in-class experiments and lectures using the computer as well as educating students to thinking in a "what-if" manner in a mathematical setting.

This on-going experiment has generally been successful; each semester provides an improved approximation of effective use of computers in the classroom. The Fall 1990 version of this project will consist of a single computer screen-projector format, instead of each student at a personal computer, and will be extended to the calculus courses for mathematics majors. Materials for such a format will be available by September at a nominal fee to cover copying and mailing.

Univ. of Wisc. Ctr.—Waukesha

Contact:

WALTER SADLER, Department of Mathematics, University of Wisconsin Center—Waukesha Campus, 1500 University Drive, Waukesha, WI 53188. PHONE: 414-521-5448. E-MAIL: sadler-@csd4.milw.wisc.edu.

Institutional Data:

The University of Wisconsin Center—Waukesha County is the largest of the 13 two-year University of Wisconsin campuses. There are 2160 full and part-time students (1478 FTE). The Mathematics Department, consisting of 10 full-time and 2 part-time faculty, has the largest teaching load at this Campus of the University (1200 students each semester), primarily in precalculus and two sequences of calculus.

Project Data:

To date only the proposer has taught in the project. Both regular and project course settings average about 25 students per class. The course has been taught three times in the present form of revision, and was taught six times in an earlier form during the late 1970's. A computer laboratory is available and recommended, but not required. In the late 1970's the principal software used was home grown Basic programs. In recent use, Epic has been used for classroom demonstrations. We use IBM XT clones. The experimenter's text *A Model Calculus* (unpublished) is used. The experimenter has been granted a semester's sabbatical to work on the project.

Project Description:

In the late 1970's, an experimental class using these materials with computer demonstrations did significantly better on modelling calculus world problems on a final exam than did a control class, and showed no differences on other questions. The most recent revision has not been tested against a control group, but class success (on comparable exams—often the same) has been as good as found in the late 1970's.

It has been hard to interest local colleagues in experimenting with this text, which has unusual emphasis on modelling, as well as nonstandard ordering of topics. Therefore the experimenter is looking for other experimental sites and is also striking out in another direction to create a "lively" dynamic computer book—a learning form that not only gives a new way of implementing curricular reform, but also includes a significant new teaching method that is independent of the classical classroom setting. Use of notebooks on a Computer Algebra System allows the integration of a calculus revision text with a super calculator mathematics tool to provide a dynamic computer textbook. This project will test the viability of dissemination and the cost-effectiveness of such calculus reform "dynabooks."

This calculus project's product is expected to impact calculus and mathematics teaching in several ways:

1. The dynabook provides an alternative for dissemination of new ideas for teaching mathematics.
2. The dynabook is easier and cheaper to duplicate and mail than a textbook.
3. Since the dynabook runs on a computer, it gives an unbiased presentation medium not encumbered by the gender and minority stereotypes often present in a classroom.
4. This freeing of the calculus and mathematics from the classroom will have obvious implications for university extension.
5. Technology has finally provided an answer to the problem of keeping up with the ever-changing knowledge base.

The written text is now available on disk in MS Word format. The dynamic textbook is under development, with sample demonstrations for the Macintosh expected by January 1991. Send a blank disk for a copy.

Univ. of Wisc.—Platteville

Contact:

FREDRIC W. TUFTE, Department of Mathematics, University of Wisconsin—Platteville, Platteville, WI 53818. PHONE: 608-342-1745. FAX: 608-342-1232.

Institutional Data:

The University of Wisconsin—Platteville is a four-year public university of approximately 5200 students. The Department of Mathematics with 22 full-time faculty supports a major in mathematics. Most of the calculus students are enrolled in the College of Engineering. The number of mathematics majors fluctuates between 70-100, with approximately 15-20 graduating each year.

Project Data:

The project consists of a one-credit supplemental course called Computational Calculus that can be taken by students that are enrolled in the regular calculus sequence. Another prerequisite is knowledge of a high-level programming language. About 450 students are enrolled in the regular 3-semester calculus sequence, and the project enrollment has been 20-25 students per semester. The course has been taught for 5 semesters.

Students do most of their homework in an IBM PC laboratory using programs they have written and graphics utilities which are furnished (Master Grapher). The project originated with a small curriculum development grant funded by the College of Arts and Sciences. The materials used in the course were produced locally. One faculty member has been involved with the project to-date.

Project Description:

The one-credit Computational Calculus course meets for 50 minutes each week for one semester. The classroom in which it is held contains an IBM PC, small monochrome monitor, and 2 large color monitors mounted above the blackboard. A major portion of each class period is spent with students demonstrating their problem solutions. The instructor rarely demonstrates programs or solutions.

The Mathematics Department has not had available a lab in which it could hold classes. As a result most students have had to do their work in a general university computer lab available to all students at all times. This lab contains IBM PC's, several printers, and has a consultant on duty.

Calculus students have difficulty in bridging the gap from algorithmic and computational proficiency to conceptual understanding and the ability to apply calculus concepts. They have great difficulty in understanding the notation of mathematics and consequently in reading mathematics. Our hypothesis is that by programming numerical procedures corresponding to the central ideas of calculus (limits, derivatives, Riemann integrals), students must pay attention to the language of mathematics and the definitions of these concepts. Using the programs which they have written, as well as graphing utilities that are provided, we require that students reconcile numeric, graphic, and algebraic representations.

Two conceptually oriented exams are given. They require students to understand the symbolic descriptions of limits, derivatives and Riemann integrals, and to interpret those descriptions in terms of the symbolic, graphic, and numeric representations of the functions. Students do not use computers on their exams. We require that all programs be turned in on a disk at the completion of the course.

The response from students taking the course has been extremely favorable. Most wonder why it is not a part of the regular calculus course. The most common comment: "I can work the problems in the regular calculus course, but this course taught me what derivatives and integrals really are." Preliminary evidence indicates that students have increased conceptual understandings.

The Mathematics Department has recently gained approval and funding to establish a computer laboratory within the Department. Experience with the calculus project made a significant contribution to the Department's argument for this laboratory.

The materials used in the course can be obtained from the author for $4 for duplicating and postage.

Ursinus College

Contact:

NANCY HAGELGANS and WILLIAM E. ROSENTHAL, Department of Mathematics and Computer Science, Ursinus College, Collegeville, PA 19426. PHONE: 215-489-4111.

Institutional Data:

Ursinus College is a private four-year liberal arts college enrolling about 1200 full-time students, all undergraduates. The faculty of the Department of Mathematics and Computer Science consists of 8 full-time members. In recent years the typical graduating class has included 12-18 majors from the Department.

Project Data:

There are two distinct calculus experiments at Ursinus: Humanistic Calculus, taught by Rosenthal and Calculus I with Derive, taught by Hagelgans.

Humanistic Calculus: During the academic year 1989-1990, 28 students enrolled in the project course, and 180 students took one of Ursinus's two standard introductory calculus courses. The project course has been taught for two semesters (Spring only). A computer laboratory is not required. No software or hardware is used. The text

is an unpublished book by Rosenthal entitled *The Satanic Calculus* written specifically for use in the project course.

Calculus I with Derive: During the Fall Semester of 1989, 50 students enrolled in the project course while 49 students enrolled in the standard Calculus I course. The project course has been taught for one semester. A computer laboratory is required; the only software used was Derive. The hardware consisted of Leading Edge Model Ds with Hercules graphics monochrome monitors. The Model Ds are IBM XT clones, each with 640K of memory. The computer laboratory contains 23 computers linked in a Novell network. The textbook was *Calculus* by Stewart.

Project Description:

Humanistic Calculus is inspired by the idea that *all* college students, including and especially those who have been depressed and repressed by conventional mathematics pedagogy, can benefit by a vernacular-based calculus course rooted in an historical, foundational, and philosophical perspective. Its intended audience consists of humanities and social science majors who must complete one mathematics course as a college graduation requirement. The standard secondary-school curriculum has left most of these persons bereft of curiosity, a sense of self-efficacy, and even mathematical ability. Humanistic Calculus offers its participants one final opportunity to recover something of value from their schooling in mathematics.

This is accomplished within a theoretical framework in which knowledge is construed as socially and subjectively constructed by human agents acting from concrete experience. Students are invited to exercise their subjectivity and create their own knowledge by a critical reading of an English-language text in tandem with a dialogue-based learning process. The conversations occur in many settings, including appointments with the instructor, journals, and the non-authoritarian class discussions that replace lectures. Two undergraduate peer tutors, positioned midway between the students' fears and the instructor's authority, help to mitigate the former and mediate the latter.

Humanistic Calculus's content is the calculus, treated from the perspective of discovering methods for measuring physical objects and processes. The course commences with finding the area inside any circle; the ancient Greek approach then spawns a unifying metaphor for finding unknown quantities:

create a sequence of approximations progressing toward the quantity, then discover its limit. In all such investigations, including those involving continuously changing quantities, only sequences are employed.

The support of Ursinus's faculty and administration has been instrumental in enabling Humanistic Calculus to become a valued and permanent course offering. An information packet, which includes two published papers on the course, excerpts from student journals, and the introductory chapter of the textbook is available from Rosenthal at a cost of $3. Copies of the entire text can be purchased for $25.

Calculus I with Derive was introduced in the Fall Semester of 1989 in two sections. Students were introduced to the computer system Derive through a sequence of assignments developed during the preceding summer. One aim was to emphasize concepts by freeing the students from computing and graphing. Other aims were to improve the class attitude with a calculus experience different from high school, and to reduce the number of withdrawals.

The class met twice a week for 1 hour 15 minutes. One class period during each of the first two weeks was spent in the microcomputer laboratory. After that, all computer work was completed as homework outside of class and there was little class time spent on Derive.

During the class sessions in the microcomputer laboratory, the students worked with partners at their own pace. The material was meant to be self-explanatory, and students were encouraged to help each other. The instructor usually was able to keep up with any problems and questions as they occurred. There was enough work planned to keep all students busy throughout the class period, and students were permitted to finish the assignments as homework.

All the assignments were graded with corrections and comments by the instructor. Throughout the semester the exercises became less structured so that students were led to develop their own examples. The grades, which counted as one-eighth of the total grade, included generous partial credit.

It is difficult to evaluate the results of the use of Derive in the calculus classes. The reduction in the withdrawal rate from 27% one year to 6% the next year might be attributed to the interest generated by the use of Derive, or to the expanded personal interaction with the instructor fostered in the computer laboratory. Additionally, students were observed talking to each other about calculus as they worked together at the computers. There was no discernible increase in the depth of understanding of calculus concepts for the class as a whole, but then the lower withdrawal rate indicated some success with weaker students.

The Department supports the use of Derive in calculus classes for prospective science and mathematics majors. Next Fall, all sections of Calculus I will use Derive in a course expanded from 3 to 4 credits. Derive exercises also will be introduced in second and third-semester calculus courses. The computer exercises are available from Hagelgans at a cost of $3.

U.S. Coast Guard Academy

Contact:

ERNEST J. MANFRED, Department of Mathematics, United States Coast Guard Academy, 15 Mohegan Avenue, New London, CT 06320. PHONE: 203-444-8394.

Institutional Data:

The United States Coast Guard Academy is one of five federal service academies. Its primary purpose is to prepare young men and women for service in the United States Coast Guard. All cadets earn a Bachelor of Science degree. 75% of which are in Engineering or Applied Sciences, while the remaining 25% are in Government and Management. There are approximately 900 cadets. The United States Coast Guard Academy is the only service academy which tenders appointments solely on the basis of an annual nationwide competition.

The Mathematics Department has twelve faculty members: four civilian, three permanent military officers, and five rotating military officers. There are approximately 100 mathematics and computer science majors. As a core requirement, all cadets are required to take a full year of calculus plus a calculus-based probability and statistics course.

Project Data:

The calculus sequence represents the core of the Academy's science and technical programs and has a significant impact on the cadet's academic success. It is extremely important that these courses provide the cadets with an in-depth understanding of the material covered. To this end, the use of computers in the calculus courses at the Academy has become commonplace.

The initiative for the use of computers in the calculus courses was modelled after several ideas which were stated in an article appearing in the September 1985 issue of *SIAM News* by Breuer and Zwas. This article introduced the notion of a computational laboratory for the calculus that would allow students to use microcomputers to test ideas and conjectures by utilizing numerical evidence. The first attempt at utilizing the laboratory involved scheduling students for a lab period once every eight meetings. The classes, consisting of no more than twenty students, would go into the laboratory and utilize True Basic Calculus package to complete prepared lab assignments. The cadets' response to the laboratory was very positive. They felt that the computer provided an opportunity to visualize and experiment with different situations thus providing a better understanding of the concepts covered in class.

In 1986, the Academic Division made the decision to require all entering freshmen to purchase a Macintosh computer. This decision, coupled with class sizes of over twenty cadets, caused the department to rethink its implementation of the computational laboratory. The scheduled lab periods were replaced by classroom demonstrations and computer projects. Software is available in the department and was written by a faculty member. All sections in the calculus are given a minimum of two computer projects a semester.

Project Description:

The following sample of projects involve the computer. These projects represent a fifth of the cadet's grade in the course. Each project is submitted in the form of a written report.

Evaluating the designs of buoy hulls: Students are given several navigational buoy hulls which have been described as solids of revolution. Equipped with a general function evaluator and a numerical integration program, they must determine if the buoy hulls float within specified design criteria.

Analyzing static stability curve: Students are given data points for the static stability curve and the heeling moment curve of a Coast Guard cutter along with a numerical integration program. Students are asked to find the permanent angle of list and the maximum angle for the ship.

Determining underwater volume of a ship: Students are given a table of offsets for a ship along with a numerical integration program and they are asked to find the underwater volume of the ship.

Each of the projects listed above requires the students to exercise their knowledge of calculus as well as considering error tolerances, relative error, and approximation techniques. The student response to such computer projects has been good. They understand the need for exactness and the importance of translating concepts to problem solving.

U.S. Military Academy

Contact:

DAVE OLWELL, Department of Mathematical Sciences, USMA, West Point, NY 10996-1786. PHONE: 914-938-4811. E-MAIL: ad7417@trotter.usma.edu.

Institutional Data:

USMA is a public four-year federal service academy. We have 4400 full-time undergraduates. All cadets complete a four-semester mathematics core curriculum, and a five-semester engineering sequence. The student body is roughly 10% female and 10% black. Our first-year cadets have a mean mathematics SAT of 646.

The Department of Mathematical Sciences has 68 full-time faculty members who are officers in the US Army. We have 20 mathematics majors in a typical graduating class of 1000.

Project Data:

We are moving this fall to a sequence of one semester of discrete dynamical systems, followed by two semesters of calculus with a "lean-and-lively" focus, and capped by a semester of probability and statistics. All our cadets who enter after this summer will take this new sequence. We will have approximately 20 faculty members teaching the sequence this fall, and that will increase to about 40 the second year. Our class size will be at most 18 cadets.

Each cadet purchases an IBM PC/386 compatible computer, as well as an HP-28S advanced scientific calculator. We support the hardware with a mandatory issue of Derive, Mathematics Plotting Package, Quattro, Minitab, and several local products. For our Discrete Dynamical Systems text, we will use *Discrete Dynamical Systems* by James Sandefur. For our two semesters of calculus, we will continue to use *Calculus, One and Several Variables* by S.L. Salas, *et al.* We have no external funding.

Project Description:

The semester of discrete dynamical systems (DDS) will progress from discrete algebra to matrix algebra to discrete dynamical systems. It will preview the most difficult concepts (for example, the limit) that underlie continuous mathematics. The DDS course will have a heavy modelling emphasis. Instructors and students will use computer software to demonstrate and solve problems in both the matrix algebra and the discrete dynamical systems sections of the course. We also will instruct our cadets on the use of Derive, Quattro, and the HP-28S in this course.

Our second and third semesters will be Calculus I and II. Calculus I includes the differential and integral calculus through techniques of integration. Calculus II introduces differential equations, parametric equations, series and sequences, and the study of multivariable calculus including 3-dimensional geometry and vectors, functions of two variables, partial derivatives, and multiple integrals. We rely heavily on computer software and advanced scientific calculators throughout the course.

Our final semester of the core curriculum is Probability and Statistics. As a service to the other departments, we will offer several different fifth-semester courses, tailored to different majors and fields of study. Other departments were briefed on the new curriculum, and the department heads unanimously concurred on our curriculum revision. All departments will continue to allow unrestricted use of the HP-28S in all classes and on all tests.

We will assess our success or failure by several means. We will compare results on our final examinations with historical results. The departments served by the core curriculum will test in their initial class meetings to see if cadets arrive better prepared for upper-division coursework. West Point currently surveys all students at various points in their cadet careers. We will include survey questions to measure attitudes toward mathematics, other sciences, and the engineering thought process to determine any changes. We will continue to track the distribution of grades, number of mathematics and operations research majors, success at the national mathematics competitions, and feedback from the other departments. Finally, our department has an "Across the Curriculum Coordination Team." This team coordinates our department approach to different issues, including core curriculum coverage, the use of Derive and HP-28S, writing requirements in our core courses, student development and growth, faculty preparation and development, and student evaluation.

U.S. Naval Academy

Contact:

HOWARD LEWIS PENN, CRAIG K. BAILEY, Department of Mathematics, United States Naval Academy, Chauvenet Hall, Annapolis, MD 21402. PHONE: 301-267-3892. E-MAIL: hlp@usna.navy.mil.

Institutional Data:

The Naval Academy is a four-year undergraduate college devoted to the mental, moral, and physical preparation of Midshipmen for service as officers in the Navy and Marine Corps. The student body consists of 4500 full-time students. The core curriculum is mainly technical. Every Midshipman is required to take three semesters of Calculus and one semester of Differential Equations. The work load for the students at the Academy is high, including an average of 18 credit hours per semester in addition to required daily participation in athletics and military activities. The Mathematics Department consists of approximately 70 full-time instructors. There are approximately 40 mathematics majors per graduating class.

Project Data:

This project has been designed for use in the standard Calculus track taken by approximately half the students. Last fall in Calculus I there were 464 students enrolled in 25 sections taught by 14 instructors; in Calculus II there were 307 students in 16 sections taught by 9 instructors. This semester there are 362 students in Calculus II registered in 24 sections taught by 13 instructors. The inclusion of computer assignments has been in effect for three years. The principal software packages used are a program produced at the Naval Academy, MPP, Microcalc, and Calc Pad. Every student has an 80286-based PC with an EGA card and color monitor. Every faculty member in the department has a similar machine, and every classroom used by the Department has a computer with a tablet for display on an overhead projector. Each student receives a copy of the programs. The text used is *Calculus with Analytic Geometry, Second Alternate Edition* by Swokowski. The project has been funded by an Instructional Development Project Grant from the Naval Academy.

Project Description:

The project consists of a series of 43 computer and supplemental assignments that have been added to the syllabi of the three Calculus courses. These syllabi are given to each student taking the course. The instructors assign these exercises as part of the homework.

There are several goals of this project. One of the most important of these is to stress the concepts of Calculus. A second goal is to reinforce the connection between the analytic and graphical representations of functions and equations. Another goal is to place additional emphasis on the numerical aspects of Calculus. A last goal is to introduce more and different applications of the Calculus.

The project has strong backing from the Department Chairman and the administration as a whole which has a strong commitment to the use of computers in all the core courses taken by Midshipmen. Copies of the program, MPP, and the computer and supplemental assignments have been sent to approximately 300 faculty members at other colleges, universities, and high schools.

The program runs on any IBM compatible with a graphics card and at least 360K memory. It makes heavy use of graphics and color. Therefore, it is recommended that the program be used on computers with an EGA or VGA card and a color monitor.

The feedback on these materials has been very positive. The program, documentation, and the collection of assignments may be obtained by sending $1 to cover the cost of the computer disks to Howard Lewis Penn. While the assignments are used as homework at the Naval Academy, they would be equally useful in a laboratory setting.

Wellesley College

Contact:

ALAN SHUCHAT or FRED SHULTZ, Department of Mathematics, Wellesley College, Wellesley, MA 02181. PHONE: 617-235-0320. E-MAIL: ashuchat@lucy.wellesley.edu or ashuchat@wellco.bitnet; fshultz@lucy.wellesley.edu or fshultz@wellco.bitnet.

Institutional Data:

Wellesley College is a four-year liberal arts college for women with approximately 2100 full-time students.

The Mathematics Department has 12 full-time faculty members. Most Wellesley students take at least one mathematics course, although there is no mathematics requirement for graduation. The Department averages about 1000 enrollments per year, one of the largest enrollments among the sciences, and each year the College graduates approximately 25 mathematics majors. Many of our majors complete majors in two departments, with mathematics-economics and mathematics-computer science as two frequent combinations.

Project Data:

Eight of our twelve full-time faculty have so far participated in the project, which is in its first year. The project includes about 450 students in 19 sections of the two-year core calculus sequence, as opposed to about 225 students in 9 non-project sections. Project and non-project sections use the same texts: for first-year Calculus, Stein's *Calculus and Analytic Geometry* is used; for Intermediate Calculus, Marsden and Tromba's *Vector Calculus* text is used; for Methods of Advanced Calculus, Marsden and Tromba and additional material on complex analysis are used. Project sections have the same profile as others, and students are not sorted into sections according to intended major or ability. Project and non-project classes normally meet two or three times a week for an average total of 175 minutes, and are taught by the faculty in sections of about 25-30 students. There are no TA's, but student homework readers are usually available.

The project sections use a new Mathematics Graphics Classroom equipped with 15 Macintosh IIcx's for up to 30 students, an instructor's IIcx attached to a ceiling-mounted Electrohome color projector, a LaserWriter NTX, and an Imagewriter II, all on an AppleTalk network with file server. The principal software for the project is Mathematica, together with Calculus Menus, a front end being developed at Wellesley with SuperCard. Additional small-scale, special-purpose software includes Analyzer*, Analyzer 3D, MacFunction, MacGrapher, and MacVector. The Graphics Classroom was established with the aid of matching grants from NSF ILI, and grants in-kind from Apple Computer and Wolfram Research.

Project Description:

Our goal is to enhance students' ability to approach mathematics in a more active and creative way, to develop their mathematical intuition and insight, especially in visualizing higher dimensions, and improve their skill in observing mathematical phenomena, making predictions, and testing hypotheses.

The Calculus Menus program we are writing gives beginning students the ability to use Mathematica to achieve these goals without needing to learn its commands or syntax. It passes the appropriate commands to Mathematica according to choices students make from standard Macintosh menus and responses they give in the dialog boxes that then appear. These enable or will enable students to do such things as graph, differentiate, and integrate functions (symbolically and numerically), study sequences and series, visualize surfaces, gradients, and level curves, and manipulate algebraic expressions and matrices. Special items include using Newton's method and Lagrange multipliers. Calculus Menus is being written primarily by Fred Shultz, with student assistance.

The project sections use the Mathematics Graphics Classroom in multiple ways. Some instructors have all class meetings there, others use it once a week, and still others use it for occasional laboratory sessions. Instructors demonstrate examples and applications that draw attention to essential parts of the theory, with active class participation. In some cases, students are given access to the room but no special meetings are held there. The room is available for doing homework. Thus, students use the Graphics Classroom to carry out experiments that motivate the introduction and discovery of new mathematical concepts, or illustrate ideas already presented by the instructor; to analyze additional examples as part of their homework assignments, and to carry out extended numeric and symbolic computations; to compute parts of in-class or take-home examinations; and to study for examinations or for individual experimentation.

Westminster College

Contact:

J.E. HALL, Department of Mathematics and Computer Science, Westminster College, New Wilmington, PA 16172. PHONE: 412-946-7284.

Institutional Data:

Westminster College is a private, four-year liberal arts college of about 1300 students. The student body is almost entirely full-time, non-commuting, non-minority. There is a small cooperative engineering program with Penn State (3-2 program), and a large department of economics and business.

The Department of Mathematics and Computer Science has nine full-time faculty, and embraces three majors: mathematics, computer science, and computer information systems. In recent years we have graduated about 20 majors, slightly more than half in mathematics (including mathematics education), the rest in the computer majors.

Project Data:

During 1989-90, one faculty member (J.E. Hall) taught in the project which involved all Calculus I students (67 students, 3 sections) in the fall term, and all Calculus II students (35 students, 2 sections) in the spring. (About 25 of these students will complete Calculus III with Hall in the fall of 1990. Those who started Calculus I in 1988-89 or in the spring of 1989-90 were not involved.)

[A new group of about the same size, again comprising all Calculus I students, will begin in the fall of 1990 with two or three teachers (J.E. Hall, W.D. Hickman, C.K. Cuff).]

A computer laboratory was used as an out-of-class resource in Semester I, and for one class per week in Semester II with two primary software tools: Framework III spreadsheet and Derive. (Some students used Basic or Pascal programs or personal calculators from simple to sophisticated.) The "Microlab" classroom is equipped with 20 IBM PC/30s with hard drives.

The textbook was *Introductory Calculus with Analytic Geometry* by Edward G. Begle (reprinted from the June 1960 printing by permission of the publisher). It contained about 2/3 of the material for Calculus I, and 1/3 of that needed for Calculus II. The balance was written by Hall (this included all laboratory materials).

There was no external funding. The only internal funding came out of Hall's hide and leisure time.

Project Description:

In courses entitled "Exploring the Mathematics of Change" and Extending the Mathematics of Change," materials of a two-semester, single-variable calculus course (for all non-business liberal arts students) were developed through group and individual problem solving activities, writing assignments, and laboratory exercises, together with traditional problem sets, quizzes, and exams.

First Semester: Three 60-minute periods per week traditional classroom; approximately one day per week on problem solving (class or group), one day per week on precalculus and calculus computational skills (review, drill and practice, testing), one day per week on underlying theory. Computer with overhead projector was used in classroom for demonstrations with Framework spreadsheet and Derive (groups completed projects outside using computer labs).

Second Semester: Two 60-minute periods per week regular classroom; one 60-minute period per week "microlab classroom" (see above); classroom periods divided between skills development and theory presentations; problem solving through ten lab exercises (10-15 minute presentations, 45-50 minutes for project teamwork).

Chief Features: Major emphasis on problem solving; subsidiary emphasis on historical perspective, social utility, and aesthetic appreciation; early use of all transcendental functions; use of technology; use of writing (group and lab projects, verbalization of principles, exam essays).

Assessment: Informal (course viewed as *evolutionary,* not *revolutionary:* many individual features used earlier by Hall). Anecdotal: students seemed better motivated and more interested, had better grasp of concepts. (Those weaker in computational skills know how to get help from Derive.) Chairman and departmental colleagues were supportive and enthusiastic.

Weaknesses: Lack of single text (but still better than huge commercial books); lack of formal lab period first term; need to steal second term lab from regular class time; too little structure for skills development component; need to use writing more systematically.

Plans for Next Year: One 45-minute lab and three 60-minute class periods; use of Dick-Patton text materials from Oregon State; involvement of more staff.

Materials Available: Course syllabi, set of major applied problems from first term, set of ten lab specs from second term. Order from the department for $5 to cover postage and handling.

Whitman College

Contact:

DOUGLAS H. UNDERWOOD, Department of Mathematics and Computer Science, Whitman College, Walla Walla, WA 99362. PHONE: 509-527-5151.

Institutional Data:

Whitman College is a private four-year liberal arts college with 1200 students, primarily residential. A large number of graduates attend graduate or professional school. The Department has 7 full-time faculty. From 10-20 majors (some combined, e.g., math-physics, math-computer science) graduate each year.

Project Data:

The special section of first-year calculus has been offered for two years by one instructor. Section size has been 15-25 students; the section is limited to students with no exposure to high school calculus. Standard section sizes are up to 40 students. No separate computer lab is required, but students must use the computer for numerical integration. We are using Microcalc on PC's this year. The text was written by Underwood. A PEW Foundation grant supported revisions for the Second Edition.

Project Description:

The intended audience is *beginning* calculus students. Enrollment has been limited, as far as possible, to students with no previous exposure to calculus in order to avoid having students skip the important reasoning steps if they already know the results.

The unique features of this project are the nature of the text and the type of work expected of the students. The text leads the students through examples which motivate and derive the standard calculus applications and formulas. Most formulas are not presented in the text. Rather, students are guided through their derivation. They are required to supply steps of algebra and reasoning. The expectation is that the student who is actively involved in this way will learn more about the meaning of calculus than the student who passively reads formulas and examples from the text.

The class meets for three 50-minute sessions per week. Students are encouraged to work in small groups outside of class. The notebooks in which they do this outside-of-class work are collected and read weekly to make sure that they are keeping up. In class students are asked to present the missing steps in derivations and examples and the solutions to exercises. Because students are expected to learn the material themselves, there is more discussion than lecturing in class. But the instructor must be alert to problems and be prepared to discuss material (including review material) as necessary.

Our final exam scheduling makes it impractical to give a common final to all sections of calculus, so comparative exam scores are not available. In spite of the demands, students rate the method highly.

The text is produced locally in TEX. As yet it has not been used elsewhere. Inquiries should be directed to Underwood.

Wilkes University

Contact:

RICHARD E. SOURS, Department of Mathematics and Computer Science, Wilkes University, Wilkes-Barre, PA 18766. PHONE: 717-824-4651.

Institutional Data:

Wilkes University is a private four-year institution with approximately 2000 full-time undergraduate students. There is an active evening and summer part-time undergraduate program. About two-thirds of the full-time students live on campus. The School of Engineering and Applied Science is large.

The Mathematics and Computer Science Department consists of 12 full-time and 12 part-time people (approximately 15 FTE's). The Department offers undergraduate majors in mathematics, computer science, and computer information systems. There are about 100 undergraduate mathematics/computer science majors. About 200 students take Calculus I each Fall semester.

Project Data:

As of Spring 1990, two people have taught Calculus I and II in the project, and one person has taught Differential Equations. For Calculus I and II, one or two sections of 25-35 students per section were involved in the project, versus 4-6 sections not involved. One of three Differential Equations sections was involved in the project. The project consists of using a computer algebra system on Macintosh computers and teaching the courses in a laboratory format. For Calculus I and II we use Analyzer* and for Differential Equations we use DiffEq. Both are part of a package (but can be obtained and used separately) called MacMath developed by people in the Cornell University Mathematics Department. We have taught the labs by meeting one class per week in a Macintosh lab, and we have taught them in a regular classroom using a Kodak Datashow projection system. We have no external funding.

Project Description:

The main thrust of our project is to get the students to discover mathematics. In Calculus I and II the discoveries are clearly not original, but they are new to the students. We are operating on the assumption that ideas which the student discovers will be learned more thoroughly, retained longer, and provide more motivation. A good computer algebra system is an excellent tool for leading students to the edges of new ideas.

Where possible the labs and demonstrations are scheduled so that they introduce new topics. We try to work through lots of examples and we ask lots of questions. The lab sessions are not traditional. There is a lot of talk between students and a lot of conjecturing. The class ends with the instructor summarizing the new ideas learned during that lab. The students are then given a computer assignment which they complete on their own. The students turn in a 2-4 page lab report which typically involves some computer-generated graphs and their analysis of the graphs and data. There are 8-10 such labs each semester.

We are in the process of writing a Laboratory Manual for a computer-based Calculus I and II course. It will be independent of textbook and software. The plan is to have 20-25 labs on various topics from calculus. Each lab will consist of a description of the examples and problems for the lab class itself plus the computer assignments for the students. The first draft should be available (from this author) during the summer of 1990.

Other than a few isolated exam questions, we have not tried to do any comparative testing of the students in this project. They enroll voluntarily and can change sections if they do not want to use the computer. Informal faculty and student feedback has been positive.

Worcester Polytechnic Institute

Contact:

W.J. HARDELL, Department of Mathematical Sciences, Worcester Polytechnic Institute, 100 Institute Road, Worcester, MA 01609. PHONE: 508-831-5406.

Institutional Data:

Worcester Polytechnic Institute offers degrees through the Ph.D. The Department of Mathematical Sciences offers the B.S. in various areas of applied mathematics. This private technological university enrolls 2600 undergraduates. WPI's mathematics department has 28 full-time faculty.

There are 13-16 mathematics majors graduating each year, and most of these have pursued programs which prepare them for employment immediately upon graduation. WPI is widely recognized for its innovative approaches to engineering education.

Project Data:

After a 1988-89 pilot program which involved two sections of calculus, full implementation of the new sequence was begun in the Fall of 1989 with all 12 sections of Calculus I. By the end of the 1989-90 academic year, 15 faculty members will have taught the project courses and some 430 students will have been in the sequence. Although no formal computer lab is required, computer-oriented problem sets are assigned. The principal pieces of software used are MicroCalc 4.04 and Calculus-Pad 1.5. An AT&T 6300 is used in the classroom and in the computer workrooms; computer images are projected in class with the MagniView 400. Our text, selected in part for its wide range of applications, is *Calculus* by Hurley.

Project Description:

What is described here is a first response to the need for calculus reform—one that can be "easily" implemented with existing texts and software. The ultimate goal is to facilitate understanding of fundamentals, to use applications to show students the power of the fundamentals, and to use computers to help accomplish these goals and open the possibility for experimentation. As early as the first day, students can be presented with an overview of the ideas of differential calculus. The Zoom feature of MicroCalc is used to enlarge a portion of a graph until that graph becomes straight. Now the ideas of tangent line, linear approximation, derivative, monotonicity, and extrema are all available for discussion. A little later $\epsilon - \delta$ can be visualized by graphing the function on $[a - \delta, a + \delta]$ by $[L - \epsilon, L + \epsilon]$ and discussing how the function runs out of this box. We introduce numerical integration (with error bounds) immediately after defining the definite integral so that a variety of applications can be considered before doing techniques of integration. A packet of some of our materials (course outlook, instructor's notes, assignments, etc.) is available from Hardell at a cost of $3.

The new course sequence is aimed at all entering freshmen. WPI's program is based on four 7-week terms per academic year (students take 3 courses per term). Calculus is a five-term sequence with four 50-minute meetings per week, and class sizes ranging from 32-40. Sophisticated calculators are not used, since PC's are closer to the professional environment students will encounter after graduation. TA's are not utilized. However, the school sponsors M*A*S*H Leaders—specially selected undergraduates who do tutoring and much more. Efforts to reform calculus have been strongly backed by the Department Chair. Department members voted for the new sequence, although some question changes in emphasis in certain topics. Support from other departments has been good; they approved the new courses and have supplied us with examples of applications which occur in their courses.

References

Resource Collections for Calculus

Contact

A. Wayne Roberts, Department of Mathematics, Macalester College, St. Paul, MN 55105. Phone: 612-696-6337. E-Mail: ikgrun@macalstr.

Institutions Involved

Associated Colleges of the Midwest: Beloit, Carleton, Coe, Colorado, Cornell, Grinnell, Knox, Lake Forest, Lawrence, Macalester, Monmouth, Ripon, St. Olaf, University of Chicago.

Great Lakes Colleges Association: Albion, Antioch, Denison, DePauw, Earlham, Hope, Kalamazoo, Kenyon, Oberlin, Ohio Wesleyan, Wabash, Wooster.

Goals

It was the idea of our consortia to identify those recurring criticisms of the calculus course that raised legitimate issues related to the content of the course, as opposed to problems of placement of students, articulation with high schools, the use of TA's, etc. With such criticisms in mind, we next asked ourselves what resources a person would need to address the issues, and then we set about the business of both collecting from others and generating ourselves those resources.

These resources are being organized into five Collections, each addressing one of the criticisms we identified. It is to be emphasized that we are not trying to design the ultimate course or to write a text. The Collections we pull together will in their totality contain far more material than could be used in any single course. Our intention is simply to assemble the best materials we can find that address the needs we have identified, and then to put them at the disposal of writers, teachers, and interested students to use as they will.

Our Collections can be characterized as follows. They will be, to borrow a phrase from our computer science friends, in the public domain. They will be ready to use—to be copied and handed out if someone chooses to use them in that way. They will certainly deal with more issues than the use of computing instruments in the teaching of calculus, but where they raise questions best addressed with the aid of technology, those questions will be raised about mathematical content in a way that is independent of the user's particular hardware or software. They will be bound in a permanent form and indexed both with a course syllabus and with a traditional index to provide for easy access.

We now turn to a description of each Collection: its title, the editor, the criticism we intend to address, and indications of both content and format.

Learning by Experiment

Editor: Anita Solow, Grinnell.

Criticism: *The insights and joy of discovery that might come to students from "playing around" are methodically sacrificed in a traditional calculus course to efficiency and perceived need to cover material quickly.*

The underlying idea of this Collection is that high-speed computational aids now allow students to perform calculations, tedious and unappealing if done by hand, to experiment, to make conjectures based upon special cases, and generally to acquire a much better "feel" for many of the major ideas that are central to the calculus.

Students might, for example, explore the derivative of the rather natural looking function $f(x) = 2^x$ rather than being led quickly to $E(x) = e^x$; once they have learned that functions of the form $g(x) =$

b^x have derivatives of the form $g'(x) = kg(x)$, they can be asked if b can be chosen so that $k = 1$.

The materials to be developed in this Collection might be likened to those found in a laboratory manual. They will be intended to facilitate practical and conceptual learning in calculus through computation, directed and open-ended exploration, and forceful encouragement to analyze and interpret the results. The materials will have standardized form, will be organized into independent modules, and will be hardware independent.

Problems for a New Technology

EDITOR: Bob Fraga, Ripon.

CRITICISM: *Most exercises in current texts deal not with ideas but with rote technique, a fact underscored for us by the realization that most of them can be solved simply by entering them into a computer if not a hand-held calculator.*

We believe there is a need for new sets of homework exercises of a type that will not be rendered trivial by the use of instruments afforded by current and coming technology. Still having the purpose of reinforcing the main concepts, these exercises must be phrased in a way that requires students to actively think about the problem at hand.

We recognize a danger here. Replacement of routine drill with incisive questions could create a course devoid of the very things that provide any sense of success and personal progress for many students. We intend, therefore, to keep in mind that the average student needs a chance to develop confidence through success with most of a homework assignment.

The problems of this Collection will be distinguished from those of the previous Collection first in their intent, reinforcing ideas discussed in lecture rather than leading to those ideas through discovery, and second, in the nature of response to the questions posed; there will be a "right answer" that can be consulted at the back of the book, as opposed to the sort of speculation encouraged by the experiments of the first Collection. The exercises of this Collection will be organized into sections corresponding to major themes that occur in our syllabi, listing for each theme a set of exercises that reinforce the main ideas.

Applications

EDITOR: Phil Straffin, Beloit.

CRITICISM: *Textbook writers have tried to include applications for an increasingly diverse clientele, resulting in superficiality, and a reduced percentage of students interested in any particular application. Students are reduced to working problems they don't understand by using illustrative examples as templates for fitting new numbers into slots.*

Those who try to respond to this criticism may concentrate on just a few applications done in some depth, or they may try to individualize assignments according to a student's interest. These, or other methods one may devise to include meaningful applications all have in common the need for expositions that give background sufficient to understand the problem. The kind of exposition we envision will leave the student familiar with the notation commonly used in the discipline, and with the confidence that he or she will be able to follow similar uses when encountered in another course.

This will be a Collection of such expositions in the natural and social sciences, and in mathematics (in which are treated traditional topics about to be dropped in order to create the lean calculus).

Several models can be cited. The UMAP modules, at their best, illustrate what we have in mind, but have always suffered from problems inherent in loose packets that must be individually ordered well in advance of when they are needed. Taken as a whole, the Rorres and Anton paperback on *Applications of Linear Algebra* serves as a positive model for us.

Projects for Individual Exploration

EDITOR: Mic Jackson, Earlham.

CRITICISM: *Calculus students are never asked to develop a series of connected ideas in a way that calls upon the students to use multiple sources, to exercise their own imagination and judgment, and ultimately, to explain their ideas in a coherently written piece of mathematical exposition.*

This Collection will contain at least thirty projects which an instructor could assign to individual students, or preferably, to small groups of students to be completed over a period of one to two weeks. The projects will be self-contained extensions built upon standard course content; each will include instructions for the student and a separate set of detailed suggestions for the teacher; there will be suggestions for both implementation and evaluation, and at least one example of a correct way that students could complete the project.

We will attempt to uniformly distribute the projects over the content of the first two semesters of single-variable calculus. Our intent is that these materials should be accessible to all students taking the class, not just honor students.

Readings for Calculus

EDITOR: Underwood Dudly, DePauw.

CRITICISM: *Students typically complete several terms of calculus without acquiring any idea of the role it has played in the development of Western thought, or any notion of what its study should contribute to their own general education.*

Sometimes, in order to liven up a lecture, we would like to refresh our memory of history or folklore connected to some topic in calculus. Sometimes we would like an article to hand to a student who questions our concern for rigor, or wonders why mathematicians should worry so much about writing, or gets interested in the history of an idea. Sometimes we would like to ask a class to read contrasting ideas on the motivations for mathematical inquiry.

This Collection will provide such readings, and more. They can come from existing material reprinted with no changes, from existing material adapted for the audience, or they could be original contributions. Common characteristics will be potential interest for students, accuracy of the stories told, and a style consistent with the stated purposes: engaging to read, containing suggestions for further reading, but not burdened with the footnotes of a scholarly treatise.

We Invite Submissions

It is our intention to make the best ideas emerging from the reform projects widely available in a permanent form. The intent of involving 26 liberal arts colleges is not to produce materials aimed exclusively at the liberal arts student, but to give us a pool of able, interested, and experienced calculus teachers who can keep abreast of a wide variety of materials, and then assess, edit, and where necessary, write. To keep the project administratively manageable, editors and the people on whom they principally rely have been drawn primarily from these colleges.

It is to be emphasized, however, that we intend to draw upon ideas contributed from any source. We will approach people who have finished pieces or embryonic ideas on which we would like to draw, and we invite anyone with an idea, be it a single problem, an experiment, a project for student exploration, or an extended article, to submit it to one of the editors listed above, or to Project Director Wayne Roberts who will forward it to the appropriate editor. Contributions accepted for our publications will of course be acknowledged.

To the extent that our project succeeds, anyone interested at any level in teaching calculus will want to keep these books on a reference shelf at arm's length from his or her desk. It is clear to us that to succeed in this way we need to draw upon good ideas from as wide a variety of sources as possible. We earnestly solicit contributions from any interested person.

A Consensus Calculus Syllabus

What follows is a preliminary version of a syllabus for a one-year calculus course prepared as part of The Associated Colleges of the Midwest and The Great Lakes Colleges Calculus Project funded by the National Science Foundation.

The attached syllabus was developed by first soliciting suggestions from the faculty members of the twenty-six participating schools, returning all of the suggestions to all of the contributors and asking them to find as much common ground as possible, and by forming a committee of active contributors to combine the suggestions into a cohesive document.

The committee has worked with two goals in mind: that the course should be lean, and that it have unifying themes. The first goal is as simple to state as it is difficult to achieve: identify the basic ideas that should go into the core of any one-year calculus course. We decided from the outset to restrict the syllabus to ideas that would fit into 32-35 class meetings in each of two terms. We thought that such a syllabus would leave time for:

- Dwelling on topics that instructors particularly want to emphasize.
- Drawing from the resource collections to include large-scale applications, individual projects, and cultural background.
- Training students in the use of electronic learning aids.
- Testing.

For the second goal, while the first calculus course has the unifying ideas of differentiations and integration linked by the fundamental theorem of calculus, many instructors find the second course a conglomeration of ideas and techniques that students find unrelated. Our suggestion is to build the second course around the ideas of precision and approximation. That is, to investigate methods that produce exact solutions, see when these methods fail or do not apply, and when that is the case to find ways to get approximate solutions and to estimate the size of the errors made. This structure would give the second course a measure of cohesiveness while simultaneously emphasizing the importance of making approximations, something that present courses do not often do but, we think, ought to.

As we worked towards the goals, we made some assumptions about the manner in which we thought the courses should be taught. They include:

- The obvious should not be proved. For example, plausibility arguments should suffice for such things as the formula for the derivative of a sum, the first derivative test, and the intermediate value theorem.

- Proofs need not be completely rigorous, provided that deficiencies are explained. For example, a full and rigorous proof of the Chain Rule is not necessary in the first course.

- Topics should be brought up only when necessary. For example, continuity does not have to be mentioned until the mean value theorem is discussed.

- Functions should not be defined solely by equations. Tables and, especially, graphs should be used continually in examples and exercises.

- Extra time should be spend on investigating topics more fully, and not on introducing additional topics.

Members of the committee will welcome any comments on this preliminary draft and suggestions for its improvement. We will also be happy to pass along to one of our five Working Groups your favorite ideas for making the calculus sequence lively as well as lean.

JEAN CALLOWAY (Kalamazoo College)
BONNIE GOLD (Wabash College)
HAROLD HANES (Earlham College)
PAUL HUMKE (St. Olaf College)
ANDREW STERRETT, Chair (Denison Univ.)
March 15, 1990

Calculus I

The Derivative and the Integral

The intent of this syllabus is that the course concentrate on ideas rather than on manipulations, since manipulations can be more conveniently carried out with a pocket calculator or a computer algebra system (CAS). The thirty-two classes in the syllabus are intended to provide enough time so that the central ideas of first-semester calculus can be discussed. Any extra time could be used for a variety of purposes, for example discussing applications, testing, or developing ideas more deeply than suggested in the syllabus. Some possible applications are included in the outline below, and the ACM-GLCA collections will provide many additional suggestions for enriching the course.

Introduction (1)

Describe geometrically the two problems that dominate first-year calculus. Begin with a special case of the area problem (for example, find the area bounded by $y = x^2$, $x = a$, and the x-axis) by calculating the limit (intuitively, of course) of upper sums that approximate the area. Then describe the tangent problem (for example, find the slope of the tangent line to $y = x^3/3$ at $x = a$) and find the limit of the appropriate difference quotient. Complete the introduction by promising that there is a relation between these apparently unrelated problems, and mention the roles of Newton and Leibniz. If a CAS system is available, the two problems can easily be finished in a single class period.

Functions and Graphs (4)
TOPICS:
- Definition; domain and range; linear and quadratic functions.
- Trigonometric functions (sine, cosine, and tangent).
- Exponential and logarithmic functions.
- Composite functions.
- Functions described by tables or graphs.

The emphasis in this part of the course should be on graphing a broader than usual collection of functions, including functions described by tables, so that a rich collection of examples and applications will be available throughout the year. One sequence of assignments could require students to graph a large number of functions by any method, with examples of each type covered so far included each day.

Note that limits and continuity are not included in this section. The idea of limits is introduced in the next section, when its lack would prevent defining and finding instantaneous rates of change. Similarly, the idea of continuity should not be brought up until it arises naturally.

The Derivative (10)

TOPICS:

- Average rates of change.
- Instantaneous rates of chance, developed intuitively.
- A study of limits, either intuitive or epsilon-delta.
- Definition and properties of the derivative.
- Derivatives of polynomials.
- Derivatives of sines and cosines.
- Derivatives of exponentials and logarithms.
- Derivatives of sums, differences, products, and quotients.
- The Chain Rule and inverse functions.

In the first class a number of applications where average rates of change arise should be discussed. For example, average velocity, average revenue, slope of a secant line, and other more specialized applications such as the average rate at which a person's body assimilates and uses calcium, the average rate of production at a plant, and so on.

Applications that could be considered include motion on a line, freely falling bodies, and related rates.

Extreme Values (8)

TOPICS:

- Extreme values; approximate graphical or numerical solutions using a calculator or CAS, if available.
- Existence theorem: a function continuous on a closed interval attains maximum and minimum values.
- Critical point theorem: a function attains extreme values only at critical points.
- Monotonicity theorem: a function with a positive derivative is increasing.
- Concavity theorem: a function with a positive second derivative is concave up.
- First derivative test for local extremes.
- Second derivative test for local extremes.
- The mean-value theorem.

In the first class, a number of interesting word problems whose solutions involve finding an extreme value should be introduced and they should be used to motivate the ideas that follow.

The mean-value theorem provides an excellent opportunity to discuss the uses of existence theorems by noting how this one leads to several important results, such as the monotonicity theorem and the fact that two functions with the same derivative differ by a constant. Problems like "Find the value of c where ..." should not be emphasized.

Applications include the exact solution of earlier word problems and finding more detailed information about graphs of functions.

Antiderivatives & Differential Equations (3)

TOPICS:

- Antiderivatives and some of their basic properties.
- An introduction to differential equations; separation of variables; constants of integration and initial conditions.

Possible applications are exponential growth, escape velocity, falling bodies, and Torricelli's Law.

The Definite Integral (6)

TOPICS:

- Riemann sums.
- Limits of Riemann sums.
- Integrability theorem; properties of definite integrals.
- The fundamental theorem of calculus.
- The derivative of integrals with variable upper limits (in two ways).

Consistent with the introduction to the derivative, the introduction to the definite integral should begin with examples where the function to be integrated is piecewise constant. Students know that distance traveled is the product of velocity and time when velocity is constant, but they do not know how to find distance when velocity varies with time. The Riemann sum provides an approximate solution. Other examples and applications include areas, volumes, work, and moments.

The fundamental theorem of calculus provides the opportunity to remind students of their first class when they were made aware of two important geometric problems and were promised that they would see a relation between them.

Calculus II

Representation of Numbers and Functions

The theme of Calculus II is the representation of numbers and functions by several methods, both exact and approximate. Topics considered are sequences as functions, improper integrals as limits of sequences of numbers, functions approximated by polynomials or represented as power series, functions described by the behavior of their derivatives, exact and approximate solution of differential equations, and area and volumes considered as integrals. Calculators and computer algebra systems make easy the use of many techniques to find both exact and approximate solutions. The constant interplay of precise and approximate representations should be the focus of this course.

Introduction (1)

Introduce several examples that illustrate the role that approximate as well as exact solutions have in mathematics. For example,

- Remind students how to find $\sin 30°$ and indicate how to approximate $\sin 31°$ with a Taylor polynomial. (The use of a CAS to graph the sine and its polynomial approximation is particularly effective.)
- Exhibit the integral or arc length (to be derived later) and discuss its limitations and how they might be overcome.
- Indicate graphically how Newton's method generates a sequence of approximations.
- Discuss how to estimate π and the importance of error analysis.

The Definite Integral Revisited (9)

TOPICS:

- The definite integral: exact values from the fundamental theorem of calculus.
- Antiderivatives: finding them by substitution, including trigonometric substitutions, and by integration by parts.
- The definite integral: approximate values by using Riemann sums and by the trapezoidal rule, with some error analysis.

New applications of the definite integral, or old applications that require new methods for finding antiderivatives or for evaluating integrals of functions that lack simple antiderivatives, should be used to motivate the ideas and techniques in this section. The approach used to introduce new applications should be consistent with that used earlier, that is, using the integral to add up generic pieces of things rather than presenting formulas to be memorized.

Integration by parts and by substitution are included here because they are important in solving differential equations and in other mathematics courses. Other antiderivatives can be found by using tables or a CAS.

Emphasis should be placed on error analysis, including the graph of an appropriate derivative, of the numerical methods used to approximate definite integrals.

Possible applications include arc length (including curves defined by parametric equations), Buffon's needle problem, surface area, numerical integration of tabular data, and Monte Carlo methods to estimate area.

Sequences and Series of Numbers (10)

SEQUENCE TOPICS:

- Infinite sequences as functions.
- Limits of sequences.
- Recursively defined sequences.
- Improper integrals, including l'Hopital's Rule.
- Limits at infinity and the asymptotic behavior of functions.

The need to study sequences of numbers should be introduced by an appropriate example. For instance, finding the area bounded by $y = 1/x$ and the x-axis over the interval $[1, \infty)$, and the amount of paint it would take to fill the infinite funnel obtained by revolving $y = 1/x$ about the x-axis. One or more of the applications listed below could also be used.

Special emphasis should be placed on the rates of growth or convergence of sequences.

Possible applications include compound interest (compounded n times a year or continuously), approximating π or e, Newton's method for finding zeros, repeating decimals as approximations to rationals, and decimal approximations to irrational numbers.

SERIES TOPICS:

- Infinite series.
- Geometric series.
- The nth term test for divergence.
- Equivalence of series, and the limit comparison test.
- The harmonic series and p-series.

Motivation for this material can come from the desire to find exact or approximate values of series that result from applications. A key idea is comparing the rate of growth of a series of positive terms with other series whose behavior is known. A starting point is the formula for the sum of a finite geometric series which allows discussion of convergence as a limit of a sequence of partial sums, and which leads to conditions under which infinite geometric series converge and diverge. Convergence of a general series can be defined and examples can show what it means for a series of be comparable to a geometric series. The ratio and integral tests can be introduced as "formalized comparisons." Students should be able to compare the rates of convergence and divergence for p-series, geometric series, and series involving factorials. Errors should be estimated in every case where that is possible.

Applications include the distance traveled by a bouncing ball, numerical values of π and e, Zeno's paradox of Achilles and the tortoise, and the multiplier effect in economics.

Sequences and Series of Functions (8)

TOPICS:

- The Mean Value Theorem revisited and its second-degree analogue.
- Taylor polynomials with remainder theorem.
- Graphical comparison of a function with its Taylor polynomials; the graph of the error function for a Taylor approximation.
- Error estimation on intervals.
- Taylor series: the general expansion and examples (sine, cosine, exponential, logarithm, the binomial theorem, and so on).
- Power series, with the ratio test applied to give domains of convergence.
- Algebraic manipulation and term-by-term integration and differentiation.

The emphasis should be on three ideas: that a sequence of functions is an extension of the notion of a sequence of numbers, that sequences of Taylor polynomials are naturally associated with differentiable functions and can be used to approximate them to prescribe tolerances, and that techniques of manipulation acquired earlier still serve in this new environment.

Students could be asked to print a small three decimal table of the sine function as an application.

Series Solutions of Differential Equations (4)

TOPICS:

- Defining functions with differential equations, for example $y'' + ky = 0$ and $y' = ky$.
- Solving homogeneous linear second-order equations with constant coefficients using power series.

Differential equations were introduced in Calculus I, where equations with separable variables were solved. This section reintroduces the idea and provides a good opportunity to introduce applications that lead to differential equations that give rise to trigonometric and exponential equations.

Calculus IIB

Calculus in a Three-Dimensional World

This syllabus was prepared as an alternative to Calculus II which was limited to functions of a single variable. Calculus IIB is intended to follow Calculus I.

The purpose of Calculus IIB is to extend the concepts of derivative and integral to three-space. After revisiting the definite integral in two-space, the concepts of double and iterated integrals are introduced. The notion of partial derivative leads to the equations of tangent planes and to the solutions of optimization problems, generalizing ideas encountered in Calculus I. Finally, the integral is shown to apply to functions defined along a curve in two- and three-space. Appropriate applications are used throughout the course to motivate the need for generalizations.

Introduction (1)

Discuss non-trivial examples that clearly exhibit the need to generalize the concepts of derivative and definite integral as they were developed in Calculus I. Encourage members of the class to identify ideas from Calculus I for which generalizations likely will lead to solutions of problems that occur in a three-dimensional world, e.g., slope of a tangent line, critical points, and the definite integral. Applications that might be discussed include average temperature over a region given readings at a discrete set of points, economics-based optimization problems, and work done by a force exerted over a closed path.

The Definite Integral Revisited (9)

TOPICS:

- The definite integral: exact values from the fundamental theorem of calculus.
- Antiderivatives: finding them by substitution, including trigonometric substitution, and by integration by parts.
- The definite integral: approximate values by using Riemann sums and by Trapezoidal Rule, with some error analysis.

New applications of the definite integral, or old applications that require new methods for finding antiderivatives or for evaluating integrals of functions that lack simple antiderivatives, should be used to motivate the ideas and techniques in this section. The approach used to introduce new applications should be consistent with that used earlier, that is, using the integral to add up generic pieces of things rather than presenting formulas to be memorized.

Integration by parts and by substitution are included here because they are important in solving differential equations and in other mathematics courses. Other antiderivatives can be found by using tables or a CAS.

Emphasis should be placed on error analysis, including the graph of an appropriate derivative, of the numerical methods used to approximate definite integrals.

Possible applications include arc length (including curves defined by parametric equations), Buffon's needle problem, surface area, numerical integration of tabular data, and Monte Carlo methods to estimate area.

The Integral in R^2 and R^3 (8)

TOPICS:

- Real valued functions of two- and three-variables; graphing, level curves.
- Definitions of double and triple integrals.
- Integrals over rectangles and boxes.
- Evaluation of double integrals over regions with curved boundaries.

The emphasis in this section is on generalizing the concept of the definite integral to functions of two- and three-variables. We have consciously restricted the integrals to rectangular coordinates. While there are many nice problems which can be solved more easily using cylindrical and spherical coordinates, time constraints make it advisable to concentrate on the ideas involved in the extension of the concepts to double and triple integrals. In addition, graphing surfaces and drawing level curves pose enough technical problems for most students so that introducing additional complications posed by different coordinate systems seems inadvisable at this point. Computer Algebra Systems with 3-D graphing capabilities can be very helpful in this section.

Appropriate applications to motivate and give practice in using the new integrals might be taken from probability and economics as well as the usual geometric mathematical applications of finding areas, volumes, and centers of mass.

The Derivative in 2- and 3-Variables (11)

TOPICS:

- Partial derivatives: definition and geometric motivation.
- Equation of the tangent plane.
- Unconstrained optimization: critical points and the second derivative test.
- Curves described by parametric equations.
- The Chain Rule.
- Extreme value theorem revisited.
- Constrained optimization.
- Lagrange multipliers.

The geometric analogies to the tangent line should be drawn and used to motivate both the definition of the tangent plane and the method of solution of the optimization problem. The terminology "derivative" should be used rather than "gradient" for the ordered tuple of partial derivatives to emphasize the analogies. Vector notation should be used only when necessary; the tangent plane is to be expressed as a generalization of the point slope form.

Parametric equations are introduced in the context of constrained optimization and only the simpler ones (circles, ellipses, squares, helices, etc.). Velocity of a point moving along a curve is used to motivate the Chain Rule.

Those who wish to present a proof of Lagrange's Theorem may want to introduce a bit of vector notation, including dot product at this point.

The extreme value theorem is stated (but not proved) as justification for constrained optimization.

Constrained optimization is done in two ways: first, comparing values of critical points with boundary values (generalizing the one dimensional case); then using Lagrange multipliers.

Applications should include velocity/speed and several significant applications of Lagrange multipliers (at least one from economics).

Integration Along Curves (6)

TOPICS:

- Definition of the Riemann integral of a real function on a curve in R^2 and R^3.
- Vector fields in R^2 and R^3 and the dot product.
- Line integrals.
- Green's Theorem and path independence.

The integral of a (continuous) function on a curve is defined as a straightforward generalization of the Riemann integral of a function of a real variable. Several exercises are worked out and the independence of a parametrization is pointed out. The present goal is to introduce line integral as an example of this integration process. However, the notions of vector field R^2 and R^3 (scalar), projection of one vector onto another, and dot product must be introduced first. Line integrals are introduced via application, e.g., perhaps to compute the work done against the wind along certain paths. Green's Theorem is given and could be treated as a two-dimensional version of the Fundamental Theorem of Calculus. Path independence, conservative vector fields, and recapturing a primitive are then discussed.

A Sample Calculus Bibliography

Apostol, T.; Chrestenson, H., *et al.*, (Eds.). (1968) *Selected Papers on Calculus.* Washington, DC: Mathematical Association of America.

Austin, J.D. (January 1979). "High school calculus and first-quarter college calculus grades." *Journal for Research in Mathematics Education,* 10:1, pp. 69-72.

Beard, R.M.; Hartley, J. (1984). *Teaching and Learning in Higher Education.* London: Harper and Row.

Boas, R.P. (1971). "Calculus as an experimental science." *American Mathematical Monthly,* 78, pp. 664-667. Reprinted in the *College Mathematics Journal,* 1 (1971) 36-39.

Campbell, P.J.; Grinstein, L.S. (1988). *Mathematics Education in Secondary Schools and Two Year Colleges: A Source Book.* New York, NY: Garland Publishers.

CASE Newsletter, c/o Don Small, Colby College, Waterville, ME 04901.

Case, B.A. (Ed.). (1989). *Keys to Improved Instruction by Teaching Assistants and Part-Time Instructors.* (MAA Notes No. 11.) Washington, DC: Mathematical Association of America.

Cohen, D. (August-September, 1982). "Modified Moore method of teaching undergraduate mathematics." *American Mathematical Monthly,* 89, pp. 473-490.

Committee on the Teaching of Undergraduate Mathematics. (1979). *College Mathematics: Suggestions on How to Teach It.* Washington, DC: Mathematical Association of America.

Committee on the Teaching of Undergraduate Mathematics. (1990). *A Source Book for College Mathematics Teaching.* Washington, DC: Mathematical Association of America.

Committee on the Undergraduate Program in Mathematics. (1989). *Reshaping College Mathematics.* (MAA Notes No. 13.) Washington, DC: Mathematical Association of America.

Davidson, N. (August-September, 1971). "The small group-discovery method as applied to calculus instruction." *American Mathematical Monthly,* 78, pp. 789-791.

Davis, R.; Vinner S. (1986). "The notion of limit: Some seemingly unavoidable misconception stages." *The Journal of Mathematical Behavior,* 5, pp. 281-303.

Demana, F.; Waits, B.; Harvey J. (Eds). (1990). *Twilight of Paper and Pencil: Proceedings of the Ohio State Technology Conference.* Reading, MA: Addison-Wesley.

Douglas, R.G. (Ed.). (1986). *Toward A Lean and Lively Calculus.* (MAA Notes No. 6.) Washington, DC: Mathematical Association of America.

Garfunkel, S.; Young, G. (1990). *Math Outside of Math.* Arlington, MA: COMAP.

Gersting, J.; Kuczkowski, J. (November, 1977). "Why and how to use small groups in the mathematics classroom." *College Mathematics Journal,* 8, pp. 270-274.

Gopen, G.D.; Smith, D.A. (1990). "What's an assignment like you doing in a course like this? Writing to learn mathematics." *College Mathe-*

matics Journal, 21, pp. 2-19.

Gore, H.; Gilmer, G. *Effective Strategies for Teaching Calculus at the College Level.* Proceedings of Conference at Morehouse College, 1989. (Contact H. Gore at Morehouse.)

Grinstein, L. (December, 1974). "Calculus by mistake." *College Mathematics Journal,* 5, pp. 49-53.

Grinstein, L.S.; Michaels, B. (Eds.). (1977). *Calculus Readings from the Mathematics Teacher.* Reston, VA: National Council of Teachers of Mathematics.

Halmos, P. (1975). "The problem of learning to teach: I. The teaching of problem solving." *American Mathematical Monthly,* 82, pp. 466-470. (The other sections of the article, by E. Moise and G. Piranian, also contain useful suggestions.)

Heid, M.K. (1983). "Calculus with muMath: Implications for curriculum reform." *The Computing Teacher,* pp. 46-49.

Hosack, J.M. (1986). "A guide to computer algebra systems." *College Mathematics Journal,* 17, pp. 434-441.

Howson, A.G.; Kahane J.-P.; P. Lauginie (Eds). (1988). *Mathematics as a Service Subject.* International Commission on Mathematical Instruction Study Series. New York, NY: Cambridge University Press.

Humke, P. "Proceedings of the St. Olaf Conference on Symbolic Computation and Calculus Instruction." Available from Department of Mathematics, St. Olaf College, Northfield, MN 55057 (include $8.00).

IREM (Institut de recherches sur l'Enseignement des Mathematiques), Strasbourg (1985). *The Influence of Computers and Informatics on Mathematics and its Teaching.* Strasbourg, France: IREM Strasbourg.

Kemeny, J. (1988). "MATHEMATICS–How computers have changed the way I teach." *Academic Computing.*

Klopfenstein, K. (February, 1977). "The personalized system of instruction in introductory calculus." *American Mathematical Monthly,* 84, pp. 120-124.

Kranes, David P.; Smith, David A. (May 1988). "A computer in the classroom: The time is right." *College Mathematics Journal,* 19:3, pp. 261-267.

Lax, P. (1990). "Calculus reform: A modest proposal." *UME Trends,* 2:2.

Mason, J.; Burton, L.; Stacey, K. (1982). *Thinking Mathematically.* New York, NY: Addison Wesley.

McKen, R.; Davidson, N. (December, 1975). "An alternative to individual instruction in mathematics." *American Mathematical Monthly,* 82, pp. 1006-1009.

Moise, E. (1965). "Activity and motivation in mathematics." *American Mathematical Monthly,* 72, pp. 407-412.

Monk, S. (1983). "Student engagement and teaching power in large classes." In C. Bouton and R.Y. Garth (Eds.), *Learning in Groups,* pp.7-12. San Francisco, CA: Jossey-Bass.

Orton, A. (1983). "Students' understanding of integration." *Educational Studies in Mathematics,* 14, pp. 1-18.

Page, W. (1990). "Computer algebra systems: Issues and inquiries." *Computers and Mathematics with Applications,* 19:6, pp. 51-69.

Page, W. (1979). "Small group strategy for enhanced learning." *American Mathematical Monthly,* 86:10, pp. 856-858.

Price, J.J. (November, 1989). "Learning mathematics through writing: Some guidelines." *College Mathematics Journal,* 20:5, pp. 393-401.

Ralston, Anthony; Young, Gail S. (Eds.). (1983). *The Future of College Mathematics: Proceedings of a Conference/Workshop on the Future of the First Two Years of College Mathematics.* New York, NY: Springer-Verlag.

Sallee, G.T. (August-September, 1979). "Teaching 200 students in a personal way." *American Mathematical Monthly,* 86, pp. 589-590.

Schoenfeld, A.H. (Ed.). (1987). *Cognitive Science and Mathematics Education.* Hillsdale, NJ: Erlbaum.

Sierpinska, A. (1988). "Epistemological remarks on functions." *Proceedings of the 12th Annual PME,* 2, pp. 568-575.

Small, D. (October 1987). "Report of the CUPM panel on calculus articulation: Problems in the transition from high school calculus to college calculus." *American Mathematical Monthly,* 94:8, pp. 776-785.

Small, D.; Hosack, J.; Lane, K. (Nov 1986). "Computer algebra systems in undergraduate instruction." *College Mathematics Journal,* 17, pp. 423-441.

Smith, D.A.; Porter, G.A.; Leinbach, L.C.; Wenger, R.H. (Eds.). (1988). *Computers and Mathematics: The Use of Computers in Undergraduate In-*

struction. (MAA Notes No. 9.) Washington, DC: Mathematical Association of America.

Smith, D.A. (1984). *Interface: Calculus and the Computer, Second Edition.* Philadelphia, PA: W.B. Saunders.

Smith, J.C. (1988). "The problems of teaching calculus to technical students and the difficulties students have in learning calculus." *Proceedings of the 3rd International Conference: Theory of Mathematics Education,* pp. 141-147.

Steen, Lynn A. (Ed). (1988). *Calculus for a New Century.* (MAA Notes No. 8.) Washington, DC: Mathematical Association of America.

Stein, S. (March/April, 1987). "The triex: Explore, extract, explain. *Teaching Thinking and Problem Solving,* 9:2, pp. 10-11.

Struik, R.; Flexer, R. (February, 1977). "Self-paced calculus: A preliminary evaluation." *American Mathematical Monthly,* 84, pp. 129-134.

Tall, D. (1985). "Understanding the calculus." *Mathematics Teaching,* 110, pp. 49-53.

Weissglass, J. (February, 1976). "Small groups: An alternative to the lecture method." *Two-Year College Mathematics Journal,* 7, pp. 15-20.

White, J. (1988) "Teaching with CAL: A mathematics teaching and learning environment." *College Mathematics Journal,* 19:5, pp. 424-443.

Wildfogel, D. (January, 1983). "A mock symposium for your calculus class." *American Mathematical Monthly,* 90, pp. 52-53.

Wilf, H. (January, 1982). "The disk with the college education." *American Mathematical Monthly,* 89, pp. 4-8.

Winkel, B.J. (forthcoming). "First year calculus students as in-class consultants." *International Journal of Mathematical Education in Science and Technology.*

Zorn, Paul. (1987). "Computing in undergraduate mathematics." *Notices of the American Mathematical Society,* 34, pp. 917-923.

OTHERS:

UME Trends

Notices of the American Mathematical Society (especially Vols. 1, No. 3, 4, 5)

"Computer Corner" of *College Mathematics Journal*

"Classroom Notes" of *American Mathematical Monthly*

Consortium for Mathematics and Its Applications (COMAP)

COMAP is a useful resource, especially for add-on modules in mathematical modelling. The original project on Undergraduate Mathematics and Its Applications (UMAP) has generated since 1976 more than 300 modules, which are distributed by COMAP. Nearly 100 of these are related to calculus. Some sample titles are "The Digestive Process of Sheep," "Epidemics," "The Human Cough," "Zipf's Law and His Efforts to Use Infinite Series in Linguistics," "Price Discrimination and Consumer Surplus," "The St. Louis Arch Problem." There are also about one dozen longer UMAP monographs. The *UMAP Journal* offers current articles and new modules in every quarterly issue, as well as an annual collection of all the UMAP modules published in the past year.

For more information about COMAP, such as a catalog of all UMAP modules, phone 617-641-2600, or write COMAP, Inc, 60 Lowell Street, Arlington, MA 02174.

History of Calculus: A Brief Bibliography

by V. Frederick Rickey

This bibliography gives my suggestions for those items which are some of the best things to read first in learning about the history of calculus.

Boyer, Carl B. (1968). *A History of Mathematics.* New York, NY: Wiley. (Reprinted in paperback by Princeton University Press, 1985. Second edition 1989 edited by Uta Merzbach.)

It contains revised bibliographies and revisions of the later chapters. Still the best textbook available. Chapters XVIII-XXII and XXV are the pertinent ones.

Edwards, C.H., Jr. (1979). *The Historical Development of the Calculus.* New York, NY: Springer-Verlag.

A very good sketch. A book worth buying.

Grabiner, Judith V. (1983). "The changing concept of change: The derivative from Fermat to Weierstrass." *Mathematics Magazine,* 56 (1983), pp. 195-203.

> Distinguishes four stages in the chronological development of the derivative: use, discovery, exploration and development, definition. In your reading try to adapt this scheme to the other concepts of the calculus. "Invention of the calculus" is explained on page 199.

Grattan-Guinness, Ivor (Ed.). (1980). *From the Calculus to Set Theory, 1630-1910: An Introductory History.* London: Duckworth. (Also available in paperback.)

> Six excellent chapters, by some of the best contemporary historians of mathematics.

Grattan-Guinness, Ivor (Ed.). (1987). *History in Mathematics Education.* Proceedings of a Workshop Held at the University of Toronto, Canada, July-August 1983. Published as No. 21 of *Cahiers d'Historie and Philosophie des Sciences.* Paris: Belin.

> This volume contains several papers that discuss the use of history in the classroom.

Kitcher, Philip. (1983). "The development of analysis: A case study." In Kitcher, Philip. *The Nature of Mathematical Knowledge.* Chapter 10, pp. 229-271. New York, NY: Oxford University Press (Paperback 1984).

> A penetrating survey, by a good philosopher of mathematics, treating Lakatos's ideas on Cauchy's famous wrong proof.

Rickey, V. Frederick. (1987). "Isaac Newton: Man, myth, and mathematics." *College Mathematics Journal,* 18, pp. 362-389.

> A survey of Newton's contributions.

Simmons, George F. (1986). *Calculus With Analytic Geometry.* New York, NY: McGraw-Hill.

> Look at this textbook for ideas on how to use the history of the calculus in the classroom. Most obvious is the long (pp. 763-848) biographical section, but there are numerous historical remarks scattered throughout. Think about what additional historical notes you would like to have, as I am in the process of writing some, and would be most happy to have your suggestions.

Struik, Dirk J. (1967). *A Concise History of Mathematics, Third Revised Edition.* Mineola, NY: Dover.

> If you have time for nothing else, read chapters V-VIII of this. Note how he pays attention to the social history.

Youschkevitch, A.P. (1981). "Leonhard Euler." In Gillispie, C.C. (Ed.). *Dictionary of Scientific Biography.* Volume 4, pp. 467-484. New York, NY: Charles Scribners Sons.

> Good summary of Euler's work, including his contributions to the calculus. The DSB is the preeminent source of bibliographical information, so if you don't know it, go to the library and have a careful look. You will find information on virtually anyone who has made a contribution to the calculus. For example, Freudenthal's article on Cauchy is also excellent.

Youschkevitch, A.P. (1976). "The concept of function up to the middle of the 19th century." *Archive for the History of Exact Sciences,* 16, pp. 37-85.

> An excellent survey concentrating on the 18th century.

NSF Undergraduate Mathematics Awards, 1985-1990

The program for Undergraduate (Calculus) Curriculum Development in Mathematics, administered by the Division of Undergraduate Science, Engineering, and Mathematics Education (USEME) in the NSF Directorate for Education and Human Resources (EHR) [formerly called Science and Engineering Education (SEE)], announced its first round of awards in the fall of 1988. There were 25 awards for $1.29 million; 19 were planning grants, 4 Multi-year grants, one Multi-year grant to help establish the newsletter *Undergraduate Mathematics Education Trends,* and one grant for a series of conferences.

In 1989 there were 18 more awards totaling $1.94 million in addition to the continuation of the Multi-

year awards of 1988. One of these was for a conference, one to the MAA to produce this book, and most of the rest were Multi-year grants. In 1990 there were 20 more awards totaling $1.41 million.

Another NSF program that has supported curricular changes related to technology is the College Science Instrumentation Program (CSIP, 1985-1988), which has become the Instrumentation and Laboratory Improvement Program (ILI, 1989-). The former was open only to nondoctoral-granting colleges and universities, while the latter is open to all universities, four-year and two-year colleges. In the initial years of CSIP, there were very few proposals in mathematics and consequently very few grants. That has changed; in the most recent round there were 31 awards made in mathematics. Generally speaking, the more proposals there are, the more awards are made.

For more information about these two programs and other NSF programs related to undergraduate mathematics education, contact the Office of Undergraduate Science, Engineering, and Mathematics Education, Room 639, National Science Foundation, Washington, DC 20550; telephone: 202-357-7051; e-mail: undergrad@nsf or undergrad@note.nsf.gov.

We list here all awards in the Calculus and CSIP/ILI programs beginning in 1985. Each list is ordered by state, with one principal investigator (even if there are two or more co-PI's), comments, the amount of the grant (for the given fiscal year), and the grant number. It should be noted that any proposal accepted for funding by NSF is public, and copies of accepted proposals are available from NSF.

NSF Calculus Awards, FY 1988

COLORADO SCHOOL OF MINES, Golden, CO 80401. PI: Richard Yeatts; [Multi-year] $74,517; No. 8813784.

ROLLINS COLLEGE, Winter Park, FL 32789. PI: J. Douglas Child; [Multi-year] $74,730; No. 8814048.

UNIVERSITY OF MIAMI, Coral Gables, FL 33124. PI: Shair Ahmad; [Workshop] $45,000; No. 8813860.

SPELMAN COLLEGE, Atlanta, GA 30314. PI: Nagambal Shah; [Planning] $50,000; No. 8813792.

UNIVERSITY OF ILLINOIS, URBANA, Urbana, IL 61801. PI: Gerald Janusz; [Planning] $45,785; No. 8813873.

PURDUE UNIVERSITY, West Lafayette, IN 47907. PI: Edward Dubinsky; [Planning] $50,011; No. 8813996.

IOWA STATE UNIVERSITY, Ames, IA 50011. PI: Elgin H. Johnston; [Planning] $49,954; No. 8813895.

BOSTON UNIVERSITY, Boston, MA 02215. PI: Robert L. Devaney; [Planning] $40,306; No. 8813865.

FIVE COLLEGES INC., Amherst, MA 01002. PI: James Callahan; [Multi-year] $141,707; No. 8814004.

HARVARD UNIVERSITY, Cambridge, MA 02138. PI: Andrew M. Gleason; [Planning] $20,372; No. 8813997.

MACALESTER COLLEGE, Saint Paul, MN 55105. PI: A. Wayne Roberts; [Planning] $62,650; No. 8813914.

DUKE UNIVERSITY, Durham, NC 27706. PI: Lawrence C. Moore; [Planning] $20,000; No. 8814083.

DARTMOUTH COLLEGE, Hanover, NH 03755. PI: Richard H. Crowell; [Planning] $50,464; No. 8814009.

UNIVERSITY OF NEW HAMPSHIRE, Durham, NH 03824. PI: Joan Ferrini-Mundy; [Planning] $40,487; No. 8814057.

NEW MEXICO STATE UNIVERSITY, Las Cruces, NM 88003. PI: Marcus S. Cohen; [Multi-year] $228,888; No. 8813904.

ITHACA COLLEGE, Ithaca, NY 14850. PI: Stephen R. Hilbert; [Planning] $50,193; No. 8814177.

RENSSELAER POLYTECHNIC INSTITUTE, Troy, NY 12180-3590. PI: William E. Boyce; [Planning] $50,000; No. 8814011.

MIAMI UNIVERSITY, Oxford, OH 45056. PI: Thomas A. Farmer; [Planning] $48,595; No. 8813786.

OREGON STATE UNIVERSITY, Corvallis, OR 97331. PI: Thomas Dick; [Planning] $27,401; No. 8813785.

COMMUNITY COLLEGE OF PHILADELPHIA, Philadelphia, PA 19107. PI: Alain Schremmer; [Planning] $40,124; No. 8814000.

PENNSYLVANIA STATE UNIVERSITY, University Park, PA 16802. PI: Mary McCammon; [Planning] $42,399; No. 8813779.

AMERICAN MATHEMATICAL SOCIETY, Providence, RI 02901. PI: James H. Voytuk; [Newsletter] $104,413; No. 8814683.

UNIVERSITY OF RHODE ISLAND, Kingston, RI 02881. PI: Edmund A. Lamagna; [Planning] $51,350; No. 8814017.

FURMAN UNIVERSITY, Greenville, SC 29613. PI: Daniel C. Sloughter; [Planning] $22,476; No. 8813781.

WASHINGTON STATE UNIVERSITY, Pullman, WA 99164. PI: Michael E. Moody; [Planning] $29,716; No. 8814131.

NSF Calculus Awards, FY 1989

UNIVERSITY OF ARIZONA, Tucson, AZ 85721. PI: David O. Lomen; $104,806; No. 8953930.

UNIVERSITY OF CALIFORNIA, BERKELEY, Berkeley, CA 94720. PI: Marcia Linn; [Conference] $42,898; No. 8953974.

MATHEMATICAL ASSOC. OF AMERICA, Washington, DC 20036. PI: Marcia Sward; [*CRAFTY book*] $41,450; No. 8953912.

UNIVERSITY OF ILLINOIS, URBANA, Urbana, IL 61801. PI: J. Jerry Uhl; $25,000; No. 8953906.

BUTLER UNIVERSITY, Indianapolis, IN 46208. PI: Judith H. Morrel; $27,000; No. 8953948.

IOWA STATE UNIVERSITY, Ames, IA 50011. PI: Elgin H. Johnston; $63,600; No. 8953949.

UNIVERSITY OF IOWA, Iowa City, IA 52242. PI: Keith D. Stroyan; $65,000; No. 8953937.

HARVARD UNIVERSITY, Cambridge, MA 02138. PI: Andrew M. Gleason; $346,245; No. 8953923.

MACALESTER COLLEGE, Saint Paul, MN 55105. PI: A. Wayne Roberts; $199,203; No. 8953947.

MERIDIAN JUNIOR COLLEGE, Meridian, MS 39301. PI: Wanda Dixon; [*Planning*] $25,000; No. 8953931.

DARTMOUTH COLLEGE, Hanover, NH 03755. PI: Richard H. Crowell; $289,171; No. 8953908.

CUNY BOROUGH OF MANHATTAN COMMUNITY COLLEGE, New York, NY 10007. PI: Patricia R. Wilkinson; $50,000; No. 8953949.

NAZARETH COLLEGE OF ROCHESTER, Rochester, NY 14610. PI: Ronald W. Jorgensen; $78,232; No. 8953926.

RENSSELAER POLYTECHNIC INSTITUTE, Troy, NY 12180-3590. PI: William E. Boyce; $55,000; No. 8953904.

DUKE UNIVERSITY, Durham, NC 27706. PI: Lawrence C. Moore; $198,522; No. 8953961.

OREGON STATE UNIVERSITY, Corvallis, OR 97331-5503. PI: Thomas Dick; $84,219; No. 8953938.

BROWN UNIVERSITY, Providence, RI 02912. PI: Thomas Banchoff; $81,500; No. 8820503.

UNIVERSITY OF RHODE ISLAND, Kingston, RI 02881. PI: Edmund A. Lamagna; $161,535; No. 8953939.

NSF Calculus Awards, FY 1990

UNIVERSITY OF ARIZONA, Tucson, AZ 85721. PI: David O. Lomen; [*Software Development and Student Projects, Two Years*] $80,000; No. 9053431.

CAL POLY STATE UNIVERSITY, San Luis Obispo, CA 93407. Michael E. Colvin. [*Consortium, Conferences, One Year*] $46,996; No. 9053404.

GOLDEN WEST COMMUNITY COLLEGE, Huntington Beach, CA 92647. PI: David A. Horowitz; [*Simulation of Differential Equations, One Year*] $36,000; No. 9053390.

STETSON UNIVERSITY, Deland, FL 32720. PI: Gareth Williams; [*Projects and Software for Linear Algebra, One Year*] $24,964; No. 9053365.

UNIVERSITY OF ILLINOIS, Urbana, IL 61801. PI: Jerry Uhl; [*Mathematica Notebook, Two Years*] $87,501; No. 9053372.

PURDUE UNIVERSITY, West Lafayette, IN 47907. PI: Edward L. Dubinsky; [*ISETL, Research in Student Learning Processes, Three Years*] $220,000; No. 9053432.

IOWA STATE UNIVERSITY, Ames, IA 50011. PI: Leslie Hogben; [*Engineering/Physical Science Second Year Calculus and Differential*] [*Equations, One Year*] $48,241; No. 905342.

UNIVERSITY OF IOWA, Iowa City, IA 52242. PI: Keith Stroyan; [*NeXT and Mathematica in Accelerated Calculus, Two Years*] $64,000; No. 9053383.

BOWDOIN COLLEGE, Brunswick, ME 04011. PI: William H. Barker; [*Mathematica and Duke's Project CALC, Two Years*] $35,000; No. 9053397.

COLBY COLLEGE, Waterville, ME 04901. PI: Donald Small; [*Computer Algebra System Workshops, One Year*] $84,875; No. 9053427.

UNIVERSITY OF MARYLAND, College Park, MD 20742. PI: David Lay; [*Workshop on Linear Algebra Curriculum, One Year*] $43,992; No. 9053422.

CONSORTIUM FOR MATHEMATICS AND ITS APPLICATIONS (COMAP), Arlington, MA 02174. PI: Frank Giordano; [*Video Applications Modules, One Year*] $101,851; No. 9053407.

WORCESTER POLYTECHNIC INSTITUTE, Worcester, MA 01609. PI: William Farr; [*Early Multivariable, Team Projects, Two Years*] $56,981; No. 9053430.

UNIVERSITY OF MICHIGAN–DEARBORN, Dearborn, MI 48128. PI: David James; [*Laboratory Materials, Two Years*] $57,500; No. 9053385.

ST. OLAF COLLEGE, Northfield, MN 55057. PI: Arnold Ostebee; [*First-year Calculus Curriculum, Two Years*] $49,977; No. 9053363.

NEW MEXICO STATE UNIVERSITY, Las Cruces, NM 88003. PI: David Pengelley; [*Student Projects in Calculus and Differential Equations, Two Years*] $120,000; No. 9053387.

CORNELL UNIVERSITY, Ithaca, NY 14853. PI: George R. Livesay; [*Software for Differential Equations and Multivariable Calculus, Two Years*] $93,000; No. 9053426.

ITHACA COLLEGE, Ithaca, NY 14850. PI: Stephen R. Hilbert; [*Student Projects, Especially Applied,*

Two Years] $86,175; No. 9053416.

GETTYSBURG COLLEGE, Gettysburg, PA 17325. PI: Carl L. Leinbach; [*Workshops to Develop Computer Algebra Based Laboratories, Two Years*] $59,225; No. 9053401.

TEXAS A&I UNIVERSITY, Kingsville, TX 78363. PI: Alvin J. Kay; [*Project CALC Test Site*] $17,664; No. 9053364.

CSIP, FY 1985

LOYOLA MARYMOUNT UNIVERSITY, Los Angeles, CA 90045. PI: Michael Cullen; [*Variety of Courses*] $23,577; No. 8551189.

TOWSON STATE UNIVERSITY, Towson, MD 21204. PI: E. Sharon Jones; [*Laboratory for Modeling, Numerical Analysis, Statistics*] $25,366; No. 8551368.

COLLEGE OF THE HOLY CROSS, Worcester, MA 01610. PI: Melvin C. Tews; [*Computer Graphics for Calculus*] $16,242; No. 8551683.

UNIVERSITY OF NORTH CAROLINA AT CHARLOTTE, Charlotte, NC 28223. PI: David E. Nixon; [*Minitab and Statistics*] $50,000; No. 8551529.

CSIP, FY 1986

CALIFORNIA STATE UNIVERSITY AT HAYWARD, Hayward, CA 94542. PI: Richard Kakigi; [*Statistics*] $50,000; No. 8650824.

SAN JOSE STATE UNIVERSITY, San Jose, CA 95192. PI: Jane M. Day; [*Laboratory for Applied and Computational Mathematics*] $25,140; No. 8650753.

ST. OLAF COLLEGE, Northfield, MN 55057. PI: Paul Zorn; [*Computer Algebra (SMP) for Calculus*] $36,157; No. 8650912.

CSIP, FY 1987

GRINNELL COLLEGE, Grinnell, IA 50112. PI: Eugene A. Herman; [*Laboratory*] $50,000; No. 8750715.

MOUNT HOLYOKE COLLEGE, South Hadley, MA 01075. PI: Robert J. Weaver; [*Courses for Majors*] $36,525; No. 8750848.

CARLETON COLLEGE, Northfield, MN 55057. PI: Jack Goldfeather; [*Computer Graphics*] $39,430; No. 8750099.

SIENA COLLEGE, Loudonville, NY 12211. PI: Russell Dubisch; [*SMP for Math and Physics Courses*] $28,610; No. 8750276.

DENISON UNIVERSITY, Granville, OH 43023. PI: Zaven Karian; [*MAPLE and Calculus*] $36,000; No. 8750137.

UNIVERSITY OF HOUSTON, DOWNTOWN, Houston, TX 77002. PI: Carol Jones; [*Business Statistics*] $42,000; No. 8750532.

ILI (Nondoctoral), FY 1988

CALIFORNIA STATE UNIVERSITY AT LOS ANGELES, Los Angeles, CA 90032. PI: P.K. Subramian; [*Numerical Laboratory*] $50,449; No. 8852276.

UNIVERSITY OF PACIFIC, Stockton, CA 95211. PI: Walter Zimmerman; [*Computer Graphics in Calculus*] $36,000; No. 8851255.

SONOMA STATE UNIVERSITY, Rohnert Park, CA 94928. PI: Jean B. Chan; [*Includes Calculus*] $8,270; No. 8852470.

UNIVERSITY OF SAN FRANCISCO, San Francisco, CA 94117. PI: Stanley D. Nel; [*Laboratory including Calculus*] $42,897; No. 8852453.

SAN JOSE STATE UNIVERSITY, San Jose, CA 95192. PI: Leslie V. Foster; [*Modeling, Numerical Analysis*] $48,182; No. 8852808.

WHITTIER COLLEGE, Whittier, CA 90606. PI: Raymond Smith; [*A Laboratory for Calculus*] $13,813; No. 8851727.

ROLLINS COLLEGE, Winter Park, FL 32789. PI: J. Douglas Child; [*MAPLE*] $46,528; No. 8852244.

STETSON UNIVERSITY, Deland, FL 32720. PI: Gareth Williams; [*Campus Network for Calculus, etc.*] $30,390; No. 8851567.

ROSE-HULMAN INSTITUTE OF TECHNOLOGY, Terre Haute, IN 47803. PI: Robert J. Lopez; [*MAPLE Laboratory for Calculus and Differential Equations*] $100,000; No. 8851339.

VALPARAISO UNIVERSITY, Valparaiso, IN 46383. PI: Joel P. Lehman; [*Laboratory*] $20,696; No. 8852197.

LEXINGTON COMMUNITY COLLEGE, Lexington, KY 40506. PI: Lillie F. Crowley; [*Laboratory for Calculus*] $28,983; No. 8851483.

MOORHEAD STATE UNIVERSITY, Moorhead, MN 56560. PI: Christine E. McLaren; [*Statistics*] $29,157; No. 8851944.

ST. OLAF COLLEGE, Northfield, MN 55057. PI: Alan Magnuson; [*Classroom, All Courses*] $14,680; No. 8851796.

CUNY BOROUGH OF MANHATTAN COMMUNITY COLLEGE, New York, NY 10019. PI: Patricia R. Wilkinson; [*Calculus and Precalculus*] $37,421; No. 8851194.

APPALACHIAN STATE UNIVERSITY, Boone, NC 28608. PI: William C. Bauldry; [*Laboratory for Calculus and Discrete Mathematics*] $45,680; No. 8853085.

CONNORS STATE COLLEGE, Warner, OK 74403. PI: Betty Lou Acord; [*Laboratory for Remediation*] $13,609; No. 8852310.

ST. MARY'S UNIVERSITY, San Antonio, TX 78284. PI: Stillman E. Sims; [*Laboratory for New Applied Mathematics Program*] $45,448; No. 8852311.

TEXAS A&M UNIVERSITY, Kingsville, TX 78363. PI: Dwight Goode; [*Laboratory for Calculus*] $19,213; No. 8852083.

ILI (Doctoral), FY 1989

UNIVERSITY OF ARIZONA, Tucson, AZ 85721. PI: David Lovelock; [*Undergraduate Resource Laboratory*] $75,000; No. 8851907.

UNIVERSITY OF CONNECTICUT, Storrs, CT 06268. PI: James Hurley; [*Main-Track Calculus*] $42,000; No. 8852395.

GEORGIA INSTITUTE OF TECHNOLOGY, Atlanta, GA 30332. PI: William Kammerer; [*Differential Equations*] $41,000; No. 8852537.

PURDUE UNIVERSITY, West Lafayette, IN 47907. PI: Edward Dubinsky; [*Discrete Mathematics and Calculus Courses*] $47,500; No. 8852360.

IOWA STATE UNIVERSITY, Ames, IA 50010. PI: Jerold Mathews; [*Laboratory for Calculus*] $80,000; No. 8851659.

BOSTON UNIVERSITY, Boston, MA 02215. PI: Robert Devaney; [*Computer Graphics and Dynamics Lab*] $51,500; No. 8851613.

COLUMBIA UNIVERSITY, New York, NY 10027. PI: Henry Pinkham; [*Laboratory*] $32,500; No. 8851363.

ILI (Nondoctoral), FY 1989

GLENDALE COMMUNITY, Glendale, CA 91208. PI: George J. Witt; [*Calculus*] $24,395; No. 8950704.

UNIVERSITY OF CENTRAL FLORIDA, Orlando, FL 32816. PI: Paul Somerville; [*Statistics*] $46,500; No. 8951299.

ARMSTRONG STATE COLLEGE, Savannah, GA 31419. PI: Anne L. Hudson; [*Mathematica*] $34,013; No. 8950421.

EASTERN ILLINOIS UNIVERSITY, Charleston, IL 61920. PI: Anthony J. Schaeffer; [*MAPLE and Calculus*] $53,211; No. 8950506.

EARLHAM COLLEGE, Richmond, IN 47374. PI: Michael B. Jackson; [*MAPLE and Calculus*] $29,426; No. 8951335.

NORTHERN KENTUCKY UNIVERSITY, Highland Heights, KY 41076. PI: Bart Braden; [*Mathematica and Calculus*] $50,600; No. 8951219.

COLBY COLLEGE, Waterville, ME 04901. PI: Don Small; [*MAPLE in Calculus, etc.*] $44,873; No. 8951620.

MOUNT HOLYOKE COLLEGE, South Hadley, MA 01075. PI: Donal B. O'Shea; [*Laboratory in Advanced Courses*] $83,739; No. 8951358.

SMITH COLLEGE, Northampton, MA 01063. PI: James Callahan; [*Calculus Laboratory*] $42,268; No. 8951485.

WELLESLEY COLLEGE, Wellesley, MA 02181. PI: Alan H. Shuchat; [*Mathematica, Graphics Laboratory for Calculus*] $43,405; No. 8950662.

ITHACA COLLEGE, Ithaca, NY 14850. PI: Diane D. Schwartz; [*Mathematica Upper Division Courses*] $78,000; No. 8950797.

VASSAR COLLEGE, Poughkeepsie, NY 12601. PI: John A. Feroe; [*Mathematica in Calculus with Discrete Mathematics*] $56,978; No. 8950938.

KENYON COLLEGE, Gambier, OH 43022. PI: James E. White; [*Calculus using PI-Developed System CAL*] $53,663; No. 8951715.

MARIETTA COLLEGE, Marietta, OH 45750. PI: John R. Michel; [*Mathematica and Calculus*] $36,554; No. 8951470.

WITTENBERG UNIVERSITY, Springfield, OH 45501. PI: Alan Stickney; [*Calculus and Introductory Programming*] $62,832; No. 8952097.

FURMAN UNIVERSITY, Greenville, SC 29613. PI: Daniel C. Sloughter; [*Calculus Laboratory*] $26,683; No. 8952215.

TRINITY UNIVERSITY, San Antonio, TX 78284. PI: Phoebe T. Judson; [*MAPLE in Calculus, etc.*] $30,908; No. 8951036.

UNIVERSITY OF TEXAS AT SAN ANTONIO, San Antonio, TX 78285. PI: Betty Trairs; [*Calculus Laboratory*] $43,300; No. 8951071.

MARSHALL UNIVERSITY, Huntington, WV 25755. PI: Gerald E. Rubin; [*Calculus and Statistics Laboratory*] $44,706; No. 8950975.

ILI (Doctoral), FY 1990

UNIVERSITY OF IOWA, Iowa City, IA 52240. PI: Keith D. Stroyan; [*Accelerated Calculus Laboratory*] $30,000; No. 8951562.

UNIVERSITY OF CHICAGO, Chicago, IL 60637. PI: Michael J. O'Donnell; [*Mathematical Sciences Curriculum for the Humanities and Social Sciences*] $19,500; No. 8950775.

KANSAS STATE UNIVERSITY, Manhattan, KS 66506. PI: Qisu Zou; [*Numerical Analysis*] $21,000; No. 8951571.

UNIVERSITY OF NEBRASKA, Lincoln, NE 68588. PI: Thomas S. Shores; [*Advanced Laboratory for Mathematics Experimentation*] $51,000; No. 8952235.

UNIVERSITY OF NEW HAMPSHIRE AT DURHAM, Durham, NH 03824. PI: Loren D. Meeker; [*Modeling and Computation in Applied Mathematics*] $10,000; No. 8951520.

DUKE UNIVERSITY, Durham, NC 27706. PI: Lawrence C. Moore; [*Laboratories for Application, Motivated Calculus*] $60,000; No. 8951909.

BROWN UNIVERSITY, Providence, RI 02912. PI: Ulf Grenander; [*Laboratory for Experimental Mathematics*] $19,400; No. 8952446.

CLEMSON UNIVERSITY, Clemson, SC 29634. PI: Donald R. LaTorre; [*Calculus and Rest of First Two-Year Sequence*] $48,000; No. 8950571.

UNIVERSITY OF SOUTH CAROLINA AT COLUMBIA, Columbia, SC 29208. PI: David P. Sumner; [*The HyperMath Classroom*] $45,000; No. 8950650.

UNIVERSITY OF VERMONT, Burlington, VT 05405. PI: James W. Burgmeier; [*Classroom Computers and Calculus*] $21,000; No. 8951878.

UNIVERSITY OF WISCONSIN AT MADISON, Madison, WI 53706. PI: J.R. Smart; [*Laboratory*] $70,000; No. 8951698.

UNIVERSITY OF WISCONSIN AT MILWAUKEE, Milwaukee, WI 53201. PI: David H. Schultz; [*Computational Mathematics Laboratory*] $44,000; No. 8951440.

ILI (Nondoctoral), FY 1990

LIVINGSTON UNIVERSITY, Livingston, AL 35470. PI: Julie E. Massey; [*Calculus and Precalculus with Derive*] $33,983; No. 9050638.

NORTHERN ARIZONA UNIVERSITY, Flagstaff, AZ 86011. PI: John W. Hagood; [*Capstone Modeling Projects for Majors, Mathematica*] $21,924; No. 9051163.

CALIFORNIA STATE UNIVERSITY, CHICO, Chico, CA 95929. PI: LaDawn Haws; [*Mathematica for Calculus and LOGO for Prospective K-6 Teachers*] $52,077; No. 9051194.

GOLDEN WEST COLLEGE, Huntington Beach, CA 92647. PI: John H. Wadmus; [*Calculus and Precalculus with Mathematica*] $57,024; No. 9051185.

SAN JOSE STATE UNIVERSITY, San Jose, CA 95192. PI: Michael J. Beeson; [*Calculus, College Algebra with MathPert*] $37,717; No. 9050894.

GEORGIA SOUTHERN COLLEGE, Statesboro, GA 30460. PI: Arthur G. Sparks; [*Calculus with Mathematica*] $48,968; No. 9052022.

KNOX COLLEGE, Galesburg, IL 61401. PI: Dennis Schneider; [*Calculus with Mathematica*] $45,647; No. 9050757.

BATES COLLEGE, Lewiston, ME 04240. PI: Shepley Ross; [*Dynamical Systems and Fractals*] $24,514; No. 9050684.

BOWDOIN COLLEGE, Brunswick, ME 04011. PI: William H. Barker; [*Calculus with Mathematica and Project CALC*] $63,787; No. 9052224.

LOYOLA COLLEGE, Baltimore, MD 21210. PI: Anne L. Young; [*Team Projects in Required Sophomore Courses for Majors*] $23,399; No. 9050571.

MONTGOMERY COLLEGE, ROCKVILLE, Rockville, MD 20850. PI: Judy E. Ackerman; COURSE FOR GENERAL EDUCATION REQUIREMENTS WITH DERIVE $32,387; No. 9050550.

GREENFIELD COMMUNITY COLLEGE, Greenfield, MA 01301. PI: Ira Rubenzahl; [*Calculus and Precalculus with MathCad*] $54,970; No. 9050505.

COLLEGE OF HOLY CROSS, Worcester, MA 01610. PI: Melvin C. Tews; [*General Laboratory*] $18,993; No. 9051318.

ST. OLAF COLLEGE, Northfield, MN 55057. PI: Arnold Ostebee; [*Laboratory for Advanced Courses*] $99,623; No. 9050637.

UNIVERSITY OF MINNESOTA–MORRIS, Morris, MN 56267. PI: Michael F. O'Reilly; [*Calculus with Mathematica*] $26,075; No. 9050797.

MERIDIAN COMMUNITY COLLEGE, Meridian, MS 39301. PI: Wanda Dixon; [*Calculus with HP-28S Calculators and Derive*] $42,453; No. 9051605.

XAVIER UNIVERSITY, Cincinnati, OH 45207. PI: Raymond J. Collins; CALCULUS WITH MICRO-CALC $33,000; No. 9051989.

LANGSTON UNIVERSITY, Langston, OK 73050. PI: Julia E. Massey; [*Calculus and Precalculus with Derive*] $33,193; No. 9050672.

CHRISTIAN BROTHERS COLLEGE, Memphis, TN 38104. PI: Arthur A. Yanushka; [*Calculus with MicroCalc*] $24,182; No. 9052185.

STATE TECHNICAL INSTITUTE AT MEMPHIS,

Memphis, TN 38134. PI: Robert O. Armstrong; [*Laboratory for Careers in High Technology with Derive*] $16,035; No. 9050755.

TENNESSEE TECHNOLOGICAL UNIV., Cookeville, TN 37203. PI: John J. Selden; [*Calculus and Mathematica*] $86,400; No. 9052083.

COLLEGE OF ST. THOMAS, St. Paul, MN 55105. PI: John T. Kemper; [*Projects Provided by Local Industry*] $26,000; No. 9050562.

RIPON COLLEGE, Ripon, WI 54971. PI: Robert Fraga; [*Calculus with Mathematica and Statistics with Minitab*] $45,454; No. 9050406.

Spotlight on Calculus: Minicourses and Meetings

MAA Minicourses, 1987-1990

January 1987, San Antonio

#6 *Using microcomputer software in teaching calculus,* David A. Smith and David P. Kraines, Duke University.

#10 *Integrating history into undergraduate mathematics,* Judith V. Grabiner, Pitzer College.

#12 *TrueBASIC in freshman calculus,* James F. Hurley, University of Connecticut.

#15 *Constructing placement examinations,* John Kenelly, Clemson University (Committee on Placement Examinations; see also January meetings 1988, 1989, 1990).

August 1987, Salt Lake City

#2 *Using computer spreadsheet programs in calculus, differential equations, and combinatorics,* Donald R. Snow, Brigham Young University.

#4 *A survey of educational software,* David P. Kraines, Duke University, and Vivian Kraines, Meredith College (see also August, 1988 and January, 1990).

#6 *A calculus lab course using MicroCalc,* Harley Flanders, University of Michigan, Ann Arbor.

January 1988, Atlanta

#1 *Using computer graphics to enhance the teaching and learning of calculus and precalculus mathematics,* Franklin R. Demana and Bert K. Waits, Ohio State University.

#2 *Computer software for differential equations,* Howard Lewis Penn and James Buchanan, United States Naval Academy.

#4 *Teaching calculus with an HP-28S symbol manipulating calculator,* John Kenelly, Clemson University, and Thomas Tucker, Colgate University (see also August 1988, January 1989, August 1989).

#9 *Constructing placement examinations,* John Harvey, University of Wisconsin (COPE; see January 1987).

#12 *Using computer algebra systems in undergraduate mathematics,* Paul Zorn, St. Olaf College.

August 1988, Providence

#3 *A survey of educational software,* David P. Kraines and Vivian Kraines (see also August 1987).

#5 *Teaching calculus with an HP-28S,* Kenelly (see January 1988).

#6 *An introduction to MATLAB,* David R. Hill, Temple University.

January 1989, Phoenix

#2 *Computer graphics in calculus and precalculus,* Franklin R. Demana and Bert K. Waits, Ohio State University (see January, 1988).

#3 *Using history in teaching calculus,* V. Frederick Rickey, Bowling Green State University (see also January, 1990).

#5 *Writing in mathematics courses,* George D. Gopen and David A. Smith, Duke University (see also January, 1990).

#12 *muMath workshop,* Wade Ellis Jr., West Valley College.

#13 *Applications of the HP-28S for more experienced users,* Thomas Tucker, Colgate University (see January, 1988).

#14 *Instituting a mathematics placement program,* Billy Rhoades, Indiana University (COPE; see January 1987).

August 1989, Boulder

#4 *Faculty-managed programs that produce minority mathematics majors,* Ray Shiflett, California

Polytechnic University, Pomona, and Uri Treisman, University of California at Berkeley.

#5 *Starting, funding, and sustaining mathematics laboratories,* Stavros N. Busenberg, Harvey Mudd College (see also January, 1990).

#7 *An HP-28S short course for nearly inexperienced users,* Jerold Mathews, Iowa State University.

#8 *Applications of the HP-28S for experienced users,* Thomas Tucker, Colgate University (see January, 1989).

January 1990, Louisville

#4 *Lagrange first year calculus,* Francesca Schremmer, West Chester University and Alain Schremmer, Community College of Philadelphia.

#7 *DERIVE workshop,* Wade Ellis Jr., West Valley College.

#8 *Using history in teaching calculus,* V. Frederick Rickey, Bowling Green State University (see January, 1989).

#9 *Computer graphics in calculus and precalculus,* Franklin R. Demana and Bert K. Waits, Ohio State University (see January, 1988).

#11 *Writing in mathematics courses,* George D. Gopen and David A. Smith, Duke University (see January, 1989).

#13 *A survey of educational software,* Virginia E. Knight and Vivian Yoh Kraines, Meredith College (see August, 1987).

#14 *Instituting a mathematics placement program,* Linda Boyd, DeKalb College (COPE, see January 1987).

#15 *Mathematica and college teaching,* Stan Wagon, Smith College.

#16 *Starting, finding, and sustaining mathematics laboratories,* Stavros N. Busenberg, Harvey Mudd College (see August, 1989).

#17 *The informed consumer's instructional guide to graphing calculators,* Iris B. Fetta and John Kenelly, Clemson University.

August 1990, Columbus

#4 *A calculus laboratory using Mathematica,* Michael Barry, Benjamin Haytock, and Richard McDermot, Allegheny College.

#5 *Using history in teaching calculus,* V. Frederick Rickey, Bowling Green State University (see January 1990 and January 1989).

#7 *Exploring mathematics with the NeXT computer,* Charles G. Fleming and Judy D. Halchin, Eastern Illinois University.

#8 *A mathematician's introduction to the HP-48SX calculator for first-time users,* John Kenelly and Don LaTorre, Clemson University.

#9 *Starting, funding, and sustaining mathematics laboratories,* James E. White, Kenyon College.

#10 *CAS laboratory projects for calculus,* Carl Leinbach, Gettysburg College.

#11 *Producing mathematics courseware with Mathematica: Calculus and mathematics,* Don Brown, Horacio Porter, and Jerry Uhl, University of Illinois.

#13 *Spreadsheet based mathematical topics for nonmathematics majors,* V.S. Ramamurti, University of North Florida.

Calculus Panels & Paper Sessions

August 1987, Salt Lake City

In search of the lean and lively calculus. Contributed paper session organized by Katherine A. Franklin, Los Angeles Pierce College.

January 1988, Atlanta

Symposium on Computer Algebra Systems. Organizer, Warren Page, New York City Technical College.

January 1989, Phoenix

What is happening with calculus revision? Contributed paper session organized by John Kenelly, Clemson University, and Thomas Tucker, Colgate University (see also August, 1989).

Calculus initiatives—an update. Panel organized by Ronald Douglas, SUNY Stony Brook, and Thomas Tucker, Colgate University.

August 1989, Boulder

Calculus revision. Contributed paper session organized by Thomas Tucker, Colgate University (see January, 1989).

Pedagogical uses of symbolic computer systems.
Contributed paper session organized by Arnold Ostebee, St. Olaf College.

The role of the computer in calculus reform. Panel
organized by the Committee on Computers in
Mathematics Education (CCIME).

January 1990, Louisville

Calculus revision.   Poster session organized by
Thomas Tucker, Colgate University (CRAFTY).

Classic classroom calculus problems. Contributed
paper session organized by Anthony Barcellos,
American River College.

Computers in the classroom: the time is right. Contributed paper session organized by Vivian Kraines,
Meredith College, and David Kraines, Duke University.

Providing computer resources for mathematics.
Panel organized by David Smith, Duke University.

Calculus for the twenty-first century. Panel organized by Sheldon Gordon, Suffolk County Community College.

Computer algebra systems as teaching tools. Organized by Donald Small, Colby College.

Calculus texts in a time of reform. Panel sponsored
by the Textbook Authors Association.

August 1990, Columbus

*What students learn in the symbolic computing en-
vironment.* Poster session organized by Joan Hundhausen, Colorado School of Mines.

*The pedagogical impact of computer algebra systems
on college mathematics curricula.* Panel sponsored
by CUPM Subcommittee on Symbolic Computer
Systems.

Calculus reform today: An overview. Panel sponsored by CRAFTY.

Conferences and Workshops

Sloan Conference on Calculus Instruction, Tulane
University, January 1986. PROCEEDINGS: *Toward
a Lean and Lively Calculus,* MAA Notes 6.

Colloquium, National Academy of Sciences, October 1987. PROCEEDINGS: *Calculus for a New Century: A Pump, Not a Filter,* MAA Notes 8.

*Conferences on Technology in Collegiate Mathe-
matics,* Ohio State University, November 1988
and November 1989. PROCEEDINGS: (1988 Conference) *Twilight of Paper and Pencil,* Addison-
Wesley, 1989.

Boston Workshops for Mathematics Faculty, Wellesley College, August 1988, June 1989. CONTACT:
Gil Strang, MIT.

Meetings of NSF Calculus Principal Investigators,
University of Minnesota, November 1988, and Duke
University, November 1989.   CONTACT: John
(Spud) Bradley, NSF.

Calculus Conference, Ithaca College, April 1989.
CONTACT: Stephen Hilbert, Ithaca College.

Spring Conference on the First Two Years, April
1989, University of Hartford.   PROCEEDINGS:
Teaching the Mathematical Core, available. CON-
TACT: Bob Decker or John Williams, University of
Hartford.

Conference on Strategies for Teaching Calculus,
April 1989, Morehouse College. CONTACT: Henry
Gore, Morehouse College.

Calculus Conference, May 1989, University of
Maine. CONTACT: Grattan Murphy, University of
Maine.

*Seventeenth Annual Mathematics and Statistics
Conference: Issues in the Teaching of Calculus,* October 1989, Miami University (Ohio), proceedings
available. CONTACT: Fred Gass, Miami University.

*Conference on Symbolic Computation and Calculus
Instruction,* St. Olaf College, October 1989, proceedings available. CONTACT: St. Olaf College
(include $8.00).

Conference on Calculus Revision, William Jewell
College, January 1990. CONTACT: Darrel Thoman,
William Jewell College.

*Workshops on Computer Algebra Systems as Teach-
ing Tools,* Mississippi State, Colby College, and St.
Olaf College, June and July 1990. CONTACT: Don
Small, Colby College.

*California Calculus Consortium Summer Confer-
ence,* Cal Poly San Luis Obispo, July 1990. CON-
TACT: Michael Colvin or Donald Hartig, Cal Poly
San Luis Obispo.

Graphing Calculators

Eight graphing calculators are currently available: Casio's fx-7000G, fx-7500G, fx-8000G, and fx-8500G; Hewlett-Packard's HP-28S and HP-48SX; Sharp's EL-5200; and Texas Instruments' TI-81. The TI-81 and HP-48SX have just become available in 1990. All of these models have multiline text displays that allow expressions to be entered, evaluated, and edited while all the information is on the screen. All are fully programmable and have "continuous memories" that retain programs when the calculator is turned off. All can display the graph of one or more functions, and each has ways of changing the viewing rectangle (e.g. zooming in or out) and displaying x, y-coordinates of points on the graph. All can be programmed easily to plot polar equations and parametric equations (for the TI-81 and HP-48SX this is built in). All have some sort of built-in statistical package. All but the Casio models allow matrices and matrix operations. Finally, all have (or soon will have) special overhead models (prices \$250 – \$500) for classroom demonstrations.

In what follows, we give a very brief sketch of each calculator. For a more detailed comparison, see the article "Graphing Calculators: Comparisons and Recommendations" by Demana, Dick, Harvey, Kenelly, Musser, and Waits *The Computing Teacher,* April 1990, pp 24-31). Much of this information given here is taken directly from that article.

Casio: Street prices of all models are \$60-\$75. All but the fx-7500G are 6.5×3.25 inches and $\frac{1}{2}$ to $\frac{11}{16}$ inches thick and have a one-face keyboard that is somewhat cluttered since many keys have three functions. Keys have a "rubber feel." The case is durable with an inexpensive plastic cover, but the screen is unprotected and can be damaged in backpacks. The fx-7500G has a smaller hard case (4.9 $\times$ 2.7 inches closed) that opens to a less cluttered two-face keyboard; the keys, however, are touch-sensitive surface "keys" and it is hard to know when you have made positive contact. The Casio screens are 2 inches wide by 1.4 inches high, 96 pixels by 64 pixels. User memory (RAM) is between 3K and 8K with 10 program addresses; ROM is 11K to 38K.

The Casio models can graph $y = 2\sin x$ with 6 key pushes in 5 to 8 seconds. The plotter "interpolates" with vertical columns of dots joining adjacent points on the graph; this makes graphs look

nice and continuous but leads to spurious effects such as steep lines joining points on opposite sides of a vertical asymptote. There is no built-in root finder; roots are found by tracing along a graph and zooming. There is no numerical integrator either. Display notation in expressions is nonstandard for exponents.

Sharp EL-5200: The street price is around \$80. The size is 5.2×2.9 inches closed. The soft plastic case opens to a two-face keyboard; the left side has dual function plastic keys while the right side has touch-sensitive surface "keys." Durability of the hinge is questionable and the manual recommends not to fold the right-hand keyboard back, even though it is easily opened beyond 180 degrees. The screen is 2 inches wide by .7 inches high, 96 by 32 pixels, the smallest of any of the graphing calculator screens (although slightly higher than the HP-28S). User memory is about 9K with 53 program addresses; ROM is 138K.

The Sharp model graphs $y = 2\sin x$ with 5 button pushes in about 12 seconds. Unlike the other graphing calculators, the graph is not produced dynamically but rather the screen shows the word COMPUTING until the entire graph is displayed. Like the Casio models, the Sharp plotter always interpolates pixels. Graphs can be scrolled up or down, right or left, with cursor controls. There is a built-in root finder based on bisection. There is no numerical integrator. Display notation in expressions is nonstandard for exponents.

Texas Instruments TI-81: The street price will be about \$75 - \$80. The size is 3.1×6.7 inches. The case has a hard plastic sliding cover to protect the one-face keyboard and screen. The keys are large, hard plastic with a crisp, positive feedback. Because of the use of "pull down" menus, the keyboard is remarkably uncluttered. The screen is 2 inches wide by 1.4 inches high, 96 by 64 pixels, about the same as the Casio models. The user memory is 32K with 37 program addresses; ROM is 32K.

The TI-81 graphs $y = 2\sin x$ with 5 button pushes in 8 seconds. The user has a choice of interpolating mode or dot mode (one pixel per column as in the HP-28S). Graphs can be scrolled. Zoom-in rectangles are displayed on the screen. There is a built-in parametric equation plotter. There is no built-in root finder or numerical integrator, but there is a numerical differentiator. Display notation for expressions is standard, although mostly one line without superscripts.

Hewlett-Packard: The street price of the HP-28S is $175 and of the HP-48SX about $275. The HP-28S has a 6.5 × 3.8 sturdy case that opens to a two-faced keyboard of dual function, firm, "click" keys. Because of pull down menus the keyboard is uncluttered. The HP-48SX has a single keyboard like a traditional calculator and consequently each key has up to 6 functions. The calculator is 7 × 3.1 inches with thickness varying from 1.25 to .75 inches; it comes with a soft, padded, protective carrying case. The screen of the HP-28S is 2.6 inches wide by .6 inches high, 137 by 32 pixels (the disproportionate width is for menu key labels). The HP-48SX screen is 2.5 inches wide by 1.5 inches high, 131 by 64 pixels. It is the largest of any of the screens and the only one using the Supertwist display found in laptop computers.

Both models have 32K of user memory with unlimited program addresses; the HP-28S has 128K of ROM and the HP-48SX 256 of ROM. However, the HP-48SX has two slots for either interchangeable program modules (ROM) or additional (RAM) user memory up to 256K. In addition, the HP-48SX can be accessorized to interface fully with a PC, including screen dumps into and out of the calculator. The HP-48SX also communicates by infrared signals with other HP-48SX calculators within a distance of about two feet.

The HP models graph $y = 2\sin x$ with 10 or 12 keystrokes in about 10 seconds. The HP-28S plotter does not interpolate; the user of the HP-48SX can choose either interpolating or dot plotting. The built-in plotter of the HP-28S allows at most two functions to be graphed simultaneously and cannot toggle between graph and test; it is not difficult, however, to customize the plotting environment with short programs to emulate features on the other calculators, such as toggling, elaborate windowing, multiple function plotting. The HP-48SX has an elaborate built-in plotting environment with menu key labels on screen below the graph; features include scrolling, line and circle drawing, unlimited multiple function plots, parametric and polar curve plotting.

Both models have very efficient built-in root-finders and numerical integrators. Even more impressive, there is symbolic manipulation with significant calculus applications. Both have symbolic differentiation, and the HP-48SX has some modest symbolic antidifferentiation. Both machines have a surprisingly powerful, structured programming language (vaguely reminiscent of FORTH), with full editing and debugging capability, and an operating system that includes an infinite stack, directories of ROM programs, and a tree of directories and subdirectories of user memory. Display notation is standard, one line without superscripts, but the HP-48SX also has a full screen "equation" editor for creating and editing multiline expressions involving superscripts, fractions, and special symbols like integral or summation signs.

Software for College Mathematics

This list of software for college mathematics is reproduced from a database maintained for The College Mathematics Journal *by Jon Wilkin of Northern Virginia Community College. Longer reviews of many of the software packages listed here can be found in* The College Mathematics Journal, *the* Notices of the American Mathematical Society, *and other mathematics and computer periodicals.*

Software in this list is arranged by general areas of predominant use, approximately paralleling standard courses:

*General
Algebra*

*Linear Algebra
Geometry*

*Trigonometry
Graphing Software
Solving Equations
Logic
Discrete Mathematics
Calculus*

*Fractals
Statistics
Problem Solving
Advanced Mathematics
Tech. Word Processing*

Each item lists the title, author (in brackets), publisher, cost, and system requirements, where known. Complete addresses and phone numbers for publishers can be found in a special list at the end of the software lists.

Caution: *Because software versions, costs, and companies change rapidly, readers must understand that any list of this sort will be somewhat dated.*

Current information on price and availability will be available from the publishers.

General

Algebraic Proposer. [Schwartz, Judah L.] True BASIC. $99.95. IBM, Mac.

Arithmetic. $49.95. IBM PC, Mac, Amiga, Atari ST.

Arithmetic Skillbuilder. [Hamilton & Owen] Addison-Wesley. IBM.

Arithmetric Interactive Instruction. [Hackworth] H&H. $560. Apple II.

Baffles. [Spain, James] CONDUIT. $50. Apple.

Baffles II. [Spain, James; Ridge, John] CONDUIT. $50. IBM.

Basic Math Competency Skill Building. [Conlon, M.] Educational Activities. $325. Apple II, IBM, C-64, TRS-80.

C Math Functions. [Novack] Kern. $37. IBM.

CHIPendale. [Davis, James A.] True BASIC. $49.95. IBM PC, Mac.

College Compentencies. [Gilbert, Helen; Howland, J.D.] H & H. $315. Apple II.

Language of Math. Krell. $299.95. Apple II.

LCSI Logo II. [LCSI] LCSI. $99. Apple II.

Lin Tek. [Fraleigh, John B.] Addison-Wesley. IBM.

Logowriter Primary. [LCSI] Lcsi. $169. IBM, Apple II.

Macsyma. Symbolics. $1250 for academics. IBM, Mac.

Magic Math Plus. [Ecker, Michael W.] Recreational Mathemagical Software. $30 (Apple), $37 (IBM), $40 (Tandy). Apple II, IBM, Tandy.

Math Anxiety Reduction. [Hackworth, R. D.] H & H. $250. Apple II.

Math Library. [Hundal] Kern. $65. IBM.

Math Mastery. Cerebic Institute. $49.95. Apple II.

Math Practice & Problem Solver, Arithmetic. H&N Software. $80. IBM.

Math Review. [Novak, Robert] Bergwall Educational Software. $99. Apple II.

MathCAD 2.0. MathSoft. $495. Mac.

MathCAD 2.5. MathSoft. $495. IBM.

Mathdisk One. [Kimberling, Clark] Mathematics Software Co.. $40. Apple II.

Mathematica. Wolfram Research. $695 and up. IBM 386, Mac.

MathStation. MathSoft. SUN 3, SUN 4.

Mirror on the Mind: Strategy. [Finzer, William & Resek, Diane] Addison-Wesley. $55. Apple.

REDUCE 3.3. Northwest Computer Algorithms. $495. IBM, Mac, Atari, VAX, SUN/Unix.

The Scientific Desk Analysis System. C. Abaci, Inc. $225. IBM.

The Scientific Desk Library. C. Abaci, Inc. $375. IBM (DOS, RT, large-scale), Mac.

Algebra

100 Mathematics Programs. [Kimberling, Clark] Mathematics Software Co.. $70. IBM.

88 Individual lessons in mathematics. GP Publishing, Inc.. $49.50 per lesson. IBM.

Algebra. Queue. $39.95. Apple II, IBM.

Algebra 3.0. [Kemeny, J. ; Bogart, R.] True BASIC. $49.95. IBM, Mac.

Algebra Arcade. [Mick, Dennis et al] COMPress. $50. Apple II, C-64.

Algebra Arcade. Queue. $49.95. Apple II.

Algebra Drill and Practice II. [Detmer, R. C., Smullen, III, S.W.] CONDUIT. $125. Apple, IBM.

Algebra Drill and Practice III. [Detmer, R. C., Wells, David] CONDUIT. $125. Apple, IBM.

Algebra Football. Queue. $29.95. Apple II.

Algebra I. True BASIC. $49.95. Amagi, Atari.

Algebra Pack. Queue. $65. Apple II.

Algebra Package. [Dodge, W., Enns, D.] D.C. Heath. $100. Apple II.

Algebra Word Problems. Queue. $39.95 (Apple), $49.95 (IBM). Apple II, IBM.

Algebra Word Problems II. Queue. $34.95. Apple II.

Algebra Word Problems III. Queue. $34.95. Apple II.

Algebra Word Problems IV. Queue. $34.95. Apple II.

Algebraic Proposer. [Schwartz, Judah L.] True BASIC. $99.95. IBM, Mac.

Balance. Queue. $49.95. Apple II, Commodore.

Balance!. [Braun, Ludwig] HRM. $49.95. Apple II, C-64.

Basic Verbal Problems. [Kimberling, Clark] Mathematics Software Co.. $40. IBM.

Beginning Algebra. Queue. $74.95. Apple II, Atari.

Cactusplot: A Mathematics Utility, v.5. [Losse] Cactusplot Company. $60. Apple II, IBM.

CAL: The Mathematics Teaching and Learning Environment, Version II. [White, James] Bluejay Lispware. $249. IBM.

Calculus Graphics. [Yuster, Thomas] Polygonal. $50. IBM.

Comprehensive Test. Courses by Computers. $39.95. Apple, IBM, Tandy.

Computer Tutor. Houghton-Mifflin. IBM, Apple II.

Conic Sections. [Luckas] Edutech. $60. Apple II.

Core Drill and Practice. [Duignan, Wendy] H & H. $315. Apple II.

Counting, including Permutations and Combinations. [Kimberling, Clark] Mathematics Software Co.. $40. IBM.

Derive. Soft Warehouse, Inc. $200. IBM.

Discovery Algebra: Exponents. [Halsey] Wisc-Ware. $100. IBM.

Distance: D = RT. [Kimberling, Clark] Mathematics Software Co.. $40. IBM.

Distributive Law and Grouping Symbols. [Kimberling, Clark] Mathematics Software Co.. $40. IBM.

Electronic Blackboard Algebra 1.1. [O'Farrell, Richard *et al*] COMPress. $95. Apple II.

Electronic Blackboard: Algebra. Queue. $95. Apple II.

Electronic Study Guide for Precalculus Algebra. [Steinbach, R., Lundsford, D.] COMPress. $275. Apple II.

Equations and Inequalities. Courses by Computers. $49.95. Apple II, IBM, Tandy.

Equations and Inequalities. Queue. $50. Apple II.

Equations I. Queue. $39.95. Apple II.

Equations II. Queue. $39.95. Apple II.

Eureka: The Solver. Borland. $167 (IBM), $195 (Mac). IBM, Mac.

Expansion of Expressions. Queue. $39.95. Apple II.

Expert Algebra Tutor 2.0. [Ovchinnikov, Sergei] TUSOFT. $400. IBM.

Expert Algebra Tutor I. [Ovchinnikov] Scott Foresman/Little, Brown. $250. IBM.

Expert Algebra Tutor II. [Ovchinnikov] Scott Foresman/Little, Brown. $250. IBM.

Exponential & Logarithmic Functions. Queue. $34.95. Apple II.

Exponential and Logarithmic Functions. Queue. $50. Apple II.

Exponents. [Kimberling, Clark] Mathematics Software Co.. $40. IBM.

Exponents, Roots, & Radicals. Courses by Computers. $49.95. Apple II, IBM, Tandy.

Expression Writer. Queue. $49.95. Apple II.

Factoring. Courses by Computers. $49.95. Apple II, IBM, Tandy.

First Year Algebra I, II. [Olson, V., Ridge, J. *et al*] CONDUIT. $240 each, $380 pair. IBM PC.

Fractions. [Kimberling, Clark] Mathematics Software Co.. $40. IBM.

Functions. Queue. $50. Apple II.

Functions. Dynacomp. $30. IBM, C-64.

Fundamental Concepts of Algebra. Queue. $50. Apple II.

GP Mathematics Series. GP Publishing, Inc.. IBM.

Graph-Calc. [Harris, Mark] Queue. $75. IBM.

Graphing. [Hessemer] Educational Activities. $59.95. Apple II, TRS 80, Atari, C-64.

Graphing Linear Equations. [Hessemer] Educational Activities. $149. Apple II.

Graphing Lines on a Plane. Courses by Computers. $49.95. Apple II, IBM, Tandy.

Graphs of Lines and Hyperbola. Queue. $39.95. Apple II.

Greatest Common Factor/LCM. Queue. $39.95. Apple II.

Guess My Rule. [Barclay, Tim] HRM. $49.95. Apple II.

Honors Algebra. Queue. $39.95. Apple II, IBM.

I-APL. I-APL. IBM.

Instructional Software for Inter. Algebra. [Bittinger & Keedy] Addison-Wesley. Apple II.

Instructional Software for Intro. Algebra. [Bittinger & Keedy] Addison-Wesley. Apple II.

Intllectual SAT Algebra. Queue. $49.95. Apple II.

Interactive Lessons I, II, III. [Hackworth, R. D.; Howland, J.D.] H & H. $315. Apple II, IBM (set I only).

Interest: I = PRT. [Kimberling, Clark] Mathematics Software Co.. $40. IBM.

Intro Algebra Instructional Software. [Keedy & Bittinger] Addison-Wesley. $150. Apple.

Linear Equations. Queue. $39.95. Apple II.

Macsyma. Symbolics. $1250 for academics. IBM, Mac.

MAPLE 4.2.1. Brooks-Cole. $695. IBM, Mac.

MATH PAC. [Graves] Harcourt Brace Jovanovich. $20. IBM.

Math Practice & Problem Solver, Algebra. H&N Software. $120.

Mathdisk Three. [Kimberling, Clark] Mathematics Software Co.. $40. Apple II.

Mathematica. Wolfram Research. $695 and up. IBM 386, Mac.

Mathematics with Applications Package. [Dodge, W.; Enns, D.] D.C. Heath. $100. Apple II.

Mathgrapher. [Cohen, Stephen] HRM. $49.95. Apple, C-64.

Mathomatic. Dynacomp. $40. IBM.

MathView Professional. BrainPower. $250. Mac.

Matrix Master. [Kimberling, Clark] Mathematics Software Co.. $50. IBM.

Mixed Verbal Problems: Easy. [Kimberling, Clark] Mathematics Software Co.. $40. IBM.

Mixed Verbal Problems: Hard. [Kimberling, Clark] Mathematics Software Co.. $40. IBM.

Mixed Verbal Problems: Medium. [Kimberling, Clark] Mathematics Software Co.. $40. IBM.

Mixtures: Liquids, Coins,.... [Kimberling, Clark] Mathematics Software Co.. $40. IBM.

muMath-80. [Rich, Stoutmeyer, & Dickey] Dickey. $40. Apple II.

PC Math Package. Dynacomp. $50. Apple II.

Polynomial & Rational Functions, Conic Sections. Queue. $50. Apple II.

Polynomial Pal. [Weinstein, M.] Polygonal. $20. IBM.

Polynomials. Courses by Computers. $49.95. Apple II, IBM, Tandy.

Pre-Calculus. [Kemeny, J. & Bogart, R.] True BASIC. $49.95. IBM PC, Mac, Amiga, Atari.

Precalculus/Calculus. Tecmath Educational Consultants. $250. Apple II.

Quadratic Equations. Courses by Computers. $49.95. Apple II, IBM, Tandy.

Quadratic Equations I (Factoring). Queue. $39.95. Apple II.

Quadratic Equations II (formula). Queue. $39.95. Apple II.

Ratio, Proportion, Percent. [Kimberling, Clark] Mathematics Software Co.. $40. IBM.

Rational Expressions. Courses by Computers. $49.95. Apple II, IBM, Tandy.

Read & Solve Math Problems. [Edison & Schwartz] Educational Activities. $359. Apple II.

Real Number System. Queue. $39.95. Apple II.

REDUCE 3.3. Northwest Computer Algorithms. $495. IBM, Mac, Atari, VAX, SUN/Unix.

Signed Number Operations. Courses by Computers. $49.95. Apple II, IBM, Tandy.

Simple Inequality. Queue. $39.95. Apple II.

Simultaneous Equations. Queue. $39.95. Apple II.

Super Math Package. [Dodge, W.; Enns, D.] D. C. Heath. $250. Apple II.

Symbols and Sets. Queue. $39.95. Apple II.

SYNTEL Math Toolbox. Dynacomp. $40. IBM PC.

Systems of Equations. Courses by Computers. $49.95. Apple II, IBM, Tandy.

Systems of Equations and Inequalities. Queue. $50. Apple II.

The Function Game. Queue. $39.95. Apple II.

Topics in Algebra. [Hogben, Anne] H & H. $150. IBM.

Turbo Pascal Numerical Methods Toolbox. Borland. $100. IBM, Mac.

Visual Vectors. Kern. $75. Apple II, IBM.

Visualizing Algebra: the Function Analyzer. Sunburst. IBM.

Work: Time on Job, Tank Filling,.... [Kimberling, Clark] Mathematics Software Co.. $40. IBM.

Algebra Mentor: CBI for Ele Algebra. [Miller] Brooks-Cole. IBM, Apple II.

Algebra Mentor: Computer Based Instruction. [Miller] Brooks/Cole. Apple II, IBM.

Interactive Algebra. Tecmath Educational Consultants. $200. Apple II.

Plot Pak Precalculus Tutorials. [Mowbray] Brooks/Cole. $30. IBM.

Sin 'n Conics. Pedagoguery Software. Apple II, IBM.

Solving Algebra Word Problems. [Hoffman, Dale] Brooks/Cole. $35. Apple II.

Trigonometry

88 Individual lessons in mathematics. GP Publishing, Inc.. $49.50 per lesson. IBM.

Analytic Trigonometry. Queue. $60. Apple II.

Cactusplot: A Mathematics Utility. [Losse] Cactusplot Company. $60. Apple II, IBM.

CAL: The Mathematics Teaching and Learning Environment, Version II. [White, James] Bluejay Lispware. $249. IBM.

Complex Numbers. Queue. $60. Apple II.

Derive. Soft Warehouse, Inc. $200. IBM.

Discovery Learning in Trigonometry. [Kelly, John C.] CONDUIT. $75. Apple, IBM PC.

Electronic Blackboard Trig 1.1. [O'Farrell, Richard, et al] COMPress. $50. Apple II.

Electronic Blackboard: Trigonometry. Queue. $50. Apple II.

Electronic Study Guide for Precalculus Algebra. [Steinbach, R., Lundsford, D.] COMPress. $275. Apple II.

Eureka: The Solver. Borland. $167 (IBM), $195 (Mac). IBM, Mac.

Getting Ready for Trigonometry. Queue. $60. Apple II.

GP Mathematics Series. GP Publishing, Inc.. IBM.

Graph-Calc. [Harris, Mark] Queue. $75. IBM.

Graphing Trigonometric Functions. [Mazzarella & Schiller] Bergwall Educational Software. $49. Apple II.

Honors Trigonometry. Queue. $39.95. Apple II, IBM.

I-APL. I-APL. IBM.

Introduction to Trigonometry. Queue. $49.95. Apple II.

MathView Professional. BrainPower. $250. Mac.

Oblique Triangles & Vectors. Queue. $60. Apple II.

Pre-Calculus. [Kemeny, J. & Bogart, R.] True BASIC. $49.95. IBM PC, Mac, Amiga, Atari.

REDUCE 3.3. Northwest Computer Algorithms. $495. IBM, Mac, Atari, VAX, SUN/Unix.

RSI Trigonometry. Queue. $39.95. Apple II, IBM.

SEITrigonometry. Queue. $65. Apple II, IBM.

Super Math Package. [Dodge, W.; Enns, D.] D. C. Heath. $250. Apple II.

SYNTEL Math Toolbox. Dynacomp. $40. IBM PC.

Trig 1.0. [Kemeny, John] True BASIC. $49.95. Amagi, Atari.

Trigonometric Functions. Queue. $60. Apple II.

Trigonometric Graphing. Queue. $60. Apple II.

Trigonometry. Queue. $34.95. Apple II.

Trigonometry of A Right Triangle. [Mazzarella, John & Schiller, R.] Bergwall Educational Software. $169. Apple II.

Vectors. Queue. $34.95. Apple II.

Macsyma. Symbolics. $1250 for academics. IBM, Mac.

Mathematica. Wolfram Research. $695 and up. IBM 386, Mac.

Precalculus/Calculus. Tecmath Educational Consultants. $250. Apple II.

Sin 'n Conics. Pedagoguery Software. Apple II, IBM.

Trig Pak - Trigonometry Tutorials. [Mowbray, John] Brooks/Cole. IBM.

Graphing Software

3-D Graphics Toolkit. True BASIC. $69.95. IBM PC, Mac, Amiga, Atari ST.

4 Dimensional Hypercube. Lascaux. IBM, Mac.

88 Individual lessons in mathematics. GP Publishing, Inc.. $49.50 per lesson. IBM.

Advanced Math Graphics. Dynacomp. $40. Apple II, IBM.

ARBPLOT. [Brown, Austin R.; Harris, Mark] CONDUIT. $125. Apple II.

ArtPack. Zepher. $29.95. IBM.

C Graphics. [Novack] Kern. $37. IBM.

Cactusplot Student Supplement. [Losse] Addison-Wesley. $10. Apple, IBM.

Cactusplot: A Mathematics Utility. [Losse] Cactusplot Company. $60. Apple II, IBM.

CAL: The Mathematics Teaching and Learning Environment, Version II. [White, James] Bluejay Lispware. $249. IBM.

CALC-87 muMATH enhancements. [Freese & Stegenga] R & D Software. $45. IBM.

Calcaide. [Chang] Harcourt. IBM.

Calculus Graphics. [Yuster, Thomas] Polygonal. $50. IBM.

Chalkboard Graphics Tool Box I. [Scharf, Fred] Scharf Systems, Inc. $59.95. Apple II.

College Package. [Dodge, W.; Enns, D.] D.C. Heath. $125. Apple II.

Computer Activities for Calculus. D.C. Heath. $250. IBM.

Computer Graphing Experiments. [Lund, C.; Anderson, E.] Addison-Wesley. $220. Apple.

Conic Sections. [Luckas] Edutech. $60. Apple II.

Curve Fit Utility. Dynacomp. $30. IBM.

CURVES. [Bridger, Mark] Bridge Software. $50. IBM.

DATASURF. [Bridger, Mark] Bridge Software. $125. IBM.

Derive. Soft Warehouse, Inc. $200. IBM.

Electronic Blackboard: Function Plotter. Queue. $50. Apple II.

Electronic Blackboard: Function Plotter. [O'Farrell, Richard] COMPress. $50. Apple II.

Equator. [Palmer & Glade] Pulse Research. $59. IBM.

Eureka: The Solver. Borland. $167 (IBM), $195 (Mac). IBM, Mac.

f(z). [Lapidus] Lascaux. $59.95. IBM, Mac.

Fields & operators. Lascaux. IBM, Mac.

FractalMagic. [Bolme, Mark W.] Sintar. $35. IBM, Apple II, Atari, Mac.

Functions and Curves. [Kimberling, Clark] Mathematics Software Co.. $50. IBM.

GP Mathematics Series. GP Publishing, Inc.. IBM.

Graph Challenge. [Harris, Mark] COMPress. $75. Apple II.

Graph-Calc. [Harris, Mark] Queue. $75. IBM.

Graphics Calculator. [Mick, D.] CONDUIT. $75. Apple.

Graphing. [Hessemer] Educational Activities. $59.95. Apple II, TRS 80, Atari, C-64.

Graphing Equations. [Dugdale, Sharon, Kibbey, David] CONDUIT. $60. Apple.

Graphing Linear Equations. [Hessemer] Educational Activities. $149. Apple II.

Graphing Lines on a Plane. Courses by Computers. $49.95. Apple II, IBM, Tandy.

Graphing Trigonometric Functions. [Mazzarella & Schiller] Bergwall Educational Software. $49. Apple II.

GraphMaster. Zepher. $39.95. IBM.

Green Globe and Graphing Equations. [Dugdale, Sharon; Kibbey, David] Sunburst. $65. Apple II, IBM, Tandy 1000.

Hypercard Graph. [Snow, Dennis M.] Snow. Mac.

I-APL. I-APL. IBM.

Interpreting Graphs. [Dugdale, Sharon; Kibbey, David] Sunburst. $65. Apple II, IBM, Tandy 1000.

Interpreting Graphs (including Escape!). [Dugdale, Sharon, Kibbey, David] CONDUIT. $50. Apple.

KaleidoScope. [Bolme, Mark W.] Sintar. $25. IBM.

M.P.P.. [Penn *et al*] Penn. IBM.

MacFunction. [Lewis, Harry; Tecosky, Jeff] True BASIC. $49.95. Mac.

Mandelbrot Fractals. MACE. $10. IBM.

Master Grapher/3D Grapher. [Demana, Franklin; Waits, Bert] Addison-Wesley. $20. Apple, Mac, IBM.

MATH 3D. [Foley, Robert] Bergwall Educational Software. $49. Apple II.

Math Pack III - Graphs. Micro-ED. $39. Apple II.

Math Utilities. [Bridger, Mark] Bridge Software. $125. IBM.

Mathematics with Applications Package. [Dodge, W.; Enns, D.] D.C. Heath. $100. Apple II.

Mathgrapher. [Cohen, Stephen] HRM. $49.95. Apple, C-64.

Mathgrapher: A Complete Graphing Utility. Queue. $49.95. Apple II.

MathView Professional. BrainPower. $250. Mac.

Microcomputer Graph & Elem Cal Concpts. [Sandstrom, Ron] Sandstrom. Apple II, IBM, Mac.

Microcomputer Graphics. [Myers, Roy E.] Addison-Wesley. $15. Apple, IBM PC.

Omni-Fit. Dynacomp. $50. IBM.

PC Graphics. Dynacomp. $50. IBM.

PC Math Package. Dynacomp. $50. Apple II.

Pre-Calculus. [Kemeny, J. & Bogart, R.] True BASIC. $49.95. IBM PC, Mac, Amiga, Atari.

Super Math Package. [Dodge, W.; Enns, D.] D. C. Heath. $250. Apple II.

Supergraph. Queue. $59.95. Apple II.

Supergraph. Ventura Educational Systems. $59.95. Apple II.

SUPERPLOT. [Steketee, Scott] Edusoft. $50. Apple II.

Surface. [Smith, David A.; Myers, Roy E.] CONDUIT. $65. Apple.

Surfaces for Multivariate Calculus. [Myers, Roy E.] CONDUIT. $65. Apple.

SURFS. [Bridger, Mark] Bridge Software. $50. IBM.

Symmetry. Queue. $34.95. Apple II.

SYNTEL Math Toolbox. Dynacomp. $40. IBM PC.

Tecmath. Technical Education Cousultants. $250. Apple II.

Turbo Pascal Numerical Methods Toolbox. Borland. $100. IBM, Mac.

CALCULUS PAD. [Bell, I.; Davis, J.; Rice, S.] Brooks/Cole. $35. IBM.

GrafEq. Pedagoguery Software. Apple II.

Graphitti. [Best] DC Heath. IBM, MAC.

GyroGrapics v2.2. [Johnson] Cipher Systems. $75. IBM.

Macsyma. Symbolics. $1250 for academics. IBM, Mac.

Mathematica. Wolfram Research. $695 and up. IBM 386, Mac.

One-variable Function Plotter. [Brooks] Prentice Hall. $34.95. IBM.

Plot Pak - Precalculus Tutorials. [Mowbray] Brooks/Cole. $30. IBM.

Precalculus/Calculus. Tecmath Educational Consultants. $250. Apple II.

100 Mathematics Programs. [Kimberling, Clark] Mathematics Software Co.. $70. IBM.

Solving Equations

88 Individual lessons in mathematics. GP Publishing, Inc.. $49.50 per lesson. IBM.

Cactusplot: A Mathematics Utility. [Losse] Cactusplot Company. $60. Apple II, IBM.

CAL: The Mathematics Teaching and Learning Environment, Version II. [White, James] Bluejay Lispware. $249. IBM.

Derive. Soft Warehouse, Inc. $200. IBM.

Eureka: The Solver. Borland. $167 (IBM), $195 (Mac). IBM, Mac.

GP Mathematics Series. GP Publishing, Inc.. IBM.

I-APL. I-APL. IBM.

Mathomatic. Dynacomp. $40. IBM.

MathView Professional. BrainPower. $250. Mac.

REDUCE 3.3. Northwest Computer Algorithms. $495. IBM, Mac, Atari, VAX, SUN/Unix.

SEQS. CET Research Group, Ltd. $125. IBM.

SEQS 3.0 Student version. CET Research Group, Ltd. $25. IBM.

SOLVE 1. [Holden] Holden. $35. IBM.

Solving Equations & Finding Roots. [Hoffman, Dale T.] Queue. $50. Apple II.

Tecmath. Technical Education Cousultants. $250. Apple II.

Turbo Pascal Numerical Methods Toolbox. Borland. $100. IBM, Mac.

Macsyma. Symbolics. $1250 for academics. IBM, Mac.

Mathematica. Wolfram Research. $695 and up. IBM 386, Mac.

Plot Pak - Precalculus Tutorials. [Mowbray] Brooks/Cole. $30. IBM.

Logic

Barbara the Syllogizer. [Wengert] Wisc-Ware. $25. IBM.

Comp-U-Solve. [Engel] Educational Activities. $109. Apple II, TRS 80, C-64.

Elementary Logic level I. Courses by Computers. $49.95.

Elementary Logic level II. Courses by Computers. $49.95. Apple II, IBM, Tandy.

LogicLab - Programs for Math Logic. [Keisler et al] Wisc-Ware. $50. IBM.

Philo the Logician. [Wengert] Wisc-Ware. $50. IBM.

PREDCALC - Predicate Logic. [Keisler] Wisc-Ware. $50. IBM w/ windows.

Reasoning, the Logical Process. [Ouding, Connie] MCE. $69.95. Apple II.

Solving Logical Problems. Krell. $70. Apple, IBM, Tandy.

Discrete Mathematics

Algebraic Proposer. [Schwartz, Judah L.] True BASIC. $99.95. IBM, Mac.

Counting, including Permutations and Combinations. [Kimberling, Clark] Mathematics Software Co.. $40. IBM.

Discrete Mathematics. True BASIC. $49.95. IBM, Mac, Amagi, Atari.

Discrete Probability, Part I. [Trumbo, Bruce E.] Queue. $50. Apple II.

Discrete Probability, Part II. [Trumbo, Bruce E.] Queue. $50. Apple II.

Finitepak. Addison-Wesley. IBM.

Interactive Finite Mathematics. [Avery & Soler] The Math Lab.

Probability Theory. [Kemeny, John G.] True BASIC. $49.95. IBM, Mac, Amagi, Atari.

Solving Logical Problems. Krell. $70. Apple, IBM, Tandy.

The Scientific Desk Analysis System. C. Abaci, Inc. $225. IBM.

The Scientific Desk Library. C. Abaci, Inc. $375. IBM (DOS, RT, large-scale), Mac.

Discrete Simulation. [Curry et al] Holden-Day. $49.95. IBM.

Macsyma. Symbolics. $1250 for academics. IBM, Mac.

Mathematica. Wolfram Research. $695 and up. IBM 386, Mac.

Calculus

100 Mathematics Programs. [Kimberling, Clark] Mathematics Software Co.. $70. IBM.

88 Individual lessons in mathematics. GP Publishing, Inc.. $49.50 per lesson. IBM.

Advanced Placement Calculus. Queue. $89.95. Apple II.

Algebraic Proposer. [Schwartz, Judah L.] True BASIC. $99.95. IBM, Mac.

Cactusplot: A Mathematics Utility, v.5. [Losse] Cactusplot Company. $60. Apple II, IBM.

CAL: The Mathematics Teaching and Learning Environment, Version II. [White, James] Bluejay Lispware. $249. IBM.

CALC-87 muMATH enhancements. [Freese & Stegenga] R & D Software. $45. IBM.

Calcaide. [Chang] Harcourt. IBM.

Calculus. Broderbund. $100. Mac.

Calculus. Queue. $34.95. Apple II.

Calculus. [Kemeny, John G.] True BASIC. $49.95. IBM, Mac, Atari, Amiga.

Calculus and the Computer. [Oberle, W.] Addison-Wesley. Apple, IBM.

Calculus Package. [Dodge, W.; Enns, D.] D.C. Heath. $150. Apple II.

Calculus Student's Toolkit. [Finney, Rose L. *et al*] Addison-Wesley. $20. Apple, IBM.

Calculus Toolkit, The. [Finney, Ross L. *et al*] Addison-Wesley. $250. Apple, IBM.

College Package. [Dodge, W.; Enns, D.] D.C. Heath. $125. Apple II.

Computer Activities for Calculus. D.C. Heath. $250. IBM.

Computer Applications for Finite Math & Calculus. [Coscia, D.R.] Scott, Foresman/Little,Brown. $23.44. Apple, IBM.

Computer Explorations in Calculus. [Stroyan, K.D.] Harcourt. $12. IBM.

Derive. Soft Warehouse, Inc. $200. IBM.

Eureka: The Solver. Borland. $167 (IBM), $195 (Mac). IBM, Mac.

Exploring Calculus. [Fraleigh, John B.; Pakula, Lewis I.] Addison-Wesley. IBM.

Fields & Operators. Lascaux. IBM, Mac.

GP Mathematics Series. GP Publishing, Inc.. IBM.

Graph-Calc. [Harris, Mark] Queue. $75. IBM.

Honors Calculus. Queue. $39.95. Apple II, IBM.

I-APL. I-APL. IBM.

Introductory Calculus Series. Microphys. $200. Apple II, IBM, Tandy, Commodore 64/128/PET.

M.P.P.. [Penn *et al*] Penn. IBM.

Macsyma. Symbolics. $1250 for academics. IBM, Mac.

MAPLE 4.2.1. Brooks-Cole. $695. IBM, Mac.

Math Pack I - Calculus & Diff Equations. Micro-Ed. $39. Apple II.

MathCAD,v 2.5. MathSoft Inc. $495, $40 for student version. IBM.

Mathdisk Four. [Kimberling, Clark] Mathematics Software Co.. $40. Apple II.

Mathematica. Wolfram Research. $695 and up. IBM 386, Mac.

Mathematics with Applications Package. [Dodge, W.; Enns, D.] D.C. Heath. $100. Apple II.

MathView Professional. BrainPower. $250. Mac.

Microcomputer Graph & Elem Cal Concpts. [Sandstrom, Ron] Sandstrom. Apple II, IBM, Mac.

Multiple Integration. Dynacomp. $20. IBM, C-64.

muMath-80. [Rich, Stoutmeyer, & Dickey] Dickey. $40. Apple II.

Omni-Fit. Dynacomp. $50. IBM.

Practice Problems in Calculus. [Wells, David M.] COMPress. $50. Apple II, IBM.

Practice Problems in Calculus. [Wells, David M.] Queue. $50. Apple II, IBm.

REDUCE 3.3. Northwest Computer Algorithms. $495. IBM, Mac, Atari, VAX, SUN/Unix.

Riemann Integral, The. [Keyton, Michael] Micro Power & Light. $30. Apple II.

RSI Calculus. Queue. $39.95. Apple II, IBM.

Solving Equations & Finding Roots. [Hoffman, Dale T.] Queue. $50. Apple II.

Student Edition of MathCAD 2.0. Addison-Wesley. $36.20. IBM.

Super Math Package. [Dodge, W.; Enns, D.] D. C. Heath. $250. Apple II.

Surface. [Smith, David A.; Myers, Roy E.] CONDUIT. $65. Apple.

Surfaces for Multivariate Calculus. [Myers, Roy E.] CONDUIT. $65. Apple.

SYNTEL Math Toolbox. Dynacomp. $40. IBM PC.

Tecmath. Technical Education Cousultants. $250. Apple II.

The Function Game. Queue. $39.95. Apple II.

The Scientific Desk Analysis System. C. Abaci, Inc. $225. IBM.

The Scientific Desk Library. C. Abaci, Inc. $375. IBM (DOS, RT, large-scale), Mac.

The Scientific Desk Student Pack. C. Abaci, Inc. $95. IBM.

Turbo Pascal Numerical Methods Toolbox. Borland. $100. IBM, Mac.

CALCULUS PAD 1.5. [Bell, I.; Davis, J.; Rice, S.] Brooks/Cole. $35. IBM.

Calculus T/L. [Child] Brooks-Cole. MAC.

Graphitti. [Best] DC Heath. IBM, MAC.

Interactive Calculus. [Avery] DC Heath. IBM, Apple II.

Interactive Computer Applications Package. [Meitler] Dellen.

MATHPACK. McGraw-Hill.

MATHPATH. [Bergeman] Saunders. IBM.

Precalculus/Calculus. Tecmath Educational Consultants. $250. Apple II.

Work-Out. [Mazur] DC Heath. IBM.

Linear Algebra

88 Individual lessons in mathematics. GP Publishing, Inc.. $49.50 per lesson. IBM.

CLR Hyperarrays. Clear Lake Research. $65. Mac.

Derive. Soft Warehouse, Inc. $200. IBM.

Eureka: The Solver. Borland. $167 (IBM), $195 (Mac). IBM, Mac.

I-APL. I-APL. IBM.

LINDO. [Schrage, Linus] Scientific Press. $45 (student edition). IBM, Mac.

Linear Algebra. [Fraleigh, John B.] Addison-Wesley. IBM PC.

Linear Algebra. Dynacomp. $30. IBM, C-64.

Linear Algebra Toolkit. [Wilde, C.] Addison-Wesley. IBM.

Linear and Non-Linear Programming. Lionheart. $145. IBM, Mac.

LP. [Wassyng] Wisc-Ware. $50. IBM.

Math Pack II - Matrix & Linear Algebra. Micro-Ed. $39. Apple II.

MathView Professional. BrainPower. $250. Mac.

MATLAB. [Little & Moler] The Math Works. $395. IBM, Mac, VAX.

Matrix 100. Dynacomp. $80. IBM.

Matrix 100. [Thapa & Laventhold] Stanford Business Software. $80. IBM.

Matrix Algebra Software. [Wong, P. K.] Wong, P. K.. $10 (Apple II), $15 (IBM). Apple II, IBM.

Matrix Calculator. [Hogben, L.; Henzel, I. R.] CONDUIT. $85. IBM.

Matrix Master. [Kimberling, Clark] Mathematics Software Co.. $50. IBM.

Matrix Reducer. [Turner, V. Lee] Creative Communication. $20. Apple II.

Matrix Workshop 1.0. Puma Software. $295, $99 (students). Mac.

MatrixPad. [Orzech, Morris] D. C. Heath. IBM PC.

Mr. Matrix. [Weinstein, Michael] Polygonal. $20. IBM.

muMath-80. [Rich, Stoutmeyer, & Dickey] Dickey. $40. Apple II.

REDUCE 3.3. Northwest Computer Algorithms. $495. IBM, Mac, Atari, VAX, SUN/Unix.

SOLVE 1. [Holden] Holden. $35. IBM.

The Scientific Desk Analysis System. C. Abaci, Inc. $225. IBM.

The Scientific Desk Library. C. Abaci, Inc. $375. IBM (DOS, RT, large-scale), Mac.

The Scientific Desk Student Pack. C. Abaci, Inc. $95. IBM.

Three Dim'l Graphics: An Appl. of Linear Algebra. [Smith, David A.] Smith. Apple II, IBM (DOS 3.3).

Turbo Pascal Numerical Methods Toolbox. Borland. $100. IBM, Mac.

Macsyma. Symbolics. $1250 for academics. IBM, Mac.

Mathematica. Wolfram Research. $695 and up. IBM 386, Mac.

Matrix Operations. Lionheart. $125. IBM, Mac.

MAX - The MatriX Algebra Calculator. [Herman, E. A.; Jepsen, C. H.] Brooks/Cole. $35. IBM.

Geometry

100 Mathematics Programs. [Kimberling, Clark] Mathematics Software Co.. $70. IBM.

Angles of a Circle. [Mazzarella & Schiller] Bergwall Educational Software. $169. Apple II.

Circles and Conics. Queue. $49.95.

Coordinate Geometry. [Mazzarella & Schiller] Bergwall Educational Software. $169. Apple II.

Elements of Geometry. Queue. $49.95. Apple II.

Geometric Constructor, The. [Kimberling, Clark] Mathematics Software Co.. $130. IBM.

Geometric Supposer - Circles. [Schwartz, J.; Yerushalmy, M.] Sunburst. $99. Apple II, IBM, Tandy 1000.

Geometric Supposer - Quadrilaterals. [Schwartz, J.; Yerushalmy, M.] Sunburst. $99. Apple II, IBM, Tandy 1000.

Geometric Supposer - Triangles. [Schwartz, J.; Yerushalmy, M.] Sunburst. $99. Apple II, IBM, Tandy 1000.

Geometry. Broderbund. $100 (Mac), $80 (Apple II). Apple II, Mac.

Geometry. Queue. $39.95. Apple II, IBM.

Geometry & Microcomputers. [Oehmke, R. H.; Stroyan, K. D.] CONDUIT. $85. Apple.

Geometry Alive!. [Dylan] Educational Activities. $159. Apple II, IBM, TRS 80.

Geometry: Concepts & Proofs. Educational Design. $40. Apple II.

GP Mathematics Series. GP Publishing, Inc.. IBM.

Honors Geometry. Queue. $39.95. Apple II, IBM.

Intllectual SAT Geometry. Queue. $49.95. Apple II.

Mathdisk Two. [Kimberling, Clark] Mathematics Software Co.. $40. Apple II.

Parallel Lines & Triangles. Queue. $49.95. Apple II.

Proportions in Geometry. [Mazzarella & Schiller] Bergwall Educational Software. $169. Apple II.

Ratios & Right Triangles. Queue. $49.95. Apple II.

REDUCE 3.3. Northwest Computer Algorithms. $495. IBM, Mac, Atari, VAX, SUN/Unix.

Special Polygons, Congruent Triangles, Area/Perimeter. Queue. $49.95. Apple II.

The Function Game. Queue. $39.95. Apple II.

Theorem Generator. National Collegiate Software. $25. IBM.

Macsyma. Symbolics. $1250 for academics. IBM, Mac.

Fractals

ChaosPlus. EduTech. $75. Apple II.

Desktop Fractal Design System. [Barnsley] Academic Press. $40. IBM.

DiscoverForm. [Carlson] Secondary Dynamical Systems. $25. MAC.

Fractal Magic. Sintar Software. $25. IBM, Mac, Apple II, Amiga.

Fractal-D. [Slice] Exeter. $75. IBM.

FractalMagic. [Bolme, Mark W.] Sintar. $35. IBM, Apple II, Atari, Mac.

FractaSketch. Dynamic Software. $30. MAC.

Julia. [Koch] Secondary Dynamical Systems. MAC.

Mandelbrot Explorer. AAmygdala DOS-ware. $30. IBM.

MandelMovie. Dynamic Software. $30. MAC.

Peanut Software. [Parris] Peanut Software. IBM.

Statistics

20/20 Statistics. [Bergeman & Scott] Saunders. $25. Apple II, IBM.

44 Probability & Statistics Programs. [Kimberling, Clark] Mathematics Software Co.. $50. IBM.

A-STAT. [Grandon] National Collegiate Software. $35. IBM.

ANOVA. Dynacomp. $44. Apple, IBM, TRS-80, C64.

APP STATS & GRAPHS. StatSoft. $99. Apple II.

BIOM-pc Applied Biostatistics. [F. James Rohlf] Exeter. $75. IBM.

BIOMLAB. [F. James Rohlf] Exeter. $90. IBM.

Chance Encounters. [Holtzman, David] Educational Activities. $100. Apple II.

Chi-Square Analysis. [Trumbo, Bruce E.] Queue. $50. Apple II.

Chi-Square Analysis of Contingency Tables. [Trumbo, Bruce E.] COMPress. $50. Apple II.

CLR ANOVA. [Lane & Kluger] Clear Lake Research. $100. Mac.

CLR StatCalculator. Clear Lake Research. $35. Mac.

Combinations. Courses by Computers. $49.95. Apple II, IBM, Tandy.

Completely Randomized Designs. [Trumbo, Bruce E.] COMPress. $50. Apple II.

Completely Randomized Designs (One-Way ANOVA). [Trumbo, Bruce E.] Queue. $50. Apple II.

Continuous Distributions. [Trumbo, Bruce E.] Queue. $50. Apple II.

Continuous Probability Distributions. [Trumbo, Bruce E.] COMPress. $50. Apple II.

CRUNCH Stat Package, ver. 3.1. Crunch Software. $495, $99 (student version). IBM.

CSS - Complete Statistical System. StatSoft. $495. IBM.

Curve Fitter. Interactive Microware. $105. Apple II, IBM.

Data Analysis. [Elberfeld] EduTech. $35. Apple II.

Data Plotting Software for Micros. Kern. $50. Apple II, IBM.

Decision Analysis Techniques. Lionheart. $145. IBM, Mac.

Descriptive Statistics. Lionheart. $145. IBM, Mac.

Descriptive Statistics. Queue. $65. Apple II.

Descriptive Statistics level I. Courses by Computers. $49.95. Apple II.

Descriptive Statistics level II. Courses by Computers. $49.95. Apple II.

Discrete Probability, Part I. [Trumbo, Bruce E.] Queue. $50. Apple II.

Discrete Probability, Part II. [Trumbo, Bruce E.] Queue. $50. Apple II.

ELF - The Statistical Package. [Weiss, E.] Winchendon. $100. Apple II, IBM.

Eureka: The Solver. Borland. $167 (IBM), $195 (Mac). IBM, Mac.

Experimental Statistics. Lionheart. $145. IBM, Mac.

Exploratory Data Analysis. [Velleman, Paul; Hoaglin, David] CONDUIT. $165. Apple.

Exploring Statistics with the IBM PC. [Doane, David P.] Addison-Wesley. $30. IBM.

FILESTAT. [Strange, H. R.; Innes, A. H.] COMPress. $150. Apple II.

Filestat: A Clear Path to Statistical Solutions. Queue. $150. Apple II.

Goodness-of-Fit. Walonick. $195. IBM.

Graphical Methods for Exploring Multivariate Data. [Trumbo, Bruce E.] COMPress. $50. Apple II.

Graphical Methods for Exploring Multivariate Data. [Trumbo, Bruce E.] Queue. $50. Apple II.

Graphitti. [Kador, S.] SERAPHIM. $5. IBM, Apple II.

Intro Statistics Software Package. [Frankenberger, W.; Blakemore, T.] Addison-Wesley. $20. Apple.

Kwikstat 2.00. [Elliot] TexaSoft. $49. IBM.

LINGEN. [Christian, S.; Tucker, E.] Seraphim. $5. IBM, Apple II.

MacSS - Macintosh Statistical System. StatSoft. $119. Mac.

Mathdisk Five. [Kimberling, Clark] Mathematics Software Co.. $40. Apple II.

MathView Professional. BrainPower. $250. Mac.

Microstat-II. Ecosoft Inc. $395, $50 (student version). IBm.

MINITAB. Minitab, Inc. $395. IBM, Mac.

MSUSTAT. [Lund, Richard E.] Research & Development Institute Inc. $193. IBM.

Multilinear Regression. Dynacomp. $25. Apple II, IBM, TRS-80, C64, CP/M.

Multiple Regression & Correlation. Dynacomp. $40. Apple II, IBM.

Multivariate Analysis. Lionheart. $145. IBM, Mac.

Mystat. SYSTAT, Inc.. $5. IBM, Mac.

Nonlinear Parametric Estimation. [Nash & Walker-Smith] Marcel Dekker. $75. IBM.

NSCC 5.02. NCSS. $99, $39 (student). IBM.

NTSYS-pc 1.5 Applied Biostatistics. [F. James Roulf] Exeter. $110. IBM.

Numerical Optimzation. [Glasser, L.] SERAPHIM. $4. IBM, Apple II.

Permutations. Courses by Computers. $49.95. Apple II, IBM, Tandy.

Probability and Statistics Programs. Microphys. $200. Apple II, IBM, Tandy.

Probability and Statistics Demos & Tutorials. [Trumbo, Bruce E.] COMPress. $135. Apple II.

Probability level I. Courses by Computers. $49.95. Apple II, IBM, Tandy.

Probability level II. Courses by Computers. $49.95. Apple II, IBM, Tandy.

Probability Theory. [Kemeny, John G.] True BASIC. $49.95. IBM, Mac, Amagi, Atari.

PROSTAT. [Ward, Charles & Reeves, James] COMPress. $125. IBM.

Prostat. [Ward & Reeves] Queue. $125. IBM.

RAMAS/a Applied Biomathematics. [Scott Ferson, et al] Exeter. $195. IBM.

Regression Analysis II. Dynacomp. $40. IBM, C-64.

Regression I, II. Dynacomp. $24 each. Apple, IBM PC, TRS-80, C-64.

Research Methods: Main Effects & Interactions. [Fazio, R. & Backler, M.] CONDUIT. $50. Apple, IBM.

SAMP: Survey Sampling. [Gilber, G. Nigel] CONDUIT. $70. Apple, IBM.

Shareware Epistat. [Gustafson, Tracy L.] Epistat Services. $25. IBM.

SPSS PC+ 3.0. SPSS. $795. IBM.

Stat-Packets. Walonick. $25. IBM PC.

Stata Release 2.05. Computing Resource Center. $125 (academic), $55 (student). IBM, Unix.

Statistical Consultant 2.06. National Collegiate Software. $25. IBM.

Statistics & Graph. Queue. $39.95. Apple II.

Statistics Demos and Tutorials. [Trumbo, Bruce E.] COMPRess. $135. Apple II.

Statistics Package. [Marshall, Thomas] Micro Power & Light. $150. Apple II.

Statistics Software for Micros. Kern. $75. Apple II, IBM.

Statistix 3.0. NH Analytical Software. $129 (Apple), $179 (IBM). Apple II, IBM.

Statmaster. [Levy, et al] Scott, Foresman/Little Brown. $262.50. IBM.

StatPac. [Walonick, David] Walonick. $495. IBM.

StatPac Gold. Walonick. $595. IBM.

Stats+ - Statistical System. StatSoft. $149. IBM.

Stattest. Dynacomp. $34. Apple, IBM PC, C64, TRS-80.

StatView. Brainpower. $50. Mac.

StatView 512+. Brainpower. $300. Mac.

StatView II. Abacus. $495. MAc II family, Mac SE.

Student Edition of MINITAB. Addison-Wesley. $41.20. IBM.

The SAS System on personal computers. SAS Institue. IBM.

The Scientific Desk Analysis System. C. Abaci, Inc. $225. IBM.

The Scientific Desk Library. C. Abaci, Inc. $375. IBM (DOS, RT, large-scale), Mac.

The Scientific Desk Student Pack. C. Abaci, Inc. $95. IBM.

TRUE EPISTAT. [Gustafson, Tracy L.] Epistat Services. $319. IBM.

True STAT. [Kurtz, Thomas E.] True BASIC. $49.95. IBM PC, Mac, Amiga.

Turbo Pascal Numerical Methods Toolbox. Borland. $100. IBM, Mac.

Visual Statistics. Kern. $65. Apple II, IBM.

Which Statistic. National Collegiate Software. $25. IBM, Mac.

WormStat. [Wolford] Small Business Computers of New England. $19.95. Mac.

Advanced Simulation & Statistics Package. [Lewis *et al*] Brooks/Cole. $65. IBM.

Advanced Simulation & Statistics Package. [Lewis *et al*] Brooks/Cole. $65. IBM.

GASP - Graphical Aids for Stochastic Processes. [Fisch, Bob; Griffeath, David] Brooks/Cole. $100. IBM.

Introductory Statistics: A Microcomputer Approach. [Elzey & Cloward] Brooks/Cole. $28. IBM, Apple II.

Macsyma. Symbolics. $1250 for academics. IBM, Mac.

Mathematica. Wolfram Research. $695 and up. IBM 386, Mac.

StaTable. Brooks-Cole. IBM.

Statcalc. Zephyr. $40. Apple II, IBM.

Statistician's MACE. MACE. $155. IBM.

Statistics with Stata. [Hamilton] Brooks-Cole. IBM.

TIMESLAB: Time Series Analysis Lab. [Newton] Brooks-Cole. IBM.

Problem Solving

88 Individual lessons in mathematics. GP Publishing, Inc.. $49.50 per lesson. IBM.

Comp-U-Solve. [Engel] Educational Activities. $109. Apple II, TRS 80, C-64.

Decision Analysis Techniques. Lionheart. $145. IBM, Mac.

I-APL. I-APL. IBM.

MSC/mate. McNeal-Schwendler. $95. IBM.

Optimization. Lionheart. $145. IBM, Mac.

SEQS. CET Research Group, Ltd. $125. IBM.

Discrete Simulation. [Curry *et al*] Holden-Day. $49.95. IBM.

GAMS. [Brooke *et al*] Scientific Press. $75 (student edition). IBM.

Macsyma. Symbolics. $1250 for academics. IBM, Mac.

Mathematica. Wolfram Research. $695 and up. IBM 386, Mac.

Advanced Mathematics

Derive. Soft Warehouse, Inc. $200. IBM.

Equator. [Palmer & Glade] Pulse Research. $59. IBM.

Exploring Small Groups. [Geissinger, Ladnor] Harcourt. $15. IBM.

Fractal-D. [Slice] Exeter. $75. IBM.

GALOIS Algebra Package. [Lidl, R.; Matthews, R. W.; Wells, R.] U. of Tasmania. $115. IBM.

Group Velocity. [Lane, Eric T.] CONDUIT. $60. Apple.

I-APL. I-APL. IBM.

Macsyma. Symbolics. $1250 for academics. IBM, Mac.

MAPLE 4.2.1. Brooks-Cole. $695. IBM, Mac.

MathCAD, v 2.5. MathSoft Inc. $495 ($40 for student version). IBM.

Mathematica. Wolfram Research. $695 and up. IBM 386, Mac.

MathView Professional. BrainPower. $250. Mac.

REDUCE 3.3. Northwest Computer Algorithms. $495. IBM, Mac, Atari, VAX, SUN/Unix.

Student Edition of MathCAD 2.0. Addison-Wesley. $36.20. IBM.

The Scientific Desk Analysis System. C. Abaci, Inc. $225. IBM.

The Scientific Desk Library. C. Abaci, Inc. $375. IBM (DOS, RT, large-scale), Mac.

Technical Word Processing

ChiWriter. Horstmann. $150. IBM.

EXP. [Smith & Smith] Brooks-Cole. $295. IBM.

MathWriter 2.0. Brooks-Cole. MAC.

T3. TCI Software Research. $595. IBM.

Publishers

ABACUS CONCEPTS, INC., 1984 Bonita Ave, Berkeley, CA 94704.

ACADEMIC PRESS, 1250 Sixth Ave, San Diego, CA 92101-4311. TEL: 800-321-5068

ADDISON-WESLEY, Route 128, Reading, MA 01867. TEL: 617-944-3700

ADDISON-WESLEY PUBLISHING CO., Reading, MA 01867. TEL: 617-944-3700

AMYGDALA DOS-WARE, Box 219, San Cristobal, NM 87564. TEL: 505-586-0197

BERGWALL EDUCATIONAL SOFTWARE, INC, 106 Charles Lingbergh Blvd, Uniondale, NY 11553. TEL: 800-645-3565

BLUEJAY LISPWARE, PO Box 1904, Gambier, OH 43022.

BORLAND INTERNATIONAL, 4585 Scotts Valley Drive, Scotts Valley, CA 95066. TEL: 408-438-8400

BRAINPOWER, INC., 24009 Venture Blvd., Suite 250, Calabasas, CA 91302. TEL: 818-884-6911

BRIDGE SOFTWARE, PO Box 118, New Town Branch, Boston, MA 02258. TEL: 617-527-1585

BRODERBUND SOFTWARE, 17 Paul Drive, San Rafael, CA 94903-2101. TEL: 415-492-3200

BROOKS-COLE, 511 Forest Lodge Road, Pacific Grove, CA 93950. TEL: 800-354-9706

C. ABACI, INC, 208 St. Mary's Street, Raleigh, NC 27605. TEL: 919-832-4847

CACTUSPLOT COMPANY, 4712 E. Osborn, Phoenix, AZ 85018. TEL: 602-945-1667

CEREBIC INSTITUTE, PO Box 9, Milton, NC 27305-0009. TEL: 804-822-7026

CET RESEARCH GROUP, LTD, PO Box 2029, Norman, OK 73070-2029. TEL: 405-360-5464

CIPHER SYSTEMS, 717 Willow Street, Stillwater, OK 74075. TEL: 405-377-4432

CLEAR LAKE RESEARCH, 2476 Bolsover, Suite 343, Houston, TX 77005. TEL: 713-523-7842

COMPRESS/QUEUE, 562 Boston Ave, Bridgeport, CT 06610.

COMPUTING RESOURCE CENTER, 1640 Fifth Street, Santa Monica, CA 90401. TEL: 213-470-4341

CONDUIT, The Univ. of Iowa, Oakdale Campus, Iowa City, IA 52242. TEL: 319-335-4100

COURSES BY COMPUTERS, INC, PO Box 830, State College, PA 16804. TEL: 814-234-2210

CREATIVE COMMUNICATION SYSTEMS, 2007 Trailpine Court, Norman, OK 73064.

CRUNCH SOFTWARE CORP., 5335 College Ave., Suite 27, Oakland, CA 94618-1416. TEL: 415-420-8660

D. C. HEATH AND COMPANY, 125 Spring Street, Lexington, MA 02173.

DELLEN (DIVISION OF MACMILLAN),

DICKEY, DR. ED, College of Education, USC, Columbia, SC 29208.

DYNACOMP, 178 Phillips Road, Webster, NY 14580.

DYNAMIC SOFTWARE, PO Box 7534, Santa Cruz, CA 95061. TEL: 408-425-8619

E. DAVID & ASSOCIATES, 27 Russett Lane, Storrs, CT 06268.

ECOSOFT, INC, 6413 N. College Ave, Indianapolis, IN 46220. TEL: 317-255-6476

EDUCATIONAL ACTIVITIES, INC, 1937 Grand Avenue, Baldwin, NY 11510. TEL: 800-645-3739

EDUCATIONAL DESIGN, INC., 47 West 13th St., New York, NY 10011. TEL: 800-225-7900

EDUSOFT, PO Box 2560, Berkeley, CA 94702. TEL: 800-EDUSOFT

EDUTECH, 1927 Culver Rd., Rochester, NY 14609.

EPISTAT SERVICES, 2011 Cap Rock Circle, Richardson, TX 75080-3417. TEL: 214-680-1376

EXETER PUBLISHING, LTD, Building B, 100 N. Country Rd, Setauket, NY 11733. TEL: 516-689-7838

GP PUBLISHING, INC., 5829 Banneker Road, Columbia, MD 21044. TEL: 800-638-3838

H&H PUBLISHING CO., INC., 1231 Kapp Drive, Clearwater, FL 34625-2116.

H&N SOFTWARE, PO Box 4067, Bricktown, NJ 08723.

HARCOURT BRACE JOVANOVICH, 7555 Caldwell Avenue, Chicago, IL 60648.

HARCOURT BRACE JOVANOVICH, INC; COLLEGE DEPARTMENT, 1250 Sixth Avenue, San Diego, CA 92101. TEL: 619-699-6227

HOLDEN, HERBERT, Math Dept Gonzaga University, Spokane, Wa 99258.

HOLDEN-DAY, INC, 4432 Telegraph Ave., Oakland, CA 94609. TEL: 415-428-9400

HORSTMANN SOFTWARE, PO Box 1807, San Jose, CA 95109-1807. TEL: 800-736-8886

HOUGHTON MIFFLIN COMPANY, One Beacon Street, Boston, MA 02108. TEL: 800-732-3223

HRM/DIVISIO9N OF QUEUE, 338 Commerce Drive, Fairfield, CT 06430. TEL: 800-232-2224

I-APL, EDWARD CHERLIN, 6611 Linville Drive, Weed, CA 96094. TEL: 916-938-4684

INTERACTIVE MICROWARE, INC, PO Box 139, State College, PA 16804-0139. TEL: 814-238-8294

KAKANER, DAVID K., Room A151, Tech Bldg, NBS, Washington, DC 20234.

KERN INTERNATIONAL, INC, 575 Washington Street, PO Box 308, Pembroke, MA 02359. TEL: 617-826-0095

KRELL SOFTWARE CORP, Flowerfield Bldg 7, Suite 1D, St. James, NY 11780. TEL: 800-245-7355

LASCAUX GRAPHICS, 3220 Steuben Ave, Bronx, NY 10467. TEL: 212-654-7429

LCSI, 3300 Cote Vertu Road, Suite 201, Montreal, Quebec, CANADA H4R 2B7. TEL: 514-331-7090

LIONHEART PRESS, PO Box 379, Alburg, VT 05440. TEL: 514-933-4918

MACE, INC, 2313 Center Ave., Madison, WI 53704. TEL: 608-244-3331

MARCEL DEKKER, INC, 270 Madison Avenue, New York, NY 10016.

MATHEMATICS SOFTWARE COMPANY, 419 S. Boeke Road, Evansville, In 47714.

MATHSOFT INC, 201 Broadway, Cambridge, MA 02139. TEL: 617-577-1017

MATHSOFT, ATTN JEAN M. HENRY, one Kendall Square, 200, Cambridge, MA 02139.

MCE, 157 S. Kalamazoo Mall, Suite 250, Kalamazoo, MI 49007.

MCGRAW-HILL, New York, NY

MCNEAL-SCHWENDLER, 815 Colorado Blvd, Los Angeles, CA 90041.

MICRO POWER & LIGHT Co., 12810 Hillcrest Road, Suite 120, Dallas, TX 75230.

MICRO-ED, 31 Marshall Street, Edison, NJ 08817.

MICROPHYS PROGRAMS, 1737 West 2nd Street, Brooklyn, NY 11223. TEL: 718-375-5151

MINITAB, INC, 3081 Enterprise Drive, State College, PA 16801. TEL: 814-238-4383

NATIONAL COLLEGIATE SOFTWARE CLEARINGHOUSE, DUKE UNIVERSITY PRESS, 6697 College Station, Durham, NC 27708.

NCSS, 865 East 400 North, Kaysville, UT 84037. TEL: 801-546-0445

NH ANALYTICAL SOFTWARE, PO Box 13204, Roseville, MN 55113. TEL: 612-631-2852

NORTHWEST COMPUTER ALGORITHMS, PO Box 1747, Novato, CA 94948. TEL: 415-897-1302

PEANUT SOFTWARE c/o R PARRIS, Phillips Exeter Academy, Exeter, NH 03833.

PEDAGOGUERY SOFTWARE, 4446 Lazelle Avenue, Terrace, BC, Canada V8G 1R8.

PENN, HOWARD, Math Dept., USNA, Annapolis, MD 21402.

POLYGONAL PUBLISHING HOUSE, 210 Broad Strret, Washington, NJ 007882.

PRENTICE HALL, SIMON & SCHUSTER, Route 59 at Brook Hill Drive, West Nyack, NY 10995. TEL: 201-767-5937

PULSE RESEARCH, P.O.Box 696, Shelburne, VT 05482. TEL: 802-985-2928

PUMA SOFTWARE INC, PO Box 35373, Albuquerque, NM 87176. TEL: 505-265-5270

QUEUE, INC, 338 Commerce Drive, Fairfield, CT 06430. TEL: 800-232-2224

R & D SOFTWARE, Mathematics, University of Hawaii, Honolulu, HI 96822.

RECREATIONAL MATHEMAGICAL SOFTWARE, 129 Carol Drive, Clarks Summit, PA 18411. TEL: 717-586-2784

RESEARCH & DEVELOPMENT INSTITITUE INC, Montana State University, Bozeman, Montana 59717-0002. TEL: 406-994-3271

SANDSTROM, RON, Fort Hays SU, 600 Park Street, Fort Hays, KS 67601-4099.

SAS INSTITUTE, INC, Box 8000, Cary, NC 27512-8000. TEL: 919-467-8000

SAUNDERS PUBLISHING Co., West Washington Square, Philadelphia, PA 19105.

SCHARF SYSTEMS, INC, 17 Hemlock Drive, North Caldwell, NJ 07006. TEL: 201-403-9787

SCIENTIFIC PRESS, 507 Seaport Court, Redwood City, CA 94063-2731. TEL: 415-2577

SCOTT FORESMAN/LITTLE BROWN, 1900 East Lake Avenue, Glenview, IL 60025. TEL: 312-729-3000

SECONDARY DYNAMICAL SYSTEMS COLLABORATIVE, Box 991, Groton, MA 01450. TEL: 508-448-3363 x570

SERAPHIM, Dept of Chemistry, U of Wisconsin, Madison, WI 53706.

SERVOSOFTWARE, P.O. Box 72, West Covina, CA 91793.

SINTAR SOFTWARE, P.O. Box 3746, Bellevue, WA 98009. TEL: 206-455-4130

SMALL BUSINESS COMPUTERS OF NEW ENGLAND, PO Box 397, Amherst, MA 03031. TEL: 603-673-0228

SMITH, DAVID A., Dept of Mathematics, Duke U., Durham, NC 27706.

SNOW, DENNIS M., Math Dept, 203 Comp Ctr & Math Bldg, Notre Dame, IN 46556.

SOFT WAREHOUSE, INC, 3615 Harding Avenue, Suite 505, Honolulu, Hawaii 96816. TEL: 808-734-5801

SPENCER ORGANIZATION, PO Box 248, Westwood, NJ 07675. TEL: 201-666-6011

SPRINGER-VERLAG, 175 Fifth Ave, New Yoprk, NY 10010.

SPSS, INC, 444 N. Michigan Ave, Chicago, IL 60611.

STANFORD BUSINESS SOFTWARE, INC, 2672 Bayshore Parkway, Suite 304, Mountain View, CA 94043. TEL: 415-424-9499

STATSOFT, 2325 East 13th Street, Tulsa, OK 74104. TEL: 918-583-4149

SUNBURST COMMUNICATIONS, 39 Washington Avenue, Pleasantville, NY 10570-2898. TEL: 800-431-1934

SYMBOLICS, 8 New England Executive Park, East Burlington, MA 01803. TEL: 617-221-1250

SYSTAT, INC., 1800 Sherman Avenue, Evanston, IL 60201. TEL: 312-864-5670

TCI SOFTWARE RESEARCH, 1190 Foster Road, Las Cruces, NM 88001. TEL: 800-874-2383

TECHNICAL EDUCATION CONSULTANTS, 76 North Broadway, Suite 4009, Hicksville, NY 11801. TEL: 516-681-1773

TEXASOFT, PO Box 1169, Cedar Hill, TX 75104. TEL: 214-291-2115

THE MATH LAB, 10893 Leavelsey Place, Cupertino, CA 95014.

THE MATH WORKS, INC, 21 Eliot Street, South Natick, MA 01760. TEL: 508-653-1415

TRUE BASIC, INC, 12 Commerce Ave., West Lebanon, NH 03784. TEL: 800-TR-BASIC (sales), 603-298-5655 (support)

TUSOFT, PO Box 9979, Berkeley, CA 94709.

U. OF TASMANIA, Math Dept., Box 252, GPO, Hobart, Tasmania 7001, Australia

VENTURA EDUCATIONAL SYSTEMS, 3440 Brokenhill St., Newbury Park, CA 91320. TEL: 805-499-1407

WADSWORTH PUBLISHING COMPANY, Ten Davis Street, Belmont, CA 94002. TEL: 415-595-2350

WALONICK ASSOCIATES, 6500 Nicollect Avenue South, Minneapolis, MN 55423. TEL: 800-328-4907

WATCOM PRODUCTS, INC., 415 Phillip Street, Waterloo, Ontario, Canada N2L 3X2.

WATERLOO MAPLE SOFTWARE, 160 Columbia Street West, Waterloo, Ontario, CANADA N2L 3L3. TEL: 519-747-2373

WILEY, New York, NY

WINCHENDON GROUP, PO Box 10339, Alexandria, VA 22310. TEL: 703-960-2587

WISC-WARE - ACADEMIC COMPUTING CENTER, Univ. of Wisconsin-Madison - 1210 W. Dayton St., Madison, WI 53706. TEL: 800-543-3201

WKM ASSOCIATES, PO Box 585, Claymont, DE 19703.

WOLFRAM RESEARCH INC, PO Box 6059, Champaign, IL 61826. TEL: 217-398-0700

WONG, P. K., Michigan State University - Mathematics, East Lansing, MI 48824.

ZEPHER SERVICES, 306 S. Homewood Ave, Pittsburgh, PA 15208. TEL: 800-533-6666

The Advanced Placement Program in Calculus

The Advanced Placement Program was introduced by the College Board in the 1950's to give talented high school students an opportunity to earn college credit for courses taken in high school. There are AP courses in about twenty subject areas. The syllabus for each course is given in great detail in the AP course description booklet for the subject area and is controlled and updated every two years by an AP subject committee, usually composed of three college and three secondary school teachers. The crediting mechanism is a three-hour exam given once each year in May. The exam is usually about half multiple choice and half essay (in test jargon, "free response"). The exams are then graded in June, over a 6-day period, by a large number of hard working college and high school teachers. The construction, administration, and the grading of each exam is handled by Educational Testing Service. In 1989, there were 455,966 exams (89,261 in mathematics or computer science) taken by 309,751 students or about 17 exams for every 100 graduating high school seniors. The volume has grown by 230 percent in the last decade.

There are two AP Calculus exams, called "AB" and "BC". The AB exam covers a little more than the standard first semester college calculus course: differentiation including all the usual applications, integration including elementary techniques up to simple integration by parts and applications to areas and solids of revolution, both exponential/logarithmic functions and trigonometric functions. The BC course covers somewhat more than the first two semesters of a college course: all of the AB material plus further techniques of integration, parametric equations, polar coordinates, series of constants, Taylor series, and a little differential equations. The course at present includes no multivariable topics.

The AP exam grades range from 1 (lowest) to 5 (highest). The vast majority of colleges and universities in this country give at least one semester of credit for a 4 or 5 on the AB exam, and at least two semesters credit for a 4 or 5 on the BC exam. Most schools extend this crediting policy to 3's as well, and some even consider 2's. Students taking an AP course are not required to take the AP exam, and, because of the $60 price tag, many students choose not to, especially if they expect to do poorly. From survey data, generally fewer than half the students enrolled in an AP calculus course take the exams. Even after this self-selection, about 30% to 40% of the students score a 1 or 2 on the AB exam, and about 20% to 25% score a 1 or 2 on the BC exam (AB exams outnumber BC exams about four to one). Thus, a typical AB course may end up with fewer than one out of three enrolled students getting a 3 or above on the AP exam.

One of the most important aspects of the AP program is that the free response parts of every test are completely public. In addition, the multiple-choice sections are released about once every five years. The result is that every AP teacher has a pool of AP exam questions to draw on, and it is these questions which shape a typical AP course. These questions range from straightforward applications of standard techniques to quite unusual problems of theoretical interest. The entire 1985 AB exam appears in the exam section of *Calculus for a New Century* (MAA Notes).

Another important aspect of the AP Calculus program is the widespread involvement of the mathematical community. During the late 1970's and early 1980's, the annual reading of AP calculus exams was perhaps the only on-going, national joint venture of secondary school and college teachers. It still is one of the best means of communication between the two groups. Even at the college level, the AP program has a friendly mix of mathematicians from both doctoral and nondoctoral institutions. In fact, two of the members of the CUPM Committee on Calculus and the First Two Years (CRAFTY) have been deeply involved in the AP program. Thomas Tucker was chair from 1983 to 1987 of the AP Calculus committee that constructs the exam and oversees the program. John Kenelly

chaired the same committee from 1979 to 1983, was "Chief Reader" in charge of the annual exams from 1975 to 1979, and for 1989-1990 was Director of the entire AP program in all disciplines.

At present, the AP Calculus is considering how to allow the use of calculators on the exam. One of the largest studies ever undertaken by the AP program is going on at this time, involving thousands of students and a specially constructed multiple-choice test including a number of questions that cannot be answered without a calculator. (For more about the history of calculator use in the AP exam and other standardized tests, see the MAA publication *The Use of Calculators in the Standardized Testing of Mathematics*.)

Considering the size and influence of the AP Calculus program, it is surprising how little most college faculty members know about it. Every department should have a current copy of the *Advanced Placement Course Description in Mathematics*, as well as copies of the most recent released examination, in this case, *The Entire 1988 AP Calculus AB Examination and Key* and *The Entire 1988 AP Calculus BC Examination and Key*. The latter booklets contain not only the exams, but also data on student responses on every multiple-choice item, data on raw scores and the relation between those scores and the final grade of 1 through 5, and a point-by-point explanation of how each of the six free response questions was graded. To order these publications or to inquire about further information, write to

AP Program
P.O. Box 6670
Princeton, NJ 08541-6670

The person at Educational Testing Service most closely involved with the AP Calculus program is James Armstrong, phone 609-734-5214.